11 – 19 PROGRESSION

Endorsed for
Pearson Edexcel
Qualifications

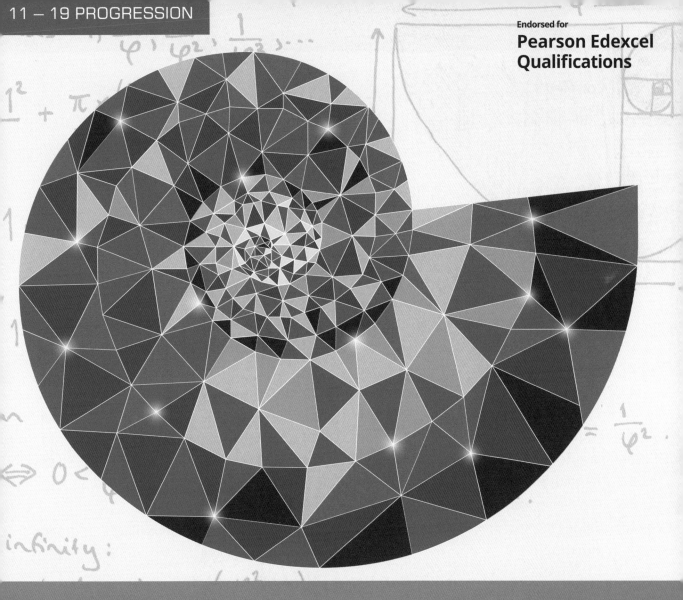

Pearson Edexcel A level Mathematics

Pure Mathematics
Year 2

Series Editor: Harry Smith
Authors: Greg Attwood, Jack Barraclough, Ian Bettison, Keith Gallick, Daniel Goldberg, Anne McAteer, Alistair Macpherson, Bronwen Moran, Joe Petran, Keith Pledger, Cong San, Harry Smith, Geoff Staley and Dave Wilkins

Contents

Overarching themes — iv
Extra online content — vi

1 Algebraic methods — 1
1.1 Proof by contradiction — 2
1.2 Algebraic fractions — 5
1.3 Partial fractions — 9
1.4 Repeated factors — 12
1.5 Algebraic division — 14
Mixed exercise 1 — 19

2 Functions and graphs — 22
2.1 The modulus function — 23
2.2 Functions and mappings — 27
2.3 Composite functions — 32
2.4 Inverse functions — 36
2.5 $y = |f(x)|$ and $y = f(|x|)$ — 40
2.6 Combining transformations — 44
2.7 Solving modulus problems — 48
Mixed exercise 2 — 53

3 Sequences and series — 59
3.1 Arithmetic sequences — 60
3.2 Arithmetic series — 63
3.3 Geometric sequences — 66
3.4 Geometric series — 70
3.5 Sum to infinity — 73
3.6 Sigma notation — 76
3.7 Recurrence relations — 79
3.8 Modelling with series — 83
Mixed exercise 3 — 86

4 Binomial expansion — 91
4.1 Expanding $(1 + x)^n$ — 92
4.2 Expanding $(a + bx)^n$ — 97
4.3 Using partial fractions — 101
Mixed exercise 4 — 104

Review exercise 1 — 107

5 Radians — 113
5.1 Radian measure — 114
5.2 Arc length — 118
5.3 Areas of sectors and segments — 122
5.4 Solving trigonometric equations — 128
5.5 Small angle approximations — 133
Mixed exercise 5 — 135

6 Trigonometric functions — 142
6.1 Secant, cosecant and cotangent — 143
6.2 Graphs of sec x, cosec x and cot x — 145
6.3 Using sec x, cosec x and cot x — 149
6.4 Trigonometric identities — 153
6.5 Inverse trigonometric functions — 158
Mixed exercise 6 — 162

7 Trigonometry and modelling — 166
7.1 Addition formulae — 167
7.2 Using the angle addition formulae — 171
7.3 Double-angle formulae — 174
7.4 Solving trigonometric equations — 177
7.5 Simplifying $a \cos x \pm b \sin x$ — 181
7.6 Proving trigonometric identities — 186
7.7 Modelling with trigonometric functions — 189
Mixed exercise 7 — 192

8 Parametric equations — 197
8.1 Parametric equations — 198
8.2 Using trigonometric identities — 202
8.3 Curve sketching — 206
8.4 Points of intersection — 209
8.5 Modelling with parametric equations — 213
Mixed exercise 8 — 220

Review exercise 2	225

9	**Differentiation**	231
9.1	Differentiating $\sin x$ and $\cos x$	232
9.2	Differentiating exponentials and logarithms	235
9.3	The chain rule	237
9.4	The product rule	241
9.5	The quotient rule	243
9.6	Differentiating trigonometric functions	246
9.7	Parametric differentiation	250
9.8	Implicit differentiation	253
9.9	Using second derivatives	257
9.10	Rates of change	261
	Mixed exercise 9	265

10	**Numerical methods**	273
10.1	Locating roots	274
10.2	Iteration	278
10.3	The Newton–Raphson method	282
10.4	Applications to modelling	286
	Mixed exercise 10	289

11	**Integration**	293
11.1	Integrating standard functions	294
11.2	Integrating $f(ax + b)$	296
11.3	Using trigonometric identities	298
11.4	Reverse chain rule	300
11.5	Integration by substitution	303
11.6	Integration by parts	307
11.7	Partial fractions	310
11.8	Finding areas	313
11.9	The trapezium rule	318
11.10	Solving differential equations	322
11.11	Modelling with differential equations	326
11.12	Integration as the limit of a sum	329
	Mixed exercise 11	330

12	**Vectors**	337
12.1	3D coordinates	338
12.2	Vectors in 3D	340
12.3	Solving geometric problems	344
12.4	Application to mechanics	348
	Mixed exercise 12	349

Review exercise 3	352
Exam-style practice: Paper 1	358
Exam-style practice: Paper 2	361
Answers	365
Index	423

Overarching themes

The following three overarching themes have been fully integrated throughout the Pearson Edexcel AS and A level Mathematics series, so they can be applied alongside your learning and practice.

1. Mathematical argument, language and proof
- Rigorous and consistent approach throughout
- Notation boxes explain key mathematical language and symbols
- Dedicated sections on mathematical proof explain key principles and strategies
- Opportunities to critique arguments and justify methods

2. Mathematical problem solving

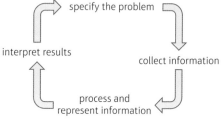

The Mathematical Problem-solving cycle

specify the problem → collect information → process and represent information → interpret results →

- Hundreds of problem-solving questions, fully integrated into the main exercises
- Problem-solving boxes provide tips and strategies
- Structured and unstructured questions to build confidence
- Challenge boxes provide extra stretch

3. Mathematical modelling
- Dedicated modelling sections in relevant topics provide plenty of practice where you need it
- Examples and exercises include qualitative questions that allow you to interpret answers in the context of the model
- Dedicated chapter in Statistics & Mechanics Year 1/AS explains the principles of modelling in mechanics

Finding your way around the book

Access an online digital edition using the code at the front of the book.

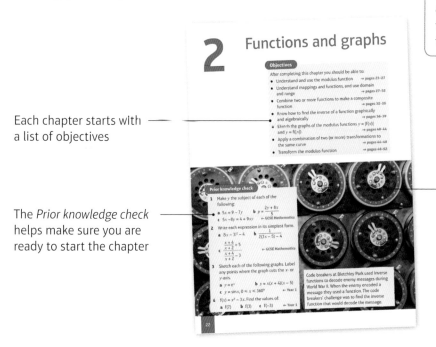

Each chapter starts with a list of objectives

The *Prior knowledge check* helps make sure you are ready to start the chapter

The real world applications of the maths you are about to learn are highlighted at the start of the chapter with links to relevant questions in the chapter

Overarching themes

Exercise questions are carefully graded so they increase in difficulty and gradually bring you up to exam standard

Problem-solving boxes provide hints, tips and strategies, and *Watch out* boxes highlight areas where students often lose marks in their exams

Exercises are packed with exam-style questions to ensure you are ready for the exams

Exam-style questions are flagged with Ⓔ

Problem-solving questions are flagged with Ⓟ

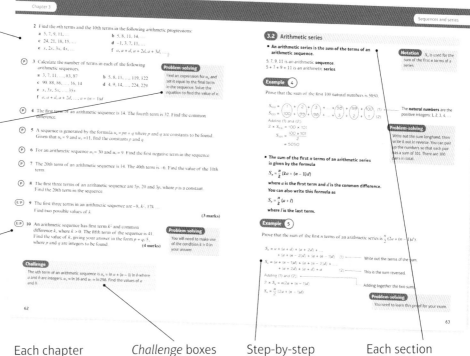

Each chapter ends with a *Mixed exercise* and a *Summary of key points*

Challenge boxes give you a chance to tackle some more difficult questions

Step-by-step worked examples focus on the key types of questions you'll need to tackle

Each section begins with explanation and key learning points

Every few chapters a *Review exercise* helps you consolidate your learning with lots of exam-style questions

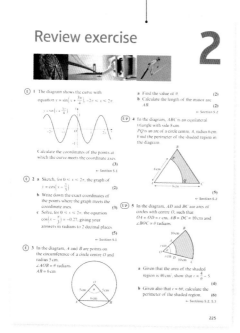

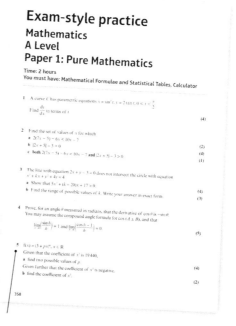

Two A level practice papers at the back of the book help you prepare for the real thing

v

Extra online content

Whenever you see an *Online* box, it means that there is extra online content available to support you.

SolutionBank

SolutionBank provides a full worked solution for every question in the book.

Online Full worked solutions are available in SolutionBank.

Download all the solutions as a PDF or quickly find the solution you need online

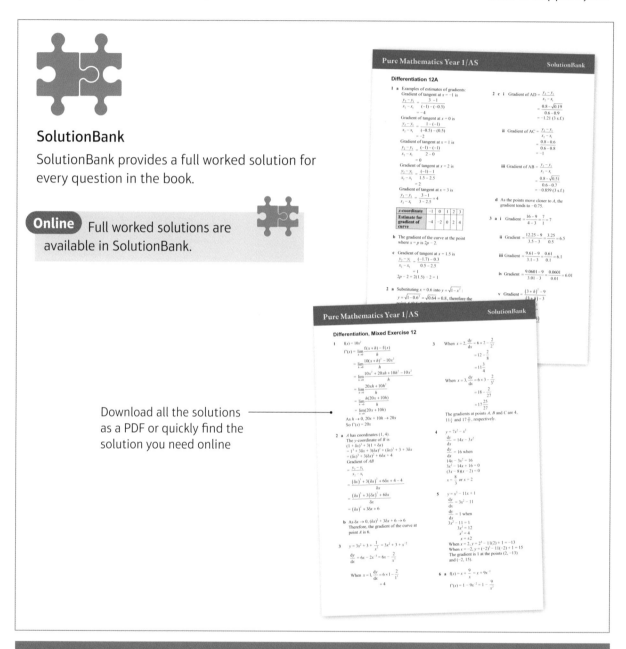

Access all the extra online content for free at:

www.pearsonschools.co.uk/p2maths

You can also access the extra online content by scanning this QR code:

Extra online content

Use of technology

Explore topics in more detail, visualise problems and consolidate your understanding. Use pre-made GeoGebra activities or Casio resources for a graphic calculator.

Online Find the point of intersection graphically using technology.

GeoGebra-powered interactives

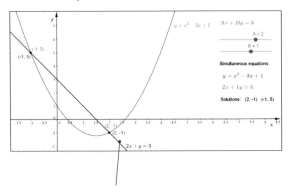

Interact with the maths you are learning using GeoGebra's easy-to-use tools

Graphic calculator interactives

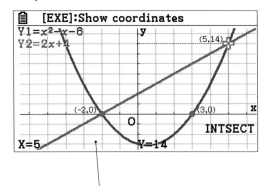

Explore the maths you are learning and gain confidence in using a graphic calculator

Calculator tutorials

Our helpful video tutorials will guide you through how to use your calculator in the exams. They cover both Casio's scientific and colour graphic calculators.

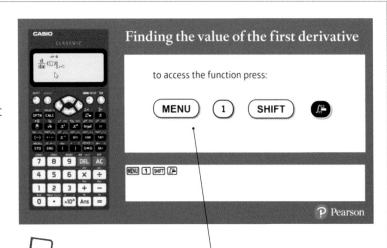

Online Work out each coefficient quickly using the ^nC_r and power functions on your calculator.

Step-by-step guide with audio instructions on exactly which buttons to press and what should appear on your calculator's screen

vii

Published by Pearson Education Limited, 80 Strand, London WC2R 0RL.

www.pearsonschoolsandfecolleges.co.uk

Copies of official specifications for all Pearson qualifications may be found on the website: qualifications.pearson.com

Text © Pearson Education Limited 2017
Edited by Tech-Set Ltd, Gateshead
Typeset by Tech-Set Ltd, Gateshead
Original illustrations © Pearson Education Limited 2017
Cover illustration Marcus@kja-artists

The rights of Greg Attwood, Jack Barraclough, Ian Bettison, Keith Gallick, Daniel Goldberg, Anne McAteer, Alistair Macpherson, Bronwen Moran, Joe Petran, Keith Pledger, Cong San, Harry Smith, Geoff Staley and Dave Wilkins to be identified as authors of this work have been asserted by them in accordance with the Copyright, Designs and Patents Act 1988.

First published 2017

25
18

British Library Cataloguing in Publication Data
A catalogue record for this book is available from the British Library

ISBN 978 1 292 18340 4

Copyright notice
All rights reserved. No part of this publication may be reproduced in any form or by any means (including photocopying or storing it in any medium by electronic means and whether or not transiently or incidentally to some other use of this publication) without the written permission of the copyright owner, except in accordance with the provisions of the Copyright, Designs and Patents Act 1988 or under the terms of a licence issued by the Copyright Licensing Agency, Barnards Inn 86 Fetter Lane, London EC4A 1EN (www.cla.co.uk). Applications for the copyright owner's written permission should be addressed to the publisher.

Printed in the UK by Bell and Bain Ltd, Glasgow

Acknowledgements
The authors and publisher would like to thank the following individuals and organisations for permission to reproduce photographs:

(Key: b-bottom; c-centre; l-left; r-right; t-top)

Alamy Stock Photo: Hans Kristian Olsen 197, 225r, Prisma Bildagentur AG 22, 107cl; **Fotolia.com:** Mita Stock Images 1, 107l, Nico_Vash 142, 225cl, vvoe 166, 225cr; **Getty Images:** Anthony Bradshaw 273, 352cl, mikedabell 231, 352l, Westend61 336, 352r; **Science Photo Library Ltd:** SPL / Millard H. Sharp 293, 352cr; **Shutterstock.com:** DeReGe 113, 225l, Michelangelus 59, 107cr, OliverSved 91, 107r

All other images © Pearson Education

A note from the publisher
In order to ensure that this resource offers high-quality support for the associated Pearson qualification, it has been through a review process by the awarding body. This process confirms that this resource fully covers the teaching and learning content of the specification or part of a specification at which it is aimed. It also confirms that it demonstrates an appropriate balance between the development of subject skills, knowledge and understanding, in addition to preparation for assessment.

Endorsement does not cover any guidance on assessment activities or processes (e.g. practice questions or advice on how to answer assessment questions), included in the resource nor does it prescribe any particular approach to the teaching or delivery of a related course.

While the publishers have made every attempt to ensure that advice on the qualification and its assessment is accurate, the official specification and associated assessment guidance materials are the only authoritative source of information and should always be referred to for definitive guidance.

Pearson examiners have not contributed to any sections in this resource relevant to examination papers for which they have responsibility.

Examiners will not use endorsed resources as a source of material for any assessment set by Pearson.

Endorsement of a resource does not mean that the resource is required to achieve this Pearson qualification, nor does it mean that it is the only suitable material available to support the qualification, and any resource lists produced by the awarding body shall include this and other appropriate resources.

Pearson has robust editorial processes, including answer and fact checks, to ensure the accuracy of the content in this publication, and every effort is made to ensure this publication is free of errors. We are, however, only human, and occasionally errors do occur. Pearson is not liable for any misunderstandings that arise as a result of errors in this publication, but it is our priority to ensure that the content is accurate. If you spot an error, please do contact us at resourcescorrections@pearson.com so we can make sure it is corrected.

Algebraic methods

1

Objectives

After completing this chapter you should be able to:
- Use proof by contradiction to prove true statements → pages 2–5
- Multiply and divide two or more algebraic fractions → pages 5–7
- Add or subtract two or more algebraic fractions → pages 7–8
- Convert an expression with linear factors in the denominator into partial fractions → pages 9–11
- Convert an expression with repeated linear factors in the denominator into partial fractions → pages 12–13
- Divide algebraic expressions → pages 14–17
- Convert an improper fraction into partial fraction form → pages 17–18

Prior knowledge check

1 Factorise each polynomial:
   a $x^2 - 6x + 5$ b $x^2 - 16$
   c $9x^2 - 25$ ← Year 1, Section 1.3

2 Simplify fully the following algebraic fractions.
   a $\dfrac{x^2 - 9}{x^2 + 9x + 18}$ b $\dfrac{2x^2 + 5x - 12}{6x^2 - 7x - 3}$
   c $\dfrac{x^2 - x - 30}{-x^2 + 3x + 18}$ ← Year 1, Section 7.1

3 For any integers n and m, decide whether the following will always be odd, always be even, or could be either.
   a $8n$ b $n - m$
   c $3m$ d $2n - 5$
   ← Year 1, Section 7.6

You can use proof by contradiction to prove that there is an infinite number of prime numbers. Very large prime numbers are used to encode chip and pin transactions. → Example 4, page 3

Chapter 1

1.1 Proof by contradiction

A **contradiction** is a disagreement between two statements, which means that both cannot be true. Proof by contradiction is a powerful technique.

- **To prove a statement by contradiction you start by assuming it is not true. You then use logical steps to show that this assumption leads to something impossible (either a contradiction of the assumption, or a contradiction of a fact you know to be true). You can conclude that your assumption was incorrect, and the original statement was true.**

Notation A statement that asserts the falsehood of another statement is called the negation of that statement.

Example 1

Prove by contradiction that there is no greatest odd integer.

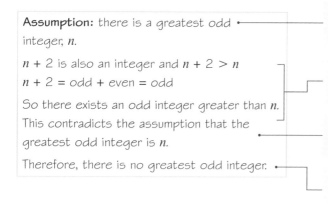

Assumption: there is a greatest odd integer, n.

$n + 2$ is also an integer and $n + 2 > n$

$n + 2 = $ odd + even = odd

So there exists an odd integer greater than n.

This contradicts the assumption that the greatest odd integer is n.

Therefore, there is no greatest odd integer.

Begin by assuming the original statement is false. This is the negation of the original statement.

You need to use logical steps to reach a contradiction. Show all of your working.

The existence of an odd integer greater than n contradicts your initial assumption.

Finish your proof by concluding that the original statement must be true.

Example 2

Prove by contradiction that if n^2 is even, then n must be even.

Assumption: there exists a number n such that n^2 is even but n is odd.

n is odd so write $n = 2k + 1$

$n^2 = (2k + 1)^2 = 4k^2 + 4k + 1$
$= 2(2k^2 + 2k) + 1$

So n^2 is odd.

This contradicts the assumption that n^2 is even.

Therefore, if n^2 is even then n must be even.

This is the negation of the original statement.

You can write any odd number in the form $2k + 1$ where k is an integer.

All multiples of 2 are even numbers, so 1 more than a multiple of 2 is an odd number.

Finish your proof by concluding that the original statement must be true.

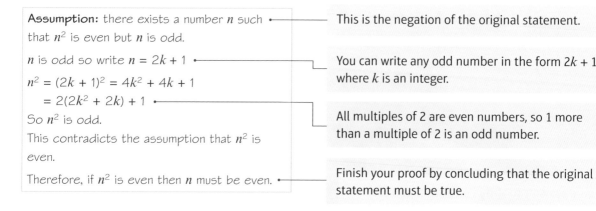

- **A rational number can be written as $\frac{a}{b}$, where a and b are integers.**

- **An irrational number cannot be expressed in the form $\frac{a}{b}$, where a and b are integers.**

Notation $\mathbb{Q}$ is the set of all rational numbers.

Algebraic methods

Example 3

Prove by contradiction that $\sqrt{2}$ is an irrational number.

Assumption: $\sqrt{2}$ is a rational number. — Begin by assuming the original statement is false.

Then $\sqrt{2} = \dfrac{a}{b}$ for some integers, a and b. — This is the definition of a rational number.

Also assume that this fraction cannot be reduced further: there are no common factors between a and b. — If a and b did have a common factor you could just cancel until this fraction was in its simplest form.

So $2 = \dfrac{a^2}{b^2}$ or $a^2 = 2b^2$ — Square both sides and make a^2 the subject.

This means that a^2 must be even, so a is also even. — We proved this result in Example 2.

If a is even, then it can be expressed in the form $a = 2n$, where n is an integer.
So $a^2 = 2b^2$ becomes $(2n)^2 = 2b^2$ which means $4n^2 = 2b^2$ or $2n^2 = b^2$.

This means that b^2 must be even, so b is also even. — Again using the result from Example 2.

If a and b are both even, they will have a common factor of 2. — All even numbers are divisible by 2.

This contradicts the statement that a and b have no common factors.
Therefore $\sqrt{2}$ is an irrational number. — Finish your proof by concluding that the original statement must be true.

Example 4

Prove by contradiction that there are infinitely many prime numbers.

Assumption: there is a finite number of prime numbers. — Begin by assuming the original statement is false.

List all the prime numbers that exist:
$p_1, p_2, p_3, \ldots, p_n$ — This is a list of all possible prime numbers.

Consider the number
$N = p_1 \times p_2 \times p_3 \times \ldots \times p_n + 1$ — This new number is one more than the product of the existing prime numbers.

When you divide N by any of the prime numbers $p_1, p_2, p_3, \ldots, p_n$ you get a remainder of 1. So none of the prime numbers $p_1, p_2, p_3, \ldots, p_n$ is a factor of N.

So N must either be prime or have a prime factor which is not in the list of all possible prime numbers.
This is a contradiction. — This contradicts the assumption that the list $p_1, p_2, p_3, \ldots, p_n$ contains all the prime numbers.

Therefore, there is an infinite number of prime numbers. — Conclude your proof by stating that the original statement must be true.

Chapter 1

Exercise 1A

1. Select the statement that is the negation of 'All multiples of three are even'.
   - A All multiples of three are odd.
   - B At least one multiple of three is odd.
   - C No multiples of three are even.

2. Write down the negation of each statement.
   - a All rich people are happy.
   - b There are no prime numbers between 10 million and 11 million.
   - c If p and q are prime numbers then $(pq + 1)$ is a prime number.
   - d All numbers of the form $2^n - 1$ are either prime numbers or multiples of 3.
   - e At least one of the above four statements is true.

3. Statement: If n^2 is odd then n is odd.
   - a Write down the negation of this statement.
   - b Prove the original statement by contradiction.

4. Prove the following statements by contradiction.
   - a There is no greatest even integer.
   - b If n^3 is even then n is even.
   - c If pq is even then at least one of p and q is even.
   - d If $p + q$ is odd then at least one of p and q is odd.

5. a Prove that if ab is an irrational number then at least one of a and b is an irrational number. **(3 marks)**

   b Prove that if $a + b$ is an irrational number then at least one of a and b is an irrational number. **(3 marks)**

   c A student makes the following statement:

   If $a + b$ is a rational number then at least one of a and b is a rational number.

   Show by means of a counterexample that this statement is not true. **(1 mark)**

6. Use proof by contradiction to show that there exist no integers a and b for which $21a + 14b = 1$.

   Hint Assume the opposite is true, and then divide both sides by the highest common factor of 21 and 14.

7. a Prove by contradiction that if n^2 is a multiple of 3, n is a multiple of 3. **(3 marks)**

   Hint Consider numbers in the form $3k + 1$ and $3k + 2$.

   b Hence prove by contradiction that $\sqrt{3}$ is an irrational number. **(3 marks)**

Algebraic methods

(P) 8 Use proof by contradiction to prove the statement:
'There are no integer solutions to the equation
$x^2 - y^2 = 2$'

Hint You can assume that x and y are positive, since $(-x)^2 = x^2$.

(E/P) 9 Prove by contradiction that $\sqrt[3]{2}$ is irrational. **(5 marks)**

(E/P) 10 This student has attempted to use proof by contradiction to show that there is no least positive rational number:

> **Assumption:** There is a least positive rational number.
> Let this least positive rational number be n.
> As n is rational, $n = \dfrac{a}{b}$ where a and b are integers.
> $n - 1 = \dfrac{a}{b} - 1 = \dfrac{a-b}{b}$
> Since a and b are integers, $\dfrac{a-b}{b}$ is a rational number that is less than n.
> This contradicts the statement that n is the least positive rational number.
> Therefore, there is no least positive rational number.

Problem-solving
You might have to analyse student working like this in your exam. The question says, 'the error', so there should only be one error in the proof.

a Identify the error in the student's proof. **(1 mark)**
b Prove by contradiction that there is no least positive rational number. **(5 marks)**

1.2 Algebraic fractions

Algebraic fractions work in the same way as numeric fractions. You can simplify them by cancelling common factors and finding common denominators.

- **To multiply fractions, cancel any common factors, then multiply the numerators and multiply the denominators.**

Example 5

Simplify the following products:

a $\dfrac{3}{5} \times \dfrac{5}{9}$ b $\dfrac{a}{b} \times \dfrac{c}{a}$ c $\dfrac{x+1}{2} \times \dfrac{3}{x^2 - 1}$

a $\dfrac{\cancel{3}^1}{\cancel{5}_1} \times \dfrac{\cancel{5}^1}{\cancel{9}_3} = \dfrac{1 \times 1}{1 \times 3} = \dfrac{1}{3}$ — Cancel any common factors and multiply numerators and denominators.

b $\dfrac{\cancel{a}^1}{b} \times \dfrac{c}{\cancel{a}_1} = \dfrac{1 \times c}{b \times 1} = \dfrac{c}{b}$ — Cancel any common factors and multiply numerators and denominators.

c $\dfrac{x+1}{2} \times \dfrac{3}{x^2 - 1} = \dfrac{x+1}{2} \times \dfrac{3}{(x+1)(x-1)}$ — Factorise $(x^2 - 1)$.

$= \dfrac{\cancel{x+1}^1}{2} \times \dfrac{3}{\cancel{(x+1)}_1(x-1)}$ — Cancel any common factors and multiply numerators and denominators.

$= \dfrac{3}{2(x-1)}$

5

Chapter 1

- **To divide two fractions, multiply the first fraction by the reciprocal of the second fraction.**

Example 6

Simplify:

a $\dfrac{a}{b} \div \dfrac{a}{c}$

b $\dfrac{x+2}{x+4} \div \dfrac{3x+6}{x^2-16}$

---

a $\dfrac{a}{b} \div \dfrac{a}{c} = \dfrac{\cancel{a}^1}{b} \times \dfrac{c}{\cancel{a}_1}$ — Multiply the first fraction by the reciprocal of the second fraction. Cancel the common factor a.

$= \dfrac{1 \times c}{b \times 1}$ — Multiply numerators and denominators.

$= \dfrac{c}{b}$

b $\dfrac{x+2}{x+4} \div \dfrac{3x+6}{x^2-16}$

$= \dfrac{x+2}{x+4} \times \dfrac{x^2-16}{3x+6}$ — Multiply the first fraction by the reciprocal of the second fraction.

$= \dfrac{x+2}{x+4} \times \dfrac{(x+4)(x-4)}{3(x+2)}$ — Factorise as far as possible.

$= \dfrac{\cancel{x+2}^1}{\cancel{x+4}^1} \times \dfrac{\cancel{(x+4)}^1(x-4)}{3\cancel{(x+2)}^1}$ — Cancel any common factors and multiply numerators and denominators.

$= \dfrac{x-4}{3}$

Exercise 1B

1 Simplify:

a $\dfrac{a}{d} \times \dfrac{a}{c}$
b $\dfrac{a^2}{c} \times \dfrac{c}{a}$
c $\dfrac{2}{x} \times \dfrac{x}{4}$
d $\dfrac{3}{x} \div \dfrac{6}{x}$
e $\dfrac{4}{xy} \div \dfrac{x}{y}$
f $\dfrac{2r^2}{5} \div \dfrac{4}{r^3}$

2 Simplify:

a $(x+2) \times \dfrac{1}{x^2-4}$
b $\dfrac{1}{a^2+6a+9} \times \dfrac{a^2-9}{2}$
c $\dfrac{x^2-3x}{y^2+y} \times \dfrac{y+1}{x}$

d $\dfrac{y}{y+3} \div \dfrac{y^2}{y^2+4y+3}$
e $\dfrac{x^2}{3} \div \dfrac{2x^3-6x^2}{x^2-3x}$
f $\dfrac{4x^2-25}{4x-10} \div \dfrac{2x+5}{8}$

g $\dfrac{x+3}{x^2+10x+25} \times \dfrac{x^2+5x}{x^2+3x}$
h $\dfrac{3y^2+4y-4}{10} \div \dfrac{3y+6}{15}$
i $\dfrac{x^2+2xy+y^2}{2} \times \dfrac{4}{(x-y)^2}$

(E/P) 3 Show that $\dfrac{x^2-64}{x^2-36} \div \dfrac{64-x^2}{x^2-36} = -1$ **(4 marks)**

(E/P) 4 Show that $\dfrac{2x^2-11x-40}{x^2-4x-32} \times \dfrac{x^2+8x+16}{6x^2-3x-45} \div \dfrac{8x^2+20x-48}{10x^2-45x+45} = \dfrac{a}{b}$ and find the values of the constants a and b, where a and b are integers. **(4 marks)**

Algebraic methods

5 a Simplify fully $\dfrac{x^2 + 2x - 24}{2x^2 + 10x} \times \dfrac{x^2 - 3x}{x^2 + 3x - 18}$ **(3 marks)**

> **Hint** Simplify and then solve the logarithmic equation.
> ← Year 1, Section 14.6

b Given that
$\ln((x^2 + 2x - 24)(x^2 - 3x)) = 2 + \ln((2x^2 + 10x)(x^2 + 3x - 18))$ find x in terms of e. **(4 marks)**

6 $f(x) = \dfrac{2x^2 - 3x - 2}{6x - 8} \div \dfrac{x - 2}{3x^2 + 14x - 24}$

> **Hint** Differentiate each term separately.
> ← Year 1, Section 12.5

a Show that $f(x) = \dfrac{2x^2 + 13x + 6}{2}$ **(4 marks)**

b Hence differentiate $f(x)$ and find $f'(4)$. **(3 marks)**

- To add or subtract two fractions, find a common denominator.

Example 7

Simplify the following:

a $\dfrac{1}{3} + \dfrac{3}{4}$ **b** $\dfrac{a}{2x} + \dfrac{b}{3x}$ **c** $\dfrac{2}{x + 3} - \dfrac{1}{x + 1}$ **d** $\dfrac{3}{x + 1} - \dfrac{4x}{x^2 - 1}$

a

$\dfrac{1}{3} + \dfrac{3}{4}$

$\times \dfrac{4}{4} \qquad \times \dfrac{3}{3}$

$= \dfrac{4}{12} + \dfrac{9}{12}$ ———— The lowest common multiple of 3 and 4 is 12.

$= \dfrac{13}{12}$

b $\dfrac{a}{2x} + \dfrac{b}{3x}$ ———— The lowest common multiple of $2x$ and $3x$ is $6x$.

$= \dfrac{3a}{6x} + \dfrac{2b}{6x}$ ———— Multiply the first fraction by $\dfrac{3}{3}$ and the second fraction by $\dfrac{2}{2}$

$= \dfrac{3a + 2b}{6x}$

c $\dfrac{2}{(x + 3)} - \dfrac{1}{(x + 1)}$ ———— The lowest common multiple is $(x + 3)(x + 1)$, so change both fractions so that the denominators are $(x + 3)(x + 1)$.

$= \dfrac{2(x + 1)}{(x + 3)(x + 1)} - \dfrac{1(x + 3)}{(x + 3)(x + 1)}$

$= \dfrac{2(x + 1) - 1(x + 3)}{(x + 3)(x + 1)}$ ———— Subtract the numerators.

$= \dfrac{2x + 2 - 1x - 3}{(x + 3)(x + 1)}$ ———— Expand the brackets.

$= \dfrac{x - 1}{(x + 3)(x + 1)}$ ———— Simplify the numerator.

Chapter 1

d $\dfrac{3}{x+1} - \dfrac{4x}{x^2-1}$

$= \dfrac{3}{x+1} - \dfrac{4x}{(x+1)(x-1)}$ ← Factorise $x^2 - 1$ to $(x+1)(x-1)$.

$= \dfrac{3(x-1)}{(x+1)(x-1)} - \dfrac{4x}{(x+1)(x-1)}$ ← The LCM of $(x+1)$ and $(x+1)(x-1)$ is $(x+1)(x-1)$.

$= \dfrac{3(x-1) - 4x}{(x+1)(x-1)}$

$= \dfrac{-x-3}{(x+1)(x-1)}$ ← Simplify the numerator: $3x - 3 - 4x = -x - 3$.

Exercise 1C

1 Write as a single fraction:

a $\dfrac{1}{3} + \dfrac{1}{4}$ **b** $\dfrac{3}{4} - \dfrac{2}{5}$ **c** $\dfrac{1}{p} + \dfrac{1}{q}$ **d** $\dfrac{3}{4x} + \dfrac{1}{8x}$ **e** $\dfrac{3}{x^2} - \dfrac{1}{x}$ **f** $\dfrac{a}{5b} - \dfrac{3}{2b}$

2 Write as a single fraction:

a $\dfrac{3}{x} - \dfrac{2}{x+1}$ **b** $\dfrac{2}{x-1} - \dfrac{3}{x+2}$ **c** $\dfrac{4}{2x+1} + \dfrac{2}{x-1}$

d $\dfrac{1}{3}(x+2) - \dfrac{1}{2}(x+3)$ **e** $\dfrac{3x}{(x+4)^2} - \dfrac{1}{x+4}$ **f** $\dfrac{5}{2(x+3)} + \dfrac{4}{3(x-1)}$

3 Write as a single fraction:

a $\dfrac{2}{x^2+2x+1} + \dfrac{1}{x+1}$ **b** $\dfrac{7}{x^2-4} + \dfrac{3}{x+2}$ **c** $\dfrac{2}{x^2+6x+9} - \dfrac{3}{x^2+4x+3}$

d $\dfrac{2}{y^2-x^2} + \dfrac{3}{y-x}$ **e** $\dfrac{3}{x^2+3x+2} - \dfrac{1}{x^2+4x+4}$ **f** $\dfrac{x+2}{x^2-x-12} - \dfrac{x+1}{x^2+5x+6}$

(E) 4 Express $\dfrac{6x+1}{x^2+2x-15} - \dfrac{4}{x-3}$ as a single fraction in its simplest form. **(4 marks)**

5 Express each of the following as a fraction in its simplest form.

a $\dfrac{3}{x} + \dfrac{2}{x+1} + \dfrac{1}{x+2}$ **b** $\dfrac{4}{3x} - \dfrac{2}{x-2} + \dfrac{1}{2x+1}$ **c** $\dfrac{3}{x-1} + \dfrac{2}{x+1} + \dfrac{4}{x-3}$

(E) 6 Express $\dfrac{4(2x-1)}{36x^2-1} + \dfrac{7}{6x-1}$ as a single fraction in its simplest form. **(4 marks)**

(E/P) 7 $g(x) = x + \dfrac{6}{x+2} + \dfrac{36}{x^2-2x-8}$, $x \in \mathbb{R}, x \neq -2, x \neq 4$

a Show that $g(x) = \dfrac{x^3 - 2x^2 - 2x + 12}{(x+2)(x-4)}$ **(3 marks)**

b Using algebraic long division, or otherwise, further show that $g(x) = \dfrac{x^2 - 4x + 6}{x-4}$ **(3 marks)**

8

1.3 Partial fractions

- **A single fraction with two distinct linear factors in the denominator can be split into two separate fractions with linear denominators. This is called splitting it into partial fractions.**

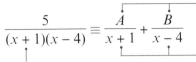

$$\frac{5}{(x+1)(x-4)} \equiv \frac{A}{x+1} + \frac{B}{x-4}$$

- A and B are constants to be found.
- The denominator contains two linear factors: $(x+1)$ and $(x-4)$.
- The expression is rewritten as the sum of two partial fractions.

Links Partial fractions are used for binomial expansions → **Chapter 4** and integration. → **Chapter 11**

There are two methods to find the constants A and B: by **substitution** and by **equating coefficients**.

Example 8

Split $\dfrac{6x-2}{(x-3)(x+1)}$ into partial fractions by: **a** substitution **b** equating coefficients.

a
$$\frac{6x-2}{(x-3)(x+1)} \equiv \frac{A}{x-3} + \frac{B}{x+1}$$

Set $\dfrac{6x-2}{(x-3)(x+1)}$ identical to $\dfrac{A}{x-3} + \dfrac{B}{x+1}$

$$\equiv \frac{A(x+1) + B(x-3)}{(x-3)(x+1)}$$

Add the two fractions.

$$6x - 2 \equiv A(x+1) + B(x-3)$$
$$6 \times 3 - 2 = A(3+1) + B(3-3)$$
$$16 = 4A$$
$$A = 4$$

To find A substitute $x = 3$.
This value of x eliminates B from the equation.

$$6 \times (-1) - 2 = A(-1+1) + B(-1-3)$$
$$-8 = -4B$$
$$B = 2$$

To find B substitute $x = -1$.
This value of x eliminates A from the equation.

$$\therefore \frac{6x-2}{(x-3)(x+1)} \equiv \frac{4}{x-3} + \frac{2}{x+1}$$

b
$$\frac{6x-2}{(x-3)(x+1)} \equiv \frac{A}{x-3} + \frac{B}{x+1}$$

$$\equiv \frac{A(x+1) + B(x-3)}{(x-3)(x+1)}$$

$$6x - 2 \equiv A(x+1) + B(x-3)$$
$$\equiv Ax + A + Bx - 3B$$

Expand the brackets.

$$\equiv (A+B)x + (A - 3B)$$

Collect like terms.

Equate coefficients of x:
$$6 = A + B \qquad (1)$$

Equate constant terms:
$$-2 = A - 3B \qquad (2)$$

You want $(A+B)x + A - 3B \equiv 6x - 2$.
Hence coefficient of x is 6, and constant term is -2.

$(1) - (2)$:
$$8 = 4B$$
$$\Rightarrow B = 2$$

Solve simultaneously.

Substitute $B = 2$ in $(1) \Rightarrow 6 = A + 2$
$$A = 4$$

Chapter 1

- **The method of partial fractions can also be used when there are more than two distinct linear factors in the denominator.**

For example, the expression $\dfrac{7}{(x-2)(x+6)(x+3)}$ can be split into $\dfrac{A}{x-2} + \dfrac{B}{x+6} + \dfrac{C}{x+3}$

The constants A, B and C can again be found by either substitution or by equating coefficients.

Watch out This method cannot be used for a repeated linear factor in the denominator. For example, the expression $\dfrac{9}{(x+4)(x-1)^2}$ cannot be rewritten as $\dfrac{A}{x+4} + \dfrac{B}{x-1} + \dfrac{C}{x-1}$ because $(x-1)$ is a repeated factor. There is more on this in the next section.

Example 9

Given that $\dfrac{6x^2 + 5x - 2}{x(x-1)(2x+1)} \equiv \dfrac{A}{x} + \dfrac{B}{x-1} + \dfrac{C}{2x+1}$, find the values of the constants A, B and C.

Let $\dfrac{6x^2 + 5x - 2}{x(x-1)(2x+1)} \equiv \dfrac{A}{x} + \dfrac{B}{x-1} + \dfrac{C}{2x+1}$ — The denominators must be x, $(x-1)$ and $(2x+1)$.

$\equiv \dfrac{A(x-1)(2x+1) + Bx(2x+1) + Cx(x-1)}{x(x-1)(2x+1)}$ — Add the fractions.

$\therefore 6x^2 + 5x - 2 \equiv A(x-1)(2x+1) + Bx(2x+1) + Cx(x-1)$ — The numerators are equal.

Let $x = 1$:
$6 + 5 - 2 = 0 + B \times 1 \times 3 + 0$
$9 = 3B$
$B = 3$

Let $x = 0$:
$0 + 0 - 2 = A \times (-1) \times 1 + 0 + 0$
$-2 = -A$
$A = 2$

Proceed by substitution OR by equating coefficients.
Here we used the method of substitution.

Let $x = -\tfrac{1}{2}$:
$\tfrac{6}{4} - \tfrac{5}{2} - 2 = 0 + 0 + C \times \left(-\tfrac{1}{2}\right) \times \left(-\tfrac{3}{2}\right)$
$-3 = \tfrac{3}{4}C$
$C = -4$

So $\dfrac{6x^2 + 5x - 2}{x(x-1)(2x+1)} \equiv \dfrac{2}{x} + \dfrac{3}{x-1} - \dfrac{4}{2x+1}$

So $A = 2$, $B = 3$ and $C = -4$. — Finish the question by listing the coefficients.

Exercise 1D

1 Express the following as partial fractions:

a $\dfrac{6x-2}{(x-2)(x+3)}$

b $\dfrac{2x+11}{(x+1)(x+4)}$

c $\dfrac{-7x-12}{2x(x-4)}$

d $\dfrac{2x-13}{(2x+1)(x-3)}$

e $\dfrac{6x+6}{x^2-9}$

Hint First factorise the denominator.

f $\dfrac{7-3x}{x^2-3x-4}$

g $\dfrac{8-x}{x^2+4x}$

h $\dfrac{2x-14}{x^2+2x-15}$

(E) 2 Show that $\dfrac{-2x-5}{(4+x)(2-x)}$ can be written in the form $\dfrac{A}{4+x}+\dfrac{B}{2-x}$ where A and B are constants to be found. **(3 marks)**

(P) 3 The expression $\dfrac{A}{(x-4)(x+8)}$ can be written in partial fractions as $\dfrac{2}{x-4}+\dfrac{B}{x+8}$.

Find the values of the constants A and B.

(E) 4 $h(x) = \dfrac{2x^2-12x-26}{(x+1)(x-2)(x+5)}, x > 2$

Given that $h(x)$ can be expressed in the form $\dfrac{A}{x+1}+\dfrac{B}{x-2}+\dfrac{C}{x+5}$, find the values of A, B and C. **(4 marks)**

(E) 5 Given that, for $x < -1$, $\dfrac{-10x^2-8x+2}{x(2x+1)(3x-2)} \equiv \dfrac{D}{x}+\dfrac{E}{2x+1}+\dfrac{F}{3x-2}$, where D, E and F are constants. Find the values of D, E and F. **(4 marks)**

6 Express the following as partial fractions:

$\dfrac{-5x^2-19x-32}{(x+1)(x+2)(x-5)}$

(P) 7 Express the following as partial fractions:

a $\dfrac{6x^2+7x-3}{x^3-x}$

b $\dfrac{8x+9}{10x^2+3x-4}$

Hint First factorise the denominator.

Challenge

Express $\dfrac{5x^2-15x-8}{x^3-4x^2+x+6}$ as a sum of fractions with linear denominators.

1.4 Repeated factors

- **A single fraction with a repeated linear factor in the denominator can be split into two or more separate fractions.**

In this case, there is a special method for dealing with the repeated linear factor.

$$\frac{2x+9}{(x-5)(x+3)^2} \equiv \frac{A}{x-5} + \frac{B}{x+3} + \frac{C}{(x+3)^2}$$

A and B and C are constants to be found.

The denominator contains three linear factors: $(x-5)$, $(x+3)$ and $(x+3)$. $(x+3)$ is a repeated linear factor.

The expression is rewritten as the sum of three partial fractions. Notice that $(x-5)$, $(x+3)$ and $(x+3)^2$ are the denominators.

Example 10

Show that $\dfrac{11x^2 + 14x + 5}{(x+1)^2(2x+1)}$ can be written in the form $\dfrac{A}{x+1} + \dfrac{B}{(x+1)^2} + \dfrac{C}{2x+1}$, where A, B and C are constants to be found.

Let
$$\frac{11x^2 + 14x + 5}{(x+1)^2(2x+1)} \equiv \frac{A}{(x+1)} + \frac{B}{(x+1)^2} + \frac{C}{(2x+1)}$$

You need denominators of $(x+1)$, $(x+1)^2$ and $(2x+1)$.

$$\equiv \frac{A(x+1)(2x+1) + B(2x+1) + C(x+1)^2}{(x+1)^2(2x+1)}$$

Add the three fractions.

Hence $11x^2 + 14x + 5$
$\equiv A(x+1)(2x+1) + B(2x+1) + C(x+1)^2$ (1)

The numerators are equal.

Let $x = -1$:
$11 - 14 + 5 = A \times 0 + B \times -1 + C \times 0$
$2 = -1B$
$B = -2$

To find B substitute $x = -1$.

Let $x = -\tfrac{1}{2}$:
$\tfrac{11}{4} - 7 + 5 = A \times 0 + B \times 0 + C \times \tfrac{1}{4}$
$\tfrac{3}{4} = \tfrac{1}{4}C$
$C = 3$

To find C substitute $x = -\tfrac{1}{2}$

$11 = 2A + C$
$11 = 2A + 3$
$2A = 8$
$A = 4$

Equate terms in x^2 in (1). Terms in x^2 are $A \times 2x^2 + C \times x^2$.

Substitute $C = 3$.

Hence $\dfrac{11x^2 + 14x + 5}{(x+1)^2(2x+1)}$

$\equiv \dfrac{4}{(x+1)} - \dfrac{2}{(x+1)^2} + \dfrac{3}{(2x+1)}$

So $A = 4$, $B = -2$ and $C = 3$.

Finish the question by listing the coefficients.

Online Check your answer using the simultaneous equations function on your calculator.

Algebraic methods

Exercise 1E

1. $f(x) = \dfrac{3x^2 + x + 1}{x^2(x + 1)}, x \neq 0, x \neq -1$

   Given that $f(x)$ can be expressed in the form $\dfrac{A}{x} + \dfrac{B}{x^2} + \dfrac{C}{x + 1}$, find the values of A, B and C. **(4 marks)**

2. $g(x) = \dfrac{-x^2 - 10x - 5}{(x + 1)^2(x - 1)}, x \neq -1, x \neq 1$

   Find the values of the constants D, E and F such that $g(x) = \dfrac{D}{x + 1} + \dfrac{E}{(x + 1)^2} + \dfrac{F}{x - 1}$ **(4 marks)**

3. Given that, for $x < 0$, $\dfrac{2x^2 + 2x - 18}{x(x - 3)^2} \equiv \dfrac{P}{x} + \dfrac{Q}{x - 3} + \dfrac{R}{(x - 3)^2}$, where P, Q and R are constants, find the values of P, Q and R. **(4 marks)**

4. Show that $\dfrac{5x^2 - 2x - 1}{x^3 - x^2}$ can be written in the form $\dfrac{C}{x} + \dfrac{D}{x^2} + \dfrac{E}{x - 1}$ where C, D and E are constants to be found. **(4 marks)**

5. $p(x) = \dfrac{2x}{(x + 2)^2}, x \neq -2$.

   Find the values of the constants A and B such that $p(x) = \dfrac{A}{x + 2} + \dfrac{B}{(x + 2)^2}$ **(4 marks)**

6. $\dfrac{10x^2 - 10x + 17}{(2x + 1)(x - 3)^2} \equiv \dfrac{A}{2x + 1} + \dfrac{B}{x - 3} + \dfrac{C}{(x - 3)^2}, x > 3$

   Find the values of the constants A, B and C. **(4 marks)**

7. Show that $\dfrac{39x^2 + 2x + 59}{(x + 5)(3x - 1)^2}$ can be written in the form $\dfrac{A}{x + 5} + \dfrac{B}{3x - 1} + \dfrac{C}{(3x - 1)^2}$ where A, B and C are constants to be found. **(4 marks)**

8. Express the following as partial fractions:

   a $\dfrac{4x + 1}{x^2 + 10x + 25}$

   b $\dfrac{6x^2 - x + 2}{4x^3 - 4x^2 + x}$

Chapter 1

1.5 Algebraic division

- An **improper algebraic fraction** is one whose numerator has a degree equal to or larger than the denominator. An improper fraction must be converted to a mixed fraction before you can express it in partial fractions.

$$\frac{x^2 + 5x + 8}{x - 2} \text{ and } \frac{x^3 + 5x - 9}{x^3 - 4x^2 + 7x - 3} \text{ are both improper fractions.}$$

The degree of the numerator is greater than the degree of the denominator.

The degrees of the numerator and denominator are equal.

Notation The **degree of a polynomial** is the largest exponent in the expression. For example, $x^3 + 5x - 9$ has degree 3.

- You can either use:
  - algebraic division
  - or the relationship $F(x) = Q(x) \times \text{divisor} + \text{remainder}$

to convert an improper fraction into a mixed fraction.

Watch out The divisor and the remainder can be numbers or functions of x.

Method 1
Use algebraic long division to show that:

$$F(x) \longrightarrow \frac{x^2 + 5x + 8}{x - 2} \equiv x + 7 + \frac{22}{x - 2} \longleftarrow \text{remainder}$$

with $Q(x) = x + 7$ and divisor $x - 2$.

Method 2
Multiply by $(x - 2)$ and compare coefficients to show that:

$$F(x) \longrightarrow x^2 + 5x + 8 \equiv (x + 7)(x - 2) + 22 \longleftarrow \text{remainder}$$

with divisor $(x - 2)$ and $Q(x) = x + 7$.

Example 11

Given that $\dfrac{x^3 + x^2 - 7}{x - 3} \equiv Ax^2 + Bx + C + \dfrac{D}{x - 3}$, find the values of A, B, C and D.

Using algebraic long division:

$$
\begin{array}{r}
x^2 + 4x + 12 \\
x - 3 \overline{\smash{)}x^3 + x^2 + 0x - 7} \\
\underline{x^3 - 3x^2} \\
4x^2 + 0x \\
\underline{4x^2 - 12x} \\
12x - 7 \\
\underline{12x - 36} \\
29
\end{array}
$$

Problem-solving
Solving this problem using algebraic long division will give you an answer in the form asked for in the question.

Algebraic methods

So $\dfrac{x^3 + x^2 - 7}{x - 3} = x^2 + 4x + 12$

with a remainder of 29.

$\dfrac{x^3 + x^2 - 7}{x - 3} = x^2 + 4x + 12 + \dfrac{29}{x - 3}$

So $A = 1$, $B = 4$, $C = 12$ and $D = 29$.

— The divisor is $(x - 3)$ so you need to write the remainder as a fraction with denominator $(x - 3)$.

— It's always a good idea to list the value of each unknown asked for in the question.

Example 12

Given that $x^3 + x^2 - 7 \equiv (Ax^2 + Bx + C)(x - 3) + D$, find the values of A, B, C and D.

Let $x = 3$:
$27 + 9 - 7 = (9A + 3B + C) \times 0 + D$
$D = 29$

Let $x = 0$:
$0 + 0 - 7 = (A \times 0 + B \times 0 + C)$
$\times (0 - 3) + D$
$-7 = -3C + D$
$-7 = -3C + 29$
$3C = 36$
$C = 12$

Compare the coefficients of x^3 and x^2.
Compare coefficients in x^3: $\quad 1 = A$
Compare coefficients in x^2: $\quad 1 = -3A + B$
$1 = -3 + B$
Therefore $A = 1$, $B = 4$, $C = 12$ and $D = 29$
and we can write
$x^3 + x^2 - 7 \equiv (x^2 + 4x + 12)(x - 3) + 29$
This can also be written as:
$\dfrac{x^3 + x^2 - 7}{x - 3} \equiv x^2 + 4x + 12 + \dfrac{29}{x - 3}$

Problem-solving

The identity is given in the form $F(x) \equiv Q(x) \times$ divisor + remainder so solve the problem by equating coefficients.

— Set $x = 3$ to find the value of D.

— Set $x = 0$ and use your value of D to find the value of C.

You can find the remaining values by equating coefficients of x^3 and x^2.
Remember there are two x^2 terms when you expand the brackets on the RHS:
x^3 terms: LHS = x^3, RHS = Ax^3
x^2 terms: LHS = x^2, RHS = $(-3A + B)x^2$

Example 13

$f(x) = \dfrac{x^4 + x^3 + x - 10}{x^2 + 2x - 3}$

Show that $f(x)$ can be written as $Ax^2 + Bx + C + \dfrac{Dx + E}{x^2 + 2x - 3}$ and find the values of A, B, C, D and E.

Chapter 1

Using algebraic long division:

$$\begin{array}{r} x^2 - x + 5 \\ x^2 + 2x - 3 \overline{\smash{\big)}\, x^4 + x^3 + 0x^2 + x - 10} \\ \underline{x^4 + 2x^3 - 3x^2} \\ -x^3 + 3x^2 + x \\ \underline{-x^3 - 2x^2 + 3x} \\ 5x^2 - 2x - 10 \\ \underline{5x^2 + 10x - 15} \\ -12x + 5 \end{array}$$

$$\frac{x^4 + x^3 + x - 10}{x^2 + 2x - 3} \equiv x^2 - x + 5 + \frac{-12x + 5}{x^2 + 2x - 3}$$

So $A = 1$, $B = -1$, $C = 5$, $D = -12$ and $E = 5$.

Watch out When you are dividing by a quadratic expression, the remainder can be a constant or a linear expression. The degree of $(-12x + 5)$ is smaller than the degree of $(x^2 + 2x - 3)$ so stop your division here. The remainder is $-12x + 5$.

Write the remainder as a fraction over the whole divisor.

Exercise 1F

(E) **1** $\dfrac{x^3 + 2x^2 + 3x - 4}{x + 1} \equiv Ax^2 + Bx + C + \dfrac{D}{x + 1}$

Find the values of the constants A, B, C and D. **(4 marks)**

(E) **2** Given that $\dfrac{2x^3 + 3x^2 - 4x + 5}{x + 3} \equiv ax^2 + bx + c + \dfrac{d}{x + 3}$ find the values of a, b, c and d. **(4 marks)**

(E) **3** $f(x) = \dfrac{x^3 - 8}{x - 2}$

Show that $f(x)$ can be written in the form $px^2 + qx + r$ and find the values of p, q and r. **(4 marks)**

(E) **4** Given that $\dfrac{2x^2 + 4x + 5}{x^2 - 1} \equiv m + \dfrac{nx + p}{x^2 - 1}$ find the values of m, n and p. **(4 marks)**

(E) **5** Find the values of the constants A, B, C and D in the following identity:
$8x^3 + 2x^2 + 5 \equiv (Ax + B)(2x^2 + 2) + Cx + D$ **(4 marks)**

(E) **6** $\dfrac{4x^3 - 5x^2 + 3x - 14}{x^2 + 2x - 1} \equiv Ax + B + \dfrac{Cx + D}{x^2 + 2x - 1}$

Find the values of the constants A, B, C and D. **(4 marks)**

(E) **7** $g(x) = \dfrac{x^4 + 3x^2 - 4}{x^2 + 1}$. Show that $g(x)$ can be written in the form $px^2 + qx + r + \dfrac{sx + t}{x^2 + 1}$
and find the values of p, q, r, s and t. **(4 marks)**

(E) **8** Given that $\dfrac{2x^4 + 3x^3 - 2x^2 + 4x - 6}{x^2 + x - 2} \equiv ax^2 + bx + c + \dfrac{dx + e}{x^2 + x - 2}$ find the values
of a, b, c, d and e. **(5 marks)**

Algebraic methods

9 Find the values of the constants A, B, C, D and E in the following identity:
$$3x^4 - 4x^3 - 8x^2 + 16x - 2 \equiv (Ax^2 + Bx + C)(x^2 - 3) + Dx + E$$ (5 marks)

10 **a** Fully factorise the expression $x^4 - 1$. (2 marks)
   b Hence, or otherwise, write the algebraic fraction $\dfrac{x^4 - 1}{x + 1}$ in the form $(ax + b)(cx^2 + dx + e)$ and find the values of a, b, c, d and e. (4 marks)

In order to express an improper algebraic fraction in partial fractions, it is first necessary to divide the numerator by the denominator. Remember an improper algebraic fraction is one where the degree of the numerator is greater than or equal to the degree of the denominator.

Example 14

Given that $\dfrac{3x^2 - 3x - 2}{(x-1)(x-2)} \equiv A + \dfrac{B}{x-1} + \dfrac{C}{x-2}$, find the values of A, B and C.

$\dfrac{3x^2 - 3x - 2}{(x-1)(x-2)} \equiv \dfrac{3x^2 - 3x - 2}{x^2 - 3x + 2}$ — Multiply out the denominator on the LHS.

$$\begin{array}{r} 3 \\ x^2 - 3x + 2 \overline{)3x^2 - 3x - 2} \\ 3x^2 - 9x + 6 \\ \hline 6x - 8 \end{array}$$

Divide the denominator into the numerator. It goes in 3 times, with a remainder of $6x - 8$.

Therefore
$\dfrac{3x^2 - 3x - 2}{(x-1)(x-2)} \equiv 3 + \dfrac{6x - 8}{x^2 - 3x + 2}$ — Write $\dfrac{3x^2 - 3x - 2}{(x-1)(x-2)}$ as a mixed fraction.

$\equiv 3 + \dfrac{6x - 8}{(x-1)(x-2)}$ — Factorise $x^2 - 3x + 2$.

Let $\dfrac{6x - 8}{(x-1)(x-2)} \equiv \dfrac{B}{(x-1)} + \dfrac{C}{(x-2)}$ — The denominators must be $(x-1)$ and $(x-2)$.

$\equiv \dfrac{B(x-2) + C(x-1)}{(x-1)(x-2)}$ — Add the two fractions.

$6x - 8 \equiv B(x-2) + C(x-1)$ — The numerators are equal.

Let $x = 2$: $12 - 8 = B \times 0 + C \times 1$ — Substitute $x = 2$ to find C.
$C = 4$

Let $x = 1$: $6 - 8 = B \times -1 + C \times 0$ — Substitute $x = 1$ to find B.
$B = 2$

$\dfrac{3x^2 - 3x - 2}{(x-1)(x-2)} \equiv 3 + \dfrac{6x - 8}{(x-1)(x-2)}$
$\equiv 3 + \dfrac{2}{(x-1)} + \dfrac{4}{(x-2)}$ — Write out the full solution.

So $A = 3$, $B = 2$ and $C = 4$. — Finish the question by listing the coefficients.

Chapter 1

Exercise 1G

1. $g(x) = \dfrac{x^2 + 3x - 2}{(x-1)(x-2)}$. Show that $g(x)$ can we written in the form $A + \dfrac{B}{x-1} + \dfrac{C}{x-2}$ and find the values of the constants A, B and C. **(4 marks)**

2. Given that $\dfrac{x^2 - 10}{(x-2)(x+1)} \equiv A + \dfrac{B}{x-2} + \dfrac{C}{x+1}$, find the values of the constants A, B and C. **(4 marks)**

3. Find the values of the constants A, B, C and D in the following identity:
$$\dfrac{x^3 - x^2 - x - 3}{x(x-1)} \equiv Ax + B + \dfrac{C}{x} + \dfrac{D}{x-1}$$
(5 marks)

4. Show that $\dfrac{-3x^3 - 4x^2 + 19x + 8}{x^2 + 2x - 3}$ can be expressed in the form $A + Bx + \dfrac{C}{(x-1)} + \dfrac{D}{(x+3)}$, where A, B, C and D are constants to be found. **(5 marks)**

5. $p(x) \equiv \dfrac{4x^2 + 25}{4x^2 - 25}$

   Show that $p(x)$ can be written in the form $A + \dfrac{B}{2x-5} + \dfrac{C}{2x+5}$, where A, B and C are constants to be found. **(4 marks)**

6. Given that $\dfrac{2x^2 - 1}{x^2 + 2x + 1} \equiv A + \dfrac{B}{x+1} + \dfrac{C}{(x+1)^2}$, find the values of the constants A, B and C. **(4 marks)**

7. By factorising the denominator, express the following as partial fractions:

   a $\dfrac{4x^2 + 17x - 11}{x^2 + 3x - 4}$ b $\dfrac{x^4 - 4x^3 + 9x^2 - 17x + 12}{x^3 - 4x^2 + 4x}$

8. Given that $\dfrac{6x^3 - 7x^2 + 3}{3x^2 + x - 10} \equiv Ax + B + \dfrac{C}{3x-5} + \dfrac{D}{x+2}$, find the values of the constants A, B, C and D. **(6 marks)**

9. $q(x) = \dfrac{8x^3 + 1}{4x^2 - 4x + 1}$

   Show that $q(x)$ can be written in the form $Ax + B + \dfrac{C}{2x-1} + \dfrac{D}{(2x-1)^2}$ and find the values of the constants A, B, C and D. **(6 marks)**

10. $h(x) = \dfrac{x^4 + 2x^2 - 3x + 8}{x^2 + x - 2}$

    Show that $h(x)$ can be written as $Ax^2 + Bx + C + \dfrac{D}{x+2} + \dfrac{E}{x-1}$ and find the values of A, B, C, D and E. **(5 marks)**

Algebraic methods

Mixed exercise 1

1. Prove by contradiction that $\sqrt{\frac{1}{2}}$ is an irrational number. **(5 marks)**

2. Prove that if q^2 is an irrational number then q is an irrational number.

3. Simplify:

   a $\dfrac{x-4}{6} \times \dfrac{2x+8}{x^2-16}$

   b $\dfrac{x^2-3x-10}{3x^2-21} \times \dfrac{6x^2+24}{x^2+6x+8}$

   c $\dfrac{4x^2+12x+9}{x^2+6x} \div \dfrac{4x^2-9}{2x^2+9x-18}$

4. a Simplify fully $\dfrac{4x^2-8x}{x^2-3x-4} \times \dfrac{x^2+6x+5}{2x^2+10x}$ **(3 marks)**

   b Given that $\ln((4x^2-8x)(x^2+6x+5)) = 6 + \ln((x^2-3x-4)(2x^2+10x))$ find x in terms of e. **(4 marks)**

5. $g(x) = \dfrac{4x^3-9x^2-9x}{32x+24} \div \dfrac{x^2-3x}{6x^2-13x-5}$

   a Show that $g(x)$ can be written in the form $ax^2 + bx + c$, where a, b and c are constants to be found. **(4 marks)**

   b Hence differentiate $g(x)$ and find $g'(-2)$. **(3 marks)**

6. Express $\dfrac{6x+1}{x-5} + \dfrac{5x+3}{x^2-3x-10}$ as a single fraction in its simplest form. **(4 marks)**

7. $f(x) = x + \dfrac{3}{x-1} - \dfrac{12}{x^2+2x-3}$, $x \in \mathbb{R}$, $x > 1$

   Show that $f(x) = \dfrac{x^2+3x+3}{x+3}$ **(4 marks)**

8. $f(x) = \dfrac{x-3}{x(x-1)}$

   Show that $f(x)$ can be written in the form $\dfrac{A}{x} + \dfrac{B}{x-1}$ where A and B are constants to be found. **(3 marks)**

9. $\dfrac{-15x+21}{(x-2)(x+1)(x-5)} \equiv \dfrac{P}{x-2} + \dfrac{Q}{x+1} + \dfrac{R}{x-5}$

   Find the values of the constants P, Q and R. **(4 marks)**

10. Show that $\dfrac{16x-1}{(3x+2)(2x-1)}$ can be written in the form $\dfrac{D}{3x+2} + \dfrac{E}{2x-1}$ and find the values of the constants D and E. **(4 marks)**

11 $\dfrac{7x^2 + 2x - 2}{x^2(x+1)} \equiv \dfrac{A}{x} + \dfrac{B}{x^2} + \dfrac{C}{x+1}$

Find the values of the constants A, B and C. **(4 marks)**

12 $h(x) = \dfrac{21x^2 - 13}{(x+5)(3x-1)^2}$

Show that $h(x)$ can be written in the form $\dfrac{D}{x+5} + \dfrac{E}{(3x-1)} + \dfrac{F}{(3x-1)^2}$ where D, E and F are constants to be found. **(5 marks)**

13 Find the values of the constants A, B, C and D in the following identity:

$x^3 - 6x^2 + 11x + 2 \equiv (x-2)(Ax^2 + Bx + C) + D$ **(5 marks)**

14 Show that $\dfrac{4x^3 - 6x^2 + 8x - 5}{2x+1}$ can be put in the form $Ax^2 + Bx + C + \dfrac{D}{2x+1}$.

Find the values of the constants A, B, C and D. **(5 marks)**

15 Show that $\dfrac{x^4 + 2}{x^2 - 1} \equiv Ax^2 + Bx + C + \dfrac{D}{x^2 - 1}$ where A, B, C and D are constants to be found. **(5 marks)**

16 $\dfrac{x^4}{x^2 - 2x + 1} \equiv Ax^2 + Bx + C + \dfrac{D}{x-1} + \dfrac{E}{(x-1)^2}$

Find the values of the constants A, B, C, D and E. **(5 marks)**

17 $h(x) = \dfrac{2x^2 + 2x - 3}{x^2 + 2x - 3}$

Show that $h(x)$ can be written in the form $A + \dfrac{B}{x+3} + \dfrac{C}{x-1}$ where A, B and C are constants to be found. **(5 marks)**

18 Given that $\dfrac{x^2 + 1}{x(x-2)} \equiv P + \dfrac{Q}{x} + \dfrac{R}{x-2}$, find the values of the constants P, Q and R. **(5 marks)**

19 Given that $f(x) = 2x^3 + 9x^2 + 10x + 3$:

a show that -3 is a root of $f(x)$

b express $\dfrac{10}{f(x)}$ as partial fractions.

Challenge

The line L meets the circle C with centre O at exactly one point, A.

Prove by contradiction that the line L is perpendicular to the radius OA.

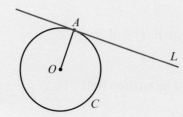

Hint In a right-angled triangle, the side opposite the right-angle is always the longest side.

Summary of key points

1. To prove a statement by contradiction you start by assuming it is **not true**. You then use logical steps to show that this assumption leads to something impossible (either a contradiction of the assumption or a contradiction of a fact you know to be true). You can conclude that your assumption was incorrect, and the original statement **was true**.

2. A rational number can be written as $\frac{a}{b}$, where a and b are integers.

   An irrational number cannot be expressed in the form $\frac{a}{b}$, where a and b are integers.

3. To multiply fractions, cancel any common factors, then multiply the numerators and multiply the denominators.

4. To divide two fractions, multiply the first fraction by the reciprocal of the second fraction.

5. To add or subtract two fractions, find a common denominator.

6. A single fraction with two distinct linear factors in the denominator can be split into two separate fractions with linear denominators. This is called splitting it into **partial fractions**:
$$\frac{5}{(x+1)(x-4)} = \frac{A}{x+1} + \frac{B}{x-4}$$

7. The method of partial fractions can also be used when there are more than two distinct linear factors in the denominator:
$$\frac{7}{(x-2)(x+6)(x+3)} = \frac{A}{x-2} + \frac{B}{x+6} + \frac{C}{x+3}$$

8. A single fraction with a repeated linear factor in the denominator can be split into two or more separate fractions:
$$\frac{2x+9}{(x-5)(x+3)^2} = \frac{A}{x-5} + \frac{B}{x+3} + \frac{C}{(x+3)^2}$$

9. An improper algebraic fraction is one whose numerator has a degree equal to or larger than the denominator. An improper fraction must be converted to a mixed fraction before you can express it in partial fractions.

10. You can either use:
    - algebraic division
    - or the relationship F(x) = Q(x) × divisor + remainder

    to convert an improper fraction into a mixed fraction.

2 Functions and graphs

Objectives

After completing this chapter you should be able to:

- Understand and use the modulus function → pages 23–27
- Understand mappings and functions, and use domain and range → pages 27–32
- Combine two or more functions to make a composite function → pages 32–35
- Know how to find the inverse of a function graphically and algebraically → pages 36–39
- Sketch the graphs of the modulus functions $y = |f(x)|$ and $y = f(|x|)$ → pages 40–44
- Apply a combination of two (or more) transformations to the same curve → pages 44–48
- Transform the modulus function → pages 48–52

Prior knowledge check

1 Make y the subject of each of the following:
   a $5x = 9 - 7y$ **b** $p = \dfrac{2y + 8x}{5}$
   c $5x - 8y = 4 + 9xy$ ← GCSE Mathematics

2 Write each expression in its simplest form.
   a $(5x - 3)^2 - 4$ **b** $\dfrac{1}{2(3x - 5) - 4}$
   c $\dfrac{\dfrac{x+4}{x+2} + 5}{\dfrac{x+4}{x+2} - 3}$ ← GCSE Mathematics

3 Sketch each of the following graphs. Label any points where the graph cuts the x- or y-axis.
   a $y = e^x$ **b** $y = x(x + 4)(x - 5)$
   c $y = \sin x$, $0 \leq x \leq 360°$ ← Year 1

4 $f(x) = x^2 - 3x$. Find the values of:
   a f(7)   **b** f(3)   **c** f(−3)   ← Year 1

Code breakers at Bletchley Park used inverse functions to decode enemy messages during World War II. When the enemy encoded a message they used a function. The code breakers' challenge was to find the inverse function that would decode the message.

Functions and graphs

2.1 The modulus function

The modulus of a number a, written as $|a|$, is its **non-negative** numerical value.
So, for example, $|5| = 5$ and also $|-5| = 5$.

- **A modulus function is, in general, a function of the type $y = |f(x)|$.**
  - **When $f(x) \geq 0$, $|f(x)| = f(x)$**
  - **When $f(x) < 0$, $|f(x)| = -f(x)$**

Notation The modulus function is also known as the **absolute value** function. On a calculator, the button is often labelled 'Abs'.

Example 1

Write down the values of
a $|-2|$ b $|6.5|$ c $\left|\dfrac{1}{3} - \dfrac{4}{5}\right|$

a $|-2| = 2$ — The positive numerical value of -2 is 2.

b $|6.5| = 6.5$ — 6.5 is a positive number.

c $\left|\dfrac{1}{3} - \dfrac{4}{5}\right| = \left|\dfrac{5}{15} - \dfrac{12}{15}\right| = \left|-\dfrac{7}{15}\right| = \dfrac{7}{15}$ — Work out the value inside the modulus.

Example 2

$f(x) = |2x - 3| + 1$

Write down the values of
a $f(5)$ b $f(-2)$ c $f(1)$

a $f(5) = |2 \times 5 - 3| + 1$
  $= |7| + 1 = 7 + 1 = 8$

b $f(-2) = |2(-2) - 3| + 1$
  $= |-7| + 1 = 7 + 1 = 8$

c $f(1) = |2 \times 1 - 3| + 1$
  $= |-1| + 1 = 1 + 1 = 2$

Watch out The modulus function acts like a pair of brackets. Work out the value inside the modulus function first.

Online Use your calculator to work out values of modulus functions.

- **To sketch the graph of $y = |ax + b|$, sketch $y = ax + b$ then reflect the section of the graph below the x-axis in the x-axis.**

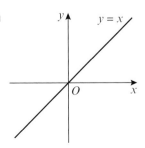 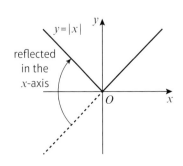

Example 3

Sketch the graph of $y = |3x - 2|$.

Online Explore graphs of f(x) and |f(x)| using technology.

Step 1
Sketch the graph of $y = 3x - 2$.
(Ignore the modulus.)

Step 2
For the part of the line below the x-axis (the negative values of y), reflect in the x-axis. For example, this will change the y-value −2 into the y-value 2.

You could check your answer using a table of values:

x	−1	0	1	2
$y = \|3x - 2\|$	5	2	1	4

Example 4

Solve the equation $|2x - 1| = 5$.

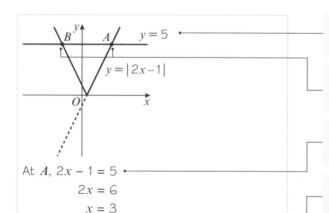

At A, $2x - 1 = 5$
$2x = 6$
$x = 3$

At B, $-(2x - 1) = 5$
$-2x + 1 = 5$
$2x = -4$
$x = -2$

The solutions are $x = 3$ and $x = -2$.

Start by sketching the graphs of $y = |2x - 1|$ and $y = 5$.

The graphs intersect at two points, A and B, so there will be two solutions to the equation.

A is the point of intersection on the original part of the graph.

B is the point of intersection on the reflected part of the graph.

Notation The function inside the modulus is called the **argument** of the modulus. You can solve modulus equations algebraically by considering the positive argument and the negative argument separately.

Functions and graphs

Example 5

Solve the equation $|3x - 5| = 2 - \frac{1}{2}x$.

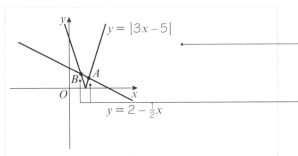

Online Explore intersections of straight lines and modulus graphs using technology.

First draw a sketch of the line $y = |3x - 5|$ and the line $y = 2 - \frac{1}{2}x$.

The sketch shows there are two solutions, at A and B, the points of intersection.

At A: $3x - 5 = 2 - \frac{1}{2}x$

$\frac{7}{2}x = 7$

$x = 2$

At B: $-(3x - 5) = 2 - \frac{1}{2}x$

$-3x + 5 = 2 - \frac{1}{2}x$

$-\frac{5}{2}x = -3$

$x = \frac{6}{5}$

The solutions are $x = 2$ and $x = \frac{6}{5}$

This is the solution on the original part of the graph.

When $f(x) < 0$, $|f(x)| = -f(x)$, so $-(3x - 5) = 2 - \frac{1}{2}x$ gives you the second solution.

This is the solution on the reflected part of the graph.

Example 6

Solve the inequality $|5x - 1| > 3x$.

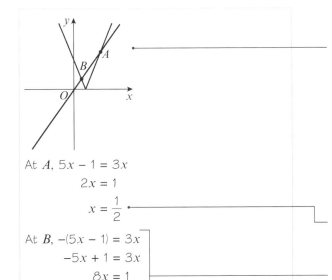

At A, $5x - 1 = 3x$

$2x = 1$

$x = \frac{1}{2}$

At B, $-(5x - 1) = 3x$

$-5x + 1 = 3x$

$8x = 1$

$x = \frac{1}{8}$

First draw a sketch of the line $y = |5x - 1|$ and the line $y = 3x$.

Solve the equation $|5x - 1| = 3x$ to find the x-coordinates of the points of intersection, A and B.

This is the intersection on the original part of the graph.

Consider the negative argument to find the point of intersection on the reflected part of the graph.

The points of intersection are $x = \frac{1}{2}$ and $x = \frac{1}{8}$

So the solution to $|5x - 1| > 3x$ is $x < \frac{1}{8}$ or $x > \frac{1}{2}$

Problem-solving

Look at the sketch to work out which values of x satisfy the inequality. $y = |5x - 1|$ is above $y = 3x$ when $x > \frac{1}{2}$ or $x < \frac{1}{8}$. You could write the solution in set notation as $\{x : x > \frac{1}{2}\} \cup \{x : x < \frac{1}{8}\}$.

Exercise 2A

1 Write down the values of
  a $\left|\frac{3}{4}\right|$
  b $|-0.28|$
  c $|3 - 11|$
  d $\left|\frac{5}{7} - \frac{3}{8}\right|$
  e $|20 - 6 \times 4|$
  f $|4^2 \times 2 - 3 \times 7|$

2 $f(x) = |7 - 5x| + 3$. Write down the values of:
  a $f(1)$
  b $f(10)$
  c $f(-6)$

3 $g(x) = |x^2 - 8x|$. Write down the values of:
  a $g(4)$
  b $g(-5)$
  c $g(8)$

4 Sketch the graph of each of the following. In each case, write down the coordinates of any points at which the graph meets the coordinate axes.
  a $y = |x - 1|$
  b $y = |2x + 3|$
  c $y = |4x - 7|$
  d $y = \left|\frac{1}{2}x - 5\right|$
  e $y = |7 - x|$
  f $y = |6 - 4x|$
  g $y = -|x|$
  h $y = -|3x - 1|$

  Hint $y = -|x|$ is a reflection of $y = |x|$ in the x-axis. ← Year 1, Chapter 4

5 $g(x) = \left|4 - \frac{3}{2}x\right|$ and $h(x) = 5$
  a On the same axes, sketch the graphs of $y = g(x)$ and $y = h(x)$.
  b Hence solve the equation $\left|4 - \frac{3}{2}x\right| = 5$.

6 Solve:
  a $|3x - 1| = 5$
  b $\left|\frac{x - 5}{2}\right| = 1$
  c $|4x + 3| = -2$
  d $|7x - 3| = 4$
  e $\left|\frac{4 - 5x}{3}\right| = 2$
  f $\left|\frac{x}{6} - 1\right| = 3$

7 a On the same diagram, sketch the graphs $y = -2x$ and $y = \left|\frac{1}{2}x - 2\right|$.
  b Solve the equation $-2x = \left|\frac{1}{2}x - 2\right|$.

(E) 8 Solve $|3x - 5| = 11 - x$. **(4 marks)**

9 a On the same set of axes, sketch $y = |6 - x|$ and $y = \frac{1}{2}x - 5$.
  b State with a reason whether there are any solutions to the equation $|6 - x| = \frac{1}{2}x - 5$.

10 A student attempts to solve the equation $|3x + 4| = x$. The student writes the following working:

$$\begin{aligned} 3x + 4 &= x & & & -(3x + 4) &= x \\ 4 &= -2x & \text{or} & & -3x - 4 &= x \\ x &= -2 & & & -4 &= 4x \\ & & & & x &= -1 \end{aligned}$$

Solutions are $x = -2$ and $x = -1$.

Explain the error made by the student.

11 a On the same diagram, sketch the graphs of $y = -|3x + 4|$ and $y = 2x - 9$.

b Solve the inequality $-|3x + 4| < 2x - 9$.

12 Solve the inequality $|2x + 9| < 14 - x$. **(4 marks)**

13 The equation $|6 - x| = \frac{1}{2}x + k$ has exactly one solution.

a Find the value of k. **(2 marks)**

b State the solution to the equation. **(2 marks)**

Problem-solving

The solution must be at the vertex of the graph of the modulus function.

Challenge

$f(x) = |x^2 + 9x + 8|$ and $g(x) = 1 - x$

a On the same axes, sketch graphs of $y = f(x)$ and $y = g(x)$.

b Use your sketch to find all the solutions to $|x^2 + 9x + 8| = 1 - x$.

2.2 Functions and mappings

A **mapping** transforms one set of numbers into a different set of numbers. The mapping can be described in words or through an algebraic equation. It can also be represented by a graph.

- **A mapping is a function if every input has a distinct output. Functions can either be one-to-one or many-to-one.**

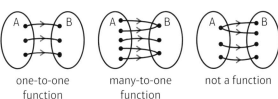

one-to-one function   many-to-one function   not a function

Many mappings can be made into functions by changing the domain. Consider $y = \sqrt{x}$:

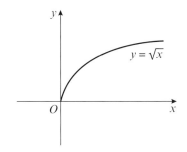

Notation The **domain** is the set of all possible inputs for a mapping.

The **range** is the set of all possible outputs for the mapping.

If the domain were all of the real numbers, ℝ, then $y = \sqrt{x}$ would not be a function because values of x less than 0 would not be mapped anywhere.

If you restrict the domain to $x \geqslant 0$, every element in the domain is mapped to exactly one element in the range.

We can write this function together with its domain as $f(x) = \sqrt{x},\ x \in \mathbb{R},\ x \geqslant 0$.

Notation You can also write this function as:
$f: x \mapsto \sqrt{x},\ x \in \mathbb{R},\ x \geqslant 0$

Example 7

For each of the following mappings:
i State whether the mapping is one-to-one, many-to-one or one-to-many.
ii State whether the mapping is a function.

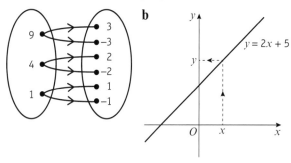
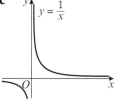
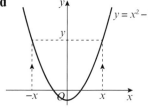

a i Every element in set A gets mapped to two elements in set B, so the mapping is **one-to-many**.
   ii The mapping is not a function.

b i Every value of x gets mapped to one value of y, so the mapping is **one-to-one**.
   ii The mapping is a function.

c i The mapping is **one-to-one**.
   ii $x = 0$ does not get mapped to a value of y so the mapping is not a function.

d i On the graph, you can see that x and $-x$ both get mapped to the same value of y. Therefore, this is a **many-to-one** mapping.
   ii The mapping is a function.

— You couldn't write down a single value for f(9).

— For a mapping to be a function, every input in the domain must map onto exactly one output.

— The mapping in part **c** could be a function if $x = 0$ were omitted from the domain. You could write this as a function as $f(x) = \frac{1}{x},\ x \in \mathbb{R},\ x \neq 0$.

Example 8

Find the range of each of the following functions:
a $f(x) = 3x - 2$, domain $\{x = 1, 2, 3, 4\}$
b $g(x) = x^2$, domain $\{x \in \mathbb{R},\ -5 \leqslant x \leqslant 5\}$
c $h(x) = \frac{1}{x}$, domain $\{x \in \mathbb{R},\ 0 < x \leqslant 3\}$

State if the functions are one-to-one or many-to-one.

Functions and graphs

a f(x) = 3x − 2, {x = 1, 2, 3, 4}

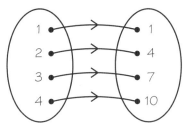

Range of f(x) is {1, 4, 7, 10}.
f(x) is one-to-one.

The domain contains a finite number of elements, so you can draw a mapping diagram showing the whole function.

b g(x) = x^2, {−5 ⩽ x ⩽ 5}

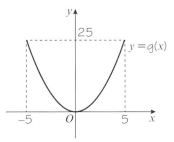

Range of g(x) is 0 ⩽ g(x) ⩽ 25.
g(x) is many-to-one.

The domain is the set of all the x-values that correspond to points on the graph. The range is the set of y-values that correspond to points on the graph.

c h(x) = $\frac{1}{x}$, {x ∈ ℝ, 0 < x ⩽ 3}

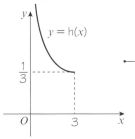

Range of h(x) is h(x) ⩾ $\frac{1}{3}$
h(x) is one-to-one.

Calculate h(3) = $\frac{1}{3}$ to find the minimum value in the range of h. As x approaches 0, $\frac{1}{x}$ approaches ∞, so there is no maximum value in the range of h.

Example 9

The function f(x) is defined by

f: x ↦ $\begin{cases} 5 - 2x, & x < 1 \\ x^2 + 3, & x \geqslant 1 \end{cases}$

a Sketch y = f(x), and state the range of f(x).
b Solve f(x) = 19.

Notation This is an example of a **piecewise-defined function**. It consists of two parts: one linear (for x < 1) and one quadratic (for x ⩾ 1).

Online Explore graphs of functions on a given domain using technology.

a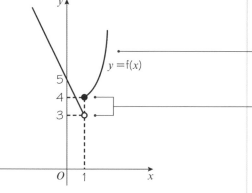

The range is the set of values that y takes and therefore f(x) > 3.

b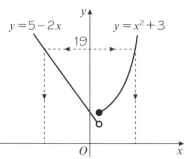

The positive solution is where
$x^2 + 3 = 19$
$x^2 = 16$
$x = \pm 4$
$x = 4$

The negative solution is where
$5 - 2x = 19$
$-2x = 14$
$x = -7$

The solutions are $x = 4$ and $x = -7$.

Watch out Although the graph jumps at $x = 1$, the function is still defined for all real values of x:
$f(0.9) = 5 - 2(0.9) = 3.2$
$f(1) = (1)^2 + 3 = 4$

Sketch the graph of $y = 5 - 2x$ for $x < 1$, and the graph of $y = x^2 + 3$ for $x \geq 1$.

$f(1)$ lies on the quadratic curve, so use a solid dot on the quadratic curve, and an open dot on the line.

Note that $f(x) \neq 3$ at $x = 1$
so $f(x) > 3$
not $f(x) \geq 3$

There are 2 values of x such that $f(x) = 19$.

Problem-solving

Use $x^2 + 3 = 19$ to find the solution when $x \geq 1$ and use $5 - 2x = 19$ to find the solution when $x < 1$.

Ignore $x = -4$ because the function is only equal to $x^2 + 3$ for $x \geq 1$.

Exercise 2B

1 For each of the following functions:
  i draw the mapping diagram
  ii state if the function is one-to-one or many-to-one
  iii find the range of the function.
  a $f(x) = 5x - 3$, domain $\{x = 3, 4, 5, 6\}$
  b $g(x) = x^2 - 3$, domain $\{x = -3, -2, -1, 0, 1, 2, 3\}$
  c $h(x) = \dfrac{7}{4 - 3x}$, domain $\{x = -1, 0, 1\}$

2 For each of the following mappings:
  i State whether the mapping is one-to-one, many-to-one or one-to-many.
  ii State whether the mapping could represent a function.

a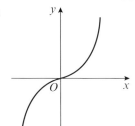
b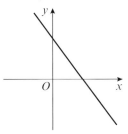
c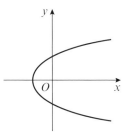
d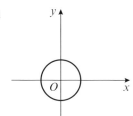
e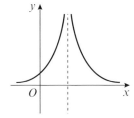
f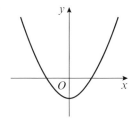

3 Calculate the value(s) of a, b, c and d given that:
  a $p(a) = 16$ where p: $x \mapsto 3x - 2$, $x \in \mathbb{R}$
  b $q(b) = 17$ where q: $x \mapsto x^2 - 3$, $x \in \mathbb{R}$
  c $r(c) = 34$ where r: $x \mapsto 2(2^x) + 2$, $x \in \mathbb{R}$
  d $s(d) = 0$ where s: $x \mapsto x^2 + x - 6$, $x \in \mathbb{R}$

4 For each function:
  i represent the function on a mapping diagram, writing down the elements in the range
  ii state whether the function is one-to-one or many-to-one.
    a $f(x) = 2x + 1$ for the domain $\{x = 1, 2, 3, 4, 5\}$
    b g: $x \mapsto \sqrt{x}$ for the domain $\{x = 1, 4, 9, 16, 25, 36\}$
    c $h(x) = x^2$ for the domain $\{x = -2, -1, 0, 1, 2\}$
    d j: $x \mapsto \dfrac{2}{x}$ for the domain $\{x = 1, 2, 3, 4, 5\}$
    e $k(x) = e^x + 3$ for the domain $\{x = -2, -1, 0, 1, 2\}$

Notation Remember, $\sqrt{x}$ means the positive square root of x.

5 For each function:
  i sketch the graph of $y = f(x)$
  ii state the range of $f(x)$
  iii state whether $f(x)$ is one-to-one or many-to-one.
    a f: $x \mapsto 3x + 2$ for the domain $\{x \geq 0\}$
    b $f(x) = x^2 + 5$ for the domain $\{x \geq 2\}$
    c f: $x \mapsto 2\sin x$ for the domain $\{0 \leq x \leq 180\}$
    d f: $x \mapsto \sqrt{x + 2}$ for the domain $\{x \geq -2\}$
    e $f(x) = e^x$ for the domain $\{x \geq 0\}$
    f $f(x) = 7\log x$, for the domain, $\{x \in \mathbb{R}, x > 0\}$

6 The following mappings f and g are defined on all the real numbers by
$$f(x) = \begin{cases} 4 - x, & x < 4 \\ x^2 + 9, & x \geq 4 \end{cases} \qquad g(x) = \begin{cases} 4 - x, & x < 4 \\ x^2 + 9, & x > 4 \end{cases}$$
  a Explain why $f(x)$ is a function and $g(x)$ is not.   b Sketch $y = f(x)$.
  c Find the values of: i $f(3)$ ii $f(10)$ d Solve $f(a) = 90$.

Chapter 2

P 7 The function s is defined by
$$s(x) = \begin{cases} x^2 - 6, & x < 0 \\ 10 - x, & x \geq 0 \end{cases}$$

a Sketch $y = s(x)$.
b Find the value(s) of a such that $s(a) = 43$.
c Solve $s(x) = x$.

Problem-solving
The solutions of $s(x) = x$ are the values in the domain that get mapped to themselves in the range.

E/P 8 The function p is defined by
$$p(x) = \begin{cases} e^{-x}, & -5 \leq x < 0 \\ x^3 + 4, & 0 \leq x \leq 4 \end{cases}$$

a Sketch $y = p(x)$. **(3 marks)**
b Find the values of a, to 2 decimal places, such that $p(a) = 50$. **(4 marks)**

E/P 9 The function h has domain $-10 \leq x \leq 6$, and is linear from $(-10, 14)$ to $(-4, 2)$ and from $(-4, 2)$ to $(6, 27)$.

a Sketch $y = h(x)$. **(2 marks)**
b Write down the range of $h(x)$. **(1 mark)**
c Find the values of a, such that $h(a) = 12$. **(4 marks)**

Problem-solving
The graph of $y = h(x)$ will consist of two line segments which meet at $(-4, 2)$.

P 10 The function g is defined by $g(x) = cx + d$, $x \in \mathbb{R}$ where c and d are constants to be found. Given $g(3) = 10$ and $g(8) = 12$ find the values of c and d.

P 11 The function f is defined by $f(x) = ax^3 + bx - 5$, $x \in \mathbb{R}$ where a and b are constants to be found. Given that $f(1) = -4$ and $f(2) = 9$, find the values of the constants a and b.

E/P 12 The function h is defined by $h(x) = x^2 - 6x + 20$ and has domain $x \geq a$. Given that $h(x)$ is a one-to-one function find the smallest possible value of the constant a. **(6 marks)**

Problem-solving
First complete the square for $h(x)$.

2.3 Composite functions

Two or more functions can be combined to make a new function. The new function is called a **composite function**.

- **fg(x) means apply g first, then apply f.**
- **fg(x) = f(g(x))**

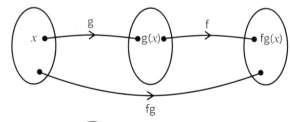

Watch out The order in which the functions are combined is important: $fg(x)$ is not necessarily the same as $gf(x)$.

Example 10

Given $f(x) = x^2$, $x \in \mathbb{R}$ and $g(x) = x + 1$, $x \in \mathbb{R}$, find:
a $fg(1)$ **b** $gf(3)$ **c** $ff(-2)$

32

a fg(1) = f(1 + 1) ———————————— g(1) = 1 + 1
       $= 2^2$
       $= 4$ ———————————————————— $f(2) = 2^2$

b gf(3) = g(3^2) ———————————————
       = g(9)
       = 9 + 1 ————————————————— $f(3) = 3^2$
       = 10
                                    g(9) = 9 + 1

c ff(−2) = f((−2)^2) ———————————
       = f(4)
       = 4^2 ———————————————— $f(−2) = (−2)^2$
       = 16
                                    $f(4) = 4^2$

Example 11

The functions f and g are defined by f(x) = $3x + 2$, $x \in \mathbb{R}$ and g(x) = $x^2 + 4$, $x \in \mathbb{R}$. Find:

a the function fg(x)
b the function gf(x)
c the function f^2(x)

Notation f^2(x) is ff(x).

d the values of b such that fg(b) = 62.

a fg(x) = f($x^2 + 4$) ————————— g acts on x first, mapping it to $x^2 + 4$.
       = 3($x^2 + 4$) + 2
       = $3x^2 + 14$, $x \in \mathbb{R}$ —————————— f acts on the result.

b gf(x) = g($3x + 2$) ————————— Simplify answer.
       = $(3x + 2)^2 + 4$
       = $9x^2 + 12x + 8$, $x \in \mathbb{R}$ —————— f acts on x first, mapping it to $3x + 2$.

                                    g acts on the result.

c f^2(x) = f($3x + 2$) ————————
       = 3($3x + 2$) + 2
       = $9x + 8$, $x \in \mathbb{R}$ ————————— f maps x to $3x + 2$.

                                    f acts on the result.

d   fg(x) = $3x^2 + 14$, $x \in \mathbb{R}$
    If   fg(b) = 62 ————————————
    then $3b^2 + 14 = 62$
          $b^2 = 16$ Set up and solve an equation in b.
          $b = \pm 4$

Watch out When you are asked to find a function you should give the equation **and** the domain.

Example 12

The functions f and g are defined by
f: $x \mapsto |2x - 8|$, $x \in \mathbb{R}$
g: $x \mapsto \dfrac{x + 1}{2}$, $x \in \mathbb{R}$

a Find fg(3).   **b** Solve fg(x) = x.

a $fg(3) = f\left(\dfrac{3+1}{2}\right)$ ⎯⎯⎯ $g(3) = \left(\dfrac{3+1}{2}\right)$

$= f(2)$

$= |2 \times 2 - 8|$ ⎯⎯⎯ $f(2) = |2 \times 2 - 8|$

$= |-4|$

$= 4$

b First find $fg(x)$:

$fg(x) = f\left(\dfrac{x+1}{2}\right)$ ⎯⎯⎯ g acts on x first, mapping it to $\dfrac{x+1}{2}$

$= \left|2\left(\dfrac{x+1}{2}\right) - 8\right|$ ⎯⎯⎯ f acts on the result.

$= |x - 7|$ ⎯⎯⎯ Simplify the answer.

$fg(x) = x, \ x \in \mathbb{R}$

$|x - 7| = x$

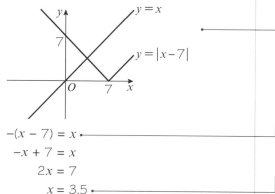

Draw a sketch of $y = |x - 7|$ and $y = x$.

The sketch shows there is only one solution to the equation $|x - 7| = x$ and that it occurs on the reflected part of the graph.

$-(x - 7) = x$ ⎯⎯⎯ When $f(x) < 0$, $|f(x)| = -f(x)$. The solution is on the reflected part of the graph so use $-(x-7)$.

$-x + 7 = x$

$2x = 7$

$x = 3.5$ ⎯⎯⎯ This is the x-coordinate at the point of intersection marked on the graph.

Exercise 2C

1 Given the functions $p(x) = 1 - 3x, x \in \mathbb{R}$, $q(x) = \dfrac{x}{4}, x \in \mathbb{R}$ and $r(x) = (x-2)^2, x \in \mathbb{R}$, find:

a $pq(-8)$ **b** $qr(5)$ **c** $rq(6)$ **d** $p^2(-5)$ **e** $pqr(8)$

2 Given the functions $f(x) = 4x + 1, x \in \mathbb{R}$, $g(x) = x^2 - 4$, $x \in \mathbb{R}$ and $h(x) = \dfrac{1}{x}, x \in \mathbb{R}, x \neq 0$, find the functions:

a $fg(x)$ **b** $gf(x)$ **c** $gh(x)$

d $fh(x)$ **e** $f^2(x)$

Hint Give the domain for each function. The domains of gh and fh are the same as the domain of h.

(E) 3 The functions f and g are defined by

$f(x) = 3x - 2, x \in \mathbb{R}$ $g(x) = x^2, x \in \mathbb{R}$

a Find $fg(x)$. **(2 marks)**

b Solve $fg(x) = gf(x)$. **(4 marks)**

(E) 4 The functions p and q are defined by

$p(x) = \dfrac{1}{x-2}, x \in \mathbb{R}, x \neq 2$ $q(x) = 3x + 4, x \in \mathbb{R}$

a Show that $qp(x) = \dfrac{ax+b}{cx+d}, x \in \mathbb{R}, x \neq 2$, where a, b, c and d are integers to be found. **(3 marks)**

b Solve $qp(x) = 16$. **(3 marks)**

5 The functions f and g are defined by:

f: $x \mapsto |9 - 4x|$, $x \in \mathbb{R}$

g: $x \mapsto \dfrac{3x - 2}{2}$, $x \in \mathbb{R}$

a Find fg(6). **(2 marks)**

b Solve fg(x) = x. **(5 marks)**

6 Given $f(x) = \dfrac{1}{x + 1}$, $x \neq -1$

a Prove that $f^2(x) = \dfrac{x + 1}{x + 2}$, $x \neq -1$, $x \neq -2$ **b** Find $f^3(x)$.

7 The functions s and t are defined by

s(x) = 2^x, $x \in \mathbb{R}$

t(x) = x + 3, $x \in \mathbb{R}$

a Find st(x).

b Find ts(x).

c Solve st(x) = ts(x), leaving your answer in the form $\dfrac{\ln a}{\ln b}$

Hint Rearrange the equation in part **c** into the form $2^x = k$ where k is a real number, then take natural logs of both sides. ← Year 1, Section 14.5

8 Given $f(x) = e^{5x}$, $x \in \mathbb{R}$ and $g(x) = 4\ln x$, $x \in \mathbb{R}$, $x > 0$ find in its simplest form:

a gf(x) **(2 marks)**

b fg(x) **(2 marks)**

9 The functions p and q are defined by

p: $x \mapsto \ln(x + 3)$, $x \in \mathbb{R}$, $x > -3$

q: $x \mapsto e^{3x} - 1$, $x \in \mathbb{R}$

Hint The range of p will be the set of possible inputs for q in the function qp.

a Find qp(x) and state its range. **(3 marks)**

b Find the value of qp(7). **(1 mark)**

c Solve qp(x) = 124. **(3 marks)**

10 The function t is defined by

t: $x \mapsto 5 - 2x$, $x \in \mathbb{R}$

Solve the equation $t^2(x) - (t(x))^2 = 0$. **(5 marks)**

Problem-solving

You need to work out the intermediate steps for this problem yourself, so plan your answer before you start. You could start by finding an expression for tt(x).

11 The function g has domain $-5 \leq x \leq 14$ and is linear from (−5, −8) to (0, 12) and from (0, 12) to (14, 5).

A sketch of the graph of y = g(x) is shown in the diagram.

a Write down the range of g. **(1 mark)**

b Find gg(0). **(2 marks)**

The function h is defined by h: $x \mapsto \dfrac{2x - 5}{10 - x}$, $x \in \mathbb{R}$, $x \neq 10$

c Find gh(7). **(2 marks)**

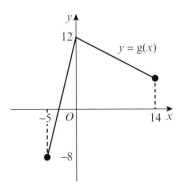

35

2.4 Inverse functions

The **inverse** of a function performs the opposite operation to the original function. It takes the elements in the range of the original function and maps them back into elements of the domain of the original function. For this reason, inverse functions only exist for one-to-one functions.

- Functions $f(x)$ and $f^{-1}(x)$ are inverses of each other. $ff^{-1}(x) = f^{-1}f(x) = x$.
- The graphs of $y = f(x)$ and $y = f^{-1}(x)$ are reflections of each another in the line $y = x$.
- The domain of $f(x)$ is the range of $f^{-1}(x)$.
- The range of $f(x)$ is the domain of $f^{-1}(x)$.

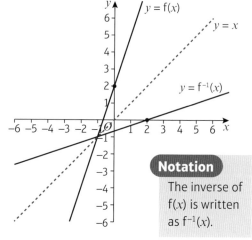

Notation
The inverse of $f(x)$ is written as $f^{-1}(x)$.

Example 13

Find the inverse of the function $h(x) = 2x^2 - 7$, $x \geq 0$.

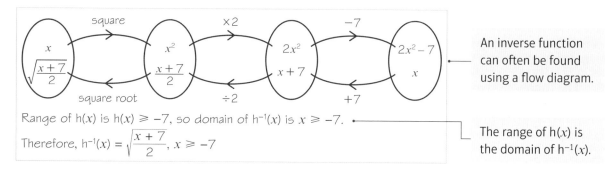

Range of $h(x)$ is $h(x) \geq -7$, so domain of $h^{-1}(x)$ is $x \geq -7$.

Therefore, $h^{-1}(x) = \sqrt{\dfrac{x + 7}{2}}$, $x \geq -7$

An inverse function can often be found using a flow diagram.

The range of $h(x)$ is the domain of $h^{-1}(x)$.

Example 14

Find the inverse of the function $f(x) = \dfrac{3}{x - 1}$, $x \in \mathbb{R}$, $x \neq 1$ by changing the subject of the formula.

Let $y = f(x)$

$$y = \dfrac{3}{x - 1}$$

$$y(x - 1) = 3$$

$$yx - y = 3$$

$$yx = 3 + y$$

$$x = \dfrac{3 + y}{y}$$

Range of $f(x)$ is $f(x) \neq 0$, so domain of $f^{-1}(x)$ is $x \neq 0$.

Therefore $f^{-1}(x) = \dfrac{3 + x}{x}$, $x \neq 0$

$f(4) = \dfrac{3}{4 - 1} = \dfrac{3}{3} = 1$

$f^{-1}(1) = \dfrac{3 + 1}{1} = \dfrac{4}{1} = 4$

You can rearrange to find an inverse function. Start by letting $y = f(x)$.

Rearrange to make x the subject of the formula.

Define $f^{-1}(x)$ in terms of x. Remember to include the domain.

Check to see that at least one element works. Try 4. Note that $f^{-1}f(4) = 4$.

Example 15

The function, $f(x) = \sqrt{x-2}$, $x \in \mathbb{R}$, $x \geq 2$.

a State the range of $f(x)$. **b** Find the function $f^{-1}(x)$ and state its domain and range.
c Sketch $y = f(x)$ and $y = f^{-1}(x)$ and the line $y = x$.

a The range of $f(x)$ is $y \in \mathbb{R}$, $y \geq 0$. ——— $f(2) = 0$. As x increases from 2, $f(x)$ also increases without limit, so the range is $f(x) \geq 0$, or $y \geq 0$.

b $y = \sqrt{x-2}$
$y^2 = x - 2$
$x^2 = y - 2$
$y = x^2 + 2$ ——— Interchange x and y.
The inverse function is $f^{-1}(x) = x^2 + 2$.
The domain of $f^{-1}(x)$ is $x \in \mathbb{R}$, $x \geq 0$. ——— Always write your function in terms of x.
The range of $f^{-1}(x)$ is $y \in \mathbb{R}$, $y \geq 2$.

The range of $f(x)$ is the same as the domain of $f^{-1}(x)$.

c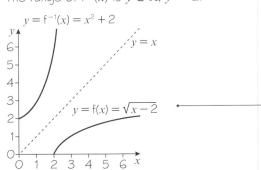

The range of $f^{-1}(x)$ is the same as the domain of $f(x)$.

The graph of $f^{-1}(x)$ is a reflection of $f(x)$ in the line $y = x$. This is because the reflection transforms y to x and x to y.

Example 16

The function $f(x)$ is defined by $f(x) = x^2 - 3$, $x \in \mathbb{R}$, $x \geq 0$.

a Find $f^{-1}(x)$. **b** Sketch $y = f^{-1}(x)$. **c** Solve the equation $f(x) = f^{-1}(x)$.

a Let $y = f(x)$
$y = x^2 - 3$
$y + 3 = x^2$ ——— Change the subject of the formula.
$x = \sqrt{y + 3}$
$f^{-1}(x) = \sqrt{x + 3}$, $x \in \mathbb{R}$, $x \geq -3$ ——— The range of the original function is $f(x) \geq -3$.

b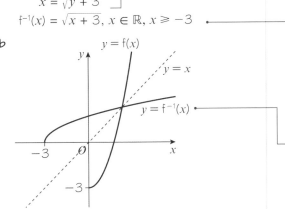

Online Explore functions and their inverses using technology.

First sketch $f(x)$. Then reflect $f(x)$ in the line $y = x$.

c When f(x) = f^{-1}(x)
f(x) = x
$x^2 - 3 = x$
$x^2 - x - 3 = 0$
So $x = \dfrac{1 + \sqrt{13}}{2}$

Problem-solving

y = f(x) and y = f^{-1}(x) intersect on the line $y = x$. This means that the solution to f(x) = f^{-1}(x) is the same as the solution to f(x) = x.

From the graph you can see that the solution must be positive, so ignore the negative solution to the equation.

Exercise 2D

1 For each of the following functions f(x):
  i state the range of f(x)
  ii determine the equation of the inverse function f^{-1}(x)
  iii state the domain and range of f^{-1}(x)
  iv sketch the graphs of y = f(x) and y = f^{-1}(x) on the same set of axes.

  a f: $x \mapsto 2x + 3$, $x \in \mathbb{R}$
  b f: $x \mapsto \dfrac{x+5}{2}$, $x \in \mathbb{R}$
  c f: $x \mapsto 4 - 3x$, $x \in \mathbb{R}$
  d f: $x \mapsto x^3 - 7$, $x \in \mathbb{R}$

2 Find the inverse of each function:
  a f(x) = $10 - x$, $x \in \mathbb{R}$
  b g(x) = $\dfrac{x}{5}$, $x \in \mathbb{R}$
  c h(x) = $\dfrac{3}{x}$, $x \neq 0$, $x \in \mathbb{R}$
  d k(x) = $x - 8$, $x \in \mathbb{R}$

Notation Two of these functions are **self-inverse**. A function is self-inverse if f^{-1}(x) = f(x). In this case ff(x) = x.

(P) 3 Explain why the function g: $x \mapsto 4 - x$, $\{x \in \mathbb{R}, x > 0\}$ is not identical to its inverse.

4 For each of the following functions g(x) with a restricted domain:
  i state the range of g(x)
  ii determine the equation of the inverse function g^{-1}(x)
  iii state the domain and range of g^{-1}(x)
  iv sketch the graphs of y = g(x) and y = g^{-1}(x) on the same set of axes.

  a g(x) = $\dfrac{1}{x}$, $\{x \in \mathbb{R}, x \geq 3\}$
  b g(x) = $2x - 1$, $\{x \in \mathbb{R}, x \geq 0\}$
  c g(x) = $\dfrac{3}{x - 2}$, $\{x \in \mathbb{R}, x > 2\}$
  d g(x) = $\sqrt{x - 3}$, $\{x \in \mathbb{R}, x \geq 7\}$
  e g(x) = $x^2 + 2$, $\{x \in \mathbb{R}, x > 2\}$
  f g(x) = $x^3 - 8$, $\{x \in \mathbb{R}, x \geq 2\}$

(E) 5 The function t(x) is defined by
t(x) = $x^2 - 6x + 5$, $x \in \mathbb{R}$, $x \geq 5$
Find t^{-1}(x).

Hint First complete the square for the function t(x).

(5 marks)

(E/P) 6 The function m(x) is defined by m(x) = $x^2 + 4x + 9$, $x \in \mathbb{R}$, $x > a$, for some constant a.
  a State the least value of a for which m^{-1}(x) exists. **(4 marks)**
  b Determine the equation of m^{-1}(x). **(3 marks)**
  c State the domain of m^{-1}(x). **(1 mark)**

Functions and graphs

7 The function h(x) is defined by $h(x) = \dfrac{2x+1}{x-2}$, $\{x \in \mathbb{R}, x \neq 2\}$.

  a What happens to the function as x approaches 2?
  b Find $h^{-1}(3)$.
  c Find $h^{-1}(x)$, stating clearly its domain.
  d Find the elements of the domain that get mapped to themselves by the function.

8 The functions m and n are defined by
  m: $x \mapsto 2x + 3$, $x \in \mathbb{R}$
  n: $x \mapsto \dfrac{x-3}{2}$, $x \in \mathbb{R}$

  a Find $nm(x)$
  b What can you say about the functions m and n?

(P) 9 The functions s and t are defined by
  $s(x) = \dfrac{3}{x+1}$, $x \neq -1$
  $t(x) = \dfrac{3-x}{x}$, $x \neq 0$

  Show that the functions are inverses of each other.

(E/P) 10 The function f(x) is defined by $f(x) = 2x^2 - 3$, $\{x \in \mathbb{R}, x < 0\}$.
  Find:
  a $f^{-1}(x)$ **(4 marks)**
  b the value(s) of a for which $f(a) = f^{-1}(a)$. **(4 marks)**

(E) 11 The functions f and g are defined by
  f: $x \mapsto e^x - 5$, $x \in \mathbb{R}$
  g: $x \mapsto \ln(x-4)$, $x > 4$

  a State the range of f. **(1 mark)**
  b Find f^{-1}, the inverse function of f. **(3 marks)**
  c On the same axes, sketch the curves with equation $y = f(x)$ and $y = f^{-1}(x)$, giving the coordinates of all the points where the curves cross the axes. **(4 marks)**
  d Find g^{-1}, the inverse function of g. **(3 marks)**
  e Solve the equation $g^{-1}(x) = 11$, giving your answer to 2 decimal places. **(3 marks)**

(E/P) 12 The function f is defined by
  f: $x \mapsto \dfrac{3(x+2)}{x^2+x-20} - \dfrac{2}{x-4}$, $x > 4$

  a Show that f: $x \mapsto \dfrac{1}{x+5}$, $x > 4$. **(4 marks)**
  b Find the range of f. **(2 marks)**
  c Find $f^{-1}(x)$. **(4 marks)**

2.5 $y = |f(x)|$ and $y = f(|x|)$

- **To sketch the graph of $y = |f(x)|$:**
  - Sketch the graph of $y = f(x)$.
  - Reflect any parts where $f(x) < 0$ (parts below the x-axis) in the x-axis.
  - Delete the parts below the x-axis.

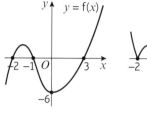
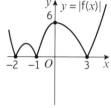

- **To sketch the graph of $y = f(|x|)$:**
  - Sketch the graph of $y = f(x)$ for $x \geqslant 0$.
  - Reflect this in the y-axis.

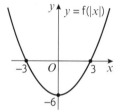

Example 17

$f(x) = x^2 - 3x - 10$

a Sketch the graph of $y = f(x)$.
b Sketch the graph of $y = |f(x)|$.
c Sketch the graph of $y = f(|x|)$.

a $f(x) = x^2 - 3x - 10 = (x - 5)(x + 2)$
$f(x) = 0$ implies $(x - 5)(x + 2) = 0$
So $x = 5$ or $x = -2$
$f(0) = -10$

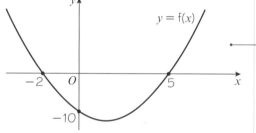

The graph of $y = x^2 - 3x - 10$ cuts the x-axis at -2 and 5.

The graph cuts the y-axis at -10.

This is the sketch of $y = x^2 - 3x - 10$.
The sketch includes the points where the graph intercepts the coordinate axes.
A sketch does not have to be to scale.

b $y = |f(x)| = |x^2 - 3x - 10|$

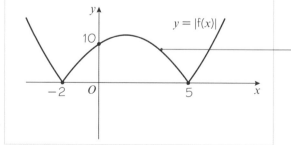

Online Explore graphs of modulus functions using technology.

Reflect the part of the curve where $y = f(x) < 0$ (the negative values of y) in the x-axis.

Functions and graphs

c $y = f(|x|) = |x|^2 - 3|x| - 10$

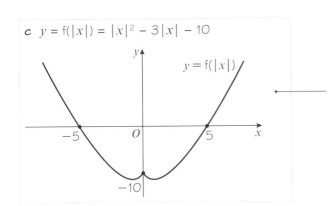

Reflect the part of the curve where $x \geqslant 0$ (the positive values of x) in the y-axis.

Example 18

$g(x) = \sin x, \; -360° \leqslant x \leqslant 360°$

a Sketch the graph of $y = g(x)$.
b Sketch the graph of $y = |g(x)|$.
c Sketch the graph of $y = g(|x|)$.

a
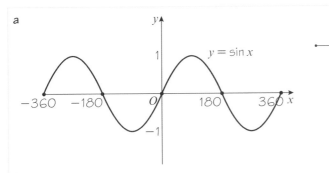

The graph is periodic and passes through the origin, (±180, 0) and (±360, 0).

← Year 1, Section 9.5

b
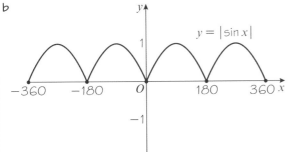

Reflect the part of the curve below the x-axis in the x-axis.

c
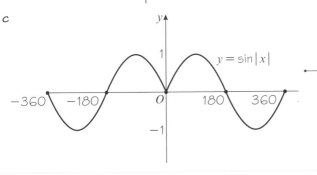

Reflect the part of the curve where $x \geqslant 0$ in the y-axis.

41

Chapter 2

Example 19

The diagram shows the graph of $y = h(x)$, with five points labelled.

Sketch each of the following graphs, labelling the points corresponding to A, B, C, D and E, and any points of intersection with the coordinate axes.

a $y = |h(x)|$
b $y = h(|x|)$

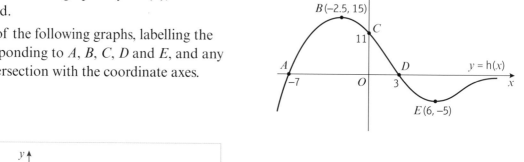

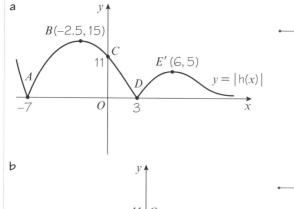

a The parts of the curve below the x-axis are reflected in the x-axis.

The points A, B, C and D are unchanged.

The point E was reflected, so the new coordinates are $E'(6, 5)$.

b The part of the curve to the right of the y-axis is reflected in the y-axis.

The old points A and B had negative x-values so they are no longer part of the graph.

The points C, D and E are unchanged.

There is a new point of intersection with the x-axis at $(-3, 0)$.

The point E was reflected, so the new coordinates are $E'(-6, -5)$.

Exercise 2E

1 $f(x) = x^2 - 7x - 8$

  a Sketch the graph of $y = f(x)$.
  b Sketch the graph of $y = |f(x)|$.
  c Sketch the graph of $y = f(|x|)$.

2 $g: x \mapsto \cos x, -360° \leq x \leq 360°$

  a Sketch the graph of $y = g(x)$.
  b Sketch the graph of $y = |g(x)|$.
  c Sketch the graph of $y = g(|x|)$.

3 $h: x \mapsto (x-1)(x-2)(x+3)$

  a Sketch the graph of $y = h(x)$.
  b Sketch the graph of $y = |h(x)|$.
  c Sketch the graph of $y = h(|x|)$.

Functions and graphs

(P) 4 The function k is defined by $k(x) = \dfrac{a}{x^2}$, $a > 0$, $x \in \mathbb{R}$, $x \neq 0$.

    a Sketch the graph of $y = k(x)$.

    b Explain why it is not necessary to sketch $y = |k(x)|$ and $y = k(|x|)$.

    The function m is defined by $m(x) = \dfrac{a}{x^2}$, $a < 0$, $x \in \mathbb{R}$, $x \neq 0$.

    c Sketch the graph of $y = m(x)$.

    d State with a reason whether the following statements are true or false.

      i $|k(x)| = |m(x)|$ ii $k(|x|) = m(|x|)$ iii $m(x) = m(|x|)$

(E) 5 The diagram shows the graph of $y = p(x)$ with 5 points labelled.

Sketch each of the following graphs, labelling the points corresponding to A, B, C, D and E, and any points of intersection with the coordinate axes.

    a $y = |p(x)|$ **(3 marks)**

    b $y = p(|x|)$ **(3 marks)**

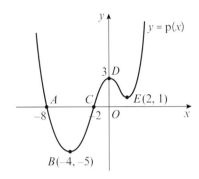

(E) 6 The diagram shows the graph of $y = q(x)$ with 7 points labelled.

Sketch each of the following graphs, labelling the points corresponding to A, B, C, D and E, and any points of intersection with the coordinate axes.

    a $y = |q(x)|$ **(4 marks)**

    b $y = q(|x|)$ **(3 marks)**

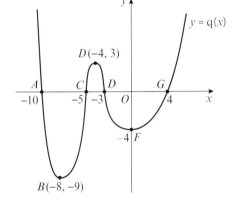

7 $k(x) = \dfrac{a}{x}$, $a > 0$, $x \neq 0$

    a Sketch the graph of $y = k(x)$.

    b Sketch the graph of $y = |k(x)|$.

    c Sketch the graph of $y = k(|x|)$.

8 $m(x) = \dfrac{a}{x}$, $a < 0$, $x \neq 0$

    a Sketch the graph of $y = m(x)$.

    b Describe the relationship between $y = |m(x)|$ and $y = m(|x|)$.

9 $f(x) = e^x$ and $g(x) = e^{-x}$

    a Sketch the graphs of $y = f(x)$ and $y = g(x)$ on the same axes.

    b Explain why it is not necessary to sketch $y = |f(x)|$ and $y = |g(x)|$.

    c Sketch the graphs of $y = f(|x|)$ and $y = g(|x|)$ on the same axes.

43

Chapter 2

E/P 10 The function f(x) is defined by

$$f(x) = \begin{cases} -2x - 6, & -5 \leqslant x < -1 \\ (x + 1)^2, & -1 \leqslant x \leqslant 2 \end{cases}$$

a Sketch f(x) stating its range. (5 marks)
b Sketch the graph of $y = |f(x)|$. (3 marks)
c Sketch the graph of $y = f(|x|)$. (3 marks)

Problem-solving

A piecewise function like this does not have to be continuous. Work out the value of both expressions when $x = -1$ to help you with your sketch.

2.6 Combining transformations

You can use combinations of the following transformations of a function to sketch graphs of more complicated transformations.

- $f(x + a)$ **is a translation by the vector** $\begin{pmatrix} -a \\ 0 \end{pmatrix}$
- $f(x) + a$ **is a translation by the vector** $\begin{pmatrix} 0 \\ a \end{pmatrix}$
- $f(-x)$ **reflects f(x) in the y-axis.**
- $-f(x)$ **reflects f(x) in the x-axis.**
- $f(ax)$ **is a horizontal stretch of scale factor** $\frac{1}{a}$
- $af(x)$ **is a vertical stretch of scale factor** a

Links You can think of $f(-x)$ and $-f(x)$ as stretches with scale factor -1. ← Year 1, Sections 4.6, 4.7

Example 20

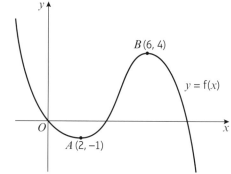

The diagram shows a sketch of the graph of $y = f(x)$. The curve passes through the origin O, the point $A(2, -1)$ and the point $B(6, 4)$.

Sketch the graphs of:

a $y = 2f(x) - 1$
b $y = f(x + 2) + 2$
c $y = \frac{1}{4}f(2x)$
d $y = f(3x - 12)$

In each case, find the coordinates of the images of the points O, A and B.

a $y = 2f(x) - 1$

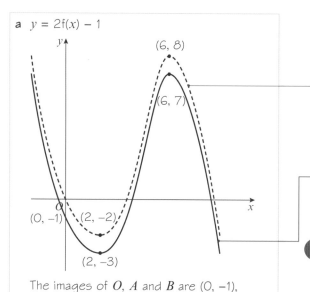

The images of O, A and B are $(0, -1)$, $(2, -3)$ and $(6, 7)$ respectively.

Apply the stretch first. The dotted curve is the graph of $y = 2f(x)$, which is a vertical stretch with scale factor 2.

Next apply the translation. The solid curve is the graph of $y = 2f(x) - 1$, as required. This is a translation of $y = 2f(x)$ by vector $\begin{pmatrix} 0 \\ -1 \end{pmatrix}$.

Watch out The order is important. If you applied the transformations in the opposite order you would have the graph of $y = 2(f(x) - 1)$ or $y = 2f(x) - 2$.

b $y = f(x + 2) + 2$

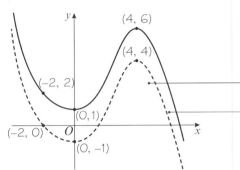

The images of O, A and B are $(-2, 2)$, $(0, 1)$ and $(4, 6)$ respectively.

Apply the translation **inside** the brackets first. The dotted curve is the graph of $y = f(x + 2)$, which is a translation of $y = f(x)$ by vector $\begin{pmatrix} -2 \\ 0 \end{pmatrix}$.

Next apply the translation **outside** the brackets. The solid curve is the graph of $y = f(x + 2) + 2$, as required. This is a translation of $y = f(x + 2)$ by vector $\begin{pmatrix} 0 \\ 2 \end{pmatrix}$.

c $y = \frac{1}{4}f(2x)$

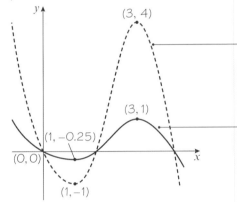

The images of O, A and B are $(0, 0)$, $(1, -0.25)$ and $(3, 1)$ respectively.

Apply the stretch **inside** the brackets first. The dotted curve is the graph of $y = f(2x)$, which is a horizontal stretch with scale factor $\frac{1}{2}$

Then apply the stretch **outside** the brackets. The solid curve is the graph of $y = \frac{1}{4}f(2x)$, as required. This is a vertical stretch of $y = f(2x)$ with scale factor $\frac{1}{4}$

Problem-solving

To sketch a transformation of the form $y = f(ax + b)$ rewrite it as $y = f\left(a\left(x + \frac{b}{a}\right)\right)$.
Carry out the horizontal stretch first, then the translation.

d $y = f(3x - 12)$
$y = f(3(x - 4))$

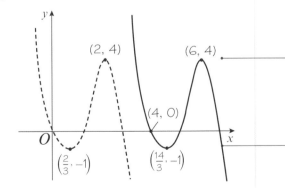

The images of O, A and B are $(4, 0)$, $\left(\frac{14}{3}, -1\right)$ and $(6, 4)$ respectively.

Write $y = f(3x - 12)$ as $y = f(3(x - 4))$, and apply the horizontal stretch. The dotted curve is the graph of $y = f(3x)$, which is a horizontal stretch by scale factor $\frac{1}{3}$.

Then apply the translation. The solid curve is the graph of $y = f(3(x - 4))$, as required. This is a translation of $y = f(3x)$ by vector $\begin{pmatrix} 4 \\ 0 \end{pmatrix}$.

Example 21

$f(x) = \ln x$, $x > 0$

Sketch the graphs of

a $y = 2f(x) - 3$ **b** $y = |f(-x)|$

Show, on each diagram, the point where the graph meets or crosses the x-axis.
In each case, state the equation of the asymptote.

a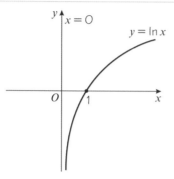

Online Explore combinations of transformations using technology.

Problem-solving

You have not been asked to sketch $y = f(x)$ in this question, but it is a good idea to do this before sketching transformations of this graph.
Sketch $y = f(x)$, labelling its asymptote and the coordinates of the point where it crosses the x-axis. ← Year 1, Section 14.3

$2\ln x - 3 = 0$

$\ln x = \dfrac{3}{2}$

$x = e^{\frac{3}{2}}$

$= 4.48$ (3 s.f.)

The graph $y = 2\ln x - 3$ will cross the x-axis at $(4.48, 0)$.

Solve this equation to find the x-intercept of $y = 2f(x) - 3$.

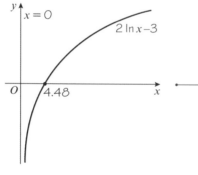

The original graph underwent a vertical stretch by a scale factor of 2 and then a vertical translation by vector $\begin{pmatrix} 0 \\ -3 \end{pmatrix}$.

b The graph of $y = f(-x)$ is a reflection of $y = f(x)$ in the y-axis.

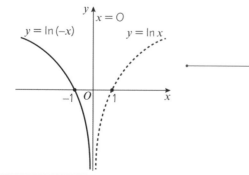

The original graph is first reflected in the y-axis.
The x-intercept becomes $(-1, 0)$.
The asymptote is unchanged.

Functions and graphs

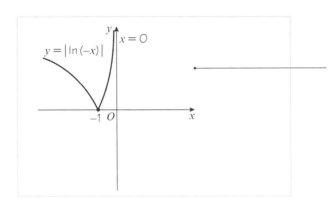

To sketch the graph of $y = |f(-x)|$ reflect any negative y-values of $y = f(-x)$ in the x-axis.

Exercise 2F

1 The diagram shows a sketch of the graph $y = f(x)$. The curve passes through the origin O, the point $A(-2, -2)$ and the point $B(3, 4)$.
On separate axes, sketch the graphs of:

a $y = 3f(x) + 2$
b $y = f(x - 2) - 5$
c $y = \frac{1}{2}f(x + 1)$
d $y = -f(2x)$
e $y = |f(x)|$
f $y = |f(-x)|$

In each case find the coordinates of the images of the points O, A and B.

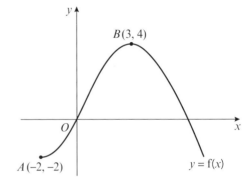

2 The diagram shows a sketch of the graph $y = f(x)$. The curve has a maximum at the point $A(-1, 4)$ and crosses the axes at the points $(0, 3)$ and $(-2, 0)$.
On separate axes, sketch the graphs of:

a $y = 3f(x - 2)$
b $y = \frac{1}{2}f\left(\frac{1}{2}x\right)$
c $y = -f(x) + 4$
d $y = -2f(x + 1)$
e $y = 2f(|x|)$
f $y = f(2x - 6)$

For each graph, find, where possible, the coordinates of the maximum or minimum and the coordinates of the intersection points with the axes.

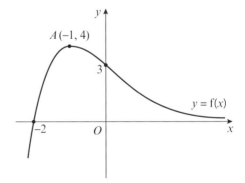

3 The diagram shows a sketch of the graph $y = f(x)$. The lines $x = 2$ and $y = 0$ (the x-axis) are asymptotes to the curve.
On separate axes, sketch the graphs of:

a $y = 3f(x) - 1$
b $y = f(x + 2) + 4$
c $y = -f(2x)$
d $y = f(|x|)$

For each part, state the equations of the asymptotes and the new coordinates of the point A.

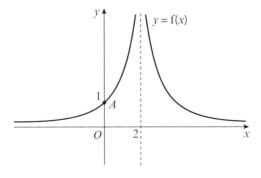

Chapter 2

E 4 The function g is defined by
g: $x \mapsto (x - 2)^2 - 9$, $x \in \mathbb{R}$.

a Draw a sketch of the graph of $y = g(x)$, labelling the turning point and the x- and y-intercepts. **(3 marks)**

b Write down the coordinates of the turning point when the curve is transformed as follows:

  i $2g(x - 4)$ **(2 marks)**
  ii $g(2x)$ **(2 marks)**
  iii $|g(x)|$ **(2 marks)**

c Sketch the curve with equation $y = g(|x|)$. On your sketch show the coordinates of all turning points and all x- and y-intercepts. **(4 marks)**

5 $h(x) = 2\sin x$, $-180° \leq x \leq 180°$.

a Sketch the graph of $y = h(x)$.

b Write down the coordinates of the minimum, A, and the maximum, B.

c Sketch the graphs of:

  i $h(x - 90°) + 1$ ii $\frac{1}{4}h\left(\frac{1}{2}x\right)$ iii $\frac{1}{2}|h(-x)|$

In each case find the coordinates of the images of the points O, A and B.

E/P 6 The diagram shows the curve with equation $y = f(x)$.

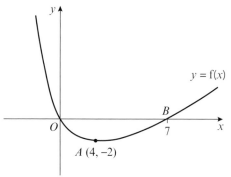

Find the coordinates of the points to which O, A and B are transformed on the curve with equation $y = f\left(\frac{1}{3}x + 2\right)$. **(3 marks)**

2.7 Solving modulus problems

You can use combinations of transformations together with $|f(x)|$ and $f(|x|)$ and an understanding of domain and range to solve problems.

Example 22

Given the function $t(x) = 3|x - 1| - 2$, $x \in \mathbb{R}$,

a sketch the graph of the function

b state the range of the function

c solve the equation $t(x) = \frac{1}{2}x + 3$.

48

Functions and graphs

Problem-solving

Use transformations to sketch the graph of $y = 3|x - 1| - 2$.

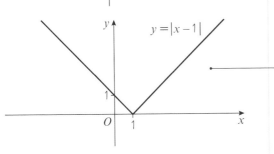

Sketch the graph of $y = |x|$.

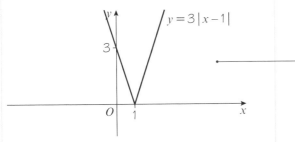

Step 1
Horizontal translation by vector $\begin{pmatrix} 1 \\ 0 \end{pmatrix}$.

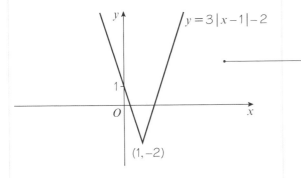

Step 2
Vertical stretch, scale factor 3.

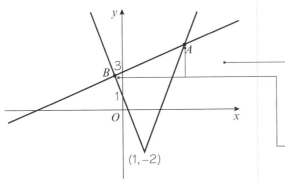

Step 3
Vertical translation by vector $\begin{pmatrix} 0 \\ -2 \end{pmatrix}$.

b The range of the function $t(x)$ is $y \in \mathbb{R}$, $y \geq -2$.

The graph has a minimum at $(1, -2)$.

c

First draw a sketch of $y = 3|x - 1| - 2$ and the line $y = \frac{1}{2}x + 3$.

The sketch shows there are two solutions, at A and B, the points of intersection.

At A, $3(x - 1) - 2 = \frac{1}{2}x + 3$

$3x - 5 = \frac{1}{2}x + 3$

$\frac{5}{2}x = 8$

$x = \frac{16}{5}$ ──── This is the solution on the original part of the graph.

At B, $-3(x - 1) - 2 = \frac{1}{2}x + 3$ ──── When $f(x) < 0$, $|f(x)| = -f(x)$, so use $-(3x - 1) - 2$ to find the solution on the reflected part of the graph.

$-3x + 3 - 2 = \frac{1}{2}x + 3$

$-\frac{7}{2}x = 2$

$x = -\frac{4}{7}$ ──── This is the solution corresponding to point B on the sketch.

The solutions are $x = \frac{16}{5}$ and $x = -\frac{4}{7}$.

Example 23

The function f is defined by f: $x \mapsto 6 - 2|x + 3|$.
A sketch of the graph of the function is shown in the diagram.
a State the range of f.
b Give a reason why f^{-1} does not exist.
c Solve the inequality $f(x) > 5$.

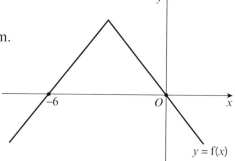

a The range of $f(x)$ is $f(x) \leq 6$. ──── The greatest value $f(x)$ can take is 6.

b $f(x)$ is a many-to-one function.
Therefore, f^{-1} does not exist.

For example, $f(0) = f(-6) = 0$.

c $f(x) = 5$ at the points A and B.
$f(x) > 5$ between the points A and B.

Problem-solving

Only one-to-one functions have inverses.

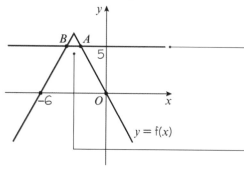

Add the line $y = 5$ to the graph of $y = f(x)$.

Between the points A and B, the graph of $y = f(x)$ is above the line $y = 5$.

At A, $6 - 2(x + 3) = 5$

$-2(x + 3) = -1$

$x + 3 = \frac{1}{2}$

$x = -\frac{5}{2}$ ──── This is the solution on the original part of the graph.

At B, $6 - (-2(x + 3)) = 5$

$2(x + 3) = -1$

$x + 3 = -\frac{1}{2}$

$x = -\frac{7}{2}$

When $f(x) < 0$, $|f(x)| = -f(x)$, so use the negative argument, $-2(x + 3)$.

This is the solution on the reflected part of the graph.

The solution to the inequality $f(x) > 5$ is

$-\frac{7}{2} < x < -\frac{5}{2}$

Online Explore the solution using technology.

Exercise 2G

P 1 For each function
  i sketch the graph of $y = f(x)$
  ii state the range of the function.
  a $f: x \mapsto 4|x| - 3$, $x \in \mathbb{R}$
  b $f(x) = \frac{1}{3}|x + 2| - 1$, $x \in \mathbb{R}$
  c $f(x) = -2|x - 1| + 6$, $x \in \mathbb{R}$
  d $f: x \mapsto -\frac{5}{2}|x| + 4$, $x \in \mathbb{R}$

Hint For part **b** transform the graph of $y = |x|$ by:
- A translation by vector $\begin{pmatrix} -2 \\ 0 \end{pmatrix}$
- A vertical sketch with scale factor $\frac{1}{3}$
- A translation by vector $\begin{pmatrix} 0 \\ -1 \end{pmatrix}$

2 Given that $p(x) = 2|x + 4| - 5$, $x \in \mathbb{R}$,
  a sketch the graph of $y = p(x)$
  b shade the region of the graph that satisfies $y \geq p(x)$.

P 3 Given that $q(x) = 6 - |3x + 9|$, $x \in \mathbb{R}$,
  a sketch the graph of $y = q(x)$
  b shade the region of the graph that satisfies $y \leq q(x)$.

4 The function f is defined as
  $f: x \mapsto 4|x + 6| + 1$, $x \in \mathbb{R}$.
  a Sketch the graph of $y = f(x)$.
  b State the range of the function.
  c Solve the equation $f(x) = -\frac{1}{2}x + 1$.

5 Given that $g(x) = -\frac{5}{2}|x - 2| + 7$, $x \in \mathbb{R}$,
  a sketch the graph of $y = g(x)$
  b state the range of the function
  c solve the equation $g(x) = x + 1$.

6 The functions m and n are defined as
  m$(x) = -2x + k$, $x \in \mathbb{R}$
  n$(x) = 3|x - 4| + 6$, $x \in \mathbb{R}$
  where k is a constant.

  Problem-solving
  m$(x) =$ n(x) has no real roots means that $y =$ m(x) and $y =$ n(x) do not intersect.

  The equation m$(x) =$ n(x) has no real roots.
  Find the range of possible values for the constant k. **(4 marks)**

7 The functions s and t are defined as
  s$(x) = -10 - x$, $x \in \mathbb{R}$
  t$(x) = 2|x + b| - 8$, $x \in \mathbb{R}$
  where b is a constant.
  The equation s$(x) =$ t(x) has exactly one real root. Find the value of b. **(4 marks)**

8 The function h is defined by
  h$(x) = \frac{2}{3}|x - 1| - 7$, $x \in \mathbb{R}$
  The diagram shows a sketch of the graph $y =$ h(x).
  a State the range of h. **(1 mark)**
  b Give a reason why h^{-1} does not exist. **(1 mark)**
  c Solve the inequality h$(x) < -6$. **(4 marks)**
  d State the range of values of k for which the equation h$(x) = \frac{2}{3}x + k$ has no solutions. **(4 marks)**

9 The diagram shows a sketch of part of the graph $y =$ h(x), where h$(x) = a - 2|x + 3|$, $x \in \mathbb{R}$.
  The graph intercepts the y-axis at $(0, 4)$.
  a Find the value of a. **(2 marks)**
  b Find the coordinates of P and Q. **(3 marks)**
  c Solve h$(x) = \frac{1}{3}x + 6$. **(4 marks)**

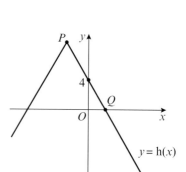

10 The diagram shows a sketch of part of the graph $y =$ m(x), where m$(x) = -4|x + 3| + 7$, $x \in \mathbb{R}$.
  a State the range of m. **(1 mark)**
  b Solve the equation m$(x) = \frac{3}{5}x + 2$. **(4 marks)**
  Given that m$(x) = k$, where k is a constant, has two distinct roots
  c state the set of possible values for k. **(4 marks)**

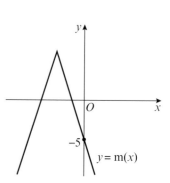

Functions and graphs

Challenge

1 The functions f and g are defined by
$f(x) = 2|x - 4| - 8, x \in \mathbb{R}$
$g(x) = x - 9, x \in \mathbb{R}$

The diagram shows a sketch of the graphs of $y = f(x)$ and $y = g(x)$.

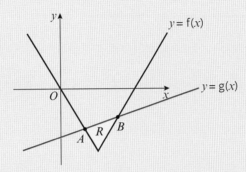

a Find the coordinates of the points A and B.
b Find the area of the region R.

2 The functions f and g are defined as:
$f(x) = -|x - 3| + 10, x \in \mathbb{R}$
$g(x) = 2|x - 3| + 2, x \in \mathbb{R}$
Show that the area of the shaded region is $\dfrac{64}{3}$.

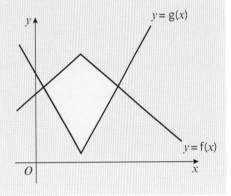

Mixed exercise 2

1 **a** On the same axes, sketch the graphs of $y = 2 - x$ and $y = 2|x + 1|$.
  b Hence, or otherwise, find the values of x for which $2 - x = 2|x + 1|$.

(E/P) 2 The equation $|2x - 11| = \dfrac{1}{2}x + k$ has exactly two distinct solutions.
Find the range of possible values of k. **(4 marks)**

(E/P) 3 Solve $|5x - 2| = -\dfrac{1}{4}x + 8$. **(4 marks)**

(E/P) 4 **a** On the same set of axes, sketch $y = |12 - 5x|$ and $y = -2x + 3$. **(3 marks)**
  b State with a reason whether there are any solutions to the equation
  $|12 - 5x| = -2x + 3$ **(2 marks)**

53

5 For each of the following mappings:
  i state whether the mapping is one-to-one, many-to-one or one-to-many
  ii state whether the mapping could represent a function.

a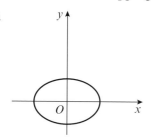
b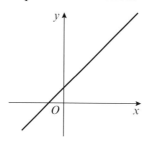
c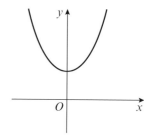
d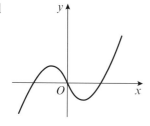
e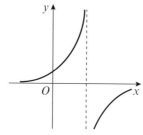
f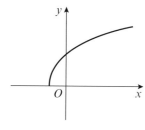

E 6 The function f(x) is defined by

$$f(x) = \begin{cases} -x, & x \leq 1 \\ x - 2, & x > 1 \end{cases}$$

  a Sketch the graph of f(x) for $-2 \leq x \leq 6$. (4 marks)
  b Find the values of x for which $f(x) = -\dfrac{1}{2}$ (3 marks)

E 7 The functions p and q are defined by
  p: $x \mapsto x^2 + 3x - 4$, $x \in \mathbb{R}$
  q: $x \mapsto 2x + 1$, $x \in \mathbb{R}$
  a Find pq(x). (2 marks)
  b Solve pq(x) = qq(x). (3 marks)

E 8 The function g(x) is defined as $g(x) = 2x + 7$, $\{x \in \mathbb{R}, x \geq 0\}$.
  a Sketch $y = g(x)$ and find the range. (3 marks)
  b Find $y = g^{-1}(x)$. (3 marks)
  c Sketch $y = g^{-1}(x)$ on the same axes as $y = g(x)$, stating the relationship between the two graphs. (2 marks)

E 9 The function f is defined by

$$f: x \mapsto \frac{2x + 3}{x - 1}, \quad \{x \in \mathbb{R}, x > 1\}$$

  a Find $f^{-1}(x)$. (4 marks)
  b State the range of $f^{-1}(x)$ (1 mark)

10 The functions f and g are given by

$$f: x \mapsto \frac{x}{x^2 - 1} - \frac{1}{x + 1}, \quad x \in \mathbb{R}, x > 1$$

$$g: x \mapsto \frac{2}{x}, \quad x \in \mathbb{R}, x > 0$$

a Show that $f(x) = \frac{1}{(x-1)(x+1)}$, $x \in \mathbb{R}, x > 1$ (3 marks)

b Find the range of $f(x)$. (1 mark)

c Solve $gf(x) = 70$. (4 marks)

11 The following functions $f(x)$, $g(x)$ and $h(x)$ are defined by

$$f(x) = 4(x - 2), \quad \{x \in \mathbb{R}, x \geq 0\}$$
$$g(x) = x^3 + 1, \quad \{x \in \mathbb{R}\}$$
$$h(x) = 3^x, \quad \{x \in \mathbb{R}\}$$

a Find $f(7)$, $g(3)$ and $h(-2)$.
b Find the range of $f(x)$ and the range of $g(x)$.
c Find $g^{-1}(x)$.
d Find the composite function $fg(x)$.
e Solve $gh(a) = 244$.
f Solve $f^{-1}(x) = -\frac{1}{2}$.

12 The function $f(x)$ is defined by $f: x \mapsto x^2 + 6x - 4$, $x \in \mathbb{R}, x > a$, for some constant a.

a State the least value of a for which f^{-1} exists. (4 marks)

b Given that $a = 0$, find f^{-1}. (4 marks)

13 The functions f and g are given by

$$f: x \mapsto 4x - 1, \{x \in \mathbb{R}\}$$

$$g: x \mapsto \frac{3}{2x - 1}, \left\{x \in \mathbb{R}, x \neq \frac{1}{2}\right\}$$

Find in its simplest form:

a the inverse function f^{-1} (2 marks)

b the composite function gf (3 marks)

c the values of x for which $2f(x) = g(x)$, giving your answers to 3 decimal places. (4 marks)

14 The functions f and g are given by

$$f: x \mapsto \frac{x}{x - 2}, \quad \{x \in \mathbb{R}, x \neq 2\}$$

$$g: x \mapsto \frac{3}{x}, \quad \{x \in \mathbb{R}, x \neq 0\}$$

a Find $f^{-1}(x)$. (2 marks)

b Write down the range of $f^{-1}(x)$. (1 mark)

c Calculate $gf(1.5)$. (2 marks)

d Use algebra to find the values of x for which $g(x) = f(x) + 4$. (4 marks)

15 The function $n(x)$ is defined by

$$n(x) = \begin{cases} 5 - x, & x \leq 0 \\ x^2, & x > 0 \end{cases}$$

a Find $n(-3)$ and $n(3)$.
b Solve the equation $n(x) = 50$.

16 g(x) = tan x, −180° ⩽ x ⩽ 180°

   a Sketch the graph of y = g(x).

   b Sketch the graph of y = |g(x)|.

   c Sketch the graph of y = g(|x|).

(E) **17** The diagram shows the graph of f(x).

The points A(4, −3) and B(9, 3) are turning points on the graph.

Sketch on separate diagrams, the graphs of

   a y = f(2x) + 1   **(3 marks)**

   b y = |f(x)| **(3 marks)**

   c y = −f(x − 2)   **(3 marks)**

   d $y = f\left(\frac{1}{2}x + 2\right)$ **(3 marks)**

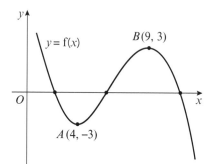

Indicate on each diagram the coordinates of any turning points on your sketch.

(E) **18** Functions f and g are defined by

      f : x ↦ 4 − x,   {x ∈ ℝ}

      g : x ↦ 3x²,   {x ∈ ℝ}

   a Write down the range of g.   **(1 mark)**

   b Solve gf(x) = 48.   **(4 marks)**

   c Sketch the graph of y = |f(x)| and hence find the values of x for which |f(x)| = 2.   **(4 marks)**

(E/P) **19** The function f is defined by f : x ↦ |2x − a|, {x ∈ ℝ}, where a is a positive constant.

   a Sketch the graph of y = f(x), showing the coordinates of the points where the graph cuts the axes.   **(3 marks)**

   b On a separate diagram, sketch the graph of y = f(2x), showing the coordinates of the points where the graph cuts the axes.   **(2 marks)**

   c Given that a solution of the equation $f(x) = \frac{1}{2}x$ is x = 4, find the two possible values of a.   **(4 marks)**

(E/P) **20 a** Sketch the graph of y = |x − 2a|, where a is a positive constant. Show the coordinates of the points where the graph meets the axes.   **(3 marks)**

   b Using algebra solve, for x in terms of a, $|x − 2a| = \frac{1}{3}x$.   **(4 marks)**

   c On a separate diagram, sketch the graph of y = a − |x − 2a|, where a is a positive constant. Show the coordinates of the points where the graph cuts the axes.   **(4 marks)**

(E/P) **21 a** Sketch the graph of y = |2x + a|, a > 0, showing the coordinates of the points where the graph meets the coordinate axes.   **(3 marks)**

   b On the same axes, sketch the graph of $y = \frac{1}{x}$ **(2 marks)**

   c Explain how your graphs show that there is only one solution of the equation

      x|2x + a| − 1 = 0   **(2 marks)**

   d Find, using algebra, the value of x for which x|2x + a| − 1 = 0.   **(3 marks)**

22 The diagram shows part of the curve with equation $y = f(x)$, where
$$f(x) = x^2 - 7x + 5\ln x + 8, \quad x > 0$$
The points A and B are the stationary points of the curve.

a Using calculus and showing your working, find the coordinates of the points A and B. **(4 marks)**

b Sketch the curve with equation $y = -3f(x - 2)$. **(3 marks)**

c Find the coordinates of the stationary points of the curve with equation $y = -3f(x - 2)$. State, without proof, which point is a maximum and which point is a minimum. **(3 marks)**

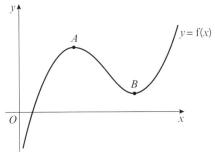

23 The function f has domain $-5 \leq x \leq 7$ and is linear from $(-5, 6)$ to $(-3, -2)$ and from $(-3, -2)$ to $(7, 18)$.
The diagram shows a sketch of the function.

a Write down the range of f. **(1 mark)**

b Find ff(-3). **(2 marks)**

c Sketch the graph of $y = |f(x)|$, marking the points at which the graph meets or cuts the axes. **(3 marks)**

The function g is defined by g: $x \mapsto x^2 - 7x + 10$.

d Solve the equation fg(x) = 2. **(3 marks)**

24 The function p is defined by
p: $x \mapsto -2|x + 4| + 10$
The diagram shows a sketch of the graph.

a State the range of p. **(1 mark)**

b Give a reason why p^{-1} does not exist. **(1 mark)**

c Solve the inequality p(x) > -4. **(4 marks)**

d State the range of values of k for which the equation $p(x) = -\dfrac{1}{2}x + k$ has no solutions. **(4 marks)**

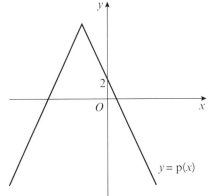

25 $f(x) = 3x^2 - 12x + 20, \quad x \in \mathbb{R}$

a Write $f(x)$ in the form $a(x + b)^2 + c$, where a, b and c are integers to be found. **(3 marks)**

b Hence find the range of the function
$$g(x) = \dfrac{1}{3x^2 - 12x + 20}, \quad x \in \mathbb{R}$$ **(2 marks)**

Challenge

a Sketch, on a single diagram, the graphs of $y = a^2 - x^2$ and $y = |x + a|$, where a is a constant and $a > 1$.

b Write down the coordinates of the points where the graph of $y = a^2 - x^2$ cuts the coordinate axes.

c Given that the two graphs intersect at $x = 4$, calculate the value of a.

Summary of key points

1. A modulus function is, in general, a function of the type $y = |f(x)|$.
   - When $f(x) \geq 0$, $|f(x)| = f(x)$
   - When $f(x) < 0$, $|f(x)| = -f(x)$

2. To sketch the graph of $y = |ax + b|$, sketch $y = ax + b$ then reflect the section of the graph below the x-axis in the x-axis.

3. A mapping is a **function** if every input has a distinct output. Functions can either be **one-to-one** or **many-to-one**.

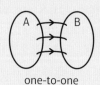
one-to-one function

many-to-one function

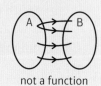
not a function

4. $fg(x)$ means apply g first, then apply f.
   $fg(x) = f(g(x))$

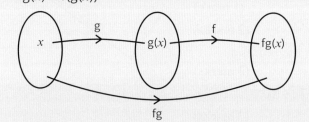

5. Functions $f(x)$ and $f^{-1}(x)$ are inverses of each other. $ff^{-1}(x) = x$ and $f^{-1}f(x) = x$.

6. The graphs of $y = f(x)$ and $y = f^{-1}(x)$ are reflections of each another in the line $y = x$.

7. The domain of $f(x)$ is the range of $f^{-1}(x)$.

8. The range of $f(x)$ is the domain of $f^{-1}(x)$.

9. To sketch the graph of $y = |f(x)|$
   - Sketch the graph of $y = f(x)$
   - Reflect any parts where $f(x) < 0$ (parts below the x-axis) in the x-axis
   - Delete the parts below the x-axis

10. To sketch the graph of $y = f(|x|)$
    - Sketch the graph of $y = f(x)$ for $x \geq 0$
    - Reflect this in the y-axis

11. $f(x + a)$ is a horizontal translation of $-a$.

12. $f(x) + a$ is a vertical translation of $+a$.

13. $f(ax)$ is a horizontal stretch of scale factor $\frac{1}{a}$

14. $af(x)$ is a vertical stretch of scale factor a.

15. $f(-x)$ reflects $f(x)$ in the y-axis.

16. $-f(x)$ reflects $f(x)$ in the x-axis.

Sequences and series

3

Objectives

After completing this chapter you should be able to:
- Find the nth term of an arithmetic sequence → **pages 60–62**
- Prove and use the formula for the sum of the first n terms of an arithmetic series → **pages 63–66**
- Find the nth term of a geometric sequence → **pages 66–70**
- Prove and use the formula for the sum of a finite geometric series → **pages 70–73**
- Prove and use the formula for the sum to infinity of a convergent geometric series → **pages 73–76**
- Use sigma notation to describe series → **pages 76–78**
- Generate sequences from recurrence relations → **pages 79–83**
- Model real-life situations with sequences and series → **pages 83–86**

Prior knowledge check

1. Write down the next three terms of each sequence.
   - **a** 2, 7, 12, 17
   - **b** 11, 8, 5, 2
   - **c** −15, −9, −3, 3
   - **d** 3, 6, 12, 24
   - **e** $\frac{1}{2}, \frac{1}{4}, \frac{1}{8}, \frac{1}{16}$
   - **f** $-\frac{1}{16}, \frac{1}{4}, -1, 4$

   ← **GCSE Mathematics**

2. Solve, giving your answers to 3 s.f.:
   - **a** $2^x = 50$
   - **b** $0.2^x = 0.0035$
   - **c** $4 \times 3^x = 78\,732$

   ← **Year 1, Section 14.6**

Sequences and series can be found in nature, and can be used to model population growth or decline, or the spread of a virus. → **Exercise 3I, Q12**

Chapter 3

3.1 Arithmetic sequences

- In an **arithmetic sequence**, the difference between consecutive terms is **constant**.

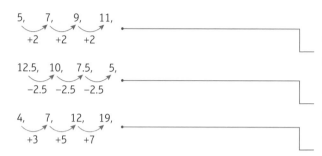

This sequence is arithmetic. The difference between consecutive terms is +2. The sequence is **increasing**.

This sequence is arithmetic. The difference between consecutive terms is −2.5. The sequence is **decreasing**.

The difference is not constant so the sequence is not arithmetic.

Notation An arithmetic sequence is sometimes called an **arithmetic progression**.

- The formula for the nth term of an arithmetic sequence is:

$$u_n = a + (n-1)d$$

where a is the first term and d is the common difference.

Notation In questions on sequences and series:
- u_n is the nth term
- a is the first term
- d is the common difference

Example 1

The nth term of an arithmetic sequence is $u_n = 55 - 2n$.

a Write down the first 5 terms of the sequence.
b Find the 99th term in the sequence.
c Find the first term in the sequence that is negative.

Online Use the table function on your calculator to generate terms in the sequence for this function, or to check an nth term.

a $u_n = 55 - 2n$
$n = 1 \rightarrow u_1 = 55 - 2(1) = 53$
$n = 2 \rightarrow u_2 = 55 - 2(2) = 51$
$n = 3 \rightarrow u_3 = 55 - 2(3) = 49$
$n = 4 \rightarrow u_4 = 55 - 2(4) = 47$
$n = 5 \rightarrow u_5 = 55 - 2(5) = 45$

b $u_{99} = 55 - 2(99) = -143$

c $55 - 2n < 0$
$-2n < -55$
$n > 27.5$
$n = 28$
$u_{28} = 55 - 2(28) = -1$

Remember, n is the position in the sequence, so for the first term substitute $n = 1$.

For the second term substitute $n = 2$.

For the 99th term substitute $n = 99$.

Problem-solving

To find the first negative term, set $u_n < 0$ and solve the inequality. n is the term number so it must be a positive integer.

Example 2

Find the nth term of each arithmetic sequence.

a 6, 20, 34, 48, 62

b 101, 94, 87, 80, 73

a $a = 6, d = 14$ — Write down the values of a and d.
$u_n = 6 + 14(n - 1)$
$u_n = 6 + 14n - 14$
$u_n = 14n - 8$

Substitute the values of a and d into the formula $a + (n - 1)d$ and simplify.

b $a = 101, d = -7$
$u_n = 101 - 7(n - 1)$
$u_n = 101 - 7n + 7$
$u_n = 108 - 7n$

Watch out If the sequence is decreasing then d is negative.

Example 3

A sequence is generated by the formula $u_n = an + b$ where a and b are constants to be found.
Given that $u_3 = 5$ and $u_8 = 20$, find the values of the constants a and b.

Problem-solving

You know two terms and there are two unknowns in the expression for the nth term. You can use this information to form two simultaneous equations. ← Year 1 Section 3.1

$u_3 = 5$, so $3a + b = 5$. (1)

$u_8 = 20$, so $8a + b = 20$. (2)

Substitute $n = 3$ and $u_3 = 5$ in $u_n = an + b$.

Substitute $n = 8$ and $u_8 = 20$ in $u_n = an + b$.

(2) − (1) gives:
$5a = 15$
$a = 3$
Substitute $a = 3$ in (1):
$9 + b = 5$
$b = -4$
Constants are $a = 3$ and $b = -4$.

Solve simultaneously.

Exercise 3A

1 For each sequence:

  i write down the first 4 terms of the sequence

  ii write down a and d.

  a $u_n = 5n + 2$ **b** $u_n = 9 - 2n$ **c** $u_n = 7 + 0.5n$ **d** $u_n = n - 10$

Chapter 3

2 Find the nth terms and the 10th terms in the following arithmetic progressions:
 a 5, 7, 9, 11, ...
 b 5, 8, 11, 14, ...
 c 24, 21, 18, 15, ...
 d −1, 3, 7, 11, ...
 e $x, 2x, 3x, 4x, ...$
 f $a, a + d, a + 2d, a + 3d, ...$

(P) **3** Calculate the number of terms in each of the following arithmetic sequences.
 a 3, 7, 11, ..., 83, 87
 b 5, 8, 11, ..., 119, 122
 c 90, 88, 86, ..., 16, 14
 d 4, 9, 14, ..., 224, 229
 e $x, 3x, 5x, ..., 35x$
 f $a, a + d, a + 2d, ..., a + (n − 1)d$

> **Problem-solving**
>
> Find an expression for u_n and set it equal to the final term in the sequence. Solve the equation to find the value of n.

(P) **4** The first term of an arithmetic sequence is 14. The fourth term is 32. Find the common difference.

(P) **5** A sequence is generated by the formula $u_n = pn + q$ where p and q are constants to be found. Given that $u_6 = 9$ and $u_9 = 11$, find the constants p and q.

(P) **6** For an arithmetic sequence $u_3 = 30$ and $u_9 = 9$. Find the first negative term in the sequence.

(P) **7** The 20th term of an arithmetic sequence is 14. The 40th term is −6. Find the value of the 10th term.

(P) **8** The first three terms of an arithmetic sequence are $5p$, 20 and $3p$, where p is a constant. Find the 20th term in the sequence.

(E/P) **9** The first three terms in an arithmetic sequence are −8, k^2, $17k$ ...
 Find two possible values of k. **(3 marks)**

(E/P) **10** An arithmetic sequence has first term k^2 and common difference k, where $k > 0$. The fifth term of the sequence is 41. Find the value of k, giving your answer in the form $p + q\sqrt{5}$, where p and q are integers to be found. **(4 marks)**

> **Problem-solving**
>
> You will need to make use of the condition $k > 0$ in your answer.

> **Challenge**
>
> The nth term of an arithmetic sequence is $u_n = \ln a + (n − 1) \ln b$ where a and b are integers. $u_3 = \ln 16$ and $u_7 = \ln 256$. Find the values of a and b.

Sequences and series

3.2 Arithmetic series

- **An arithmetic series is the sum of the terms of an arithmetic sequence.**

5, 7, 9, 11 is an arithmetic **sequence**.
5 + 7 + 9 + 11 is an arithmetic **series**.

Notation S_n is used for the sum of the first n terms of a series.

Example 4

Prove that the sum of the first 100 natural numbers is 5050.

$S_{100} = 1 + 2 + 3 + \ldots + 98 + 99 + 100$ (1)
$S_{100} = 100 + 99 + 98 + \ldots + 3 + 2 + 1$ (2)

Adding (1) and (2):
$2 \times S_{100} = 100 \times 101$
$S_{100} = \dfrac{100 \times 101}{2}$
$= 5050$

The **natural numbers** are the positive integers: 1, 2, 3, 4, …

Problem-solving

Write out the sum longhand, then write it out in reverse. You can pair up the numbers so that each pair has a sum of 101. There are 100 pairs in total.

- **The sum of the first n terms of an arithmetic series is given by the formula**

$$S_n = \frac{n}{2}(2a + (n-1)d)$$

where a is the first term and d is the common difference.
You can also write this formula as

$$S_n = \frac{n}{2}(a + l)$$

where l is the last term.

Example 5

Prove that the sum of the first n terms of an arithmetic series is $\dfrac{n}{2}(2a + (n-1)d)$.

$S_n = a + (a+d) + (a+2d) + \ldots$
$\quad + (a + (n-2)d) + (a + (n-1)d)$ (1) — Write out the terms of the sum.

$S_n = (a + (n-1)d) + (a + (n-2)d) + \ldots$
$\quad + (a + 2d) + (a + d) + a$ (2) — This is the sum reversed.

Adding (1) and (2):
$2 \times S_n = n(2a + (n-1)d)$
$S_n = \dfrac{n}{2}(2a + (n-1)d)$

— Adding together the two sums.

Problem-solving

You need to learn this proof for your exam.

Chapter 3

Example 6

Find the sum of the first 50 terms of the arithmetic series 32 + 27 + 22 + 17 + 12 + ...

$a = 32, d = -5$ — Write down a and d.

$S_{50} = \dfrac{50}{2}(2(32) + (50-1)(-5))$ — Substitute into the formula.

$S_{50} = -4525$ — Simplify.

Example 7

Find the least number of terms required for the sum of 4 + 9 + 14 + 19 + ... to exceed 2000.

4 + 9 + 14 + 19 + ... > 2000 — Always establish what you are given in a question. As you are adding on positive terms, it is easier to solve the equality $S_n = 2000$.

Using $S_n = \dfrac{n}{2}(2a + (n-1)d)$

$2000 = \dfrac{n}{2}(2 \times 4 + (n-1)5)$

$4000 = n(8 + 5n - 5)$ — Knowing $a = 4$, $d = 5$ and $S_n = 2000$, you need to find n.

$4000 = n(5n + 3)$

Substitute into $S_n = \dfrac{n}{2}(2a + (n-1)d)$.

$4000 = 5n^2 + 3n$

$0 = 5n^2 + 3n - 4000$

$n = \dfrac{-3 \pm \sqrt{9 + 80\,000}}{10}$ — Solve using the quadratic formula.

$n = 27.99$ or -28.59

28 terms are needed. — n is the number of terms, so must be a positive integer.

Exercise 3B

1 Find the sums of the following series.
  a 3 + 7 + 11 + 15 + ... (20 terms)
  b 2 + 6 + 10 + 14 + ... (15 terms)
  c 30 + 27 + 24 + 21 + ... (40 terms)
  d 5 + 1 + −3 + −7 + ... (14 terms)
  e 5 + 7 + 9 + ... + 75
  f 4 + 7 + 10 + ... + 91
  g 34 + 29 + 24 + 19 + ... + −111
  h $(x + 1) + (2x + 1) + (3x + 1) + ... + (21x + 1)$

  Hint For parts **e** to **h**, start by using the last term to work out the number of terms in the series.

2 Find how many terms of the following series are needed to make the given sums.
  a 5 + 8 + 11 + 14 + ... = 670
  b 3 + 8 + 13 + 18 + ... = 1575
  c 64 + 62 + 60 + ... = 0
  d 34 + 30 + 26 + 22 + ... = 112

  Hint Set the expression for S_n equal to the total and solve the resulting equation to find n.

Sequences and series

P 3 Find the sum of the first 50 even numbers.

P 4 Find the least number of terms for the sum of $7 + 12 + 17 + 22 + 27 + \ldots$ to exceed 1000.

P 5 The first term of an arithmetic series is 4. The sum to 20 terms is -15. Find, in any order, the common difference and the 20th term.

P 6 The sum of the first three terms of an arithmetic series is 12. If the 20th term is -32, find the first term and the common difference.

P 7 Prove that the sum of the first 50 natural numbers is 1275.

> **Problem-solving**
> Use the same method as Example 4.

P 8 Show that the sum of the first $2n$ natural numbers is $n(2n + 1)$.

P 9 Prove that the sum of the first n odd numbers is n^2.

E/P 10 The fifth term of an arithmetic series is 33. The tenth term is 68. The sum of the first n terms is 2225.
  a Show that $7n^2 + 3n - 4450 = 0$. **(4 marks)**
  b Hence find the value of n. **(1 mark)**

E/P 11 An arithmetic series is given by $(k + 1) + (2k + 3) + (3k + 5) + \ldots + 303$
  a Find the number of terms in the series in terms of k. **(1 mark)**
  b Show that the sum of the series is given by $\dfrac{152k + 46208}{k + 2}$ **(3 marks)**
  c Given that $S_n = 2568$, find the value of k. **(1 mark)**

E/P 12 a Calculate the sum of all the multiples of 3 from 3 to 99 inclusive,
  $$3 + 6 + 9 + \ldots + 99$$ **(3 marks)**
  b In the arithmetic series
  $$4p + 8p + 12p + \ldots + 400$$
  where p is a positive integer and a factor of 100,
  i find, in terms of p, an expression for the number of terms in this series.
  ii Show that the sum of this series is $200 + \dfrac{20\,000}{p}$ **(4 marks)**
  c Find, in terms of p, the 80th term of the arithmetic sequence
  $$(3p + 2), (5p + 3), (7p + 4), \ldots,$$
  giving your answer in its simplest form. **(2 marks)**

Chapter 3

E/P **13** Joanna has some sticks that are all of the same length.
She arranges them in shapes as shown opposite and has made the following 3 rows of patterns.

She notices that 6 sticks are required to make the single pentagon in the first row, 11 sticks in the second row and for the third row she needs 16 sticks.

Row 1

Row 2

Row 3

a Find an expression, in terms of n, for the number of sticks required to make a similar arrangement of n pentagons in the nth row. **(3 marks)**

Joanna continues to make pentagons following the same pattern. She continues until she has completed 10 rows.

b Find the total number of sticks Joanna uses in making these 10 rows. **(3 marks)**

Joanna started with 1029 sticks. Given that Joanna continues the pattern to complete k rows but does not have enough sticks to complete the $(k + 1)$th row:

c show that k satisfies $(5k - 98)(k + 21) \leq 0$ **(4 marks)**

d find the value of k. **(2 marks)**

Challenge

An arithmetic sequence has nth term $u_n = \ln 9 + (n - 1) \ln 3$. Show that the sum of the first n terms $= a \ln 3^{n^2+3n}$ where a is a rational number to be found.

3.3 Geometric sequences

- **A geometric sequence has a common ratio between consecutive terms.**

To get from one term to the next you **multiply** by the common ratio.

Notation A geometric sequence is sometimes called a **geometric progression**.

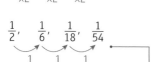

This is a geometric sequence with common ratio 2. This sequence is increasing.

This is a geometric sequence with common ratio $\frac{1}{3}$. This sequence is decreasing but will never get to zero.

Notation A geometric sequence with a common ratio $|r| < 1$ converges. This means it tends to a certain value. You call the value the **limit** of the sequence.

Here the common ratio is -2. The sequence alternates between positive and negative terms.

Notation An **alternating sequence** is a sequence in which terms are alternately positive and negative.

- **The formula for the nth term of a geometric sequence is:**

$$u_n = ar^{n-1}$$

where a is the first term and r is the common ratio.

66

Sequences and series

Example 8

Find the **i** 10th and **ii** nth terms in the following geometric sequences:
a 3, 6, 12, 24, …
b 40, −20, 10, −5, …

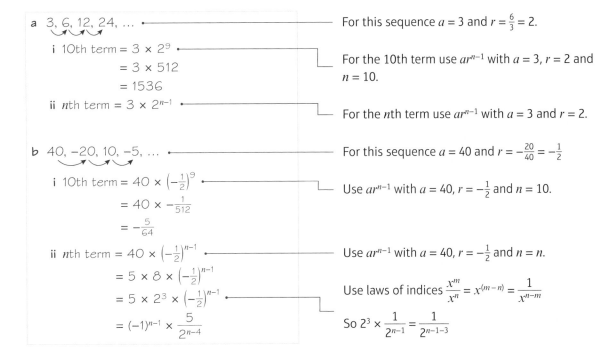

Example 9

The 2nd term of a geometric sequence is 4 and the 4th term is 8. Given that the common ratio is positive, find the exact value of the 11th term in the sequence.

nth term = ar^{n-1}, so the 2nd term is ar, and the 4th term is ar^3

$$ar = 4 \quad (1)$$
$$ar^3 = 8 \quad (2)$$

Dividing equation (2) by equation (1):

$$\frac{ar^3}{ar} = \frac{8}{4}$$
$$r^2 = 2$$
$$r = \sqrt{2}$$

Problem-solving

You can use the general term of a geometric sequence to write two equations. Solve these simultaneously to find a and r, then find the 11th term in the sequence.

You are told in the question that $r > 0$ so use the positive square root.

Substituting back into equation (1):

$a\sqrt{2} = 4$

$a = \dfrac{4}{\sqrt{2}}$

$a = 2\sqrt{2}$ — Rationalise the denominator.

nth term $= ar^{n-1}$, so

11th term $= (2\sqrt{2})(\sqrt{2})^{10}$

$= 64\sqrt{2}$ — Simplify your answer as much as possible.

Example 10

The numbers 3, x and $(x + 6)$ form the first three terms of a geometric sequence with all positive terms. Find:

a the possible values of x,   **b** the 10th term of the sequence.

a $\dfrac{u_2}{u_1} = \dfrac{u_3}{u_2}$

$\dfrac{x}{3} = \dfrac{x+6}{x}$

$x^2 = 3(x+6)$

$x^2 = 3x + 18$

$x^2 - 3x - 18 = 0$

$(x - 6)(x + 3) = 0$ — Factorise.

$x = 6$ or -3

So x is either 6 or -3, but there are no negative terms so $x = 6$. — If there are no negative terms then -3 cannot be an answer.

Problem-solving

In a geometric sequence the ratio between consecutive terms is the same, so $\dfrac{u_2}{u_1} = \dfrac{u_3}{u_2}$. Simplify the algebraic fraction to form a quadratic equation.   ← Year 1, Section 3.2

b   10th term $= ar^9$

$= 3 \times 2^9$

$= 3 \times 512$

$= 1536$

The 10th term is 1536.

Use the formula nth term $= ar^{n-1}$ with $n = 10$, $a = 3$ and $r = \dfrac{x}{3} = \dfrac{6}{3} = 2$.

Example 11

What is the first term in the geometric progression 3, 6, 12, 24, … to exceed 1 million?

nth term $= ar^{n-1}$

$= 3 \times 2^{n-1}$

We want nth term $> 1\,000\,000$

Problem-solving

Determine a and r, then write an inequality using the formula for the general term of a geometric sequence.

Sequence has $a = 3$ and $r = 2$.

Sequences and series

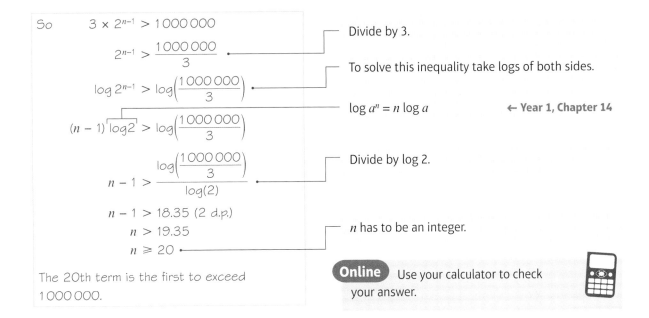

The 20th term is the first to exceed 1 000 000.

Online Use your calculator to check your answer.

Exercise 3C

1 Which of the following are geometric sequences? For the ones that are, give the value of the common ratio, r.

   a 1, 2, 4, 8, 16, 32, …
   b 2, 5, 8, 11, 14, …
   c 40, 36, 32, 28, …
   d 2, 6, 18, 54, 162, …
   e 10, 5, 2.5, 1.25, …
   f 5, −5, 5, −5, 5, …
   g 3, 3, 3, 3, 3, 3, 3, …
   h 4, −1, 0.25, −0.0625, …

2 Continue the following geometric sequences for three more terms.

   a 5, 15, 45, …
   b 4, −8, 16, …
   c 60, 30, 15, …
   d 1, $\frac{1}{4}$, $\frac{1}{16}$, …
   e 1, p, p^2, …
   f x, $-2x^2$, $4x^3$, …

(P) 3 If 3, x and 9 are the first three terms of a geometric sequence, find:

   a the exact value of x,
   b the exact value of the 4th term.

   Problem-solving

   In a geometric sequence the common ratio can be calculated by $\frac{u_2}{u_1}$ or $\frac{u_3}{u_2}$

4 Find the sixth and nth terms of the following geometric sequences.

   a 2, 6, 18, 54, …
   b 100, 50, 25, 12.5, …
   c 1, −2, 4, −8, …
   d 1, 1.1, 1.21, 1.331, …

5 The nth term of a geometric sequence is 2×5^n. Find the first and 5th terms.

6 The sixth term of a geometric sequence is 32 and the 3rd term is 4. Find the first term and the common ratio.

7 A geometric sequence has first term 4 and third term 1. Find the two possible values of the 6th term.

(E/P) 8 The first three terms of a geometric sequence are given by $8 - x$, $2x$, and x^2 respectively where $x > 0$.
   a Show that $x^3 - 4x^2 = 0$. (2 marks)
   b Find the value of the 20th term. (3 marks)
   c State, with a reason, whether 4096 is a term in the sequence. (1 mark)

(E/P) 9 A geometric sequence has first term 200 and a common ratio p where $p > 0$. The 6th term of the sequence is 40.
   a Show that p satisfies the equation $5 \log p + \log 5 = 0$. (3 marks)
   b Hence or otherwise, find the value of p correct to 3 significant figures. (1 mark)

(P) 10 A geometric sequence has first term 4 and fourth term 108. Find the smallest value of k for which the kth term in this sequence exceeds 500 000.

(P) 11 The first three terms of a geometric sequence are 9, 36, 144. State, with a reason, whether 383 616 is a term in the sequence.

> **Problem-solving**
>
> Determine the values of a and r and find the general term of the sequence. Set the number given equal to the general term and solve to find n. If n is an integer, then the number is in the sequence.

(P) 12 The first three terms of a geometric sequence are 3, −12, 48. State, with a reason, whether 49 152 is a term in the sequence.

(P) 13 Find which term in the geometric progression 3, 12, 48, … is the first to exceed 1 000 000.

3.4 Geometric series

A geometric **series** is the sum of the terms of a geometric **sequence**. 3, 6, 12, 24, … is a geometric sequence. $3 + 6 + 12 + 24 + \ldots$ is a geometric series.

■ **The sum of the first n terms of a geometric series is given by the formula**

$$S_n = \frac{a(1 - r^n)}{1 - r}, r \neq 1$$

or $S_n = \dfrac{a(r^n - 1)}{r - 1}, r \neq 1$

> **Hint** These two formulae are equivalent. It is often easier to use the first one if $r < 1$ and the second one if $r > 1$.

where a is the first term and r is the common ratio.

Sequences and series

Example 12

A geometric series has first term a and common difference r. Prove that the sum of the first n terms of this series is given by $S_n = \dfrac{a(1-r^n)}{1-r}$

Let $S_n = a + ar + ar^2 + ar^3 + \ldots ar^{n-2} + ar^{n-1}$ (1) — Multiply by r.

$rS_n = ar + ar^2 + ar^3 + \ldots ar^{n-1} + ar^n$ (2) — Subtract rS_n from S_n.

(1) − (2) gives $S_n - rS_n = a - ar^n$ — Take out the common factor.

$S_n(1-r) = a(1-r^n)$ — Divide by $(1-r)$.

$S_n = \dfrac{a(1-r^n)}{1-r}$

Problem-solving

You need to learn this proof for your exam.

Example 13

Find the sums of the following geometric series.
a $2 + 6 + 18 + 54 + \ldots$ (for 10 terms)
b $1024 - 512 + 256 - 128 + \ldots + 1$

a Series is
$2 + 6 + 18 + 54 + \ldots$ (for 10 terms) — As in all questions, write down what is given.
So $a = 2$, $r = \frac{6}{2} = 3$ and $n = 10$

So $S_{10} = \dfrac{2(3^{10} - 1)}{3 - 1} = 59048$ — When $r > 1$ it is easier to use the formula $S_n = \dfrac{a(r^n - 1)}{r - 1}$

b Series is
$1024 - 512 + 256 - 128 + \ldots + 1$
So $a = 1024$, $r = -\frac{512}{1024} = -\frac{1}{2}$
and the nth term $= 1$

$1024\left(-\frac{1}{2}\right)^{n-1} = 1$ — First solve $ar^{n-1} = 1$ to find n.

$(-2)^{n-1} = 1024$
$2^{n-1} = 1024$ — $(-2)^{n-1} = (-1)^{n-1}(2^{n-1}) = 1024$, so $(-1)^{n-1}$ must be positive and $2^{n-1} = 1024$.

$n - 1 = \dfrac{\log 1024}{\log 2}$

$n - 1 = 10$ — $1024 = 2^{10}$
$n = 11$

So $S_n = \dfrac{1024\left(1 - \left(-\frac{1}{2}\right)^{11}\right)}{1 - \left(-\frac{1}{2}\right)}$ — When $r < 1$, it is easier to use the formula $S_n = \dfrac{a(1-r^n)}{1-r}$

$= \dfrac{1024\left(1 + \frac{1}{2048}\right)}{1 + \frac{1}{2}}$

$= \dfrac{1024.5}{\frac{3}{2}} = 683$

Chapter 3

Example 14

Find the least value of n such that the sum of $1 + 2 + 4 + 8 + \ldots$ to n terms exceeds $2\,000\,000$.

Sum to n terms is $S_n = \dfrac{1(2^n - 1)}{2 - 1}$
$= 2^n - 1$

If this is to exceed $2\,000\,000$ then
$S_n > 2\,000\,000$
$2^n - 1 > 2\,000\,000$
$2^n > 2\,000\,001$ — Add 1.
$n \log 2 > \log(2\,000\,001)$ — Use laws of logs: $\log a^n = n \log a$.
$n > \dfrac{\log(2\,000\,001)}{\log(2)}$
$n > 20.9$

It needs 21 terms to exceed $2\,000\,000$. — Round n up to the nearest integer.

Problem-solving

Determine the values of a and r, then use the formula for the sum of a geometric series to form an inequality.

Exercise 3D

1 Find the sum of the following geometric series (to 3 d.p. if necessary).
   a $1 + 2 + 4 + 8 + \ldots$ (8 terms)
   b $32 + 16 + 8 + \ldots$ (10 terms)
   c $\dfrac{2}{3} + \dfrac{4}{15} + \dfrac{8}{75} + \ldots + \dfrac{256}{234\,375}$
   d $4 - 12 + 36 - 108 + \ldots$ (6 terms)
   e $729 - 243 + 81 - \ldots - \dfrac{1}{3}$
   f $-\dfrac{5}{2} + \dfrac{5}{4} - \dfrac{5}{8} \ldots - \dfrac{5}{32\,768}$

2 A geometric series has first three terms $3 + 1.2 + 0.48\ldots$ Evaluate S_{10}, giving your answer to 4 d.p.

3 A geometric series has first term 5 and common ratio $\dfrac{2}{3}$. Find the value of S_8.

(P) 4 The sum of the first three terms of a geometric series is 30.5. If the first term is 8, find possible values of r.

(P) 5 Find the least value of n such that the sum $3 + 6 + 12 + 24 + \ldots$ to n terms exceeds 1.5 million.

(P) 6 Find the least value of n such that the sum $5 + 4.5 + 4.05 + \ldots$ to n terms exceeds 45.

(E) 7 A geometric series has first term 25 and common ratio $\dfrac{3}{5}$.
   Given that the sum to k terms of the series is greater than 61,
   a show that $k > \dfrac{\log(0.024)}{\log(0.6)}$ (4 marks)
   b find the smallest possible value of k. (1 mark)

E/P 8 A geometric series has first term a and common ratio r.
The sum of the first two terms of the series is 4.48.
The sum of the first four terms is 5.1968. Find the two
possible values of r. **(4 marks)**

Problem-solving
One value will be positive and one value will be negative.

E/P 9 The first term of a geometric series is a and the common ratio is $\sqrt{3}$.
Show that $S_{10} = 121a(\sqrt{3} + 1)$. **(4 marks)**

E/P 10 A geometric series has first term a and common ratio 2. A different geometric series has first term b and common ratio 3. Given that the sum of the first 4 terms of both series is the same, show that $a = \dfrac{8}{3} b$. **(4 marks)**

E/P 11 The first three terms of a geometric series are $(k - 6)$, k, $(2k + 5)$, where k is a positive constant.
   a Show that $k^2 - 7k - 30 = 0$. **(4 marks)**
   b Hence find the value of k. **(2 marks)**
   c Find the common ratio of this series. **(1 mark)**
   d Find the sum of the first 10 terms of this series, giving your answer to the nearest whole number. **(2 marks)**

3.5 Sum to infinity

You can work out the sum of the first n terms of a geometric series. As n tends to infinity, the sum of the series is called the **sum to infinity**.

Notation You can write the sum to infinity of a geometric series as S_∞.

Consider the sum of the first n terms of the geometric series $2 + 4 + 8 + 16 + \ldots$

The terms of this series are getting larger, so as n tends to infinity, S_n also tends to infinity. This is called a **divergent** series.

Now consider the sum of the first n terms of the geometric series $1 + \tfrac{1}{2} + \tfrac{1}{4} + \tfrac{1}{8} + \tfrac{1}{16} + \ldots$

The terms of this series are getting smaller. As n tends to infinity, S_n gets closer and closer to a finite value, S_∞. This is called a **convergent** series.

- **A geometric series is convergent if and only if $|r| < 1$, where r is the common ratio.**

Hint You can also write this condition as $-1 < r < 1$.

The sum of the first n terms of a geometric series is given by $S_n = \dfrac{a(1 - r^n)}{1 - r}$

When $|r| < 1$, $\displaystyle\lim_{n \to \infty} \left(\dfrac{a(1 - r^n)}{1 - r} \right) = \dfrac{a}{1 - r}$

This is because $r^n \to 0$ as $n \to \infty$.

Notation $\displaystyle\lim_{n \to \infty}$ means 'the limit as n tends to ∞'.
You can't evaluate the expression when n is ∞, but as n gets larger the expression gets closer to a fixed (or **limiting**) value.

- **The sum to infinity of a convergent geometric series is given by $S_\infty = \dfrac{a}{1 - r}$**

Watch out You can only use this formula for a convergent series, i.e. when $|r| < 1$.

Chapter 3

Example 15

The fourth term of a geometric series is 1.08 and the seventh term is 0.233 28.

a Show that this series is convergent.

b Find the sum to infinity of the series.

a $ar^3 = 1.08$ (1)

$ar^6 = 0.233\,28$ (2)

| Use the nth term of a geometric sequence ar^{n-1} to write down 2 simultaneous equations.

Dividing (2) by (1):

$\dfrac{ar^6}{ar^3} = \dfrac{0.233\,28}{1.08}$

| Divide equation (2) by equation (1) to eliminate a.

$r^3 = 0.216$

$r = 0.6$

The series is convergent as $|r| = 0.6 < 1$.

Problem-solving

To show that a series is convergent you need to find r, then state that the series is convergent if $|r| < 1$.

b Substituting the value of r^3 into equation (1) to find a

$0.216a = 1.08$

$a = \dfrac{1.08}{0.216}$

$a = 5$

Substituting into S_∞ formula:

$S_\infty = \dfrac{a}{1-r}$

$S_\infty = \dfrac{5}{1-0.6}$

$S_\infty = 12.5$

Example 16

For a geometric series with first term a and common ratio r, $S_4 = 15$ and $S_\infty = 16$.

a Find the possible values of r.

b Given that all the terms in the series are positive, find the value of a.

a $\dfrac{a(1-r^4)}{1-r} = 15$ (1)

| $S_4 = 15$ so use the formula $S_n = \dfrac{a(1-r^n)}{1-r}$ with $n = 4$.

$\dfrac{a}{1-r} = 16$ (2)

| $S_\infty = 16$ so use the formula $S_\infty = \dfrac{a}{1-r}$ with $S_\infty = 16$.

$16(1-r^4) = 15$

$1 - r^4 = \dfrac{15}{16}$

$r^4 = \dfrac{1}{16}$

$r = \pm\dfrac{1}{2}$

| Solve equations simultaneously.

Replace $\dfrac{a}{1-r}$ by 16 in equation (1).

Take the 4th root of $\dfrac{1}{16}$

Sequences and series

b As all terms are positive, $r = +\frac{1}{2}$

$$\frac{a}{1-\frac{1}{2}} = 16$$

Substitute $r = \frac{1}{2}$ into equation (2) to find a.

$16(1 - \frac{1}{2}) = a$

$a = 8$

The first term in the series is 8.

Exercise 3E

1 For each of the following series:
  i state, with a reason, whether the series is convergent.
  ii If the series is convergent, find the sum to infinity.

  a $1 + 0.1 + 0.01 + 0.001 + \ldots$
  b $1 + 2 + 4 + 8 + 16 + \ldots$
  c $10 - 5 + 2.5 - 1.25 + \ldots$
  d $2 + 6 + 10 + 14 + \ldots$
  e $1 + 1 + 1 + 1 + 1 + \ldots$
  f $3 + 1 + \frac{1}{3} + \frac{1}{9} + \ldots$
  g $0.4 + 0.8 + 1.2 + 1.6 + \ldots$
  h $9 + 8.1 + 7.29 + 6.561 + \ldots$

2 A geometric series has first term 10 and sum to infinity 30. Find the common ratio.

3 A geometric series has first term -5 and sum to infinity -3. Find the common ratio.

4 A geometric series has sum to infinity 60 and common ratio $\frac{2}{3}$. Find the first term.

5 A geometric series has common ratio $-\frac{1}{3}$ and $S_\infty = 10$. Find the first term.

(P) 6 Find the fraction equal to the recurring decimal $0.\dot{2}\dot{3}$.

Hint $0.\dot{2}\dot{3} = \frac{23}{100} + \frac{23}{10\,000} + \frac{23}{1\,000\,000} + \ldots$

7 For a geometric series $a + ar + ar^2 + \ldots$, $S_3 = 9$ and $S_\infty = 8$, find the values of a and r.

(E/P) 8 Given that the geometric series $1 - 2x + 4x^2 - 8x^3 + \ldots$ is convergent,
  a find the range of possible values of x (3 marks)
  b find an expression for S_∞ in terms of x. (1 mark)

(E/P) 9 In a convergent geometric series the common ratio is r and the first term is 2.
  Given that $S_\infty = 16 \times S_3$,
  a find the value of the common ratio, giving your answer to 4 significant figures (3 marks)
  b find the value of the fourth term. (2 marks)

(E/P) 10 The first term of a geometric series is 30. The sum to infinity of the series is 240.
  a Show that the common ratio, r, is $\frac{7}{8}$ (2 marks)
  b Find to 3 significant figures, the difference between the 4th and 5th terms. (2 marks)
  c Calculate the sum of the first 4 terms, giving your answer to 3 significant figures. (2 marks)
  The sum of the first n terms of the series is greater than 180.
  d Calculate the smallest possible value of n. (4 marks)

Chapter 3

E/P **11** A geometric series has first term a and common ratio r. The second term of the series is $\frac{15}{8}$ and the sum to infinity of the series is 8.

    a Show that $64r^2 - 64r + 15 = 0$.   **(4 marks)**

    b Find the two possible values of r.   **(2 marks)**

    c Find the corresponding two possible values of a.   **(2 marks)**

Given that r takes the smaller of its two possible values,

    d find the smallest value of n for which S_n exceeds 7.99.   **(2 marks)**

Challenge

The sum to infinity of a geometric series is 7. A second series is formed by squaring every term in the first geometric series.

a Show that the second series is also geometric.

b Given that the sum to infinity of the second series is 35, show that the common ratio of the original series is $\frac{1}{6}$.

3.6 Sigma notation

- The Greek capital letter 'sigma' is used to signify a sum. You write it as $\sum$. You write **limits** on the top and bottom to show which terms you are summing.

This tells you that are summing the expression in brackets with $r = 1, r = 2, \ldots$ up to $r = 5$.

$$\sum_{r=1}^{5}(2r - 3) = -1 + 1 + 3 + 5 + 7$$

Substitute $r = 1, r = 2, r = 3, r = 4, r = 5$ to find the five terms in this arithmetic series.

Look at the limits carefully: they don't have to start at 1.

$$\sum_{r=3}^{7}(5 \times 2^r) = 40 + 80 + 160 + 320 + 640$$

To find the terms in this geometric series, you substitute $r = 3, r = 4, r = 5, r = 6, r = 7$.

You can write some results that you already know using sigma notation:

- $\sum_{r=1}^{n} 1 = n$

- $\sum_{r=1}^{n} r = \frac{n(n+1)}{2}$

Hint $\sum_{r=1}^{n} 1 = \underbrace{1 + 1 + 1 + \ldots + 1}_{n \text{ times}}$

Example 17

Calculate $\sum_{r=1}^{20}(4r + 1)$

$\sum_{r=1}^{20}(4r + 1) = 5 + 9 + 13 + \ldots + 81$

$a = 5, d = 4$ and $n = 20$

Problem-solving

Substitute $r = 1, 2,$ etc. to find the terms in the series.

Sequences and series

$$S = \frac{n}{2}(2a + (n-1)d)$$

$$= \frac{20}{2}(2 \times 5 + (20-1)4)$$

$$= 10(10 + 19 \times 4)$$

$$= 10 \times 86$$

$$= 860$$

Use the formula for the sum to n terms of an arithmetic series.

Substitute $a = 5$, $d = 4$ and $n = 20$ into $S = \frac{n}{2}(2a + (n-1)d)$.

Online Check your answer by using your calculator to calculate the sum of the series.

Example 18

Find the values of:

a $\sum_{k=1}^{12} 5 \times 3^{k-1}$ **b** $\sum_{k=5}^{12} 5 \times 3^{k-1}$

a $\sum_{k=1}^{12} 5 \times 3^{k-1}$

$= 5 + 15 + 45 + \ldots$

$a = 5$, $r = 3$ and $n = 12$

$$S_n = \frac{a(r^n - 1)}{r - 1}$$

$$S_{12} = \frac{5(3^{12} - 1)}{3 - 1}$$

$S_{12} = 1328600$

Substitute $k = 1$, $k = 2$ and so on to write out the first few terms of the series. This will help you determine the correct values for a, r and n.

Since $r > 1$ use the formula $S_n = \frac{a(r^n - 1)}{r - 1}$ and substitute in $a = 5$, $r = 3$ and $n = 12$.

b $\sum_{k=5}^{12} 5 \times 3^{k-1} = \sum_{k=1}^{12} 5 \times 3^{k-1} - \sum_{k=1}^{4} 5 \times 3^{k-1}$

$S_{12} = 1328600$

$$S_4 = \frac{5(3^4 - 1)}{3 - 1} = 200$$

$\sum_{k=5}^{12} 5 \times 3^{k-1} = 1328600 - 200 = 1328400$

Problem-solving

When we are summing series from k to n, we can consider the sum of the terms from 1 to n and subtract the terms from 1 to $k - 1$.

Exercise 3F

1 For each series:
  i write out every term in the series
  ii hence find the value of the sum.

  a $\sum_{r=1}^{5}(3r + 1)$ **b** $\sum_{r=1}^{6} 3r^2$ **c** $\sum_{r=1}^{5} \sin(90r°)$ **d** $\sum_{r=5}^{8} 2\left(-\frac{1}{3}\right)^r$

2 For each series:
  i write the series using sigma notation
  ii evaluate the sum.

  a $2 + 4 + 6 + 8$ **b** $2 + 6 + 18 + 54 + 162$ **c** $6 + 4.5 + 3 + 1.5 + 0 - 1.5$

Chapter 3

3 For each series:
   i find the number of terms in the series
   ii write the series using sigma notation.

   a $7 + 13 + 19 + \ldots + 157$
   b $\dfrac{1}{3} + \dfrac{2}{15} + \dfrac{4}{75} + \ldots + \dfrac{64}{46875}$
   c $8 - 1 - 10 - 19 \ldots - 127$

4 Evaluate:

   a $\displaystyle\sum_{r=1}^{20}(7 - 2r)$
   b $\displaystyle\sum_{r=1}^{10} 3 \times 4^r$
   c $\displaystyle\sum_{r=1}^{100}(2r - 8)$
   d $\displaystyle\sum_{r=1}^{\infty} 7\left(-\dfrac{1}{3}\right)^r$

(P) 5 Evaluate:

   a $\displaystyle\sum_{r=9}^{30}\left(5r - \dfrac{1}{2}\right)$
   b $\displaystyle\sum_{r=100}^{200}(3r + 4)$
   c $\displaystyle\sum_{r=5}^{100} 3 \times 0.5^r$
   d $\displaystyle\sum_{i=5}^{100} 1$

   Problem-solving
   $$\sum_{r=k}^{n} u_r = \sum_{r=1}^{n} u_r - \sum_{r=1}^{k-1} u_r$$

(E/P) 6 Find the value of $\displaystyle\sum_{r=1}^{30}(r + 2^r)$ **(3 marks)**

(E/P) 7 Find the value of $\displaystyle\sum_{r=1}^{12}(2r - 5 + 3^r)$ **(4 marks)**

   Problem-solving
   You can split up terms inside the sigma notation:
   $$\sum_{r=1}^{12}(2r - 5 + 3^r) = \sum_{r=1}^{12}(2r - 5) + \sum_{r=1}^{12} 3^r$$

8 Find in terms of k:

   a $\displaystyle\sum_{r=1}^{k} 4(-2)^r$
   b $\displaystyle\sum_{r=1}^{k}(100 - 2r)$
   c $\displaystyle\sum_{r=10}^{k}(7 - 2r)$

(P) 9 Find the value of $\displaystyle\sum_{r=10}^{\infty} 200 \times \left(\dfrac{1}{4}\right)^r$

(E/P) 10 Given that $\displaystyle\sum_{r=1}^{k}(8 + 3r) = 377$,

   a show that $(3k + 58)(k - 13) = 0$ **(3 marks)**
   b hence find the value of k. **(1 mark)**

(E/P) 11 Given that $\displaystyle\sum_{r=1}^{k} 2 \times 3^r = 59\,046$,

   a show that $k = \dfrac{\log 19\,683}{\log 3}$ **(4 marks)**
   b For this value of k, calculate $\displaystyle\sum_{r=k+1}^{13} 2 \times 3^r$. **(3 marks)**

(E/P) 12 A geometric series is given by $1 + 3x + 9x^2 + \ldots$
   The series is convergent.
   a Write down the range of possible values of x. **(3 marks)**
   Given that $\displaystyle\sum_{r=1}^{\infty}(3x)^{r-1} = 2$
   b calculate the value of x. **(3 marks)**

Challenge

Given that $\displaystyle\sum_{r=1}^{10}(a + (r - 1)d) = \sum_{r=11}^{14}(a + (r - 1)d)$, show that $d = 6a$.

3.7 Recurrence relations

If you know the rule to get from one term to the next in a sequence you can write a recurrence relation.

- **A recurrence relation of the form $u_{n+1} = f(u_n)$ defines each term of a sequence as a function of the previous term.**

For example, the recurrence relation $u_{n+1} = 2u_n + 3$, $u_1 = 6$ produces the following sequence:

6, 15, 33, 69, … $u_2 = 2u_1 + 3 = 2(6) + 3 = 15$

Watch out In order to generate a sequence from a recurrence relation like this, you need to know the **first term** of the sequence.

Example 19

Find the first four terms of the following sequences.

a $u_{n+1} = u_n + 4$, $u_1 = 7$ **b** $u_{n+1} = u_n + 4$, $u_1 = 5$

a $u_{n+1} = u_n + 4$, $u_1 = 7$
Substituting $n = 1$, $u_2 = u_1 + 4 = 7 + 4 = 11$.
Substituting $n = 2$, $u_3 = u_2 + 4 = 11 + 4 = 15$.
Substituting $n = 3$, $u_4 = u_3 + 4 = 15 + 4 = 19$.
Sequence is 7, 11, 15, 19, …

Substitute $n = 1, 2$ and 3. Use u_1 to find u_2, and then u_2 to find u_3.

b $u_{n+1} = u_n + 4$, $u_1 = 5$
Substituting $n = 1$, $u_2 = u_1 + 4 = 5 + 4 = 9$.
Substituting $n = 2$, $u_3 = u_2 + 4 = 9 + 4 = 13$.
Substituting $n = 3$, $u_4 = u_3 + 4 = 13 + 4 = 17$.
Sequence is 5, 9, 13, 17, …

This is the same recurrence formula. It produces a different sequence because u_1 is different.

Example 20

A sequence $a_1, a_2, a_3, \ldots$ is defined by
$$a_1 = p$$
$$a_{n+1} = (a_n)^2 - 1, n \geq 1$$
where $p < 0$.

a Show that $a_3 = p^4 - 2p^2$.

b Given that $a_2 = 0$, find the value of p.

c Find $\sum_{r=1}^{200} a_r$

d Write down the value of a_{199}

a $a_1 = p$
$a_2 = (a_1)^2 - 1 = p^2 - 1$
$a_3 = (a_2)^2 - 1$
$= (p^2 - 1)^2 - 1$
$= p^4 - 2p^2 + 1 - 1$
$= p^4 - 2p^2$

Use $a_2 = (a_1)^2 - 1$ and substitute $a_1 = p$.

Now substitute the expression for a_2 to find a_3.

Chapter 3

b $p^2 - 1 = 0$ ——— Set the expression for a_2 equal to zero and solve.
  $p^2 = 1$
  $p = \pm 1$ but since $p < 0$ is given, $p = -1$

c $a_1 = -1$, $a_2 = 0$, $a_3 = -1$ series alternates between -1 and 0

Since this is a recurrence relation, we can see that the sequence is going to alternate between -1 and 0. The first 200 terms will have one hundred -1s and one hundred 0s.

In 200 terms, there will be one hundred -1s and one hundred 0s.

$\sum_{r=1}^{200} a_r = -100$

d $a_{199} = -1$ as 199 is odd

Problem-solving

For an alternating series, consider the sums of the odd and even terms separately. Write the first few terms of the series. The odd terms are -1 and the even terms are 0. Only the odd terms contribute to the sum.

Exercise 3G

1 Find the first four terms of the following recurrence relationships.

  a $u_{n+1} = u_n + 3$, $u_1 = 1$
  b $u_{n+1} = u_n - 5$, $u_1 = 9$
  c $u_{n+1} = 2u_n$, $u_1 = 3$
  d $u_{n+1} = 2u_n + 1$, $u_1 = 2$
  e $u_{n+1} = \dfrac{u_n}{2}$, $u_1 = 10$
  f $u_{n+1} = (u_n)^2 - 1$, $u_1 = 2$

2 Suggest possible recurrence relationships for the following sequences. (Remember to state the first term.)

  a 3, 5, 7, 9, ...
  b 20, 17, 14, 11, ...
  c 1, 2, 4, 8, ...
  d 100, 25, 6.25, 1.5625, ...
  e 1, -1, 1, -1, 1, ...
  f 3, 7, 15, 31, ...
  g 0, 1, 2, 5, 26, ...
  h 26, 14, 8, 5, 3.5, ...

3 By writing down the first four terms or otherwise, find the recurrence formula that defines the following sequences:

  a $u_n = 2n - 1$
  b $u_n = 3n + 2$
  c $u_n = n + 2$
  d $u_n = \dfrac{n+1}{2}$
  e $u_n = n^2$
  f $u_n = 3^n - 1$

(P) 4 A sequence of terms is defined for $n \geq 1$ by the recurrence relation $u_{n+1} = ku_n + 2$, where k is a constant. Given that $u_1 = 3$,

  a find an expression in terms of k for u_2
  b hence find an expression for u_3

  Given that $u_3 = 42$:

  c find the possible values of k.

(E/P) 5 A sequence is defined for $n \geq 1$ by the recurrence relation

  $u_{n+1} = pu_n + q$, $u_1 = 2$

  Given that $u_2 = -1$ and $u_3 = 11$, find the values of p and q. **(4 marks)**

6 A sequence is given by
$$x_1 = 2$$
$$x_{n+1} = x_n(p - 3x_n)$$
where p is an integer.

a Show that $x_3 = -10p^2 + 132p - 432$. (2 marks)

b Given that $x_3 = -288$ find the value of p. (1 mark)

c Hence find the value of x_4. (1 mark)

 7 A sequence $a_1, a_2, a_3, \ldots$ is defined by
$$a_1 = k$$
$$a_{n+1} = 4a_n + 5$$

a Find a_3 in terms of k. (2 marks)

b Show that $\sum_{r=1}^{4} a_r$ is a multiple of 5. (3 marks)

- **A sequence is increasing if $u_{n+1} > u_n$ for all $n \in \mathbb{N}$.**
- **A sequence is decreasing if $u_{n+1} < u_n$ for all $n \in \mathbb{N}$.**
- **A sequence is periodic if the terms repeat in a cycle. For a periodic sequence there is an integer k such that $u_{n+k} = u_n$ for all $n \in \mathbb{N}$. The value k is called the order of the sequence.**

Notation The order of a periodic sequence is sometimes called its **period**.

- 2, 3, 4, 5… is an increasing sequence.
- −3, −6, −12, −24… is a decreasing sequence.
- −2, 1, −2, 1, −2, 1 is a periodic sequence with a period of 2.
- 1, −2, 3, −4, 5, −6… is not increasing, decreasing or periodic.

Example 21

For each sequence:
i state whether the sequence is increasing, decreasing, or periodic.
ii If the sequence is periodic, write down its order.

a $u_{n+1} = u_n + 3, u_1 = 7$ **b** $u_{n+1} = (u_n)^2, u_1 = \frac{1}{2}$ **c** $u_n = \sin(90n°)$

a 7, 10, 13, 16, … — Write out the first few terms of the sequence.

$u_{n+1} > u_n$ for all n, so the sequence is increasing.

State the condition for an increasing sequence. You could also write that $k + 3 > k$ for all numbers k.

81

b $\frac{1}{2}, \frac{1}{4}, \frac{1}{16}, \frac{1}{256}, \ldots$

$u_{n+1} < u_n$ for all n, so the sequence is decreasing.

The starting value in the sequence makes a big difference. Because $u_1 < 1$ the numbers get smaller every time you square them.

c $u_1 = \sin(90°) = 1$
$u_2 = \sin(180°) = 0$
$u_3 = \sin(270°) = -1$
$u_4 = \sin(360°) = 0$
$u_5 = \sin(450°) = 1$
$u_6 = \sin(540°) = 0$
$u_7 = \sin(630°) = -1$

The sequence is periodic, with order 4.

To find u_1 substitute $n = 1$ into $\sin(90n°)$.

Watch out *Although every even term of the sequence is 0, the period is not 2 because the odd terms alternate between 1 and −1.*

The graph of $y = \sin x$ repeats with period 360°. So $\sin(x + 360°) = \sin x$. ← **Year 1, Chapter 9**

Exercise 3H

1 For each sequence:
  i state whether the sequence is increasing, decreasing, or periodic.
  ii If the sequence is periodic, write down its order.

  a 2, 5, 8, 11, 14
  b $3, 1, \frac{1}{3}, \frac{1}{9}, \frac{1}{27}$
  c 5, 9, 15, 23, 33
  d 3, −3, 3, −3, 3

2 For each sequence:
  i write down the first 5 terms of the sequence
  ii state whether the sequence is increasing, decreasing, or periodic.
  iii If the sequence is periodic, write down its order.

  a $u_n = 20 - 3n$
  b $u_n = 2^{n-1}$
  c $u_n = \cos(180n°)$
  d $u_n = (-1)^n$
  e $u_{n+1} = u_n - 5, u_1 = 20$
  f $u_{n+1} = 5 - u_n, u_1 = 20$
  g $u_{n+1} = \frac{2}{3}u_n, u_1 = k$

3 The sequence of numbers $u_1, u_2, u_3, \ldots$ is given by $u_{n+1} = ku_n, u_1 = 5$.
Find the range of values of k for which the sequence is strictly decreasing.

(E/P) 4 The sequence with recurrence relation $u_{k+1} = pu_k + q, u_1 = 5$, where p is a constant and $q = 13$, is periodic with order 2.
Find the value of p. **(5 marks)**

(E/P) 5 A sequence has nth term $a_n = \cos(90n°), n \geq 1$.
  a Find the order of the sequence. **(1 mark)**
  b Find $\sum_{r=1}^{444} a_r$ **(2 marks)**

Sequences and series

Challenge

The sequence of numbers $u_1, u_2, u_3, \ldots$ is given by $u_{n+2} = \dfrac{1 + u_{n+1}}{u_n}$, $u_1 = a$, $u_2 = b$, where a and b are positive integers.

a Show that the sequence is periodic for all positive a and b.
b State the order of the sequence.
c In the case where $a = 2$ and $b = 9$, find $\sum_{r=1}^{100} u_r$.

> **Hint** Each term in this sequence is defined in terms of the **previous two** terms.

3.8 Modelling with series

You can model real-life situations with series. For example if a person's salary increases by the same percentage every year, their salaries each year would form a **geometric sequence** and the amount they had been paid in total over n years would be modelled by the corresponding **geometric series**.

Example 22

Bruce starts a new company. In year 1 his profits will be £20 000. He predicts his profits to increase by £5000 each year, so that his profits in year 2 are modelled to be £25 000, in year 3 £30 000, and so on. He predicts this will continue until he reaches annual profits of £100 000. He then models his annual profits to remain at £100 000.

a Calculate the profits for Bruce's business in the first 20 years.
b State one reason why this may not be a suitable model.
c Bruce's financial advisor says the yearly profits are likely to increase by 5% per annum. Using this model, calculate the profits for Bruce's business in the first 20 years.

a Year 1 $P = 20\,000$, Year 2 $P = 25\,000$, Year 3 $P = 30\,000$

    This is an arithmetic sequence as the difference is constant.

$a = 20\,000$, $d = 5000$

    Write down the values of a and d.

$u_n = a + (n-1)d$
$100\,000 = 20\,000 + (n-1)(5000)$
$100\,000 = 20\,000 + 5000n - 5000$
$85\,000 = 5000n$

    Use the nth term of an arithmetic sequence to work out n when profits will reach £100 000.

$n = \dfrac{85\,000}{5000} = 17$

    Solve to find n.

$S_{17} = \dfrac{17}{2}(2(20\,000) + (17-1)(5000))$
$= 1\,020\,000$

    You want to know how much he made overall in the 17 years, so find the sum of the arithmetic series.

$S_{20} = 1\,020\,000 + 3(100\,000)$
$= 1\,320\,000$

So Bruce's total profit after 20 years is £1 320 000.

    In the 18th, 19th and 20th year he makes £100 000 each year, so add on 3 × £100 000 to the sum of the first 17 years.

Chapter 3

b It is unlikely that Bruce's profits will increase by exactly the same amount each year.

c $a = £20\,000$, $r = 1.05$ — This is a geometric series, as to get the next term you multiply the current term by 1.05.

$$S_n = \frac{a(r^n - 1)}{r - 1}$$

$$S_{20} = \frac{20\,000(1.05^{20} - 1)}{1.05 - 1}$$ — Use the formula for the sum of the first n terms of a geometric series $S_n = \frac{a(r^n - 1)}{r - 1}$

$S_{20} = 661\,319.08$

So Bruce's total profit after 20 years is £661 319.08.

Example 23

A piece of A4 paper is folded in half repeatedly. The thickness of the A4 paper is 0.5 mm.
a Work out the thickness of the paper after four folds.
b Work out the thickness of the paper after 20 folds.
c State one reason why this might be an unrealistic model.

a $a = 0.5$ mm, $r = 2$ — This is a geometric sequence, as each time we fold the paper the thickness doubles.

After 4 folds:
$u_5 = 0.5 \times 2^4 = 8$ mm — Since u_1 is the first term (after 0 folds), u_2 is after 1 fold, so u_5 is after 4 folds.

b After 20 folds
$u_{21} = 0.5 \times 2^{20} = 524\,288$ mm

c It is impossible to fold the paper that many times so the model is unrealistic.

Problem-solving
If you have to comment on the validity of a model, always refer to the context given in the question.

Exercise 3I

1 An investor puts £4000 in an account. Every month thereafter she deposits another £200. How much money in total will she have invested at the start of **a** the 10th month and **b** the mth month?

Hint At the start of the 6th month she will have only made 5 deposits of £200.

(P) 2 Carol starts a new job on a salary of £20 000. She is given an annual wage rise of £500 at the end of every year until she reaches her maximum salary of £25 000. Find the total amount she earns (assuming no other rises), **a** in the first 10 years, **b** over 15 years and **c** state one reason why this may be an unsuitable model.

Problem-solving
This is an arithmetic series with $a = 20\,000$ and $d = 500$. First find how many years it will take her to reach her maximum salary.

Sequences and series

(P) 3 James decides to save some money during the six-week holiday. He saves 1p on the first day, 2p on the second, 3p on the third and so on.
   a How much will he have at the end of the holiday (42 days)?
   b If he carried on, how long would it be before he has saved £100?

(P) 4 A population of ants is growing at a rate of 10% a year. If there were 200 ants in the initial population, write down the number of ants after:
   a 1 year   b 2 years   c 3 years   d 10 years.

Problem-solving
This is a geometric sequence.
$a = 200$ and $r = 1.1$

(P) 5 A motorcycle has four gears. The maximum speed in bottom gear is $40\,\text{km h}^{-1}$ and the maximum speed in top gear is $120\,\text{km h}^{-1}$. Given that the maximum speeds in each successive gear form a geometric progression, calculate, in km h^{-1} to one decimal place, the maximum speeds in the two intermediate gears.

(P) 6 A car depreciates in value by 15% a year. After 3 years it is worth £11 054.25.
   a What was the car's initial price?
   b When will the car's value first be less than £5000?

Problem-solving
Use your answer to part **a** to write an inequality, then solve it using logarithms.

(E) 7 A salesman is paid commission of £10 per week for each life insurance policy that he has sold. Each week he sells one new policy so that he is paid £10 commission in the first week, £20 commission in the second week, £30 commission in the third week and so on.
   a Find his total commission in the first year of 52 weeks. **(2 marks)**
   b In the second year the commission increases to £11 per week on new policies sold, although it remains at £10 per week for policies sold in the first year. He continues to sell one policy per week. Show that he is paid £542 in the second week of his second year. **(3 marks)**
   c Find the total commission paid to him in the second year. **(2 marks)**

(E) 8 Prospectors are drilling for oil. The cost of drilling to a depth of 50 m is £500. To drill a further 50 m costs £640 and, hence, the total cost of drilling to a depth of 100 m is £1140. Each subsequent extra depth of 50 m costs £140 more to drill than the previous 50 m.
   a Show that the cost of drilling to a depth of 500 m is £11 300. **(3 marks)**
   b The total sum of money available for drilling is £76 000. Find, to the nearest 50 m, the greatest depth that can be drilled. **(3 marks)**

(E) 9 Each year, for 40 years, Anne will pay money into a savings scheme. In the first year she pays in £500. Her payments then increase by £50 each year, so that she pays in £550 in the second year, £600 in the third year, and so on.
   a Find the amount that Anne will pay in the 40th year. **(2 marks)**
   b Find the total amount that Anne will pay in over the 40 years. **(3 marks)**
   c Over the same 40 years, Brian will also pay money into the savings scheme. In the first year he pays in £890 and his payments then increase by £d each year. Given that Brian and Anne will pay in exactly the same amount over the 40 years, find the value of d. **(4 marks)**

Chapter 3

(P) 10 A virus is spreading such that the number of people infected increases by 4% a day. Initially 100 people were diagnosed with the virus. How many days will it be before 1000 are infected?

(P) 11 I invest £A in the bank at a rate of interest of 3.5% per annum. How long will it be before I double my money?

(P) 12 The fish in a particular area of the North Sea are being reduced by 6% each year due to overfishing. How long will it be before the fish stocks are halved?

(P) 13 The man who invented the game of chess was asked to name his reward. He asked for 1 grain of corn to be placed on the first square of his chessboard, 2 on the second, 4 on the third and so on until all 64 squares were covered. He then said he would like as many grains of corn as the chessboard carried. How many grains of corn did he claim as his prize?

(P) 14 A ball is dropped from a height of 10 m. It bounces to a height of 7 m and continues to bounce. Subsequent heights to which it bounces follow a geometric sequence. Find out:

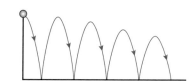

    a how high it will bounce after the fourth bounce
    b the total vertical distance travelled up to the point when the ball hits the ground for the sixth time.

(P) 15 Richard is doing a sponsored cycle. He plans to cycle 1000 miles over a number of days. He plans to cycle 10 miles on day 1 and increase the distance by 10% a day.
    a How long will it take Richard to complete the challenge?
    b What will be his greatest number of miles completed in a day?

(P) 16 A savings scheme is offering a rate of interest of 3.5% per annum for the lifetime of the plan. Alan wants to save up £20 000. He works out that he can afford to save £500 every year, which he will deposit on 1 January. If interest is paid on 31 December, how many years will it be before he has saved up his £20 000?

Mixed exercise 3

(E/P) 1 A geometric series has third term 27 and sixth term 8.
    a Show that the common ratio of the series is $\frac{2}{3}$ **(2 marks)**
    b Find the first term of the series.   **(2 marks)**
    c Find the sum to infinity of the series.   **(2 marks)**
    d Find the difference between the sum of the first 10 terms of the series and the sum to infinity. Give your answer to 3 significant figures.   **(2 marks)**

(E/P) 2 The second term of a geometric series is 80 and the fifth term of the series is 5.12.
    a Show that the common ratio of the series is 0.4.   **(2 marks)**
    Calculate:
    b the first term of the series   **(2 marks)**

c the sum to infinity of the series, giving your answer as an exact fraction (1 mark)

d the difference between the sum to infinity of the series and the sum of the first 14 terms of the series, giving your answer in the form $a \times 10^n$, where $1 \leq a < 10$ and n is an integer. (2 marks)

E/P   **3** The nth term of a sequence is u_n, where $u_n = 95\left(\frac{4}{5}\right)^n$, $n = 1, 2, 3, \ldots$

    a Find the values of u_1 and u_2. (2 marks)

    Giving your answers to 3 significant figures, calculate:

    b the value of u_{21} (1 mark)

    c $\sum_{n=1}^{15} u_n$ (2 marks)

    d the sum to infinity of the series whose first term is u_1 and whose nth term is u_n. (1 mark)

E/P   **4** A sequence of numbers $u_1, u_2, \ldots, u_n, \ldots$ is given by the formula $u_n = 3\left(\frac{2}{3}\right)^n - 1$ where n is a positive integer.

    a Find the values of u_1, u_2 and u_3. (2 marks)

    b Show that $\sum_{n=1}^{15} u_n = -9.014$ to 4 significant figures. (2 marks)

    c Prove that $u_{n+1} = \dfrac{2u_n - 1}{3}$ (2 marks)

E/P   **5** The third and fourth terms of a geometric series are 6.4 and 5.12 respectively. Find:

    a the common ratio of the series, (2 marks)

    b the first term of the series, (2 marks)

    c the sum to infinity of the series. (2 marks)

    d Calculate the difference between the sum to infinity of the series and the sum of the first 25 terms of the series. (2 marks)

E/P   **6** The price of a car depreciates by 15% per annum. Its price when new is £20 000.

    a Find the value of the car after 5 years. (2 marks)

    b Find when the value will be less than £4000. (3 marks)

E/P   **7** The first three terms of a geometric series are $p(3q + 1)$, $p(2q + 2)$ and $p(2q - 1)$, where p and q are non-zero constants.

    a Show that one possible value of q is 5 and find the other possible value. (2 marks)

    b Given that $q = 5$, and the sum to infinity of the series is 896, find the sum of the first 12 terms of the series. Give your answer to 2 decimal places. (4 marks)

E/P   **8 a** Prove that the sum of the first n terms in an arithmetic series is
$$S = \frac{n}{2}(2a + (n-1)d)$$
where a = first term and d = common difference. (3 marks)

    b Use this to find the sum of the first 100 natural numbers. (2 marks)

E/P   **9** Find the least value of n for which $\sum_{r=1}^{n}(4r - 3) > 2000$. (2 marks)

Chapter 3

E/P 10 The sum of the first two terms of an arithmetic series is 47.
The thirtieth term of this series is −62. Find:
  a the first term of the series and the common difference (3 marks)
  b the sum of the first 60 terms of the series. (2 marks)

E/P 11 a Find the sum of the integers which are divisible by 3 and lie between 1 and 400. (3 marks)
  b Hence, or otherwise, find the sum of the integers, from 1 to 400 inclusive, which are **not** divisible by 3. (2 marks)

E/P 12 A polygon has 10 sides. The lengths of the sides, starting with the shortest, form an arithmetic series. The perimeter of the polygon is 675 cm and the length of the longest side is twice that of the shortest side. Find the length of the shortest side of the polygon. (4 marks)

E/P 13 Prove that the sum of the first $2n$ multiples of 4 is $4n(2n + 1)$. (4 marks)

E/P 14 A sequence of numbers is defined, for $n \geq 1$, by the recurrence relation $u_{n+1} = ku_n - 4$, where k is a constant. Given that $u_1 = 2$:
  a find expressions, in terms of k, for u_2 and u_3. (2 marks)
  b Given also that $u_3 = 26$, use algebra to find the possible values of k. (2 marks)

E/P 15 The fifth term of an arithmetic series is 14 and the sum of the first three terms of the series is −3.
  a Use algebra to show that the first term of the series is −6 and calculate the common difference of the series. (3 marks)
  b Given that the nth term of the series is greater than 282, find the least possible value of n. (3 marks)

E/P 16 The fourth term of an arithmetic series is $3k$, where k is a constant, and the sum of the first six terms of the series is $7k + 9$.
  a Show that the first term of the series is $9 - 8k$. (3 marks)
  b Find an expression for the common difference of the series in terms of k. (2 marks)
  Given that the seventh term of the series is 12, calculate:
  c the value of k (2 marks)
  d the sum of the first 20 terms of the series. (2 marks)

E/P 17 A sequence is defined by the recurrence relation
$$a_{n+1} = \frac{1}{a_n}, \quad a_1 = p$$
  a Show that the sequence is periodic and state its order. (2 marks)
  b Find $\sum_{r=1}^{1000} a_n$ in terms of p. (2 marks)

E/P 18 A sequence $a_1, a_2, a_3, \ldots$ is defined by
$$a_1 = k$$
$$a_{n+1} = 2a_n + 6, \, n \geq 1$$
where k is an integer.

a Given that the sequence is increasing for the first 3 terms, show that $k > p$, where p is an integer to be found. **(2 marks)**
b Find a_4 in terms of k. **(2 marks)**
c Show that $\sum_{r=1}^{4} a_r$ is divisible by 3. **(3 marks)**

19 The first term of a geometric series is 130. The sum to infinity of the series is 650.
a Show that the common ratio, r, is $\frac{4}{5}$ **(3 marks)**
b Find, to 2 decimal places, the difference between the 7th and 8th terms. **(2 marks)**
c Calculate the sum of the first 7 terms. **(2 marks)**
The sum of the first n terms of the series is greater than 600.
d Show that $n > \dfrac{-\log 13}{\log 0.8}$ **(4 marks)**

20 The adult population of a town is 25 000 at the beginning of 2012.
A model predicts that the adult population of the town will increase by 2% each year, forming a geometric sequence.
a Show that the predicted population at the beginning of 2014 is 26 010. **(1 mark)**
The model predicts that after n years, the population will first exceed 50 000.
b Show that $n > \dfrac{\log 2}{\log 1.02}$ **(3 marks)**
c Find the year in which the population first exceeds 50 000. **(2 marks)**
d Every member of the adult population is modelled to visit the doctor once per year. Calculate the number of appointments the doctor has from the beginning of 2012 to the end of 2019. **(4 marks)**
e Give a reason why this model for doctors' appointments may not be appropriate. **(1 mark)**

21 Kyle is making some patterns out of squares. He has made 3 rows so far.
a Find an expression, in terms of n, for the number of squares required to make a similar arrangement in the nth row. **(2 marks)**
b Kyle counts the number of squares used to make the pattern in the kth row. He counts 301 squares. Write down the value of k. **(1 mark)**
c In the first q rows, Kyle uses a total of p squares.
 i Show that $q^2 + 2q - p = 0$. **(2 marks)**
 ii Given that $p > 1520$, find the minimum number of rows that Kyle makes. **(3 marks)**

Row 1
Row 2
Row 3

22 A convergent geometric series has first term a and common ratio r. The second term of the series is -3 and the sum to infinity of the series is 6.75.
a Show that $27r^2 - 27r - 12 = 0$. **(4 marks)**
b Given that the series is convergent, find the value of r. **(2 marks)**
c Find the sum of the first 5 terms of the series, giving your answer to 2 decimal places. **(3 marks)**

Challenge

A sequence is defined by the recurrence relation $u_{n+2} = 5u_{n+1} - 6u_n$.

a Prove that any sequence of the form $u_n = p \times 3^n + q \times 2^n$, where p and q are constants, satisfies this recurrence relation.

Given that $u_1 = 5$ and $u_2 = 12$,

b find an expression for u_n in terms of n only.

c Hence determine the number of digits in u_{100}.

Summary of key points

1. In an **arithmetic sequence**, the difference between consecutive terms is constant.

2. The formula for the nth term of an arithmetic sequence is $u_n = a + (n-1)d$, where a is the first term and d is the common difference.

3. An arithmetic series is the sum of the terms of an arithmetic sequence.

   The sum of the first n terms of an arithmetic series is given by $S_n = \frac{n}{2}(2a + (n-1)d)$, where where a is the first term and d is the common difference.

   You can also write this formula as $S_n = \frac{n}{2}(a + l)$, where l is the last term.

4. A **geometric sequence** has a **common ratio** between consecutive terms.

5. The formula for the nth term of a geometric sequence is $u_n = ar^{n-1}$, where a is the first term and r is the common ratio.

6. The sum of the first n terms of a geometric series is given by
   $$S_n = \frac{a(1 - r^n)}{1 - r}, r \neq 1 \quad \text{or} \quad S_n = \frac{a(r^n - 1)}{r - 1}, r \neq 1$$
   where a is the first term and r is the common ratio.

7. A geometric series is convergent if and only if $|r| < 1$, where r is the common ratio.

   The **sum to infinity** of a convergent geometric series is given by $S_\infty = \frac{a}{1 - r}$

8. The Greek capital letter 'sigma' is used to signify a sum. You write it as $\sum$. You write limits on the top and bottom to show which terms you are summing.

9. A recurrence relation of the form $u_{n+1} = f(u_n)$ defines each term of a sequence as a function of the previous term.

10. A sequence is **increasing** if $u_{n+1} > u_n$ for all $n \in \mathbb{N}$.

    A sequence is **decreasing** if $u_{n+1} < u_n$ for all $n \in \mathbb{N}$.

    A sequence is **periodic** if the terms repeat in a cycle. For a periodic sequence there is an integer k such that $u_{n+k} = u_n$ for all $n \in \mathbb{N}$. The value k is called the **order** of the sequence.

Binomial expansion

Objectives

After completing this chapter you should be able to:

- Expand $(1 + x)^n$ for any rational constant n and determine the range of values of x for which the expansion is valid → pages 92–97

- Expand $(a + bx)^n$ for any rational constant n and determine the range of values of x for which the expansion is valid → pages 97–100

- Use partial fractions to expand fractional expressions → pages 101–103

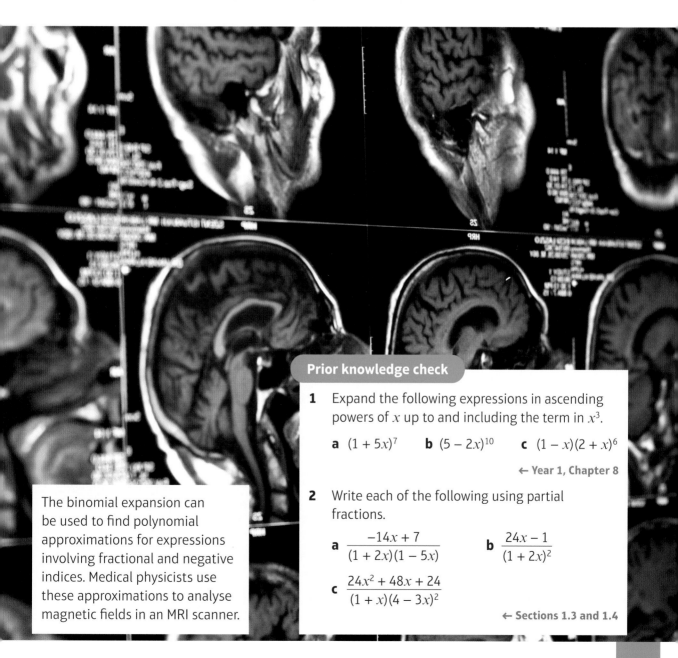

Prior knowledge check

1. Expand the following expressions in ascending powers of x up to and including the term in x^3.

   a $(1 + 5x)^7$ **b** $(5 - 2x)^{10}$ **c** $(1 - x)(2 + x)^6$

   ← Year 1, Chapter 8

2. Write each of the following using partial fractions.

   a $\dfrac{-14x + 7}{(1 + 2x)(1 - 5x)}$ **b** $\dfrac{24x - 1}{(1 + 2x)^2}$

   c $\dfrac{24x^2 + 48x + 24}{(1 + x)(4 - 3x)^2}$

   ← Sections 1.3 and 1.4

The binomial expansion can be used to find polynomial approximations for expressions involving fractional and negative indices. Medical physicists use these approximations to analyse magnetic fields in an MRI scanner.

Chapter 4

4.1 Expanding $(1 + x)^n$

If n is a natural number you can find the binomial expansion for $(a + bx)^n$ using the formula:

$$(a + b)^n = a^n + \binom{n}{1}a^{n-1}b + \binom{n}{2}a^{n-2}b^2 + \ldots + \binom{n}{r}a^{n-r}b^r + \ldots + b^n, \ (n \in \mathbb{N})$$

Hint There are $n + 1$ terms, so this formula produces a **finite** number of terms.

If n is a **fraction** or a **negative number** you need to use a different version of the binomial expansion.

- **This form of the binomial expansion can be applied to negative or fractional values of n to obtain an infinite series.**

$$(1 + x)^n = 1 + nx + \frac{n(n-1)}{2!}x^2 + \frac{n(n-1)(n-2)}{3!}x^3 + \ldots + \left(\frac{n(n-1)\ldots(n-r+1)}{r!}\right)x^r + \ldots$$

- **The expansion is valid when $|x| < 1$, $n \in \mathbb{R}$.**

When n is not a natural number, none of the factors in the expression $n(n-1)\ldots(n-r+1)$ are equal to zero. This means that this version of the binomial expansion produces an **infinite number** of terms.

Watch out This expansion is valid for any **real value** of n, but is **only** valid for values of x that satisfy $|x| < 1$, or in other words, when $-1 < x < 1$.

A finite number of terms of the binomial expansion can be used to obtain both algebraic and numerical approximations.

- **Approximations based on the binomial expansion are more accurate when:**
  - **more terms of the expansion are used**
  - **the values of x substituted into the expansion are closer to 0**

Example 1

Find the first four terms in the binomial expansion of $\dfrac{1}{1 + x}$

$\dfrac{1}{1 + x} = (1 + x)^{-1}$ — Write in index form.

$= 1 + (-1)x + \dfrac{(-1)(-2)x^2}{2!}$
$+ \dfrac{(-1)(-2)(-3)x^3}{3!} + \ldots$ — Replace n by -1 in the expansion.

$= 1 - 1x + 1x^2 - 1x^3 + \ldots$ — As n is not a positive integer, no coefficient will ever be equal to zero. Therefore, the expansion is **infinite**.

$= 1 - x + x^2 - x^3 + \ldots$

For the series to be **convergent**, $|x| < 1$.

- **The expansion of $(1 + bx)^n$, where n is negative or a fraction, is valid for $|bx| < 1$, or $|x| < \dfrac{1}{|b|}$.**

Example 2

Find the binomial expansions of

a $(1-x)^{\frac{1}{3}}$

b $\dfrac{1}{(1+4x)^2}$

up to and including the term in x^3. State the range of values of x for which each expansion is valid.

a $(1-x)^{\frac{1}{3}}$

$= 1 + (\frac{1}{3})(-x)$ — Replace n by $\frac{1}{3}$, x by $(-x)$.

$+ \dfrac{(\frac{1}{3})(\frac{1}{3}-1)(-x)^2}{2!}$ — Simplify brackets.

$+ \dfrac{(\frac{1}{3})(\frac{1}{3}-1)(\frac{1}{3}-2)(-x)^3}{3!} + \ldots$

Watch out Be careful working out whether each term should be positive or negative:
- even number of negative signs means term is positive
- odd number of negative signs means term is negative

The x^3 term here has 5 negative signs in total, so it is negative.

$= 1 + (\frac{1}{3})(-x) + \dfrac{(\frac{1}{3})(-\frac{2}{3})(-x)^2}{2}$

$+ \dfrac{(\frac{1}{3})(-\frac{2}{3})(-\frac{5}{3})(-x)^3}{6} + \ldots$

$= 1 - \frac{1}{3}x - \frac{1}{9}x^2 - \frac{5}{81}x^3 + \ldots$ — Simplify coefficients.

Expansion is valid as long as $|-x| < 1$

$\Rightarrow |x| < 1$ — Terms in expansion are $(-x)$, $(-x)^2$, $(-x)^3$.

b $\dfrac{1}{(1+4x)^2} = (1+4x)^{-2}$ — Write in index form.

$= 1 + (-2)(4x)$

$+ \dfrac{(-2)(-2-1)(4x)^2}{2!}$ — Replace n by -2, x by $4x$.

$+ \dfrac{(-2)(-2-1)(-2-2)(4x)^3}{3!} + \ldots$ — Simplify brackets.

$= 1 + (-2)(4x)$

$+ \dfrac{(-2)(-3)16x^2}{2}$ — Simplify coefficients.

$+ \dfrac{(-2)(-3)(-4)64x^3}{6} + \ldots$

— Terms in expansion are $(4x)$, $(4x)^2$, $(4x)^3$.

$= 1 - 8x + 48x^2 - 256x^3 + \ldots$

Expansion is valid as long as $|4x| < 1$

$\Rightarrow |x| < \frac{1}{4}$

Online Use technology to explore why the expansions are only valid for certain values of x.

Chapter 4

Example 3

a Find the expansion of $\sqrt{1-2x}$ up to and including the term in x^3.
b By substituting in $x = 0.01$, find a decimal approximation to $\sqrt{2}$.
c Without further calculation, state how the accuracy of this approximation could be improved.

a $\sqrt{1-2x} = (1-2x)^{\frac{1}{2}}$ — Write in index form.

$= 1 + (\frac{1}{2})(-2x)$ — Replace n by $\frac{1}{2}$, x by $(-2x)$.

$+ \dfrac{(\frac{1}{2})(\frac{1}{2}-1)(-2x)^2}{2!}$

$+ \dfrac{(\frac{1}{2})(\frac{1}{2}-1)(\frac{1}{2}-2)(-2x)^3}{3!} + \ldots$

$= 1 + (\frac{1}{2})(-2x)$

$+ \dfrac{(\frac{1}{2})(-\frac{1}{2})(4x^2)}{2!}$ — Simplify brackets.

$+ \dfrac{(\frac{1}{2})(-\frac{1}{2})(-\frac{3}{2})(-8x^3)}{6} + \ldots$

— Simplify coefficients.

$= 1 - x - \dfrac{x^2}{2} - \dfrac{x^3}{2} + \ldots$

Expansion is valid if $|-2x| < 1$ — Terms in expansion are $(-2x)$, $(-2x)^2$, $(-2x)^3$.

$\Rightarrow |x| < \frac{1}{2}$

b $\sqrt{1 - 2 \times 0.01} \approx 1 - 0.01 - \dfrac{0.01^2}{2}$ — $x = 0.01$ satisfies the validity condition $|x| < \frac{1}{2}$

$- \dfrac{0.01^3}{2}$ — Substitute $x = 0.01$ into both sides of the expansion.

$\sqrt{0.98} \approx 1 - 0.01 - 0.00005$
$\quad - 0.0000005$ — Simplify both sides. Note that the terms are getting smaller.

$\sqrt{\dfrac{98}{100}} \approx 0.9899495$ — Write 0.98 as $\dfrac{98}{100}$

$\sqrt{\dfrac{49 \times 2}{100}} \approx 0.9899495$

$\dfrac{7\sqrt{2}}{10} \approx 0.9899495$ — Use rules of surds.

$\sqrt{2} \approx \dfrac{0.9899495 \times 10}{7}$

$\sqrt{2} \approx 1.414213571$ — This approximation is accurate to 7 decimal places.

c Use more terms from the binomial expansion of $\sqrt{1-2x}$.

Example 4

$f(x) = \dfrac{2 + x}{\sqrt{1 + 5x}}$

a Find the x^2 term in the series expansion of $f(x)$.

b State the range of values of x for which the expansion is valid.

a $f(x) = (2 + x)(1 + 5x)^{-\frac{1}{2}}$ ——— Write in index form.

$(1 + 5x)^{-\frac{1}{2}} = 1 + \left(-\dfrac{1}{2}\right)(5x)$

$+ \dfrac{\left(-\frac{1}{2}\right)\left(-\frac{3}{2}\right)}{2!}(5x)^2$

$+ \dfrac{\left(-\frac{1}{2}\right)\left(-\frac{3}{2}\right)\left(-\frac{5}{2}\right)}{3!}(5x)^3 + \ldots$ ——— Find the binomial expansion of $(1 + 5x)^{-\frac{1}{2}}$

$= 1 - \dfrac{5}{2}x + \dfrac{75}{8}x^2 - \dfrac{625}{16}x^3 + \ldots$ ——— Simplify coefficients.

$f(x) = (2 + x)\left(1 - \dfrac{5}{2}x + \dfrac{75}{8}x^2 - \dfrac{625}{16}x^3 + \ldots\right)$

$2 \times \dfrac{75}{8} + 1 \times -\dfrac{5}{2} = \dfrac{65}{4}$

x^2 term is $\dfrac{65}{4}x^2$

b The expansion is valid if $|5x| < 1$

$\Rightarrow |x| < \dfrac{1}{5}$

Online Use your calculator to calculate the coefficients of the binomial expansion.

Problem-solving

There are two ways to make an x^2 term. Either $2 \times \dfrac{75}{8}x^2$ or $x \times \dfrac{5}{2}x$. Add these together to find the term in x^2.

Example 5

In the expansion of $(1 + kx)^{-4}$ the coefficient of x is 20.

a Find the value of k.

b Find the corresponding coefficient of the x^2 term.

a $(1 + kx)^{-4} = 1 + (-4)(kx) + \dfrac{(-4)(-5)}{2!}(kx)^2 + \ldots$ ——— Find the binomial expansion of $(1 + kx)^{-4}$.

$= 1 - 4kx + 10k^2x^2 + \ldots$

$-4k = 20$
$k = -5$ ——— Solve to find k.

b Coefficient of $x^2 = 10k^2 = 10(-5)^2 = 250$

Chapter 4

Exercise 4A

1 For each of the following,
   i find the binomial expansion up to and including the x^3 term
   ii state the range of values of x for which the expansion is valid.
   a $(1+x)^{-4}$
   b $(1+x)^{-6}$
   c $(1+x)^{\frac{1}{2}}$
   d $(1+x)^{\frac{5}{3}}$
   e $(1+x)^{-\frac{1}{4}}$
   f $(1+x)^{-\frac{3}{2}}$

2 For each of the following,
   i find the binomial expansion up to and including the x^3 term
   ii state the range of values of x for which the expansion is valid.
   a $(1+3x)^{-3}$
   b $\left(1+\tfrac{1}{2}x\right)^{-5}$
   c $(1+2x)^{\frac{3}{4}}$
   d $(1-5x)^{\frac{7}{3}}$
   e $(1+6x)^{-\frac{2}{3}}$
   f $\left(1-\tfrac{3}{4}x\right)^{-\frac{5}{3}}$

3 For each of the following,
   i find the binomial expansion up to and including the x^3 term
   ii state the range of values of x for which the expansion is valid.
   a $\dfrac{1}{(1+x)^2}$
   b $\dfrac{1}{(1+3x)^4}$
   c $\sqrt{1-x}$
   d $\sqrt[3]{1-3x}$
   e $\dfrac{1}{\sqrt{1+\tfrac{1}{2}x}}$
   f $\dfrac{\sqrt[3]{1-2x}}{1-2x}$

   Hint In part **f**, write the fraction as a single power of $(1-2x)$.

E/P 4 $f(x) = \dfrac{1+x}{1-2x}$

   a Show that the series expansion of $f(x)$ up to and including the x^3 term is $1 + 3x + 6x^2 + 12x^3$. **(4 marks)**
   b State the range of values of x for which the expansion is valid. **(1 mark)**

   Hint First rewrite $f(x)$ as $(1+x)(1-2x)^{-1}$.

E 5 $f(x) = \sqrt{1+3x}$, $-\tfrac{1}{3} < x < \tfrac{1}{3}$
   a Find the series expansion of $f(x)$, in ascending powers of x, up to and including the x^3 term. Simplify each term. **(4 marks)**
   b Show that, when $x = \dfrac{1}{100}$, the exact value of $f(x)$ is $\dfrac{\sqrt{103}}{10}$. **(2 marks)**
   c Find the percentage error made in using the series expansion in part **a** to estimate the value of $f(0.01)$. Give your answer to 2 significant figures. **(3 marks)**

P 6 In the expansion of $(1+ax)^{-\frac{1}{2}}$ the coefficient of x^2 is 24.
   a Find the possible values of a.
   b Find the possible coefficients of the x^3 term.

96

Binomial expansion

(P) **7** Show that if x is small, the expression $\sqrt{\dfrac{1+x}{1-x}}$ is approximated by $1 + x + \tfrac{1}{2}x^2$.

> **Notation** 'x is small' means we can assume the expansion is valid for the x values being considered, as high powers become insignificant compared to the first few terms.

(E/P) **8** $h(x) = \dfrac{6}{1+5x} - \dfrac{4}{1-3x}$

   a Find the series expansion of $h(x)$, in ascending powers of x, up to and including the x^2 term. Simplify each term. **(6 marks)**

   b Find the percentage error made in using the series expansion in part **a** to estimate the value of $h(0.01)$. Give your answer to 2 significant figures. **(3 marks)**

   c Explain why it is not valid to use the expansion to find $h(0.5)$. **(1 mark)**

(E/P) **9 a** Find the binomial expansion of $(1-3x)^{\frac{3}{2}}$ in ascending powers of x up to and including the x^3 term, simplifying each term. **(4 marks)**

   b Show that, when $x = \dfrac{1}{100}$, the exact value of $(1-3x)^{\frac{3}{2}}$ is $\dfrac{97\sqrt{97}}{1000}$. **(2 marks)**

   c Substitute $x = \dfrac{1}{100}$ into the binomial expansion in part **a** and hence obtain an approximation to $\sqrt{97}$. Give your answer to 5 decimal places. **(3 marks)**

   d Without further calculation, state how the accuracy of this approximation could be improved. **(1 mark)**

Challenge

$h(x) = \left(1 + \dfrac{1}{x}\right)^{-\frac{1}{2}}, |x| > 1$

   a Find the binomial expansion of $h(x)$ in ascending powers of x up to and including the x^2 term, simplifying each term.

   b Show that, when $x = 9$, the exact value of $h(x)$ is $\dfrac{3\sqrt{10}}{10}$.

   c Use the expansion in part **a** to find an approximate value of $\sqrt{10}$. Write your answer to 2 decimal places.

> **Hint** Replace x with $\dfrac{1}{x}$

4.2 Expanding $(a + bx)^n$

The binomial expansion of $(1 + x)^n$ can be used to expand $(a + bx)^n$ for any constants a and b.

You need to take a factor of a^n out of the expression:

$$(a + bx)^n = \left(a\left(1 + \dfrac{b}{a}x\right)\right)^n = a^n\left(1 + \dfrac{b}{a}x\right)^n$$

> **Watch out** Make sure you multiply a^n by **every term** in the expansion of $\left(1 + \dfrac{b}{a}x\right)^n$.

Chapter 4

- The expansion of $(a + bx)^n$, where n is negative or a fraction, is valid for $\left|\frac{b}{a}x\right| < 1$ or $|x| < \left|\frac{a}{b}\right|$.

Example 6

Find the first four terms in the binomial expansion of **a** $\sqrt{4 + x}$ **b** $\dfrac{1}{(2 + 3x)^2}$

State the range of values of x for which each of these expansions is valid.

a $\sqrt{4 + x} = (4 + x)^{\frac{1}{2}}$ — Write in index form.

$= \left(4\left(1 + \dfrac{x}{4}\right)\right)^{\frac{1}{2}}$ — Take out a factor of $4^{\frac{1}{2}}$.

$= 4^{\frac{1}{2}}\left(1 + \dfrac{x}{4}\right)^{\frac{1}{2}}$ — Write $4^{\frac{1}{2}}$ as 2.

$= 2\left(1 + \dfrac{x}{4}\right)^{\frac{1}{2}}$

$= 2\left(1 + \left(\dfrac{1}{2}\right)\left(\dfrac{x}{4}\right) + \dfrac{\left(\dfrac{1}{2}\right)\left(\dfrac{1}{2} - 1\right)\left(\dfrac{x}{4}\right)^2}{2!} + \dfrac{\left(\dfrac{1}{2}\right)\left(\dfrac{1}{2} - 1\right)\left(\dfrac{1}{2} - 2\right)\left(\dfrac{x}{4}\right)^3}{3!} + \ldots\right)$

— Expand $\left(1 + \dfrac{x}{4}\right)^{\frac{1}{2}}$ using the binomial expansion with $n = \dfrac{1}{2}$ and $x = \dfrac{x}{4}$

$= 2\left(1 + \left(\dfrac{1}{2}\right)\left(\dfrac{x}{4}\right) + \dfrac{\left(\dfrac{1}{2}\right)\left(-\dfrac{1}{2}\right)\left(\dfrac{x^2}{16}\right)}{2} + \dfrac{\left(\dfrac{1}{2}\right)\left(-\dfrac{1}{2}\right)\left(-\dfrac{3}{2}\right)\left(\dfrac{x^3}{64}\right)}{6} + \ldots\right)$

— Simplify coefficients.

$= 2\left(1 + \dfrac{x}{8} - \dfrac{x^2}{128} + \dfrac{x^3}{1024} + \ldots\right)$

— Multiply every term in the expansion by 2.

$= 2 + \dfrac{x}{4} - \dfrac{x^2}{64} + \dfrac{x^3}{512} + \ldots$

Expansion is valid if $\left|\dfrac{x}{4}\right| < 1$ — The expansion is infinite, and converges when $\left|\dfrac{x}{4}\right| < 1$, or $|x| < 4$.

$\Rightarrow |x| < 4$

b $\dfrac{1}{(2+3x)^2} = (2+3x)^{-2}$ ——— Write in index form.

$= \left(2\left(1+\dfrac{3x}{2}\right)\right)^{-2}$

$= 2^{-2}\left(1+\dfrac{3x}{2}\right)^{-2}$ ——— Take out a factor of 2^{-2}.

$= \dfrac{1}{4}\left(1+\dfrac{3x}{2}\right)^{-2}$ ——— Write $2^{-2} = \dfrac{1}{2^2} = \dfrac{1}{4}$

$= \dfrac{1}{4}\left(1 + (-2)\left(\dfrac{3x}{2}\right) + \dfrac{(-2)(-2-1)\left(\dfrac{3x}{2}\right)^2}{2!}\right.$

$\left. + \dfrac{(-2)(-2-1)(-2-2)\left(\dfrac{3x}{2}\right)^3}{3!} + \ldots\right)$

Expand $\left(1+\dfrac{3x}{2}\right)^{-2}$ using the binomial expansion with $n = -2$ and $x = \dfrac{3x}{2}$

$= \dfrac{1}{4}\left((1 + (-2)\left(\dfrac{3x}{2}\right) + \dfrac{(-2)(-3)\left(\dfrac{9x^2}{4}\right)}{2}\right.$

$\left. + \dfrac{(-2)(-3)(-4)\left(\dfrac{27x^3}{8}\right)}{6} + \ldots\right)$

——— Simplify coefficients.

$= \dfrac{1}{4}\left(1 - 3x + \dfrac{27x^2}{4} - \dfrac{27x^3}{2} + \ldots\right)$

——— Multiply every term by $\dfrac{1}{4}$

$= \dfrac{1}{4} - \dfrac{3}{4}x + \dfrac{27x^2}{16} - \dfrac{27x^3}{8} + \ldots$

Expansion is valid if $\left|\dfrac{3x}{2}\right| < 1$

$\Rightarrow |x| < \dfrac{2}{3}$

The expansion is infinite, and converges when $\left|\dfrac{3x}{2}\right| < 1$, $|x| < \dfrac{2}{3}$

Exercise 4B

P 1 For each of the following,

i find the binomial expansion up to and including the x^3 term

ii state the range of values of x for which the expansion is valid.

a $\sqrt{4+2x}$

b $\dfrac{1}{2+x}$

c $\dfrac{1}{(4-x)^2}$

d $\sqrt{9+x}$

e $\dfrac{1}{\sqrt{2+x}}$

f $\dfrac{5}{3+2x}$

g $\dfrac{1+x}{2+x}$

Hint Write part **g** as $1 - \dfrac{1}{x+2}$

h $\sqrt{\dfrac{2+x}{1-x}}$

Chapter 4

(E) **2** $f(x) = (5 + 4x)^{-2}$, $|x| < \dfrac{5}{4}$

Find the binomial expansion of $f(x)$ in ascending powers of x, up to and including the term in x^3. Give each coefficient as a simplified fraction. **(5 marks)**

(E) **3** $m(x) = \sqrt{4 - x}$, $|x| < 4$

    a Find the series expansion of $m(x)$, in ascending powers of x, up to and including the x^2 term. Simplify each term. **(4 marks)**

    b Show that, when $x = \dfrac{1}{9}$, the exact value of $m(x)$ is $\dfrac{\sqrt{35}}{3}$. **(2 marks)**

    c Use your answer to part **a** to find an approximate value for $\sqrt{35}$, and calculate the percentage error in your approximation. **(4 marks)**

(P) **4** The first three terms in the binomial expansion of $\dfrac{1}{\sqrt{a + bx}}$ are $3 + \dfrac{1}{3}x + \dfrac{1}{18}x^2 + \ldots$

    a Find the values of the constants a and b.

    b Find the coefficient of the x^3 term in the expansion.

(P) **5** $f(x) = \dfrac{3 + 2x - x^2}{4 - x}$

Prove that if x is sufficiently small, $f(x)$ may be approximated by $\dfrac{3}{4} + \dfrac{11}{16}x - \dfrac{5}{64}x^2$.

(E/P) **6 a** Expand $\dfrac{1}{\sqrt{5 + 2x}}$, where $|x| < \dfrac{5}{2}$, in ascending powers of x up to and including the term in x^2, giving each coefficient in simplified surd form. **(5 marks)**

    b Hence or otherwise, find the first 3 terms in the expansion of $\dfrac{2x - 1}{\sqrt{5 + 2x}}$ as a series in ascending powers of x. **(4 marks**

(E/P) **7 a** Use the binomial theorem to expand $(16 - 3x)^{\frac{1}{4}}$, $|x| < \dfrac{16}{3}$ in ascending powers of x, up to and including the term in x^2, giving each term as a simplified fraction. **(4 marks)**

    b Use your expansion, with a suitable value of x, to obtain an approximation to $\sqrt[4]{15.7}$. Give your answer to 3 decimal places. **(2 marks)**

8 $g(x) = \dfrac{3}{4 - 2x} - \dfrac{2}{3 + 5x}$, $|x| < \dfrac{1}{2}$

    a Show that the first three terms in the series expansion of $g(x)$ can be written as $\dfrac{1}{12} + \dfrac{107}{72}x - \dfrac{719}{432}x^2$. **(5 marks)**

    b Find the exact value of $g(0.01)$. Round your answer to 7 decimal places. **(2 marks)**

    c Find the percentage error made in using the series expansion in part **a** to estimate the value of $g(0.01)$. Give your answer to 2 significant figures. **(3 marks)**

4.3 Using partial fractions

Partial fractions can be used to simplify the expansions of more difficult expressions.

Links You need to be confident expressing algebraic fractions as sums of partial fractions.
← Chapter 1

Example 7

a Express $\dfrac{4 - 5x}{(1 + x)(2 - x)}$ as partial fractions.

b Hence show that the cubic approximation of $\dfrac{4 - 5x}{(1 + x)(2 - x)}$ is $2 - \dfrac{7x}{2} + \dfrac{11}{4}x^2 - \dfrac{25}{8}x^3$.

c State the range of values of x for which the expansion is valid.

a $\dfrac{4 - 5x}{(1 + x)(2 - x)} \equiv \dfrac{A}{1 + x} + \dfrac{B}{2 - x}$ —— The denominators must be $(1 + x)$ and $(2 - x)$.

$\equiv \dfrac{A(2 - x) + B(1 + x)}{(1 + x)(2 - x)}$ —— Add the fractions.

$4 - 5x \equiv A(2 - x) + B(1 + x)$ —— Set the numerators equal.

Substitute $x = 2$:
$4 - 10 = A \times 0 + B \times 3$ —— Set $x = 2$ to find B.
$-6 = 3B$
$B = -2$

Substitute $x = -1$:
$4 + 5 = A \times 3 + B \times 0$ —— Set $x = -1$ to find A.
$9 = 3A$
$A = 3$

So $\dfrac{4 - 5x}{(1 + x)(2 - x)} = \dfrac{3}{1 + x} - \dfrac{2}{2 - x}$

—— Write in index form.

b $\dfrac{4 - 5x}{(1 + x)(2 - x)} = \dfrac{3}{1 + x} - \dfrac{2}{2 - x}$

$= 3(1 + x)^{-1} - 2(2 - x)^{-1}$

Problem-solving

Use headings to keep track of your working. This will help you stay organised and check your answers.

The expansion of $3(1 + x)^{-1}$

$= 3\left(1 + (-1)x + (-1)(-2)\dfrac{x^2}{2!} + (-1)(-2)(-3)\dfrac{x^3}{3!} + \ldots\right)$

Expand $3(1 + x)^{-1}$ using the binomial expansion with $n = -1$.

$= 3(1 - x + x^2 - x^3 + \ldots)$

$= 3 - 3x + 3x^2 - 3x^3 + \ldots$

The expansion of $2(2-x)^{-1}$

$= 2\left(2\left(1-\dfrac{x}{2}\right)\right)^{-1}$

$= 2 \times 2^{-1}\left(1-\dfrac{x}{2}\right)^{-1}$ ———— Take out a factor of 2^{-1}.

$= 1 \times \left(1 + (-1)\left(-\dfrac{x}{2}\right) + \dfrac{(-1)(-2)\left(-\dfrac{x}{2}\right)^2}{2!} + \dfrac{(-1)(-2)(-3)\left(-\dfrac{x}{2}\right)^3}{3!} + \ldots\right)$

Expand $\left(1-\dfrac{x}{2}\right)^{-1}$ using the binomial expansion with $n=-1$ and $x=\dfrac{x}{2}$

$= 1 \times \left(1 + \dfrac{x}{2} + \dfrac{x^2}{4} + \dfrac{x^3}{8} + \ldots\right)$

$= 1 + \dfrac{x}{2} + \dfrac{x^2}{4} + \dfrac{x^3}{8}$

Hence $\dfrac{4-5x}{(1+x)(2-x)}$

$= 3(1+x)^{-1} - 2(2-x)^{-1}$ ———— 'Add' both expressions.

$= (3 - 3x + 3x^2 - 3x^3)$
$\quad -\left(1 + \dfrac{x}{2} + \dfrac{x^2}{4} + \dfrac{x^3}{8}\right)$

$= 2 - \dfrac{7}{2}x + \dfrac{11}{4}x^2 - \dfrac{25}{8}x^3$

The expansion is infinite, and converges when $|x|<1$.

c $\dfrac{3}{1+x}$ is valid if $|x|<1$

$\dfrac{2}{2-x}$ is valid if $\left|\dfrac{x}{2}\right|<1 \Rightarrow |x|<2$

The expansion is infinite, and converges when $\left|\dfrac{x}{2}\right|<1$ or $|x|<2$.

Watch out You need to find the range of values of x that satisfy **both** inequalities.

The expansion is valid when $|x|<1$.

Exercise 4C

P **1 a** Express $\dfrac{8x+4}{(1-x)(2+x)}$ as partial fractions.

b Hence or otherwise expand $\dfrac{8x+4}{(1-x)(2+x)}$ in ascending powers of x as far as the term in x^2.

c State the set of values of x for which the expansion is valid.

Binomial expansion

P **2 a** Express $-\dfrac{2x}{(2+x)^2}$ as partial fractions.

  b Hence prove that $-\dfrac{2x}{(2+x)^2}$ can be expressed in the form $-\dfrac{1}{2}x + Bx^2 + Cx^3$ where constants B and C are to be determined.

  c State the set of values of x for which the expansion is valid.

P **3 a** Express $\dfrac{6+7x+5x^2}{(1+x)(1-x)(2+x)}$ as partial fractions.

  b Hence or otherwise expand $\dfrac{6+7x+5x^2}{(1+x)(1-x)(2+x)}$ in ascending powers of x as far as the term in x^3.

  c State the set of values of x for which the expansion is valid.

E/P **4** $g(x) = \dfrac{12x-1}{(1+2x)(1-3x)}, \ |x| < \dfrac{1}{3}$

  Given that $g(x)$ can be expressed in the form $g(x) = \dfrac{A}{1+2x} + \dfrac{B}{1-3x}$

  a Find the values of A and B. **(3 marks)**

  b Hence, or otherwise, find the series expansion of $g(x)$, in ascending powers of x, up to and including the x^2 term. Simplify each term. **(5 marks)**

P **5 a** Express $\dfrac{2x^2+7x-6}{(x+5)(x-4)}$ in partial fractions.

  Hint First divide the numerator by the denominator.

  b Hence, or otherwise, expand $\dfrac{2x^2+7x-6}{(x+5)(x-4)}$ in ascending powers of x as far as the term in x^2.

  c State the set of values of x for which the expansion is valid.

E/P **6** $\dfrac{3x^2+4x-5}{(x+3)(x-2)} = A + \dfrac{B}{x+3} + \dfrac{C}{x-2}$

  a Find the values of the constants A, B and C. **(4 marks)**

  b Hence, or otherwise, expand $\dfrac{3x^2+4x-5}{(x+3)(x-2)}$ in ascending powers of x, as far as the term in x^2. Give each coefficient as a simplified fraction. **(7 marks)**

E/P **7** $f(x) = \dfrac{2x^2+5x+11}{(2x-1)^2(x+1)}, \ |x| < \dfrac{1}{2}$

  $f(x)$ can be expressed in the form $f(x) = \dfrac{A}{2x-1} + \dfrac{B}{(2x-1)^2} + \dfrac{C}{x+1}$

  a Find the values of A, B and C. **(4 marks)**

  b Hence or otherwise, find the series expansion of $f(x)$, in ascending powers of x, up to and including the term in x^2. Simplify each term. **(6 marks)**

  c Find the percentage error made in using the series expansion in part **b** to estimate the value of $f(0.05)$. Give your answer to 2 significant figures. **(4 marks)**

Mixed exercise 4

1. For each of the following,
   i find the binomial expansion up to and including the x^3 term
   ii state the range of values of x for which the expansion is valid.

   a $(1 - 4x)^3$
   b $\sqrt{16 + x}$
   c $\dfrac{1}{1 - 2x}$
   d $\dfrac{4}{2 + 3x}$
   e $\dfrac{4}{\sqrt{4 - x}}$
   f $\dfrac{1 + x}{1 + 3x}$
   g $\left(\dfrac{1 + x}{1 - x}\right)^2$
   h $\dfrac{x - 3}{(1 - x)(1 - 2x)}$

2. Use the binomial expansion to expand $\left(1 - \dfrac{1}{2}x\right)^{\frac{1}{2}}$, $|x| < 2$ in ascending powers of x, up to and including the term in x^3, simplifying each term. **(5 marks)**

3. a Give the binomial expansion of $(1 + x)^{\frac{1}{2}}$ up to and including the term in x^3.
   b By substituting $x = \dfrac{1}{4}$, find an approximation to $\sqrt{5}$ as a fraction.

4. The binomial expansion of $(1 + 9x)^{\frac{2}{3}}$ in ascending powers of x up to and including the term in x^3 is $1 + 6x + cx^2 + dx^3$, $|x| < \dfrac{1}{9}$.
   a Find the value of c and the value of d. **(4 marks)**
   b Use this expansion with your values of c and d together with an appropriate value of x to obtain an estimate of $(1.45)^{\frac{2}{3}}$. **(2 marks)**
   c Obtain $(1.45)^{\frac{2}{3}}$ from your calculator and hence make a comment on the accuracy of the estimate you obtained in part **b**. **(1 mark)**

5. In the expansion of $(1 + ax)^{\frac{1}{2}}$ the coefficient of x^2 is -2.
   a Find the possible values of a.
   b Find the corresponding coefficients of the x^3 term.

6. $f(x) = (1 + 3x)^{-1}$, $|x| < \dfrac{1}{3}$
   a Expand $f(x)$ in ascending powers of x up to and including the term in x^3. **(5 marks)**
   b Hence show that, for small x:
   $$\dfrac{1 + x}{1 + 3x} \approx 1 - 2x + 6x^2 - 18x^3.$$ **(4 marks)**
   c Taking a suitable value for x, which should be stated, use the series expansion in part **b** to find an approximate value for $\dfrac{101}{103}$, giving your answer to 5 decimal places. **(3 marks)**

7. When $(1 + ax)^n$ is expanded as a series in ascending powers of x, the coefficients of x and x^2 are -6 and 27 respectively.
   a Find the values of a and n. **(4 marks)**
   b Find the coefficient of x^3. **(3 marks)**
   c State the values of x for which the expansion is valid. **(1 mark)**

8 Show that if x is sufficiently small then $\dfrac{3}{\sqrt{4+x}}$ can be approximated by $\dfrac{3}{2} - \dfrac{3}{16}x + \dfrac{9}{256}x^2$.

(E) 9 a Expand $\dfrac{1}{\sqrt{4-x}}$, where $|x| < 4$, in ascending powers of x up to and including the term in x^2. Simplify each term. **(5 marks)**

b Hence, or otherwise, find the first 3 terms in the expansion of $\dfrac{1+2x}{\sqrt{4-x}}$ as a series in ascending powers of x. **(4 marks)**

(E) 10 a Find the first four terms of the expansion, in ascending powers of x, of $(2 + 3x)^{-1}$, $|x| < \frac{2}{3}$ **(4 marks)**

b Hence or otherwise, find the first four non-zero terms of the expansion, in ascending powers of x, of:

$\dfrac{1+x}{2+3x}$, $|x| < \frac{2}{3}$ **(3 marks)**

(E/P) 11 a Use the binomial theorem to expand $(4 + x)^{-\frac{1}{2}}$, $|x| < 4$, in ascending powers of x, up to and including the x^3 term, giving each answer as a simplified fraction. **(4 marks)**

b Use your answer to part **a** to obtain an approximation to $\sqrt{2}$, accurate to 4 decimal places, by substituting

  i $x = -2$ ii $x = \dfrac{1}{2}$ **(3 marks)**

c State, with a reason, which of the two possible substitutions would lead to the most accurate approximation to $\sqrt{2}$. **(1 mark)**

(E) 12 $q(x) = (3 + 4x)^{-3}$, $|x| < \frac{3}{4}$

Find the binomial expansion of $q(x)$ in ascending powers of x, up to and including the term in the x^2. Give each coefficient as a simplified fraction. **(5 marks)**

(E/P) 13 $g(x) = \dfrac{39x + 12}{(x+1)(x+4)(x-8)}$, $|x| < 1$

$g(x)$ can be expressed in the form $g(x) = \dfrac{A}{x+1} + \dfrac{B}{x+4} + \dfrac{C}{x-8}$

a Find the values of A, B and C. **(4 marks)**

b Hence, or otherwise, find the series expansion of $g(x)$, in ascending powers of x, up to and including the x^2 term. Simplify each term. **(7 marks)**

Chapter 4

14 $f(x) = \dfrac{12x+5}{(1+4x)^2}$, $|x| < \dfrac{1}{4}$

For $x \neq -\dfrac{1}{4}$, $\dfrac{12x+5}{(1+4x)^2} = \dfrac{A}{1+4x} + \dfrac{B}{(1+4x)^2}$, where A and B are constants.

a Find the values of A and B. **(3 marks)**

b Hence, or otherwise, find the series expansion of $f(x)$, in ascending powers of x, up to and including the term x^2, simplifying each term. **(6 marks)**

15 $q(x) = \dfrac{9x^2 + 26x + 20}{(1+x)(2+x)}$, $|x| < 1$

a Show that the expansion of $q(x)$ in ascending powers of x can be approximated to $10 - 2x + Bx^2 + Cx^3$ where B and C are constants to be found. **(7 marks)**

b Find the percentage error made in using the series expansion in part **a** to estimate the value of $q(0.1)$. Give your answer to 2 significant figures. **(4 marks)**

Challenge

Obtain the first four non-zero terms in the expansion, in ascending powers of x, of the function $f(x)$ where $f(x) = \dfrac{1}{\sqrt{1+3x^2}}$, $3x^2 < 1$.

Summary of key points

1 This form of the binomial expansion can be applied to negative or fractional values of n to obtain an infinite series:

$$(1+x)^n = 1 + nx + \dfrac{n(n-1)x^2}{2!} + \dfrac{n(n-1)(n-2)x^3}{3!} + \ldots + \dfrac{n(n-1)\ldots(n-r+1)x^r}{r!} + \ldots$$

The expansion is valid when $|x| < 1$, $n \in \mathbb{R}$.

2 Approximations based on the binomial expansion are more accurate when:
- more terms of the expansion are used
- the values of x substituted into the expansion are closer to 0

3 The expansion of $(1 + bx)^n$, where n is negative or a fraction, is valid for $|bx| < 1$, or $|x| < \dfrac{1}{|b|}$.

4 The expansion of $(a + bx)^n$, where n is negative or a fraction, is valid for $\left|\dfrac{b}{a}x\right| < 1$ or $|x| < \left|\dfrac{a}{b}\right|$.

Review exercise

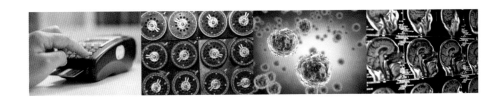

1 Prove by contradiction that there are infinitely many prime numbers. (4)

← Section 1.1

2 Prove that the equation $x^2 - 2 = 0$ has no rational solutions.
You may assume that if n^2 is an even integer then n is also an even integer. (4)

← Section 1.1

3 Express $\dfrac{4x}{x^2 - 2x - 3} + \dfrac{1}{x^2 + x}$ as a single fraction in its simplest form. (4)

← Section 1.2

4 $f(x) = 1 - \dfrac{3}{x + 2} + \dfrac{3}{(x + 2)^2}$, $x \neq -2$

a Show that $f(x) = \dfrac{x^2 + x + 1}{(x + 2)^2}$, $x \neq -2$.

b Show that $x^2 + x + 1 > 0$ for all values of x, $x \neq -2$.

c Show that $f(x) > 0$ for all values of x, $x \neq -2$.

← Section 1.2

5 Show that $\dfrac{2x - 1}{(x - 1)(2x - 3)}$ can be written in the form $\dfrac{A}{x - 1} + \dfrac{B}{2x - 3}$ where A and B are constants to be found. (3)

← Section 1.3

6 Given that
$\dfrac{3x + 7}{(x + 1)(x + 2)(x + 3)} \equiv \dfrac{P}{x + 1} + \dfrac{Q}{x + 2} + \dfrac{R}{x + 3}$
where P, Q and R are constants, find the values of P, Q and R. (4)

← Section 1.3

7 $f(x) = \dfrac{2}{(2 - x)(1 + x)^2}$, $x \neq -1, x \neq 2$.

Find the values of A, B and C such that
$f(x) = \dfrac{A}{2 - x} + \dfrac{B}{1 + x} + \dfrac{C}{(1 + x)^2}$ (4)

← Section 1.4

8 $\dfrac{14x^2 + 13x + 2}{(x + 1)(2x + 1)^2} \equiv \dfrac{A}{x + 1} + \dfrac{B}{2x + 1} + \dfrac{C}{(2x + 1)^2}$

Find the values of the constants A, B and C.

← Section 1.4

9 Given that $\dfrac{3x^2 + 6x - 2}{x^2 + 4} \equiv d + \dfrac{ex + f}{x^2 + 4}$ find the values of d, e and f. (4)

← Section 1.5

10 $p(x) = \dfrac{9 - 3x - 12x^2}{(1 - x)(1 + 2x)}$

Show that $p(x)$ can be written in the form
$A + \dfrac{B}{1 - x} + \dfrac{C}{1 + 2x}$, where A, B and C are constants to be found. (4)

← Sections 1.3, 1.5

11 Solve the inequality $|4x + 3| > 7 - 2x$. (3)

← Section 2.1

12 The function $p(x)$ is defined by
$p : x \mapsto \begin{cases} 4x + 5, & x < -2 \\ -x^2 + 4, & x \geq -2 \end{cases}$

a Sketch $p(x)$, stating its range. (3)

b Find the exact values of a such that $p(a) = -20$. (4)

← Section 2.2

107

Review exercise 1

13 The functions p and q are defined by

$p(x) = \dfrac{1}{x+4}$, $x \in \mathbb{R}$, $x \neq -4$

$q(x) = 2x - 5$, $x \in \mathbb{R}$

a Show that $qp(x) = \dfrac{ax+b}{cx+d}$, $x \in \mathbb{R}$, $x \neq -4$, where a, b, c and d are integers to be found. **(3)**

b Solve $qp(x) = 15$. **(3)**

Let $r(x) = qp(x)$.

c Find $r^{-1}(x)$. **(3)**

← Section 2.3

14 The functions f and g are defined by:

$f: x \mapsto \dfrac{x+2}{x}$, $x \in \mathbb{R}$, $x \neq 0$

$g: x \mapsto \ln(2x - 5)$, $x \in \mathbb{R}$, $x > \tfrac{5}{2}$

a Sketch the graph of f. **(3)**

b Show that

$f^2(x) = \dfrac{3x+2}{x+2}$, $x \in \mathbb{R}$, $x \neq 0$, $x \neq -2$ **(3)**

c Find the exact value of $gf\left(\tfrac{1}{4}\right)$. **(2)**

d Find $g^{-1}(x)$. **(3)**

← Section 2.3, 2.4

15 The functions p and q are defined by:

$p(x) = 3x + b$, $x \in \mathbb{R}$

$q(x) = 1 - 2x$, $x \in \mathbb{R}$

Given that $pq(x) = qp(x)$,

a show that $b = -\tfrac{2}{3}$ **(3)**

b find $p^{-1}(x)$ and $q^{-1}(x)$ **(3)**

c show that

$p^{-1}q^{-1}(x) = q^{-1}p^{-1}(x) = \dfrac{ax+b}{c}$, $x \in \mathbb{R}$,

where a, b and c are integers to be found. **(4)**

← Section 2.3, 2.4

16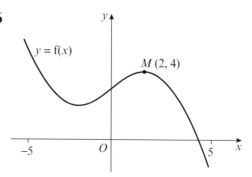

The figure shows the graph of $y = f(x)$, $-5 \leq x \leq 5$.

The point $M(2, 4)$ is the maximum turning point of the graph.

Sketch, on separate diagrams, the graphs of:

a $y = f(x) + 3$ **(2)**

b $y = |f(x)|$ **(2)**

c $y = f(|x|)$ **(2)**

d $y = f(2x - 1)$ **(2)**

Show on each graph the coordinates of any maximum turning points.

← Sections 2.5, 2.6

17 The function h is defined by

$h: x \mapsto 2(x+3)^2 - 8$, $x \in \mathbb{R}$

a Draw a sketch of $y = h(x)$, labelling the turning points and the x- and y-intercepts. **(4)**

b Write down the coordinates of the turning points on the graphs with equations:

i $y = 3h(x+2)$ **(2)**

ii $y = h(-x)$ **(2)**

iii $y = |h(x)|$ **(2)**

c Sketch the curve with equation $y = h(-|x|)$. On your sketch show the coordinates of all turning points and all x- and y-intercepts. **(4)**

← Sections 2.5, 2.6

18

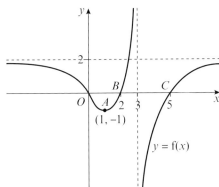

The diagram shows a sketch of the graph of $y = f(x)$.

The curve has a minimum at the point $A(1, -1)$, passes through x-axis at the origin, and the points $B(2, 0)$ and $C(5, 0)$; the asymptotes have equations $x = 3$ and $y = 2$.

a Sketch, on separate axes, the graphs of:
  i $y = |f(x)|$ (2)
  ii $y = -f(x + 1)$ (2)
  iii $y = f(-2x)$ (2)

in each case, showing the images of the points A, B and C.

b State the number of solutions to each equation.
  i $3|f(x)| = 2$ (2)
  ii $2|f(x)| = 3$ (2)

← Sections 2.6, 2.7

19 The diagram shows a sketch of part of the graph $y = q(x)$, where

$q(x) = \frac{1}{2}|x + b| - 3$, $b < 0$

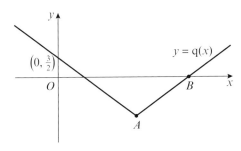

The graph cuts the y-axis at $\left(0, \frac{3}{2}\right)$.

a Find the value of b. (2)

b Find the coordinates of A and B. (3)

c Solve $q(x) = -\frac{1}{3}x + 5$. (5)

← Section 2.7

20 The function f is defined by

$f(x) = -\frac{5}{3}|x + 4| + 8$, $x \in \mathbb{R}$

The diagram shows a sketch of the graph $y = f(x)$.

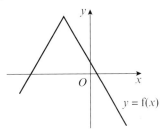

a State the range of f. (1)

b Give a reason why $f^{-1}(x)$ does not exist. (1)

c Solve the inequality $f(x) > \frac{2}{3}x + 4$. (5)

d State the range of values of k for which the equation $f(x) = \frac{5}{3}x + k$ has no solutions. (2)

← Section 2.7

21 The 4th, 5th and 6th terms in an arithmetic sequence are:

$12 - 7k$, $3k^2$, $k^2 - 10k$

a Find two possible values of k. (3)

Given that the sequence contains only integer terms,

b find the first term and the common difference. (2)

← Section 3.1

22 The 4th term of an arithmetic sequence is 72. The 11th term is 51. The sum of the first n terms is 1125.

a Show that $3n^2 - 165n + 2250 = 0$. (4)

b Find the two possible values for n. (2)

← Section 3.2

Review exercise 1

23 **a** Find, in terms of p, the 30th term of the arithmetic sequence
$(19p - 18), (17p - 8), (15p + 2), \ldots$
giving your answer in its simplest form. **(2)**

b Given $S_{31} = 0$, find the value of p. **(3)**

← Sections 3.1, 3.2

24 The second term of a geometric sequence is 256. The eighth term of the same sequence is 900. The common ratio is r, $r > 0$.

a Show that r satisfies the equation
$$6 \ln r + \ln\left(\frac{64}{225}\right) = 0 \quad \text{(3)}$$

b Find the value of r correct to 3 significant figures. **(3)**

← Section 3.3

25 The first three terms of a geometric sequence are $10, \frac{50}{6}$ and $\frac{250}{36}$.

a Find the sum to infinity of the series. **(3)**

Given that the sum to k terms of the series is greater than 55,

b show that $k > \dfrac{\log\left(\frac{1}{12}\right)}{\log\left(\frac{5}{6}\right)}$ **(4)**

c find the smallest possible value of k. **(1)**

← Sections 3.4, 3.5

26 A geometric series has first term 4 and common ratio r. The sum of the first three terms of the series is 7.

a Show that $4r^2 + 4r - 3 = 0$. **(3)**

b Find the two possible values of r. **(2)**

Given that r is positive,

c find the sum to infinity of the series. **(2)**

← Sections 3.4, 3.5

27 The fourth, fifth and sixth terms of a geometric series are x, 3 and $x + 8$.

a Find the two possible values of x and the corresponding values of the common ratio. **(4)**

Given that the sum to infinity of the series exists,

b find the first term **(1)**

c find the sum to infinity of the series. **(2)**

← Sections 3.3, 3.5

28 A sequence $a_1, a_2, a_3, \ldots$ is defined by
$a_1 = k$,
$a_{n+1} = 3a_n + 5, n \geq 1$
where k is a positive integer.

a Write down an expression for a_2 in terms of k. **(1)**

b Show that $a_3 = 9k + 20$. **(2)**

c **i** Find $\sum_{r=1}^{4} a_r$ in terms of k. **(2)**

ii Show that $\sum_{r=1}^{4} a_r$ is divisible by 10. **(2)**

← Sections 3.6, 3.7

29 At the end of year 1, a company employs 2400 people. A model predicts that the number of employees will increase by 6% each year, forming a geometric sequence.

a Find the predicted number of employees after 4 years, giving your answer to the nearest 10. **(3)**

The company expects to expand in this way until the total number of employees first exceeds 6000 at the end of a year, N.

b Show that $(N - 1)\log 1.06 > \log 2.5$ **(3)**

c Find the value of N. **(2)**

The company has a charity scheme whereby they match any employee charity contribution exactly.

d Given that the average employee charity contribution is £5 each year, find the total charity donation over the 10-year period from the end of year 1 to the end of year 10. Give your answer to the nearest £1000. **(3)**

← Section 3.8

Review exercise 1

E/P 30 A geometric series is given by
$$6 - 24x + 96x^2 - \ldots$$
The series is convergent.

a Write down a condition on x. (1)

Given that $\sum_{r=1}^{\infty} 6 \times (-4x)^{r-1} = \frac{24}{5}$

b Calculate the value of x. (5)

← Sections 3.5, 3.6

E 31 $g(x) = \dfrac{1}{\sqrt{1-x}}$

a Show that the series expansion of $g(x)$ up to and including the x^3 term is
$$1 + \frac{x}{2} + \frac{3x^2}{8} + \frac{5x^3}{16}$$ (5)

b State the range of values of x for which the expansion is valid. (1)

← Section 4.1

P 32 When $(1 + ax)^n$ is expanded as a series in ascending powers of x, the coefficients of x and x^2 are -6 and 45 respectively.

a Find the value of a and the value of n.

b Find the coefficient of x^3.

c Find the set of values of x for which the expansion is valid.

← Section 4.1

E 33 a Find the binomial expansion of $(1 + 4x)^{\frac{3}{2}}$ in ascending powers of x up to and including the x^3 term, simplifying each term. (4)

b Show that, when $x = \dfrac{3}{100}$, the exact value of $(1 + 4x)^{\frac{3}{2}}$ is $\dfrac{112\sqrt{112}}{1000}$. (2)

c Substitute $x = \dfrac{3}{100}$ into the binomial expansion in part **a** and hence obtain an approximation to $\sqrt{112}$. Give your answer to 5 decimal places. (3)

d Calculate the percentage error in your estimate to 5 decimal places. (2)

← Section 4.1

E 34 $f(x) = (1 + x)(3 + 2x)^{-3}$, $|x| < \frac{3}{2}$
Find the binomial expansion of $f(x)$ in ascending powers of x, up to and including the term in x^3. Give each coefficient as a simplified fraction. (5)

← Section 4.2

E 35 $h(x) = \sqrt{4 - 9x}$, $|x| < \frac{4}{9}$

a Find the series expansion of $h(x)$, in ascending powers of x, up to and including the x^2 term. Simplify each term. (4)

b Show that, when $x = \dfrac{1}{100}$, the exact value of $h(x)$ is $\dfrac{\sqrt{391}}{10}$. (2)

c Use the series expansion in part **a** to estimate the value of $h\left(\dfrac{1}{100}\right)$ and state the degree of accuracy of your approximation. (3)

← Section 4.2

E/P 36 Given that $(a + bx)^{-2}$ has binomial expansion $\frac{1}{4} + \frac{1}{4}x + cx^2 + \ldots$

a Find the values of the constants a, b and c. (4)

b Find the coefficient of the x^3 term in the expansion. (2)

← Section 4.2

E/P 37 $g(x) = \dfrac{3 + 5x}{(1 + 3x)(1 - x)}$, $|x| < \dfrac{1}{3}$

Given that $g(x)$ can be expressed in the form $g(x) = \dfrac{A}{1 + 3x} + \dfrac{B}{1 - x}$

a find the values of A and B. (3)

b Hence, or otherwise, find the series expansion of $f(x)$, in ascending powers of x, up to and including the x^2 term. Simplify each term. (6)

← Sections 4.1, 4.3

Review exercise 1

E/P 38 $\dfrac{3x-1}{(1-2x)^2} \equiv \dfrac{A}{1-2x} + \dfrac{B}{(1-2x)^2}, |x| < \dfrac{1}{2}$

    a Find the values of A and B. (3)

    b Hence, or otherwise, expand $\dfrac{3x-1}{(1-2x)^2}$ in ascending powers of x, as far as the term in x^3. Give each coefficient as a simplified fraction. (6)

    ← Sections 4.1, 4.3

E/P 39 $f(x) = \dfrac{25}{(3+2x)^2(1-x)}, |x| < 1$

    $f(x)$ can be expressed in the form
$$\dfrac{A}{3+2x} + \dfrac{B}{(3+2x)^2} + \dfrac{C}{1-x}$$

    a Find the values of A, B and C. (4)

    b Hence, or otherwise, find the series expansion of $f(x)$, in ascending powers of x, up to and including the term in x^2. Simplify each term. (6)

    ← Sections 4.1, 4.2, 4.3

E/P 40 $\dfrac{4x^2 + 30x + 31}{(x+4)(2x+3)} = A + \dfrac{B}{x+4} + \dfrac{C}{2x+3}$

    a Find the values of the constants A, B and C. (4)

    b Hence, or otherwise, expand $\dfrac{4x^2 + 31x + 30}{(x+4)(2x+3)}$ in ascending powers of x, as far as the term in x^2. Give each coefficient as a simplified fraction. (7)

    ← Sections 4.1, 4.2, 4.3

Challenge

1 The functions f and g are defined by
$f(x) = -3|x+3| + 15, x \in \mathbb{R}$
$g(x) = -\frac{3}{4}x + \frac{3}{2}, x \in \mathbb{R}$
The diagram shows a sketch of the graphs $y = f(x)$ and $y = g(x)$, which intersect at points A and B. M is the midpoint of AB.
The circle C, with centre M, passes through points A and B, and meets $y = f(x)$ at point P as shown in the diagram.

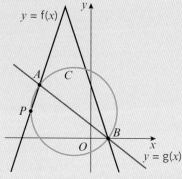

    a Find the equation of the circle.

    b Find the area of the triangle APB.

    ← Section 2.6

2 Given that $a_{n+1} = a_n + k$, $a_1 = m$ and $\sum_{i=6}^{11} a_i = \sum_{i=12}^{15} a_i$ show that $m = \frac{5}{2}k$.

    ← Section 3.6

3 The diagram shows a sketch of the functions $p(x) = |x^2 - 8x + 12|$ and $q(x) = |x^2 - 11x + 28|$.

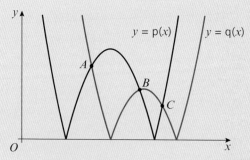

Find the exact values of the x-coordinates of the points A, B and C.

    ← Section 2.5

4 Show that $\sum_{r=1}^{40} \log_3 \left(\dfrac{2n+1}{2n-1} \right) = 4$

    ← Section 3.6

5 The graph of $y = f(x)$ is transformed to the graph of $y = f(ax + b)$ where $a \neq 1$.
Prove that any invariant points under this transformation satisfy $x = \dfrac{b}{1-a}$

Radians

5

Objectives

After completing this unit you should be able to:
- Convert between degrees and radians and apply this to trigonometric graphs and their transformations → pages 114–116
- Know exact values of angles measured in radians → pages 117–118
- Find an arc length using radians → pages 118–122
- Find areas of sectors and segments using radians → pages 122–128
- Solve trigonometric equations in radians → pages 128–132
- Use approximate trigonometric values when θ is small → pages 133–135

Prior knowledge check

1. Write down the exact values of the following trigonometric ratios.

   a $\cos 120°$ **b** $\sin 225°$ **c** $\tan(-300°)$

   d $\sin(-480°)$ ← Year 1, Chapter 10

2. Simplify each of the following expressions.

   a $(\tan\theta \cos\theta)^2 + \cos^2\theta$

   b $1 - \dfrac{1}{\cos^2\theta}$ **c** $\sqrt{1 - \dfrac{\sin\theta \cos\theta}{\tan\theta}}$

   ← Year 1, Chapter 10

3. Show that

   a $(\sin 2\theta + \cos 2\theta)^2 \equiv 1 + 2\sin 2\theta \cos 2\theta$

   b $\dfrac{2}{\sin\theta} - 2\sin\theta \equiv \dfrac{2\cos^2\theta}{\sin\theta}$

   ← Year 1, Chapter 10

4. Solve the following equations for θ in the interval $0 \leq \theta \leq 360°$, giving your answers to 3 significant figures where they are not exact.

   a $4\cos\theta + 2 = 3$ **b** $2\sin 2\theta = 1$

   c $6\tan^2\theta + 10\tan\theta - 4 = \tan\theta$

   d $10 + 5\cos\theta = 12\sin^2\theta$

   ← Year 1, Chapter 10

Radians are units for measuring angles. They are used in mechanics to describe circular motion, and can be used to work out the distances between the pods around the edge of a Ferris wheel. → Exercise 5B Q13

Chapter 5

5.1 Radian measure

So far you have probably only measured angles in degrees, with one degree representing $\frac{1}{360}$ of a complete revolution or circle.

You can also measure angles in units called **radians**. 1 radian is the angle subtended at the centre of a circle by an arc whose length is equal to the radius of the circle.

If the arc AB has length r, then $\angle AOB$ is 1 radian.

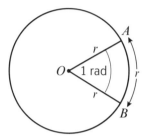

Links You always use radian measure when you are differentiating or integrating trigonometric functions. → Sections 9.1, 11.1

Notation You can write 1 radian as 1 rad.

The circumference of a circle of radius r is an arc of length $2\pi r$, so it subtends an angle of 2π radians at the centre of the circle.

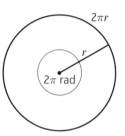

- **2π radians = 360°**
- **π radians = 180°**
- **1 radian = $\frac{180°}{\pi}$**

Hint This means that 1 radian = 57.295...°

Example 1

Convert the following angles into degrees.

a $\frac{7\pi}{8}$ rad   b $\frac{4\pi}{15}$ rad

a $\frac{7\pi}{8}$ rad
$= \frac{7}{8} \times 180°$
$= 157.5°$

b $\frac{4\pi}{15}$ rad
$= 4 \times \frac{180°}{15}$
$= 48°$

1 radian = $\frac{180°}{\pi}$, so multiply by $\frac{180°}{\pi}$

$\frac{7\pi}{8} \times \frac{180°}{\pi} = \frac{7}{8} \times 180°$

Example 2

Convert the following angles into radians. Leave your answers in terms of π.

a 150°   b 110°

a $150° = 150 \times \frac{\pi}{180}$ rad
$= \frac{5\pi}{6}$ rad

b $110° = 110 \times \frac{\pi}{180}$ rad
$= \frac{11\pi}{18}$ rad

$1° = \frac{\pi}{180}$ radians, so multiply by $\frac{\pi}{180}$

Your calculator will often give you exact answers in terms of π.

You should learn these important angles in radians:

- $30° = \frac{\pi}{6}$ radians
- $60° = \frac{\pi}{3}$ radians
- $180° = \pi$ radians
- $45° = \frac{\pi}{4}$ radians
- $90° = \frac{\pi}{2}$ radians
- $360° = 2\pi$ radians

Example 3

Find: **a** $\sin(0.3 \text{ rad})$ **b** $\cos(\pi \text{ rad})$ **c** $\tan(2 \text{ rad})$

Give your answers correct to 2 decimal places where appropriate.

Online Use your calculator to evaluate trigonometric functions in radians.

a $\sin(0.3 \text{ rad}) = 0.30$ (2 d.p.)

b $\cos(\pi \text{ rad}) = -1$

c $\tan(2 \text{ rad}) = -2.19$ (2 d.p.)

Watch out You need to make sure your calculator is in radians mode.

Example 4

Sketch the graph of $y = \sin x$ for $0 \leqslant x \leqslant 2\pi$.

If the range includes values given in terms of π, you can assume that the angle has been given in radians.

$\sin\left(\frac{\pi}{2}\right) = \sin 90° = 1$

Example 5

Sketch the graph of $y = \cos(x + \pi)$ for $0 \leqslant x \leqslant 2\pi$.

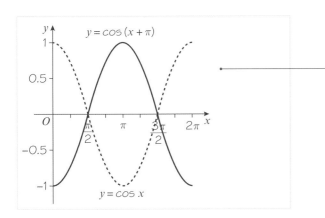

The graph of $y = \cos(x + a)$ is a translation of the graph $y = \cos x$ by the vector $\begin{pmatrix} -a \\ 0 \end{pmatrix}$.

Exercise 5A

1 Convert the following angles in radians to degrees.
   a $\dfrac{\pi}{20}$ b $\dfrac{\pi}{15}$ c $\dfrac{5\pi}{12}$ d $\dfrac{5\pi}{4}$ e $\dfrac{3\pi}{2}$ f 3π

2 Convert the following angles to degrees, giving your answer to 1 d.p.
   a 0.46 rad   b 1 rad   c 1.135 rad   d $\sqrt{3}$ rad

3 Evaluate the following, giving your answers to 3 significant figures.
   a $\sin(0.5\text{ rad})$ b $\cos(\sqrt{2}\text{ rad})$ c $\tan(1.05\text{ rad})$ d $\sin(2\text{ rad})$ e $\sin(3.6\text{ rad})$

4 Convert the following angles to radians, giving your answers as multiples of π.
   a 8°   b 10°   c 22.5°   d 30°   e 112.5°
   f 240°   g 270°   h 315°   i 330°

5 Convert the following angles to radians, giving your answers to 3 significant figures.
   a 50°   b 75°   c 100°   d 160°   e 230°   f 320°

6 Sketch the graphs of:
   a $y = \tan x$ for $0 \leqslant x \leqslant 2\pi$ b $y = \cos x$ for $-\pi \leqslant x \leqslant \pi$
   Mark any points where the graphs cut the coordinate axes.

7 Sketch the following graphs for the given ranges, marking any points where the graphs cut the coordinate axes.
   a $y = \sin(x - \pi)$ for $-\pi \leqslant x \leqslant \pi$
   b $y = \cos 2x$ for $0 \leqslant x \leqslant 2\pi$
   c $y = \tan\left(x + \dfrac{\pi}{2}\right)$ for $-\pi \leqslant x \leqslant \pi$
   d $y = \sin\dfrac{1}{3}x + 1$ for $0 \leqslant x \leqslant 6\pi$

(E/P) 8 The diagram shows the curve with equation $y = \cos\left(x - \dfrac{2\pi}{3}\right)$, $-2\pi \leqslant x \leqslant 2\pi$.

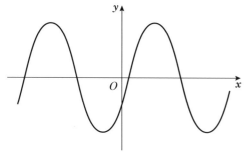

Problem-solving

Make sure you write down the coordinates of all four points of intersection with the x-axis and the coordinates of the y-intercept.

Write down the coordinates of the points at which the curve meets the coordinate axes. **(3 marks)**

Challenge

Describe all the angles, θ, in radians, that satisfy:
a $\cos\theta = 1$
b $\sin\theta = -1$
c $\tan\theta$ is undefined.

Hint You can use $n\pi$, where n is an integer, to describe any integer multiple of π.

Radians

You need to learn the exact values of the trigonometric ratios of these angles measured in radians:

- $\sin\frac{\pi}{6} = \frac{1}{2}$
- $\cos\frac{\pi}{6} = \frac{\sqrt{3}}{2}$
- $\tan\frac{\pi}{6} = \frac{1}{\sqrt{3}} = \frac{\sqrt{3}}{3}$

- $\sin\frac{\pi}{3} = \frac{\sqrt{3}}{2}$
- $\cos\frac{\pi}{3} = \frac{1}{2}$
- $\tan\frac{\pi}{3} = \sqrt{3}$

- $\sin\frac{\pi}{4} = \frac{1}{\sqrt{2}} = \frac{\sqrt{2}}{2}$
- $\cos\frac{\pi}{4} = \frac{1}{\sqrt{2}} = \frac{\sqrt{2}}{2}$
- $\tan\frac{\pi}{4} = 1$

You can use these rules to find sin, cos or tan of any positive or negative angle measured in radians using the corresponding acute angle made with the x-axis, θ.

- $\sin(\pi - \theta) = \sin\theta$
- $\sin(\pi + \theta) = -\sin\theta$
- $\sin(2\pi - \theta) = -\sin\theta$
- $\cos(\pi - \theta) = -\cos\theta$
- $\cos(\pi + \theta) = -\cos\theta$
- $\cos(2\pi - \theta) = \cos\theta$
- $\tan(\pi - \theta) = -\tan\theta$
- $\tan(\pi + \theta) = \tan\theta$
- $\tan(2\pi - \theta) = -\tan\theta$

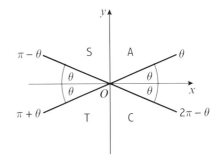

Links The CAST diagram shows you which trigonometric ratios are positive in which quadrant. You can also use the symmetry properties and periods of the graphs of sin, cos and tan to find these results. ← Year 1, Chapter 10

Example 6

Find the exact values of:

a $\cos\frac{4\pi}{3}$

b $\sin\left(\frac{-7\pi}{6}\right)$

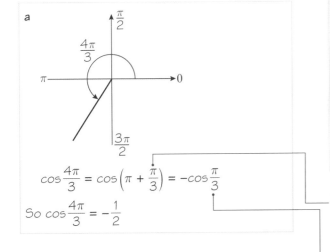

$\cos\frac{4\pi}{3} = \cos\left(\pi + \frac{\pi}{3}\right) = -\cos\frac{\pi}{3}$

So $\cos\frac{4\pi}{3} = -\frac{1}{2}$

Problem-solving

You can also use the symmetry properties of $y = \cos x$:

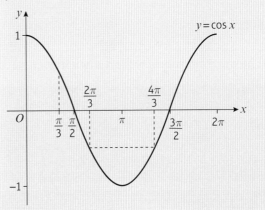

$\frac{4\pi}{3}$ is $\frac{\pi}{3}$ bigger than π.

Use $\cos(\pi + \theta) = -\cos\theta$.

b

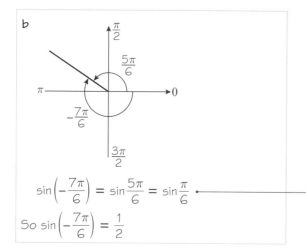

$\sin\left(-\dfrac{7\pi}{6}\right) = \sin\dfrac{5\pi}{6} = \sin\dfrac{\pi}{6}$ ——— Use $\sin(\pi - \theta) = \sin\theta$.

So $\sin\left(-\dfrac{7\pi}{6}\right) = \dfrac{1}{2}$

Exercise 5B

1 Express the following as trigonometric ratios of either $\dfrac{\pi}{6}, \dfrac{\pi}{4}$ or $\dfrac{\pi}{3}$, and hence find their exact values.

a $\sin\dfrac{3\pi}{4}$ **b** $\sin\left(-\dfrac{\pi}{3}\right)$ **c** $\sin\dfrac{11\pi}{6}$

d $\cos\dfrac{2\pi}{3}$ **e** $\cos\dfrac{5\pi}{3}$ **f** $\cos\dfrac{5\pi}{4}$

g $\tan\dfrac{3\pi}{4}$ **h** $\tan\left(-\dfrac{5\pi}{4}\right)$ **i** $\tan\dfrac{7\pi}{6}$

2 Without using a calculator, find the exact values of the following trigonometric ratios.

a $\sin\dfrac{7\pi}{3}$ **b** $\sin\left(-\dfrac{5\pi}{3}\right)$ **c** $\cos\left(-\dfrac{7\pi}{6}\right)$

d $\cos\dfrac{11\pi}{4}$ **e** $\tan\dfrac{5\pi}{3}$ **f** $\tan\left(-\dfrac{2\pi}{3}\right)$

(P) 3 The diagram shows a right-angled triangle ACD on another right-angled triangle ABC with $AD = \dfrac{2\sqrt{6}}{3}$ and $BC = 2$.

Show that $DC = k\sqrt{2}$, where k is a constant to be determined.

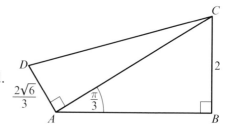

5.2 Arc length

Using radians greatly simplifies the formula for **arc length**.

- **To find the arc length l of a sector of a circle use the formula $l = r\theta$, where r is the radius of the circle and θ is the angle, in radians, contained by the sector.**

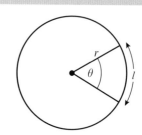

Radians

Example 7

Find the length of the arc of a circle of radius 5.2 cm, given that the arc subtends an angle of 0.8 rad at the centre of the circle.

Online Explore the arc length of a sector using GeoGebra.

Arc length = 5.2 × 0.8 = 4.16 cm — Use $l = r\theta$, with $r = 5.2$ and $\theta = 0.8$.

Example 8

An arc AB of a circle with radius 7 cm and centre O has a length of 2.45 cm. Find the angle $\angle AOB$ subtended by the arc at the centre of the circle.

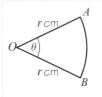

$l = r\theta$ — Use $l = r\theta$, with $l = 2.45$ and $r = 7$.

$2.45 = 7\theta$

$\dfrac{2.45}{7} = \theta$

$\theta = 0.35$ rad — Using this formula gives the angle in radians.

Example 9

An arc AB of a circle, with centre O and radius r cm, subtends an angle of θ radians at O. The perimeter of the sector AOB is P cm. Express r in terms of P and θ.

Problem-solving

When given a problem in words, it is often a good idea to sketch and label a diagram to help you to visualise the information you have and what you need to find.

$P = r\theta + 2r$

$= r(2 + \theta)$

So $r = \dfrac{P}{(2 + \theta)}$

The perimeter = arc $AB + OA + OB$, where arc $AB = r\theta$.

Factorise.

Chapter 5

Example 10

The border of a garden pond consists of a straight edge AB of length 2.4 m, and a curved part C, as shown in the diagram.
The curved part is an arc of a circle, centre O and radius 2 m.
Find the length of C.

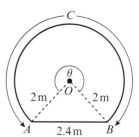

$\sin x = \dfrac{1.2}{2}$

$x = 0.6435\ldots$ rad

Acute $\angle AOB = 2x$ rad
$= 2 \times 0.6435\ldots$
$= 1.2870\ldots$ rad

So $\theta = (2\pi - 1.2870\ldots)$ rad
$= 4.9961\ldots$ rad

So length of $C = 9.99$ m (3 s.f.)

Online Explore the area of a sector using GeoGebra.

Problem-solving
Look for opportunities to use the basic trigonometric ratios rather than the more complicated cosine rule or sine rule. AOB is an isosceles triangle, so you can divide it into congruent right-angled triangles. Make sure your calculator is in radians mode.

C subtends the reflex angle θ at O, so length of $C = 2\theta$.

$\theta +$ acute $\angle AOB = 2\pi$ rad

$C = 2\theta$

Exercise 5C

1 An arc AB of a circle, centre O and radius r cm, subtends an angle θ radians at O. The length of AB is l cm.

  a Find l when: **i** $r = 6$, $\theta = 0.45$ **ii** $r = 4.5$, $\theta = 0.45$ **iii** $r = 20$, $\theta = \dfrac{3}{8}\pi$

  b Find r when: **i** $l = 10$, $\theta = 0.6$ **ii** $l = 1.26$, $\theta = 0.7$ **iii** $l = 1.5\pi$, $\theta = \dfrac{5}{12}\pi$

  c Find θ when: **i** $l = 10$, $r = 7.5$ **ii** $l = 4.5$, $r = 5.625$ **iii** $l = \sqrt{12}$, $r = \sqrt{3}$

(P) 2 A minor arc AB of a circle, centre O and radius 10 cm, subtends an angle x at O. The major arc AB subtends an angle $5x$ at O. Find, in terms of π, the length of the minor arc AB.

Notation The **minor arc** AB is the shorter arc between points A and B on a circle.

3 An arc AB of a circle, centre O and radius 6 cm, has length l cm. Given that the chord AB has length 6 cm, find the value of l, giving your answer in terms of π.

4 The sector of a circle of radius $\sqrt{10}$ cm contains an angle of $\sqrt{5}$ radians, as shown in the diagram. Find the length of the arc, giving your answer in the form $p\sqrt{q}$ cm, where p and q are integers.

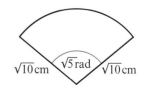

5 Referring to the diagram, find:
   a the perimeter of the shaded region when $\theta = 0.8$ radians.
   b the value of θ when the perimeter of the shaded region is 14 cm.

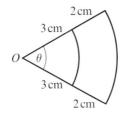

Problem-solving

The radius of the larger arc is $3 + 2 = 5$ cm.

6 A sector of a circle of radius r cm contains an angle of 1.2 radians. Given that the sector has the same perimeter as a square of area 36 cm², find the value of r.

7 A sector of a circle of radius 15 cm contains an angle of θ radians. Given that the perimeter of the sector is 42 cm, find the value of θ.

8 In the diagram AB is the diameter of a circle, centre O and radius 2 cm.
   The point C is on the circumference such that $\angle COB = \frac{2}{3}\pi$ radians.
   a State the value, in radians, of $\angle COA$. **(1 mark)**
   The shaded region enclosed by the chord AC, arc CB and AB is the template for a brooch.
   b Find the exact value of the perimeter of the brooch. **(5 marks)**

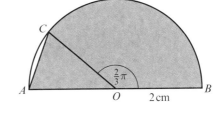

9 The points A and B lie on the circumference of a circle with centre O and radius 8.5 cm.
   The point C lies on the major arc AB. Given that $\angle ACB = 0.4$ radians, calculate the length of the minor arc AB.

10 In the diagram OAB is a sector of a circle, centre O and radius R cm, and $\angle AOB = 2\theta$ radians. A circle, centre C and radius r cm, touches the arc AB at T, and touches OA and OB at D and E respectively, as shown.
   a Write down, in terms of R and r, the length of OC. **(1 mark)**
   b Using $\triangle OCE$, show that $R \sin \theta = r(1 + \sin \theta)$. **(3 marks)**
   c Given that $\sin \theta = \frac{3}{4}$ and that the perimeter of the sector OAB is 21 cm, find r, giving your answer to 3 significant figures. **(7 marks)**

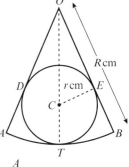

11 The diagram shows a sector AOB.
   The perimeter of the sector is twice the length of the arc AB.
   Find the size of angle AOB.

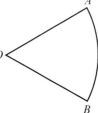

12 A circular Ferris wheel has 24 pods equally spaced on its circumference.
   Given the arc length between each pod is $\frac{3\pi}{2}$ m, and modelling each pod as a particle,
   a calculate the diameter of the Ferris wheel.
   Given that it takes approximately 30 seconds for a pod to complete one revolution,
   b estimate the speed of the pod in km/h.

Chapter 5

E/P 13 The diagram shows a triangular garden, PQR, with $PQ = 12$ m, $PR = 7$ m and $\angle QPR = 0.5$ radians. The curve SR is a small path separating the shaded patio area and the lawn, and is an arc of a circle with centre at P and radius 7 m.

Find:

a the length of the path SR (2 marks)

b the perimeter of the shaded patio, giving your answer to 3 significant figures. (4 marks)

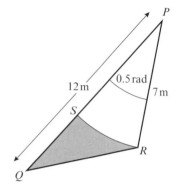

E/P 14 The shape XYZ shown is a design for an earring.

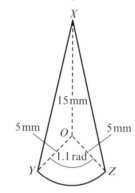

The straight lines XY and XZ are equal in length. The curve YZ is an arc of a circle with centre O and radius 5 mm. The size of $\angle YOZ$ is 1.1 radians and $XO = 15$ mm.

a Find the size of $\angle XOZ$, in radians, to 3 significant figures. (2 marks)

b Find the total perimeter of the earring, to the nearest mm. (6 marks)

5.3 Areas of sectors and segments

Using radians also greatly simplifies the formula for the area of a **sector**.

- To find the area A of a sector of a circle use the formula $A = \frac{1}{2}r^2\theta$, where r is the radius of the circle and θ is the angle, in radians, contained by the sector.

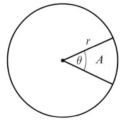

Notation A sector of a circle is the portion of a circle enclosed by two radii and an arc. The smaller area is known as the **minor** sector and the larger is known as the **major** sector.

Example 11

Find the area of the sector of a circle of radius 2.44 cm, given that the sector subtends an angle of 1.4 radians at the centre of the circle.

Area of sector $= \frac{1}{2} \times 2.44^2 \times 1.4$ — Use $A = \frac{1}{2}r^2\theta$ with $r = 2.44$ and $\theta = 1.4$.

$= 4.17$ cm² (3 s.f.)

122

Example 12

In the diagram, the area of the minor sector AOB is $28.9\,\text{cm}^2$.
Given that $\angle AOB = 0.8$ radians, calculate the value of r.

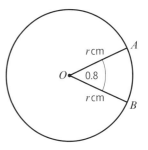

$28.9 = \dfrac{1}{2}r^2 \times 0.8 = 0.4r^2$

So $r^2 = \dfrac{28.9}{0.4} = 72.25$

$r = \sqrt{72.25} = 8.5$

— Let area of sector be $A\,\text{cm}^2$, and use $A = \dfrac{1}{2}r^2\theta$.

— Use the positive square root in this case as a length cannot be negative.

Example 13

A plot of land is in the shape of a sector of a circle of radius $55\,\text{m}$. The length of fencing that is erected along the edge of the plot to enclose the land is $176\,\text{m}$. Calculate the area of the plot of land.

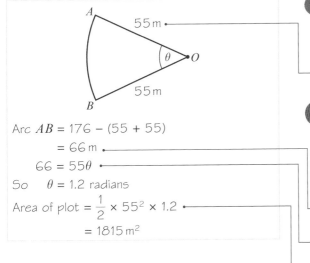

Arc $AB = 176 - (55 + 55)$
$\phantom{\text{Arc } AB} = 66\,\text{m}$
$66 = 55\theta$
So $\theta = 1.2$ radians
Area of plot $= \dfrac{1}{2} \times 55^2 \times 1.2$
$\phantom{\text{Area of plot }} = 1815\,\text{m}^2$

Online Explore the area of a segment using GeoGebra.

— Draw a diagram including all the data and let the angle of the sector be θ.

Problem-solving

In order to find the area of the sector, you need to know θ. Use the information about the perimeter to find the arc length AB.

— As the perimeter is given, first find length of arc AB.

— Use the formula for arc length, $l = r\theta$.

— Use the formula for area of a sector, $A = \dfrac{1}{2}r^2\theta$.

You can find the area of a **segment** by subtracting the area of triangle OPQ from the area of sector OPQ.

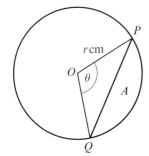

Using $\dfrac{1}{2}r^2\theta$ for the area of the sector and $\dfrac{1}{2}ab\sin\theta$ for the area of a triangle:

$A = \dfrac{1}{2}r^2\theta - \dfrac{1}{2}r^2\sin\theta$

$ = \dfrac{1}{2}r^2(\theta - \sin\theta)$

- The area of a segment in a circle of radius r is $A = \dfrac{1}{2}r^2(\theta - \sin\theta)$

Example 14

The diagram shows a sector of a circle. Find the area of the shaded segment.

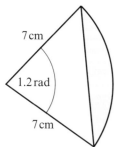

Area of segment $= \frac{1}{2} \times 7^2(1.2 - \sin 1.2)$

$= \frac{1}{2} \times 49 \times 0.26796...$

$= 6.57 \text{ cm}^2$ (3 s.f.)

Use $A = \frac{1}{2}r^2(\theta - \sin\theta)$ with $r = 7$ and $\theta = 1.2$ radians.
Make sure your calculator is in radians mode when calculating $\sin\theta$.

Example 15

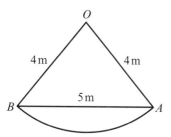

In the diagram above, OAB is a sector of a circle, radius 4 m. The chord AB is 5 m long. Find the area of the shaded segment.

Calculate angle AOB first:

$\cos\angle AOB = \dfrac{4^2 + 4^2 - 5^2}{2 \times 4 \times 4}$

$= \dfrac{7}{32}$

So $\angle AOB = 1.3502...$

Area of shaded segment

$= \frac{1}{2} \times 4^2(1.3502... - \sin 1.3502...)$

$= \frac{1}{2} \times 16 \times 0.37448...$

$= 3.00 \text{ m}^2$ (3 s.f.)

Problem-solving

In order to find the area of the segment you need to know angle AOB. You can use the cosine rule in triangle AOB, or divide the triangle into two right-angled triangles and use the trigonometric ratios.

Use the cosine rule for a non-right-angled triangle.

Watch out Use unrounded values in your calculations wherever possible to avoid rounding errors. You can use the memory function or answer button on your calculator.

Radians

Example 16

In the diagram, AB is the diameter of a circle of radius r cm, and $\angle BOC = \theta$ radians. Given that the area of $\triangle AOC$ is three times that of the shaded segment, show that $3\theta - 4\sin\theta = 0$.

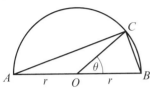

Area of segment = $\frac{1}{2}r^2(\theta - \sin\theta)$

Area of $\triangle AOC = \frac{1}{2}r^2\sin(\pi - \theta)$

$= \frac{1}{2}r^2\sin\theta$

So $\frac{1}{2}r^2\sin\theta = 3 \times \frac{1}{2}r^2(\theta - \sin\theta)$

$\sin\theta = 3(\theta - \sin\theta)$

So $3\theta - 4\sin\theta = 0$

— Area of segment = area of sector − area of triangle.

— $\angle AOB = \pi$ radians.

— Area of $\triangle AOC = 3 \times$ area of shaded segment.

Problem-solving

You might need to use circle theorems or properties when solving problems. The angle in a semicircle is a right angle so $\angle ACB = \dfrac{\pi}{2}$

Exercise 5D

1 Find the shaded area in each of the following circles. Leave your answers in terms of π where appropriate.

a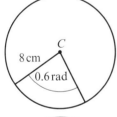

b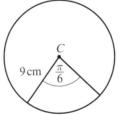

c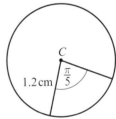

d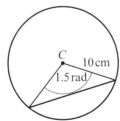

e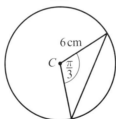

f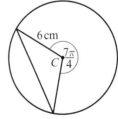

2 Find the shaded area in each of the following circles with centre C.

a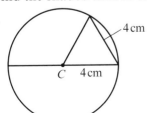

b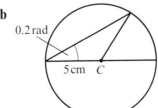

125

3 For the following circles with centre C, the area A of the shaded sector is given. Find the value of x in each case.

a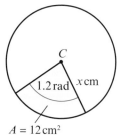

b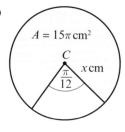

c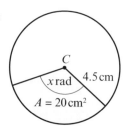

4 The arc AB of a circle, centre O and radius 6 cm, has length 4 cm.
Find the area of the minor sector AOB.

5 The chord AB of a circle, centre O and radius 10 cm, has length 18.65 cm and subtends an angle of θ radians at O.
   a Show that $\cos\theta = -0.739$ (to 3 significant figures).
   b Find the area of the minor sector AOB.

(P) 6 The area of a sector of a circle of radius 12 cm is 100 cm². Find the perimeter of the sector.

7 The arc AB of a circle, centre O and radius r cm, is such that $\angle AOB = 0.5$ radians.
Given that the perimeter of the minor sector AOB is 30 cm,
   a calculate the value of r
   b show that the area of the minor sector AOB is 36 cm²
   c calculate the area of the segment enclosed by the chord AB and the minor arc AB.

(P) 8 The arc AB of a circle, centre O and radius x cm, is such that angle $AOB = \dfrac{\pi}{12}$ radians.
Given that the arc length AB is l cm,
   a show that the area of the sector can be written as $\dfrac{6l^2}{\pi}$
   The area of the full circle is 3600π cm².
   b Find the arc length of AB.
   c Calculate the value of x.

(P) 9 In the diagram, AB is the diameter of a circle of radius r cm and $\angle BOC = \theta$ radians.
Given that the area of $\triangle COB$ is equal to that of the shaded segment, show that $\theta + 2\sin\theta = \pi$.

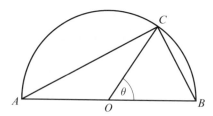

(P) 10 In the diagram, BC is the arc of a circle, centre O and radius 8 cm. The points A and D are such that $OA = OD = 5$ cm. Given that $\angle BOC = 1.6$ radians, calculate the area of the shaded region.

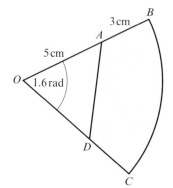

Radians

11 In the diagram, AB and AC are tangents to a circle, centre O and radius 3.6 cm. Calculate the area of the shaded region, given that $\angle BOC = \dfrac{2\pi}{3}$ radians.

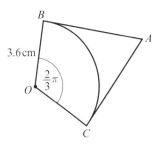

12 In the diagram, AD and BC are arcs of circles with centre O, such that $OA = OD = r$ cm, $AB = DC = 8$ cm and $\angle BOC = \theta$ radians.

a Given that the area of the shaded region is 48 cm², show that
$$r = \dfrac{6}{\theta} - 4$$ (4 marks)

b Given also that $r = 10\theta$, calculate the perimeter of the shaded region. (6 marks)

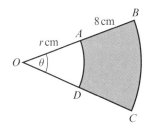

13 A sector of a circle of radius 28 cm has perimeter P cm and area A cm². Given that $A = 4P$, find the value of P.

14 The diagram shows a triangular plot of land. The sides AB, BC and CA have lengths 12 m, 14 m and 10 m respectively. The lawn is a sector of a circle, centre A and radius 6 m.

a Show that $\angle BAC = 1.37$ radians, correct to 3 significant figures.

b Calculate the area of the flowerbed.

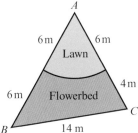

15 The diagram shows OPQ, a sector of a circle with centre O, radius 10 cm, with $\angle POQ = 0.3$ radians.

The point R is on OQ such that the ratio $OR:RQ$ is $1:3$. The region S, shown shaded in the diagram, is bounded by QR, RP and the arc PQ.

Find:

a the perimeter of S, giving your answer to 3 significant figures (6 marks)

b the area of S, giving your answer to 3 significant figures. (6 marks)

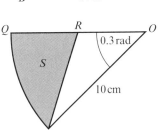

16 The diagram shows the sector OAB of a circle with centre O, radius 12 cm and angle 1.2 radians.

The line AC is a tangent to the circle with centre O, and OBC is a straight line.

The region R is bounded by the arc AB and the lines AC and CB.

a Find the area of R, giving your answer to 2 decimal places. (8 marks)

The line BD is parallel to AC.

b Find the perimeter of DAB. (5 marks)

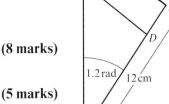

127

Chapter 5

P **17**

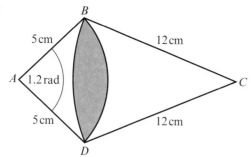

The diagram shows two intersecting sectors: ABD, with radius 5 cm and angle 1.2 radians, and CBD, with radius 12 cm.
Find the area of the overlapping section.

> **Challenge**
>
> Find an expression for the area of a sector of a circle with radius r and arc length l.

5.4 Solving trigonometric equations

In Year 1, you learned how to solve trigonometric equations in degrees. You can solve trigonometric equations in radians in the same way.

Example 17

Find the solutions of these equations in the interval $0 \leq \theta \leq 2\pi$:

a $\sin \theta = 0.3$ **b** $4 \cos \theta = 2$ **c** $5 \tan \theta + 3 = 1$

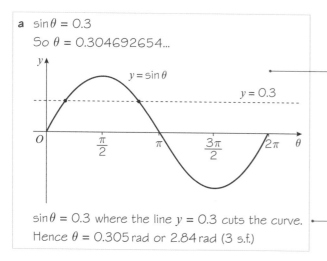

a $\sin \theta = 0.3$
So $\theta = 0.304692654...$

$\sin \theta = 0.3$ where the line $y = 0.3$ cuts the curve.
Hence $\theta = 0.305$ rad or 2.84 rad (3 s.f.)

— Draw the graph of $y = \sin \theta$ for the given interval.

— Find the first value using your calculator in radians mode.

— Since the sine curve is symmetrical in the interval $0 < \theta < \pi$, the second value is obtained by $\pi - 0.30469...$

b $4\cos\theta = 2$

$\cos\theta = \dfrac{1}{2}$

So $\theta = \dfrac{\pi}{3}$

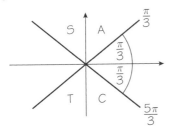

So $\theta = \dfrac{\pi}{3}$ or $\theta = 2\pi - \dfrac{\pi}{3} = \dfrac{5\pi}{3}$

c $5\tan\theta + 3 = 1$

$5\tan\theta = -2$

$\tan\theta = -0.4$

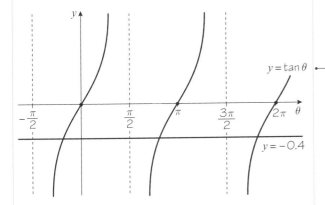

$\tan\theta = -0.4$ where the line $y = -0.4$ cuts the curve.

$\tan^{-1}(-0.4) = -0.3805...$ rad

So $\theta = 2.76108...$ rad (2.76 rad to 3 s.f.)

or $\theta = 5.90267...$ rad (5.90 rad to 3 s.f.)

Watch out When the interval is given in radians, make sure you answer in radians.

First rewrite in the form $\cos\theta = ...$

Use exact values where possible.

Putting $\dfrac{\pi}{3}$ in the four positions shown gives the angles $\dfrac{\pi}{3}, \dfrac{2\pi}{3}, \dfrac{4\pi}{3}$ and $\dfrac{5\pi}{3}$ but cosine is only positive in the 1st and 4th quadrants.

For the 2nd value, since we are working in radians, we use $2\pi - \theta$ instead of $360° - \theta$.

Draw the graph of $y = \tan\theta$ for the interval $-\dfrac{\pi}{2} < \theta < 2\pi$ since the principal value given by $\tan^{-1}(-0.4)$ is negative.

Use the symmetry and period of the tangent graph to find the required values.

Watch out Always check that your final values are within the given range; in this case $0 < \theta < 2\pi$ (remember $2\pi \approx 6.283...$)

Chapter 5

Example 18

Solve the equation $17\cos\theta + 3\sin^2\theta = 13$ in the interval $0 \leq \theta \leq 2\pi$.

$17\cos\theta + 3\sin^2\theta = 13$
$17\cos\theta + 3(1 - \cos^2\theta) = 13$
$17\cos\theta + 3 - 3\cos^2\theta = 13$
$0 = 3\cos^2\theta - 17\cos\theta + 10$
$0 = 3Y^2 - 17Y + 10$
$0 = (3Y - 2)(Y - 5)$
So $3Y - 2 = 0$ or $Y - 5 = 0$
$Y = \frac{2}{3}$ or $Y = 5$
i.e. $\cos\theta = \frac{2}{3}$ or $\cos\theta = 5$

$\cos\theta = \frac{2}{3}$

So $\theta = 0.841068...$ rad

Second solution is $2\pi - 0.841068...$
$= 5.442116...$
$\theta = 0.841$ or 5.44 (3 s.f.)

Problem-solving

Use the trigonometric identity $\sin^2\theta + \cos^2\theta \equiv 1$. Trigonometric identities work the same in radians as in degrees.

This is a quadratic so rearrange to make one side 0.

If $Y = \cos\theta$, then $Y^2 = \cos^2\theta$.

Solve the quadratic equation.

The value of $\cos\theta$ is between -1 and 1, so reject $\cos\theta = 5$.

Solve this equation to find θ.

Since the interval is given in radians, answer in radians.

Radians

Example 19

Solve the equation $\sin 3\theta = \dfrac{\sqrt{3}}{2}$, in the interval $0 \leqslant \theta \leqslant 2\pi$.

Let $X = 3\theta$ — Replace 3θ by X and solve as normal.

So $\sin X = \dfrac{\sqrt{3}}{2}$

As $X = 3\theta$, then the interval for X is $0 \leqslant X \leqslant 6\pi$ — Remember to transform the interval: $0 \leqslant \theta \leqslant 2\pi$ becomes $0 \leqslant 3\theta \leqslant 6\pi$

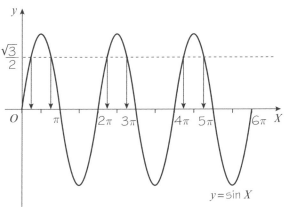

$X = \dfrac{\pi}{3}, \dfrac{2\pi}{3}, \dfrac{7\pi}{3}, \dfrac{8\pi}{3}, \dfrac{13\pi}{3}, \dfrac{14\pi}{3}$ — Remember $X = 3\theta$ so divide each value by 3.

i.e. $3\theta = \dfrac{\pi}{3}, \dfrac{2\pi}{3}, \dfrac{7\pi}{3}, \dfrac{8\pi}{3}, \dfrac{13\pi}{3}, \dfrac{14\pi}{3}$

So $\theta = \dfrac{\pi}{9}, \dfrac{2\pi}{9}, \dfrac{7\pi}{9}, \dfrac{8\pi}{9}, \dfrac{13\pi}{9}, \dfrac{14\pi}{9}$ — Always check that your solutions for θ are in the given interval for θ, in this case $0 \leqslant \theta \leqslant 2\pi$.

Exercise 5E

1 Solve the following equations for θ, in the interval $0 \leqslant \theta \leqslant 2\pi$, giving your answers to 3 significant figures where they are not exact.
   a $\cos\theta = 0.7$
   b $\sin\theta = -0.2$
   c $\tan\theta = 5$
   d $\cos\theta = -1$

2 Solve the following equations for θ, in the interval $0 \leqslant \theta \leqslant 2\pi$, giving your answers to 3 significant figures where they are not exact.
   a $4\sin\theta = 3$
   b $7\tan\theta = 1$
   c $8\tan\theta = 15$
   d $\sqrt{5}\cos\theta = \sqrt{2}$

3 Solve the following equations for θ, in the interval $0 \leqslant \theta \leqslant 2\pi$, giving your answers to 3 significant figures where they are not exact.
   a $5\cos\theta + 1 = 3$
   b $\sqrt{5}\sin\theta + 2 = 1$
   c $8\tan\theta - 5 = 5$
   d $\sqrt{7}\cos\theta - 1 = \sqrt{2}$

4 Solve the following equations for θ, giving your answers to 3 significant figures where appropriate, in the intervals indicated:
   a $\sqrt{3}\tan\theta - 1 = 0,\ -\pi \leqslant \theta \leqslant \pi$
   b $5\sin\theta = 1,\ -\pi \leqslant \theta \leqslant 2\pi$
   c $8\cos\theta = 5,\ -2\pi \leqslant \theta \leqslant 2\pi$
   d $3\cos\theta - 1 = 0.02,\ -\pi \leqslant \theta \leqslant 3\pi$
   e $0.4\tan\theta - 5 = -7,\ 0 \leqslant \theta \leqslant 4\pi$
   f $\cos\theta - 1 = -0.82,\ \dfrac{\pi}{2} \leqslant \theta \leqslant \dfrac{7\pi}{3}$

Chapter 5

5 Solve the following equations for θ, in the interval $0 \leq \theta \leq 2\pi$, giving your answers to 3 significant figures where they are not exact.

   a $5\cos 2\theta = 4$
   b $5\sin 3\theta + 3 = 1$
   c $\sqrt{3}\tan 4\theta - 5 = -4$
   d $\sqrt{10}\cos 2\theta + \sqrt{2} = 3\sqrt{2}$

6 Solve the following equations for θ, giving your answers to 3 significant figures where appropriate, in the intervals indicated.

   a $\sqrt{2}\sin 3\theta - 1 = 0$, $\quad -\pi \leq \theta \leq \pi$
   b $2\cos 4\theta = -1$, $\quad -\pi \leq \theta \leq 2\pi$
   c $8\tan 2\theta = 7$, $\quad -2\pi \leq \theta \leq 2\pi$
   d $6\cos 2\theta - 1 = 0.2$, $\quad -\pi \leq \theta \leq 3\pi$

(P) **7** Solve the following equations for θ, in the interval $0 \leq \theta \leq 2\pi$, giving your answers to 3 significant figures where they are not exact.

   a $4\cos^2 \theta = 2$
   b $3\tan^2 \theta + \tan \theta = 0$
   c $\cos^2 \theta - 2\cos \theta = 3$
   d $2\sin^2 2\theta - 5\cos 2\theta = -2$

(P) **8** Solve the following equations for θ, in the interval $0 \leq \theta \leq 2\pi$, giving your answers to 3 significant figures where they are not exact.

   a $\cos \theta + 2\sin^2 \theta + 1 = 0$
   b $10\sin^2 \theta = 3\cos^2 \theta$
   c $4\cos^2 \theta + 8\sin^2 \theta = 2\sin^2 \theta - 2\cos^2 \theta$
   d $2\sin^2 \theta - 7 + 12\cos \theta = 0$

9 Solve, for $0 \leq x < 2\pi$,

   a $\cos\left(x - \dfrac{\pi}{12}\right) = \dfrac{1}{\sqrt{2}}$
   b $\sin 3x = -\dfrac{1}{2}$
   c $\cos(2\theta + 0.2) = -0.2$, $-\dfrac{\pi}{2} \leq \theta \leq \dfrac{\pi}{2}$
   d $\tan\left(2\theta + \dfrac{\pi}{4}\right) = 1$, $0 \leq \theta \leq 2\pi$

(E/P) **10 a** Solve, for $-\pi \leq \theta < \pi$, $(1 + \tan \theta)(5\sin \theta - 2) = 0$. **(3 marks)**
   b Solve, for $0 \leq x < 2\pi$, $4\tan x = 5\sin x$. **(5 marks)**

(E) **11** Find all the solutions, in the interval $0 \leq x \leq 2\pi$, to the equation $8\cos^2 x + 6\sin x - 6 = 3$ giving each solution to one decimal place. **(6 marks)**

(E/P) **12** Find, for $0 \leq x \leq 2\pi$, all the solutions of $\cos^2 x - 1 = \dfrac{7}{2}\sin^2 x - 2$ giving each solution to one decimal place. **(6 marks)**

(E/P) **13** Show that the equation $8\sin^2 x + 4\sin x - 20 = 4$ has no solutions. **(3 marks)**

(E/P) **14 a** Show that the equation $\tan^2 x - 2\tan x - 6 = 0$ can be written as $\tan x = p \pm \sqrt{q}$ where p and q are numbers to be found. **(3 marks)**
   b Hence solve, for $0 \leq x \leq 3\pi$, the equation $\tan^2 x - 2\tan x - 6 = 0$ giving your answers to 1 decimal place where appropriate. **(5 marks)**

(E/P) **15** In the triangle ABC, $AB = 5$ cm, $AC = 4$ cm, $\angle ABC = 0.5$ radians and $\angle ACB = x$ radians.
   a Use the sine rule to find the value of $\sin x$, giving your answer to 3 decimal places. **(3 marks)**
   Given that there are two possible values of x,
   b find these values of x, giving your answers to 2 decimal places. **(3 marks)**

5.5 Small angle approximations

You can use radians to find **approximations** for the values of $\sin\theta$, $\cos\theta$ and $\tan\theta$.

- **When θ is small and measured in radians:**
  - $\sin\theta \approx \theta$
  - $\tan\theta \approx \theta$
  - $\cos\theta \approx 1 - \dfrac{\theta^2}{2}$

You can see why these approximations work by looking at the graphs of $y = \sin\theta$, $y = \cos\theta$ and $y = \tan\theta$ for values of θ close to 0.

Notation In mathematics 'small' is a relative concept. Consequently, there is not a fixed set of numbers which are small and a fixed set which are not. In this case, it is useful to think of small as being really close to 0.

Online Use technology to explore approximate values of $\sin\theta$, $\cos\theta$ and $\tan\theta$ for values of θ close to 0.

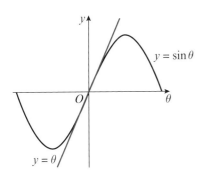
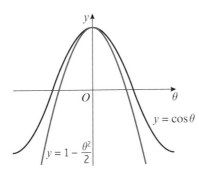
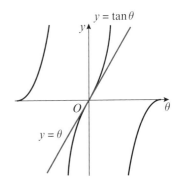

Example 20

When θ is small, find the approximate value of:

a $\dfrac{\sin 2\theta + \tan\theta}{2\theta}$

b $\dfrac{\cos 4\theta - 1}{\theta \sin 2\theta}$

a $\dfrac{\sin 2\theta + \tan\theta}{2\theta} \approx \dfrac{2\theta + \theta}{2\theta}$ ——— If $\sin\theta \approx \theta$ then $\sin 2\theta \approx 2\theta$

$= \dfrac{3\theta}{2\theta} = \dfrac{3}{2}$

When θ is small, $\dfrac{\sin 2\theta + \tan\theta}{2\theta} \approx \dfrac{3}{2}$ ——— Note that this approximation is only valid when θ is small and measured in radians.

b $\dfrac{\cos 4\theta - 1}{\theta \sin 2\theta} \approx \dfrac{1 - \dfrac{(4\theta)^2}{2} - 1}{\theta \times 2\theta}$ ——— $\cos\theta \approx 1 - \dfrac{\theta^2}{2}$ so $\cos 4\theta \approx 1 - \dfrac{(4\theta)^2}{2}$

$= \dfrac{1 - \dfrac{16\theta^2}{2} - 1}{2\theta^2}$

$= \dfrac{-\dfrac{16\theta^2}{2}}{2\theta^2} = -\dfrac{8\theta^2}{2\theta^2}$

$= -4$

Chapter 5

Example 21

a Show that, when θ is small, $\sin 5\theta + \tan 2\theta - \cos 2\theta \approx 2\theta^2 + 7\theta - 1$.

b Hence state the approximate value of $\sin 5\theta + \tan 2\theta - \cos 2\theta$ for small values of θ.

a For small values of θ:
$$\sin 5\theta + \tan 2\theta - \cos 2\theta \approx 5\theta + 2\theta - \left(1 - \frac{(2\theta)^2}{2}\right)$$
$$= 7\theta - 1 + \frac{4\theta^2}{2}$$
$$= 7\theta - 1 + 2\theta^2$$

When θ is small,
$\sin 5\theta + \tan 2\theta - \cos 2\theta \approx 2\theta^2 + 7\theta - 1$

b So, for small θ, $\sin 5\theta + \tan 2\theta - \cos 2\theta \approx -1$

Use the small angle approximations for sin, cos and tan.

When θ is small, terms in θ^2 and θ will also be small, so you can disregard the terms $2\theta^2$ and 7θ.

Exercise 5F

1 When θ is small, find the approximate values of:

a $\dfrac{\sin 4\theta - \tan 2\theta}{3\theta}$

b $\dfrac{1 - \cos 2\theta}{\tan 2\theta \sin \theta}$

c $\dfrac{3\tan \theta - \theta}{\sin 2\theta}$

2 When θ is small, show that:

a $\dfrac{\sin 3\theta}{\theta \sin 4\theta} = \dfrac{3}{4\theta}$

b $\dfrac{\cos \theta - 1}{\tan 2\theta} = -\dfrac{\theta}{4}$

c $\dfrac{\tan 4\theta + \theta^2}{3\theta - \sin 2\theta} = 4 + \theta$

3 a Find $\cos(0.244 \text{ rad})$ correct to 6 decimal places.
 b Use the approximation for $\cos \theta$ to find an approximate value for $\cos(0.244 \text{ rad})$.
 c Calculate the percentage error in your approximation.
 d Calculate the percentage error in the approximation for $\cos 0.75$ rad.
 e Explain the difference between your answers to parts **c** and **d**.

(P) **4** The percentage error for $\sin \theta$ for a given value of θ is 1%. Show that $100\theta = 101 \sin \theta$.

(E/P) **5 a** When θ is small, show that the expression $\dfrac{4\cos 3\theta - 2 + 5\sin \theta}{1 - \sin 2\theta}$ can be written as $9\theta + 2$. **(3 marks)**

 b Hence write down the value of $\dfrac{4\cos 3\theta - 2 + 5\sin \theta}{1 - \sin 2\theta}$ when θ is small. **(1 mark)**

Radians

Challenge

1 The diagram shows a right-angled triangle ABC. $\angle BAC = \theta$. An arc, CD, of the circle with centre A and radius AC has been drawn on the diagram in blue.

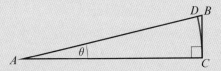

  a Write an expression for the arc length CD in terms of AC and θ.

  Given that θ is small so that, $AC = AD \approx AB$ and $CD \approx BC$,

  b deduce that $\sin\theta \approx \theta$ and $\tan\theta \approx \theta$.

2 **a** Using the binomial expansion and ignoring terms in x^4 and higher powers of x, find an approximation for $\sqrt{1-x^2}$, $|x| < 1$.

  b Hence show that for small θ, $\cos\theta \approx 1 - \dfrac{\theta^2}{2}$. You may assume that $\sin\theta \approx \theta$.

Mixed exercise 5

1 Triangle ABC is such that $AB = 5\,\text{cm}$, $AC = 10\,\text{cm}$ and $\angle ABC = 90°$. An arc of a circle, centre A and radius $5\,\text{cm}$, cuts AC at D.

  a State, in radians, the value of $\angle BAC$.

  b Calculate the area of the region enclosed by BC, DC and the arc BD.

2 The diagram shows the triangle OCD with $OC = OD = 17\,\text{cm}$ and $CD = 30\,\text{cm}$. The midpoint of CD is M. A semicircular arc A_1, with centre M is drawn, with CD as diameter. A circular arc A_2 with centre O and radius $17\,\text{cm}$, is drawn from C to D. The shaded region R is bounded by the arcs A_1 and A_2. Calculate, giving answers to 2 decimal places:

  a the area of the triangle OCD **(4 marks)**

  b the area of the shaded region R. **(5 marks)**

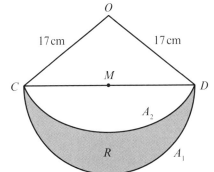

3 The diagram shows a circle, centre O, of radius $6\,\text{cm}$. The points A and B are on the circumference of the circle. The area of the shaded major sector is $80\,\text{cm}^2$. Given that $\angle AOB = \theta$ radians, where $0 < \theta < \pi$, calculate:

  a the value, to 3 decimal places, of θ **(3 marks)**

  b the length in cm, to 2 decimal places, of the minor arc AB. **(2 marks)**

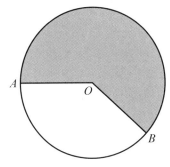

Chapter 5

E/P 4 The diagram shows a sector OAB of a circle, centre O and radius r cm.
The length of the arc AB is p cm and $\angle AOB$ is θ radians.

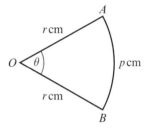

a Find θ in terms of p and r. **(2 marks)**

b Deduce that the area of the sector is $\frac{1}{2}pr$ cm². **(2 marks)**

Given that $r = 4.7$ and $p = 5.3$, where each has been measured to 1 decimal place, find, giving your answer to 3 decimal places:

c the least possible value of the area of the sector **(2 marks)**

d the range of possible values of θ. **(3 marks)**

E 5 The diagram shows a circle centre O and radius 5 cm.
The length of the minor arc AB is 6.4 cm.

a Calculate, in radians, the size of the acute angle AOB. **(2 marks)**

The area of the minor sector AOB is R_1 cm² and the area of the shaded major sector is R_2 cm².

b Calculate the value of R_1. **(2 marks)**

c Calculate $R_1 : R_2$ in the form $1:p$, giving the value of p to 3 significant figures. **(3 marks)**

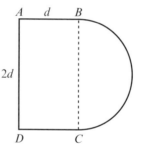

E/P 6 The diagrams show the cross-sections of two drawer handles. Shape X is a rectangle $ABCD$ joined to a semicircle with BC as diameter.
The length $AB = d$ cm and $BC = 2d$ cm.
Shape Y is a sector OPQ of a circle with centre O and radius $2d$ cm. Angle POQ is θ radians.

Given that the areas of shapes X and Y are equal,

a prove that $\theta = 1 + \frac{\pi}{4}$ **(5 marks)**

Shape X Shape Y

Using this value of θ, and given that $d = 3$, find in terms of π:

b the perimeter of shape X **(3 marks)**

c the perimeter of shape Y. **(3 marks)**

d Hence find the difference, in mm, between the perimeters of shapes X and Y. **(1 mark)**

E/P 7 The diagram shows a circle centre O and radius 6 cm.
The chord PQ divides the circle into a minor segment R_1 of area A_1 cm² and a major segment R_2 of area A_2 cm². The chord PQ subtends an angle θ radians at O.

a Show that $A_1 = 18(\theta - \sin\theta)$. **(2 marks)**

Given that $A_2 = 3A_1$,

b show that $\sin\theta = \theta - \frac{\pi}{2}$ **(4 marks)**

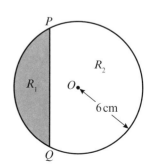

Radians

8 Triangle ABC has $AB = 9$ cm, $BC = 10$ cm and $CA = 5$ cm. A circle, centre A and radius 3 cm, intersects AB and AC at P and Q respectively, as shown in the diagram.

a Show that, to 3 decimal places,
$\angle BAC = 1.504$ radians. **(2 marks)**

b Calculate:
  i the area, in cm², of the sector APQ
  ii the area, in cm², of the shaded region $BPQC$
  iii the perimeter, in cm, of the shaded region $BPQC$. **(8 marks)**

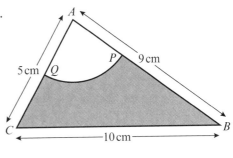

9 The diagram shows the sector OAB of a circle of radius r cm. The area of the sector is 15 cm² and $\angle AOB = 1.5$ radians.

a Prove that $r = 2\sqrt{5}$. **(2 marks)**

b Find, in cm, the perimeter of the sector OAB. **(3 marks)**

The segment R, shaded in the diagram, is enclosed by the arc AB and the straight line AB.

c Calculate, to 3 decimal places, the area of R. **(2 marks)**

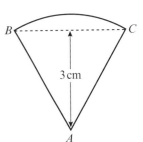

10 A badge is made in the shape of a sector ABC of a circle with centre A and radius AB, as shown in the diagram. The triangle ABC is equilateral and has perpendicular height 3 cm.

a Find, in surd form, the length of AB. **(2 marks)**

b Find, in terms of π, the area of the badge. **(2 marks)**

c Prove that the perimeter of the badge is $\dfrac{2\sqrt{3}}{3}(\pi + 6)$ cm. **(4 marks)**

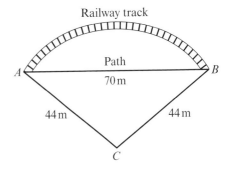

11 There is a straight path of length 70 m from the point A to the point B. The points are joined also by a railway track in the form of an arc of the circle whose centre is C and whose radius is 44 m, as shown in the diagram.

a Show that the size, to 2 decimal places, of $\angle ACB$ is 1.84 radians. **(2 marks)**

b Calculate:
  i the length of the railway track
  ii the shortest distance from C to the path
  iii the area of the region bounded by the railway track and the path. **(6 marks)**

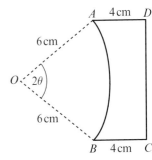

12 The diagram shows the cross-section $ABCD$ of a glass prism. $AD = BC = 4$ cm and both are at right angles to DC. AB is the arc of a circle, centre O and radius 6 cm. Given that $\angle AOB = 2\theta$ radians, and that the perimeter of the cross-section is $2(7 + \pi)$ cm,

a show that $(2\theta + 2\sin\theta - 1) = \dfrac{\pi}{3}$

b verify that $\theta = \dfrac{\pi}{6}$

c find the area of the cross-section.

137

Chapter 5

P) 13 Two circles C_1 and C_2, both of radius 12 cm have centres O_1 and O_2 respectively. O_1 lies on the circumference of C_2; O_2 lies on the circumference of C_1. The circles intersect at A and B, and enclose the region R.

 a Show that $\angle AO_1B = \dfrac{2\pi}{3}$

 b Hence write down, in terms of π, the perimeter of R.

 c Find the area of R, giving your answer to 3 significant figures.

E/P) 14 A teacher asks a student to find the area of the following sector. The attempt is shown below.

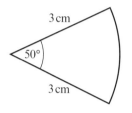

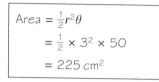

 a Identify the mistake made by the student. **(1 mark)**

 b Calculate the correct area of the sector. **(2 marks)**

15 When θ is small, find the approximate values of:

 a $\dfrac{\cos\theta - 1}{\theta \tan 2\theta}$

 b $\dfrac{2(1 - \cos\theta) - 1}{\tan\theta - 1}$

16 a When θ is small, show that the expression $\dfrac{7 + 2\cos 2\theta}{\tan 2\theta + 3}$ can be written as $3 - 2\theta$. **(3 marks)**

 b Hence write down the value of $\dfrac{7 + 2\cos 2\theta}{\tan 2\theta + 3}$ when θ is small. **(1 mark)**

E/P) 17 a When θ is small, show that the equation
$$32 \cos 5\theta + 203 \tan 10\theta = 182$$
can be written as
$$40\theta^2 - 203\theta + 15 = 0$$ **(4 marks)**

 b Hence, find the solutions of the equation
$$32 \cos 5\theta + 203 \tan 10\theta = 182$$ **(3 marks)**

 c Comment on the validity of your solutions. **(1 mark)**

P) 18 When θ is small, find the approximate value of $\cos^4 \theta - \sin^4 \theta$.

19 Solve the following equations for θ, giving your answers to 3 significant figures where appropriate, in the intervals indicated.

 a $3\sin\theta = 2, \; 0 \leqslant \theta \leqslant \pi$

 b $\sin\theta = -\cos\theta, \; -\pi \leqslant \theta \leqslant \pi$

 c $\tan\theta + \dfrac{1}{\tan\theta} = 2, \; 0 \leqslant \theta \leqslant 2\pi$

 d $2\sin^2\theta - \sin\theta - 1 = \sin^2\theta, \; -\pi \leqslant \theta \leqslant \pi$

20 a Sketch the graphs of $y = 5\sin x$ and $y = 3\cos x$ on the same axes ($0 \leq x \leq 2\pi$), marking on all the points where the graphs cross the axes.
   b Write down how many solutions there are in the given range for the equation $5\sin x = 3\cos x$.
   c Solve the equation $5\sin x = 3\cos x$ algebraically, giving your answers to 3 significant figures.

(E) 21 a Express $4\sin\theta - \cos\left(\dfrac{\pi}{2} - \theta\right)$ as a single trigonometric function. **(1 mark)**
   b Hence solve $4\sin\theta - \cos\left(\dfrac{\pi}{2} - \theta\right) = 1$ in the interval $0 \leq \theta \leq 2\pi$. Give your answers to 3 significant figures. **(3 marks)**

(E/P) 22 Find the values of x in the interval $0 < x < \dfrac{3\pi}{2}$ which satisfy the equation
$$\dfrac{\sin 2x + 0.5}{1 - \sin 2x} = 2$$ **(6 marks)**

(E/P) 23 A teacher asks two students to solve the equation $2\cos^2 x = 1$ for $-\pi \leq x \leq \pi$. The attempts are shown below.

Student A:
$\cos x = \pm\dfrac{1}{\sqrt{2}}$
Reject $-\dfrac{1}{\sqrt{2}}$ as cosine cannot be negative
$x = \dfrac{\pi}{4}$ or $x = -\dfrac{\pi}{4}$

Student B:
$2\cos^2 x = 1$
$\cos x = \pm\dfrac{1}{2}$
$x = \dfrac{\pi}{3}, -\dfrac{\pi}{3}, \dfrac{2\pi}{3}, -\dfrac{2\pi}{3}$

   a Identify the mistake made by Student A. **(1 mark)**
   b Identify the mistake made by Student B. **(1 mark)**
   c Calculate the correct solutions to the equation. **(4 marks)**

(E/P) 24 A teacher asks a student to solve the equation $2\tan 2x = 5$ for $0 \leq x \leq 2\pi$. The attempt is shown below.

$2\tan 2x = 5$
$\tan 2x = 2.5$
$2x = 1.19, 4.33$
$x = 0.595$ rad or 2.17 rad (3 s.f.)

Problem-solving
Solve the equation yourself then compare your working with the student's answer.

   a Identify the mistake made by the student. **(1 mark)**
   b Calculate the correct solutions to the equation. **(4 marks)**

(E/P) 25 a Show that the equation
$$5\sin x = 1 + 2\cos^2 x$$
can be written in the form
$$2\sin^2 x + 5\sin x - 3 = 0$$ **(2 marks)**
   b Solve, for $0 \leq x < 2\pi$,
$$2\sin^2 x + 5\sin x - 3 = 0$$ **(4 marks)**

Chapter 5

E/P 26 **a** Show that the equation
$$4\sin^2 x + 9\cos x - 6 = 0$$
can be written as
$$4\cos^2 x - 9\cos x + 2 = 0$$ (2 marks)

b Hence solve, for $0 \leqslant x < 4\pi$,
$$4\sin^2 x + 9\cos x - 6 = 0$$
giving your answers to 1 decimal place. (6 marks)

E/P 27 **a** Show that the equation
$$\tan 2x = 5 \sin 2x$$
can be written in the form
$$(1 - 5\cos 2x)\sin 2x = 0$$ (2 marks)

b Hence solve, for $0 \leqslant x \leqslant \pi$,
$$\tan 2x = 5\sin 2x$$
giving your answers to 1 decimal place where appropriate. You must show clearly how you obtained your answers. (5 marks)

E 28 **a** Sketch, for $0 \leqslant x \leqslant 2\pi$, the graph of $y = \cos\left(x + \dfrac{\pi}{6}\right)$. (2 marks)

b Write down the exact coordinates of the points where the graph meets the coordinate axes. (3 marks)

c Solve, for $0 \leqslant x \leqslant 2\pi$, the equation
$$y = \cos\left(x + \dfrac{\pi}{6}\right) = 0.65,$$
giving your answers in radians to 2 decimal places. (5 marks)

E 29 Solve, for $0 \leqslant x \leqslant \pi$, the equation
$$\sin\left(3x + \dfrac{\pi}{3}\right) = 0.45$$
giving your answers in radians to two decimal places. (5 marks)

Challenge

Use the small angle approximations to determine whether the following equations have any solutions close to $\theta = 0$. In each case, state whether each root of the resulting quadratic equation is likely to correspond to a solution of the original equation.

a $9\sin\theta\tan\theta + 25\tan\theta = 6$

b $2\tan\theta + 3 = 5\cos 4\theta$

c $\sin 4\theta = 37 - 2\cos 2\theta$

Radians

Summary of key points

1
- 2π radians = $360°$
- π radians = $180°$
- 1 radian = $\dfrac{180°}{\pi}$

2
- $30° = \dfrac{\pi}{6}$ radians
- $45° = \dfrac{\pi}{4}$ radians
- $60° = \dfrac{\pi}{3}$ radians
- $90° = \dfrac{\pi}{2}$ radians
- $180° = \pi$ radians
- $360° = 2\pi$ radians

3 You need to learn the exact values of the trigonometric ratios of these angles measured in radians.
- $\sin\dfrac{\pi}{6} = \dfrac{1}{2}$
- $\cos\dfrac{\pi}{6} = \dfrac{\sqrt{3}}{2}$
- $\tan\dfrac{\pi}{6} = \dfrac{1}{\sqrt{3}} = \dfrac{\sqrt{3}}{3}$
- $\sin\dfrac{\pi}{3} = \dfrac{\sqrt{3}}{2}$
- $\cos\dfrac{\pi}{3} = \dfrac{1}{2}$
- $\tan\dfrac{\pi}{3} = \sqrt{3}$
- $\sin\dfrac{\pi}{4} = \dfrac{1}{\sqrt{2}} = \dfrac{\sqrt{2}}{2}$
- $\cos\dfrac{\pi}{4} = \dfrac{1}{\sqrt{2}} = \dfrac{\sqrt{2}}{2}$
- $\tan\dfrac{\pi}{4} = 1$

4 You can use these rules to find sin, cos or tan of any positive or negative angle measured in radians using the corresponding acute angle made with the x-axis, θ.
- $\sin(\pi - \theta) = \sin\theta$
- $\sin(\pi + \theta) = -\sin\theta$
- $\sin(2\pi - \theta) = -\sin\theta$
- $\cos(\pi - \theta) = -\cos\theta$
- $\cos(\pi + \theta) = -\cos\theta$
- $\cos(2\pi - \theta) = \cos\theta$
- $\tan(\pi - \theta) = -\tan\theta$
- $\tan(\pi + \theta) = \tan\theta$
- $\tan(2\pi - \theta) = -\tan\theta$

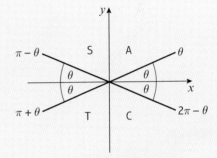

5 To find the arc length l of a sector of a circle use the formula $l = r\theta$, where r is the radius of the circle and θ is the angle, in radians, contained by the sector.

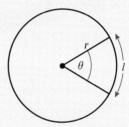

6 To find the area A of a sector of a circle use the formula $A = \tfrac{1}{2}r^2\theta$, where r is the radius of the circle and θ is the angle, in radians, contained by the sector.

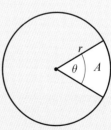

7 The area of a segment in a circle of radius r is
$$A = \tfrac{1}{2}r^2(\theta - \sin\theta)$$

8 When θ is small and measured in radians:
- $\sin\theta \approx \theta$
- $\tan\theta \approx \theta$
- $\cos\theta \approx 1 - \dfrac{\theta^2}{2}$

6 Trigonometric functions

Objectives

After completing this chapter you should be able to:

- Understand the definitions of secant, cosecant and cotangent and their relationship to cosine, sine and tangent → **pages 143–145**
- Understand the graphs of secant, cosecant and cotangent and their domain and range → **pages 145–149**
- Simplify expressions, prove simple identities and solve equations involving secant, cosecant and cotangent → **pages 149–153**
- Prove and use $\sec^2 x \equiv 1 + \tan^2 x$ and $\operatorname{cosec}^2 x \equiv 1 + \cot^2 x$ → **pages 153–157**
- Understand and use inverse trigonometric functions and their domain and ranges. → **pages 158–161**

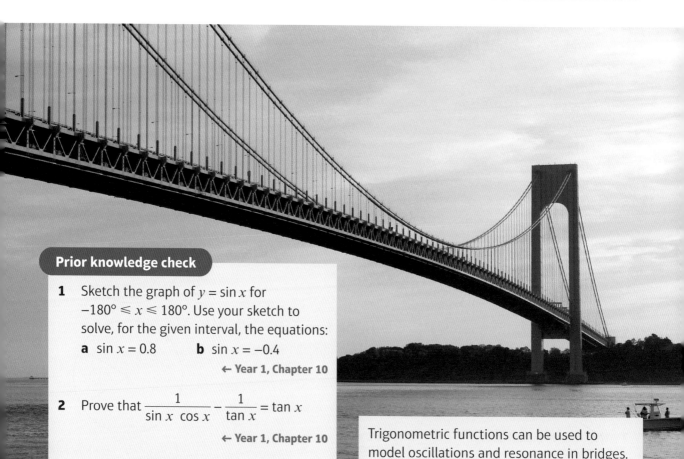

Prior knowledge check

1. Sketch the graph of $y = \sin x$ for $-180° \leq x \leq 180°$. Use your sketch to solve, for the given interval, the equations:
   a $\sin x = 0.8$ **b** $\sin x = -0.4$
   ← Year 1, Chapter 10

2. Prove that $\dfrac{1}{\sin x \, \cos x} - \dfrac{1}{\tan x} = \tan x$
   ← Year 1, Chapter 10

3. Find all the solutions in the interval $0 \leq x \leq 2\pi$ to the equation $3\sin^2(2x) = 1$.
   ← Section 5.5

Trigonometric functions can be used to model oscillations and resonance in bridges. You will use the functions in this chapter together with differentiation and integration in chapters 9 and 12.

Trigonometric functions

6.1 Secant, cosecant and cotangent

Secant (sec), cosecant (cosec) and cotangent (cot) are known as the **reciprocal** trigonometric functions.

- **sec** $x = \dfrac{1}{\cos x}$ (undefined for values of x for which $\cos x = 0$)

- **cosec** $x = \dfrac{1}{\sin x}$ (undefined for values of x for which $\sin x = 0$)

- **cot** $x = \dfrac{1}{\tan x}$ (undefined for values of x for which $\tan x = 0$)

You can also write $\cot x$ in terms of $\sin x$ and $\cos x$.

- **cot** $x = \dfrac{\cos x}{\sin x}$

Example 1

Use your calculator to write down the values of:

a $\sec 280°$ b $\cot 115°$

a $\sec 280° = \dfrac{1}{\cos 280°} = 5.76$ (3 s.f.) —— Make sure your calculator is in degrees mode.

b $\cot 115° = \dfrac{1}{\tan 115°} = -0.466$ (3 s.f.)

Example 2

Work out the exact values of:

a $\sec 210°$ b $\operatorname{cosec} \dfrac{3\pi}{4}$ —— *Exact* here means give in surd form.

a $\sec 210° = \dfrac{1}{\cos 210°}$

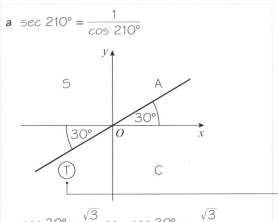

210° is in 3rd quadrant, so $\cos 210° = -\cos 30°$.

$\cos 30° = \dfrac{\sqrt{3}}{2}$ so $-\cos 30° = -\dfrac{\sqrt{3}}{2}$

So $\sec 210° = -\dfrac{2}{\sqrt{3}}$ —— Or $\sec 210° = -\dfrac{2\sqrt{3}}{3}$ if you rationalise the denominator.

143

b $\operatorname{cosec} \dfrac{3\pi}{4} = \dfrac{1}{\sin\left(\dfrac{3\pi}{4}\right)}$

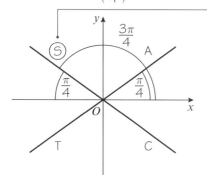

$\dfrac{3\pi}{4}$ is in the 2nd quadrant, so $\sin \dfrac{3\pi}{4} = +\sin \dfrac{\pi}{4}$

So $\operatorname{cosec} \dfrac{3\pi}{4} = \dfrac{1}{\sin\left(\dfrac{\pi}{4}\right)}$

$\sin\left(\dfrac{\pi}{4}\right) = \dfrac{1}{\sqrt{2}}$

So $\operatorname{cosec}\left(\dfrac{3\pi}{4}\right) = \sqrt{2}$

Exercise 6A

1. Without using your calculator, write down the sign of the following trigonometric ratios.
   - **a** sec 300°
   - **b** cosec 190°
   - **c** cot 110°
   - **d** cot 200°
   - **e** sec 95°

2. Use your calculator to find, to 3 significant figures, the values of:
   - **a** sec 100°
   - **b** cosec 260°
   - **c** cosec 280°
   - **d** cot 550°
   - **e** cot $\dfrac{4\pi}{3}$
   - **f** sec 2.4 rad
   - **g** cosec $\dfrac{11\pi}{10}$
   - **h** sec 6 rad

3. Find the exact values (in surd form where appropriate) of the following:
   - **a** cosec 90°
   - **b** cot 135°
   - **c** sec 180°
   - **d** sec 240°
   - **e** cosec 300°
   - **f** cot(−45°)
   - **g** sec 60°
   - **h** cosec (−210°)
   - **i** sec 225°
   - **j** cot $\dfrac{4\pi}{3}$
   - **k** sec $\dfrac{11\pi}{6}$
   - **l** cosec $\left(-\dfrac{3\pi}{4}\right)$

(P) 4. Prove that cosec $(\pi - x) \equiv$ cosec x.

(P) 5. Show that cot 30° sec 30° = 2.

(P) 6. Show that cosec $\dfrac{2\pi}{3}$ + sec $\dfrac{2\pi}{3}$ = $a + b\sqrt{3}$ where a and b are real numbers to be found.

Trigonometric functions

Challenge

The point P lies on the unit circle, centre O. The radius OP makes an acute angle of θ with the positive x-axis. The tangent to the circle at P intersects the coordinate axes at points A and B.

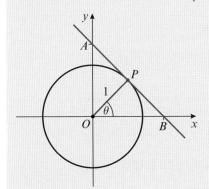

Prove that
a $OB = \sec \theta$
b $OA = \text{cosec } \theta$
c $AP = \cot \theta$

6.2 Graphs of sec x, cosec x and cot x

You can use the graphs of $y = \cos x$, $y = \sin x$ and $y = \tan x$ to sketch the graphs of their reciprocal functions.

Example 3

Sketch, in the interval $-180° \leq \theta \leq 180°$, the graph of $y = \sec \theta$.

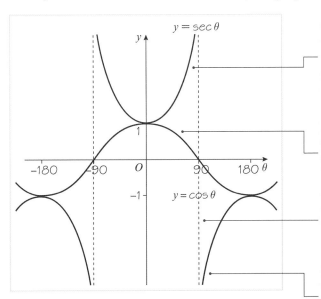

First draw the graph $y = \cos \theta$.

For each value of θ, the value of $\sec \theta$ is the reciprocal of the corresponding value of $\cos \theta$.

In particular: $\cos 0° = 1$, so $\sec 0° = 1$; and $\cos 180° = -1$, so $\sec 180° = -1$.

As θ approaches 90° from the left, $\cos \theta$ is +ve but approaches zero, and so $\sec \theta$ is +ve but becomes increasingly large.

At $\theta = 90°$, $\sec \theta$ is undefined and there is a vertical asymptote. This is also true for $\theta = -90°$.

As θ approaches 90° from the right, $\cos \theta$ is −ve but approaches zero, and so $\sec \theta$ is −ve but becomes increasingly large negative.

- The graph of $y = \sec x$, $x \in \mathbb{R}$, has symmetry in the y-axis and has period 360° or 2π radians. It has vertical asymptotes at all the values of x for which $\cos x = 0$.

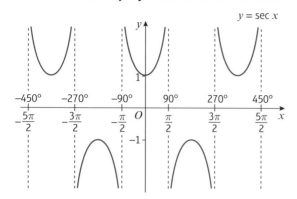

> **Notation** The domain can also be given as
> $x \in \mathbb{R}, x \neq \dfrac{(2n+1)\pi}{2}, n \in \mathbb{Z}$
>
> $\mathbb{Z}$ is the symbol used for **integers**, i.e. positive and negative whole numbers including 0.

- The domain of $y = \sec x$ is $x \in \mathbb{R}$, $x \neq 90°, 270°, 450°,\ldots$ or any odd multiple of 90°
- In radians the domain is $x \in \mathbb{R}$, $x \neq \dfrac{\pi}{2}, \dfrac{3\pi}{2}, \dfrac{5\pi}{2},\ldots$ or any odd multiple of $\dfrac{\pi}{2}$
- The range of $y = \sec x$ is $y \leqslant -1$ or $y \geqslant 1$

- The graph of $y = \operatorname{cosec} x$, $x \in \mathbb{R}$, has period 360° or 2π radians. It has vertical asymptotes at all the values of x for which $\sin x = 0$.

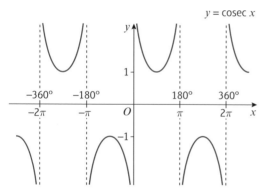

> **Notation** The domain can also be given as $x \in \mathbb{R}, x \neq n\pi, n \in \mathbb{Z}$.

- The domain of $y = \operatorname{cosec} x$ is $x \in \mathbb{R}$, $x \neq 0°, 180°, 360°,\ldots$ or any multiple of 180°
- In radians the domain is $x \in \mathbb{R}$, $x \neq 0, \pi, 2\pi,\ldots$ or any multiple of π
- The range of $y = \operatorname{cosec} x$ is $y \leqslant -1$ or $y \geqslant 1$

- The graph of $y = \cot x$, $x \in \mathbb{R}$, has period 180° or π radians. It has vertical asymptotes at all the values of x for which $\tan x = 0$.

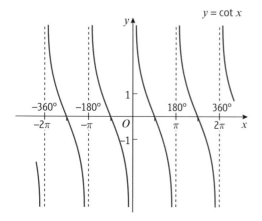

Trigonometric functions

- **The domain of $y = \cot x$ is $x \in \mathbb{R}$, $x \neq 0°$, $180°$, $360°$,... or any multiple of $180°$**
- **In radians the domain is $x \in \mathbb{R}$, $x \neq 0, \pi, 2\pi, ...$ or any multiple of π**
- **The range of $y = \cot x$ is $y \in \mathbb{R}$**

Notation The domain can also be given as $x \in \mathbb{R}, x \neq n\pi, n \in \mathbb{Z}$.

Example 4

a Sketch the graph of $y = 4\operatorname{cosec} x$, $-\pi \leqslant x \leqslant \pi$.
b On the same axes, sketch the line $y = x$.
c State the number of solutions to the equation $4\operatorname{cosec} x - x = 0$, $-\pi \leqslant x \leqslant \pi$.

a, b

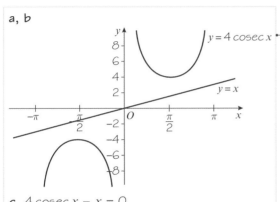

$y = 4\operatorname{cosec} x$ is a stretch of the graph of $y = \operatorname{cosec} x$, scale factor 4 in the y-direction.
You only need to draw the graph for $-\pi \leqslant x \leqslant \pi$.

c $4\operatorname{cosec} x - x = 0$
$4\operatorname{cosec} x = x$

$y = 4\operatorname{cosec} x$ and $y = x$ do not intersect for $-\pi \leqslant x \leqslant \pi$ so the equation has no solutions in the given range.

Problem-solving

The solutions to the equation $f(x) = g(x)$ correspond to the points of intersection of the graphs of $y = f(x)$ and $y = g(x)$.

Example 5

Sketch, in the interval $0 \leqslant \theta \leqslant 360°$, the graph of $y = 1 + \sec 2\theta$.

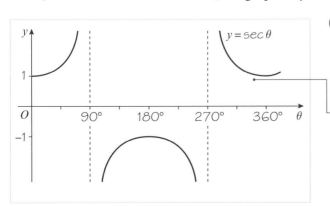

Online Explore transformations of the graphs of reciprocal trigonometric functions using technology.

Step 1
Draw the graph of $y = \sec \theta$.

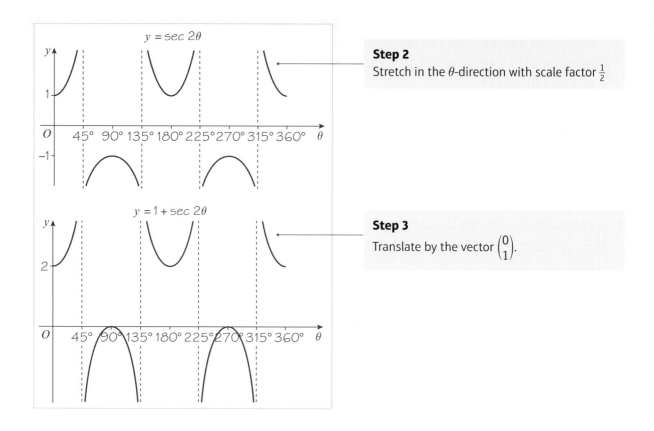

Exercise 6B

1 Sketch, in the interval $-540° \leq \theta \leq 540°$, the graphs of:
  a $y = \sec \theta$ **b** $y = \csc \theta$ **c** $y = \cot \theta$

2 a Sketch, on the same set of axes, in the interval $-\pi \leq x \leq \pi$, the graphs of $y = \cot x$ and $y = -x$.
  b Deduce the number of solutions of the equation $\cot x + x = 0$ in the interval $-\pi \leq x \leq \pi$.

3 a Sketch, on the same set of axes, in the interval $0 \leq \theta \leq 360°$, the graphs of $y = \sec \theta$ and $y = -\cos \theta$.
  b Explain how your graphs show that $\sec \theta = -\cos \theta$ has no solutions.

4 a Sketch, on the same set of axes, in the interval $0 \leq \theta \leq 360°$, the graphs of $y = \cot \theta$ and $y = \sin 2\theta$.
  b Deduce the number of solutions of the equation $\cot \theta = \sin 2\theta$ in the interval $0 \leq \theta \leq 360°$.

5 a Sketch on separate axes, in the interval $0 \leq \theta \leq 360°$, the graphs of $y = \tan \theta$ and $y = \cot(\theta + 90°)$.
  b Hence, state a relationship between $\tan \theta$ and $\cot(\theta + 90°)$.

Trigonometric functions

6 a Describe the relationships between the graphs of:

  i $y = \tan\left(\theta + \frac{\pi}{2}\right)$ and $y = \tan\theta$ **ii** $y = \cot(-\theta)$ and $y = \cot\theta$

  iii $y = \text{cosec}\left(\theta + \frac{\pi}{4}\right)$ and $y = \text{cosec}\,\theta$ **iv** $y = \sec\left(\theta - \frac{\pi}{4}\right)$ and $y = \sec\theta$

b By considering the graphs of $y = \tan\left(\theta + \frac{\pi}{2}\right)$, $y = \cot(-\theta)$, $y = \text{cosec}\left(\theta + \frac{\pi}{4}\right)$ and $y = \sec\left(\theta - \frac{\pi}{4}\right)$, state which pairs of functions are equal.

7 Sketch on separate axes, in the interval $0 \leq \theta \leq 360°$, the graphs of:

  a $y = \sec 2\theta$ **b** $y = -\text{cosec}\,\theta$ **c** $y = 1 + \sec\theta$
  d $y = \text{cosec}(\theta - 30°)$ **e** $y = 2\sec(\theta - 60°)$ **f** $y = \text{cosec}(2\theta + 60°)$
  g $y = -\cot(2\theta)$ **h** $y = 1 - 2\sec\theta$

In each case show the coordinates of any maximum and minimum points, and of any points at which the curve meets the axes.

8 Write down the periods of the following functions. Give your answers in terms of π.

  a $\sec 3\theta$ **b** $\text{cosec}\,\frac{1}{2}\theta$ **c** $2\cot\theta$ **d** $\sec(-\theta)$

9 a Sketch, in the interval $-2\pi \leq x \leq 2\pi$, the graph of $y = 3 + 5\,\text{cosec}\,x$. **(3 marks)**

  b Hence deduce the range of values of k for which the equation $3 + 5\,\text{cosec}\,x = k$ has no solutions. **(2 marks)**

10 a Sketch the graph of $y = 1 + 2\sec\theta$ in the interval $-\pi \leq \theta \leq 2\pi$. **(3 marks)**

  b Write down the θ-coordinates of points at which the gradient is zero. **(2 marks)**

  c Deduce the maximum and minimum values of $\dfrac{1}{1 + 2\sec\theta}$, and give the smallest positive values of θ at which they occur. **(4 marks)**

6.3 Using sec x, cosec x and cot x

You need to be able to simplify expressions, prove identities and solve equations involving $\sec x$, $\text{cosec}\,x$ and $\cot x$.

- $\sec x = k$ and $\text{cosec}\,x = k$ have no solutions for $-1 < k < 1$.

Example 6

Simplify:

a $\sin\theta\cot\theta\sec\theta$

b $\sin\theta\cos\theta(\sec\theta + \text{cosec}\,\theta)$

149

a $\sin\theta \cot\theta \sec\theta$

$\equiv \sin\theta \times \dfrac{\cos\theta}{\sin\theta} \times \dfrac{1}{\cos\theta}$ — Write the expression in terms of sin and cos, using $\cot\theta \equiv \dfrac{\cos\theta}{\sin\theta}$ and $\sec\theta \equiv \dfrac{1}{\cos\theta}$

$\equiv 1$

b $\sec\theta + \text{cosec}\,\theta \equiv \dfrac{1}{\cos\theta} + \dfrac{1}{\sin\theta}$ — Write the expression in terms of sin and cos, using $\sec\theta \equiv \dfrac{1}{\cos\theta}$ and $\text{cosec}\,\theta \equiv \dfrac{1}{\sin\theta}$

$\equiv \dfrac{\sin\theta + \cos\theta}{\sin\theta \cos\theta}$ — Put over common denominator.

So $\sin\theta \cos\theta (\sec\theta + \text{cosec}\,\theta) = \sin\theta + \cos\theta$ — Multiply both sides by $\sin\theta \cos\theta$.

Example 7

a Prove that $\dfrac{\cot\theta \,\text{cosec}\,\theta}{\sec^2\theta + \text{cosec}^2\,\theta} \equiv \cos^3\theta$.

b Hence explain why the equation $\dfrac{\cot\theta \,\text{cosec}\,\theta}{\sec^2\theta + \text{cosec}^2\,\theta} = 8$ has no solutions.

a Consider LHS:

The numerator $\cot\theta \,\text{cosec}\,\theta$

$\equiv \dfrac{\cos\theta}{\sin\theta} \times \dfrac{1}{\sin\theta} \equiv \dfrac{\cos\theta}{\sin^2\theta}$ — Write the expression in terms of sin and cos, using $\cot\theta \equiv \dfrac{\cos\theta}{\sin\theta}$ and $\text{cosec}\,\theta = \dfrac{1}{\sin\theta}$

The denominator $\sec^2\theta + \text{cosec}^2\,\theta$

$\equiv \dfrac{1}{\cos^2\theta} + \dfrac{1}{\sin^2\theta}$ — Write the expression in terms of sin and cos, using $\sec^2\theta \equiv \left(\dfrac{1}{\cos\theta}\right)^2 \equiv \dfrac{1}{\cos^2\theta}$ and $\text{cosec}^2\,\theta \equiv \dfrac{1}{\sin^2\theta}$

$\equiv \dfrac{\sin^2\theta + \cos^2\theta}{\cos^2\theta \sin^2\theta}$

$\equiv \dfrac{1}{\cos^2\theta \sin^2\theta}$ — Remember that $\sin^2\theta + \cos^2\theta \equiv 1$.

So $\dfrac{\cot\theta \,\text{cosec}\,\theta}{\sec^2\theta + \text{cosec}^2\,\theta}$

$\equiv \left(\dfrac{\cos\theta}{\sin^2\theta}\right) \div \left(\dfrac{1}{\cos^2\theta \sin^2\theta}\right)$

$\equiv \dfrac{\cos\theta}{\sin^2\theta} \times \dfrac{\cos^2\theta \sin^2\theta}{1}$ — Remember to invert the fraction when changing from $\div$ sign to $\times$.

$\equiv \cos^3\theta$

b Since $\dfrac{\cot\theta \,\text{cosec}\,\theta}{\sec^2\theta + \text{cosec}^2\,\theta} \equiv \cos^3\theta$ we are required to solve the equation $\cos^3\theta = 8$.

$\cos^3\theta = 8 \Rightarrow \cos\theta = 2$ which has no solutions since $-1 \leqslant \cos\theta \leqslant 1$.

Problem-solving

Write down the equivalent equation, and state the range of possible values for $\cos\theta$.

Example 8

Solve the equations

a $\sec\theta = -2.5$ **b** $\cot 2\theta = 0.6$

in the interval $0 \leq \theta \leq 360°$.

a $\dfrac{1}{\cos\theta} = -2.5$ — Substitute $\dfrac{1}{\cos\theta}$ for $\sec\theta$ and then simplify to get an equation in the form $\cos\theta = k$.

$\cos\theta = \dfrac{1}{-2.5} = -0.4$

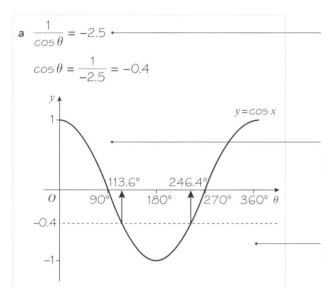

Sketch the graph of $y = \cos x$ for the given interval. The graph is symmetrical about $\theta = 180°$. Find the principal value using your calculator then subtract this from 360° to find the second solution.

You could also find all the solutions using a CAST diagram. This method is shown for part **b** below.

$\theta = 113.6°, 246.4° = 114°, 246°$ (3 s.f.) — Calculate angles from the diagram.

b $\dfrac{1}{\tan 2\theta} = 0.6$ — Substitute $\dfrac{1}{\tan 2\theta}$ for $\cot 2\theta$ and then simplify to get an equation in the form $\tan 2\theta = k$.

$\tan 2\theta = \dfrac{1}{0.6} = \dfrac{5}{3}$

Let $X = 2\theta$, so that you are solving $\tan X = \dfrac{5}{3}$, in the interval $0 \leq X \leq 720°$.

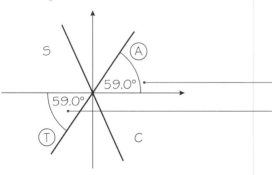

Draw the CAST diagram, with the acute angle $X = \tan^{-1}\dfrac{5}{3}$ drawn to the horizontal in the 1st and 3rd quadrants.

$X = 59.0°, 239.0°, 419.0°, 599.0°$ — Remember that $X = 2\theta$.

So $\theta = 29.5°, 120°, 210°, 300°$ (3 s.f.)

Exercise 6C

1 Rewrite the following as powers of $\sec\theta$, $\text{cosec}\,\theta$ or $\cot\theta$.

a $\dfrac{1}{\sin^3\theta}$ b $\dfrac{4}{\tan^6\theta}$ c $\dfrac{1}{2\cos^2\theta}$ d $\dfrac{1-\sin^2\theta}{\sin^2\theta}$

e $\dfrac{\sec\theta}{\cos^4\theta}$ f $\sqrt{\text{cosec}^3\,\theta\cot\theta\sec\theta}$ g $\dfrac{2}{\sqrt{\tan\theta}}$ h $\dfrac{\text{cosec}^2\,\theta\tan^2\theta}{\cos\theta}$

2 Write down the value(s) of $\cot x$ in each of the following equations.

a $5\sin x = 4\cos x$ b $\tan x = -2$ c $3\dfrac{\sin x}{\cos x} = \dfrac{\cos x}{\sin x}$

3 Using the definitions of **sec**, **cosec**, **cot** and **tan** simplify the following expressions.

a $\sin\theta\cot\theta$
b $\tan\theta\cot\theta$
c $\tan 2\theta\,\text{cosec}\,2\theta$
d $\cos\theta\sin\theta(\cot\theta + \tan\theta)$
e $\sin^3 x\,\text{cosec}\,x + \cos^3 x\sec x$
f $\sec A - \sec A\sin^2 A$
g $\sec^2 x\cos^5 x + \cot x\,\text{cosec}\,x\sin^4 x$

(P) 4 Prove that:

a $\cos\theta + \sin\theta\tan\theta \equiv \sec\theta$
b $\cot\theta + \tan\theta \equiv \text{cosec}\,\theta\sec\theta$
c $\text{cosec}\,\theta - \sin\theta \equiv \cos\theta\cot\theta$
d $(1-\cos x)(1+\sec x) \equiv \sin x\tan x$
e $\dfrac{\cos x}{1-\sin x} + \dfrac{1-\sin x}{\cos x} \equiv 2\sec x$
f $\dfrac{\cos\theta}{1+\cot\theta} \equiv \dfrac{\sin\theta}{1+\tan\theta}$

(P) 5 Solve, for values of θ in the interval $0 \leqslant \theta \leqslant 360°$, the following equations. Give your answers to 3 significant figures where necessary.

a $\sec\theta = \sqrt{2}$ b $\text{cosec}\,\theta = -3$ c $5\cot\theta = -2$ d $\text{cosec}\,\theta = 2$

e $3\sec^2\theta - 4 = 0$ f $5\cos\theta = 3\cot\theta$ g $\cot^2\theta - 8\tan\theta = 0$ h $2\sin\theta = \text{cosec}\,\theta$

(P) 6 Solve, for values of θ in the interval $-180° \leqslant \theta \leqslant 180°$, the following equations:

a $\text{cosec}\,\theta = 1$ b $\sec\theta = -3$ c $\cot\theta = 3.45$

d $2\,\text{cosec}^2\,\theta - 3\,\text{cosec}\,\theta = 0$ e $\sec\theta = 2\cos\theta$ f $3\cot\theta = 2\sin\theta$

g $\text{cosec}\,2\theta = 4$ h $2\cot^2\theta - \cot\theta - 5 = 0$

(P) 7 Solve the following equations for values of θ in the interval $0 \leqslant \theta \leqslant 2\pi$. Give your answers in terms of π.

a $\sec\theta = -1$ b $\cot\theta = -\sqrt{3}$

c $\text{cosec}\,\tfrac{1}{2}\theta = \dfrac{2\sqrt{3}}{3}$ d $\sec\theta = \sqrt{2}\tan\theta\left(\theta \neq \dfrac{\pi}{2}, \theta \neq \dfrac{3\pi}{2}\right)$

8 In the diagram $AB = 6$ cm is the diameter of the circle and BT is the tangent to the circle at B. The chord AC is extended to meet this tangent at D and $\angle DAB = \theta$.

  a Show that $CD = 6(\sec\theta - \cos\theta)$ cm. **(4 marks)**

  b Given that $CD = 16$ cm, calculate the length of the chord AC. **(3 marks)**

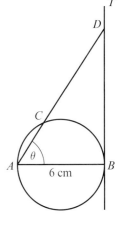

> **Problem-solving**
> AB is the diameter of the circle, so $\angle ACB = 90°$.

9 a Prove that $\dfrac{\operatorname{cosec} x - \cot x}{1 - \cos x} \equiv \operatorname{cosec} x.$ **(4 marks)**

  b Hence solve, in the interval $-\pi \leqslant x \leqslant \pi$, the equation $\dfrac{\operatorname{cosec} x - \cot x}{1 - \cos x} = 2.$ **(3 marks)**

10 a Prove that $\dfrac{\sin x \tan x}{1 - \cos x} - 1 \equiv \sec x.$ **(4 marks)**

  b Hence explain why the equation $\dfrac{\sin x \tan x}{1 - \cos x} - 1 = -\dfrac{1}{2}$ has no solutions. **(1 mark)**

11 Solve, in the interval $0 \leqslant x \leqslant 360°$, the equation $\dfrac{1 + \cot x}{1 + \tan x} = 5.$ **(8 marks)**

> **Problem-solving**
> Use the relationship $\cot x = \dfrac{1}{\tan x}$ to form a quadratic equation in $\tan x$.
> ← Year 1, Section 10.5

6.4 Trigonometric identities

You can use the identity $\sin^2 x + \cos^2 x \equiv 1$ to prove the following identities.

- **$1 + \tan^2 x \equiv \sec^2 x$**
- **$1 + \cot^2 x \equiv \operatorname{cosec}^2 x$**

> **Link** You can use the unit circle definitions of sin and cos to prove the identity $\sin^2 x + \cos^2 x \equiv 1$. ← Year 1, Section 10.5

Example 9

a Prove that $1 + \tan^2 x \equiv \sec^2 x$.

b Prove that $1 + \cot^2 x \equiv \operatorname{cosec}^2 x$.

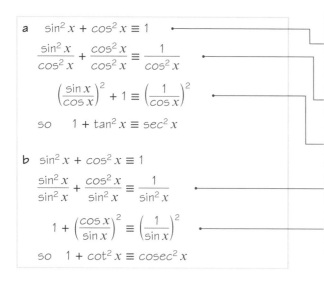

a $\sin^2 x + \cos^2 x \equiv 1$ — Unless otherwise stated, you can assume the identity $\sin^2 x + \cos^2 x \equiv 1$ in proofs involving cosec, sec and cot in your exam.

$\dfrac{\sin^2 x}{\cos^2 x} + \dfrac{\cos^2 x}{\cos^2 x} \equiv \dfrac{1}{\cos^2 x}$

$\left(\dfrac{\sin x}{\cos x}\right)^2 + 1 \equiv \left(\dfrac{1}{\cos x}\right)^2$ — Divide both sides of the identity by $\cos^2 x$.

so $1 + \tan^2 x \equiv \sec^2 x$

Use $\tan x \equiv \dfrac{\sin x}{\cos x}$ and $\sec x \equiv \dfrac{1}{\cos x}$

b $\sin^2 x + \cos^2 x \equiv 1$

$\dfrac{\sin^2 x}{\sin^2 x} + \dfrac{\cos^2 x}{\sin^2 x} \equiv \dfrac{1}{\sin^2 x}$ — Divide both sides of the identity by $\sin^2 x$.

$1 + \left(\dfrac{\cos x}{\sin x}\right)^2 \equiv \left(\dfrac{1}{\sin x}\right)^2$

Use $\cot x \equiv \dfrac{\cos x}{\sin x}$ and $\csc x \equiv \dfrac{1}{\sin x}$.

so $1 + \cot^2 x \equiv \csc^2 x$

Example 10

Given that $\tan A = -\dfrac{5}{12}$, and that angle A is obtuse, find the exact values of:

a $\sec A$ **b** $\sin A$

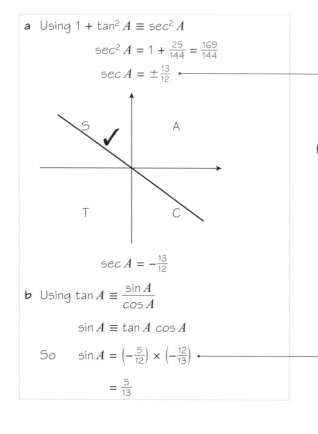

a Using $1 + \tan^2 A \equiv \sec^2 A$

$\sec^2 A = 1 + \dfrac{25}{144} = \dfrac{169}{144}$

$\sec A = \pm \dfrac{13}{12}$ — $\tan^2 A = \dfrac{25}{144}$

Problem-solving

You are told that A is obtuse. This means it lies in the second quadrant, so $\cos A$ is negative, and $\sec A$ is also negative.

$\sec A = -\dfrac{13}{12}$

b Using $\tan A \equiv \dfrac{\sin A}{\cos A}$

$\sin A \equiv \tan A \cos A$

So $\sin A = \left(-\dfrac{5}{12}\right) \times \left(-\dfrac{12}{13}\right)$ — $\cos A = -\dfrac{12}{13}$, since $\cos A = \dfrac{1}{\sec A}$

$= \dfrac{5}{13}$

Trigonometric functions

Example 11

Prove the identities:

a $\csc^4\theta - \cot^4\theta \equiv \dfrac{1+\cos^2\theta}{1-\cos^2\theta}$

b $\sec^2\theta - \cos^2\theta \equiv \sin^2\theta(1+\sec^2\theta)$

a LHS $= \csc^4\theta - \cot^4\theta$ — This is the difference of two squares, so factorise.

$\equiv (\csc^2\theta + \cot^2\theta)(\csc^2\theta - \cot^2\theta)$

$\equiv \csc^2\theta + \cot^2\theta$ — As $1 + \cot^2\theta \equiv \csc^2\theta$, so $\csc^2\theta - \cot^2\theta \equiv 1$.

$\equiv \dfrac{1}{\sin^2\theta} + \dfrac{\cos^2\theta}{\sin^2\theta}$ — Using $\csc\theta \equiv \dfrac{1}{\sin\theta}$, $\cot\theta \equiv \dfrac{\cos\theta}{\sin\theta}$

$\equiv \dfrac{1+\cos^2\theta}{\sin^2\theta}$ — Using $\sin^2\theta + \cos^2\theta \equiv 1$.

$\equiv \dfrac{1+\cos^2\theta}{1-\cos^2\theta} =$ RHS

b RHS $= \sin^2\theta + \sin^2\theta\sec^2\theta$ — Write in terms of $\sin\theta$ and $\cos\theta$.

$\equiv \sin^2\theta + \dfrac{\sin^2\theta}{\cos^2\theta}$ — Use $\sec\theta \equiv \dfrac{1}{\cos\theta}$.

$\equiv \sin^2\theta + \tan^2\theta$ — $\dfrac{\sin^2\theta}{\cos^2\theta} \equiv \left(\dfrac{\sin\theta}{\cos\theta}\right)^2 \equiv \tan^2\theta$.

$\equiv (1-\cos^2\theta) + (\sec^2\theta - 1)$

$\equiv \sec^2\theta - \cos^2\theta$

$\equiv$ LHS

Look at LHS. It is in terms of $\cos^2\theta$ and $\sec^2\theta$, so use $\sin^2\theta + \cos^2\theta \equiv 1$ and $1 + \tan^2\theta \equiv \sec^2\theta$.

Problem-solving

You can start from either the LHS or the RHS when proving an identity. Try starting with the LHS using $\cos^2\theta \equiv 1 - \sin^2\theta$ and $\sec^2\theta \equiv 1 + \tan^2\theta$.

Example 12

Solve the equation $4\csc^2\theta - 9 = \cot\theta$, in the interval $0 \leq \theta \leq 360°$.

This is a quadratic equation. You need to write it in terms of one trigonometrical function only, so use $1 + \cot^2\theta = \csc^2\theta$.

The equation can be rewritten as

$4(1 + \cot^2\theta) - 9 = \cot\theta$

So $4\cot^2\theta - \cot\theta - 5 = 0$

$(4\cot\theta - 5)(\cot\theta + 1) = 0$ — Factorise, or solve using the quadratic formula.

So $\cot\theta = \dfrac{5}{4}$ or $\cot\theta = -1$

$\therefore \tan\theta = \dfrac{4}{5}$ or $\tan\theta = -1$

For $\tan\theta = \dfrac{4}{5}$

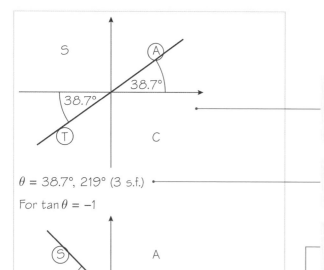

As tan θ is +ve, θ is in the 1st and 3rd quadrants. The acute angle to the horizontal is $\tan^{-1}\frac{4}{5} = 38.7°$.

θ = 38.7°, 219° (3 s.f.)

If α is the value the calculator gives for $\tan^{-1}\frac{4}{5}$, then the solutions are α and (180° + α).

For tan θ = −1

As tan θ is −ve, θ is in the 2nd and 4th quadrants. The acute angle to the horizontal is $\tan^{-1} 1 = 45°$.

If α is the value the calculator gives for $\tan^{-1}(-1)$, then the solutions are (180° + α) and (360° + α), as α is not in the given interval.

θ = 135°, 315°

Online Solve this equation numerically using your calculator.

Exercise 6D

Give answers to 3 significant figures where necessary.

1 Simplify each of the following expressions.

  a $1 + \tan^2 \frac{1}{2}\theta$
  b $(\sec \theta - 1)(\sec \theta + 1)$
  c $\tan^2 \theta (\csc^2 \theta - 1)$
  d $(\sec^2 \theta - 1) \cot \theta$
  e $(\csc^2 \theta - \cot^2 \theta)^2$
  f $2 - \tan^2 \theta + \sec^2 \theta$
  g $\dfrac{\tan \theta \sec \theta}{1 + \tan^2 \theta}$
  h $(1 - \sin^2 \theta)(1 + \tan^2 \theta)$
  i $\dfrac{\csc \theta \cot \theta}{1 + \cot^2 \theta}$
  j $(\sec^4 \theta - 2 \sec^2 \theta \tan^2 \theta + \tan^4 \theta)$
  k $4 \csc^2 2\theta + 4 \csc^2 2\theta \cot^2 2\theta$

(P) 2 Given that $\csc x = \dfrac{k}{\csc x}$, where $k > 1$, find, in terms of k, possible values of $\cot x$.

3 Given that $\cot \theta = -\sqrt{3}$, and that $90° < \theta < 180°$, find the exact values of:
  a $\sin \theta$
  b $\cos \theta$

4 Given that $\tan \theta = \frac{3}{4}$, and that $180° < \theta < 270°$, find the exact values of:
  a $\sec \theta$
  b $\cos \theta$
  c $\sin \theta$

5 Given that $\cos \theta = \frac{24}{25}$, and that θ is a reflex angle, find the exact values of:
  a $\tan \theta$
  b $\csc \theta$

6 Prove the following identities.
  a $\sec^4 \theta - \tan^4 \theta \equiv \sec^2 \theta + \tan^2 \theta$
  b $\operatorname{cosec}^2 x - \sin^2 x \equiv \cot^2 x + \cos^2 x$
  c $\sec^2 A(\cot^2 A - \cos^2 A) \equiv \cot^2 A$
  d $1 - \cos^2 \theta \equiv (\sec^2 \theta - 1)(1 - \sin^2 \theta)$
  e $\dfrac{1 - \tan^2 A}{1 + \tan^2 A} \equiv 1 - 2\sin^2 A$
  f $\sec^2 \theta + \operatorname{cosec}^2 \theta \equiv \sec^2 \theta \operatorname{cosec}^2 \theta$
  g $\operatorname{cosec} A \sec^2 A \equiv \operatorname{cosec} A + \tan A \sec A$
  h $(\sec \theta - \sin \theta)(\sec \theta + \sin \theta) \equiv \tan^2 \theta + \cos^2 \theta$

7 Given that $3 \tan^2 \theta + 4 \sec^2 \theta = 5$, and that θ is obtuse, find the exact value of $\sin \theta$.

8 Solve the following equations in the given intervals.
  a $\sec^2 \theta = 3 \tan \theta$, $0 \leq \theta \leq 360°$
  b $\tan^2 \theta - 2\sec \theta + 1 = 0$, $-\pi \leq \theta \leq \pi$
  c $\operatorname{cosec}^2 \theta + 1 = 3 \cot \theta$, $-180° \leq \theta \leq 180°$
  d $\cot \theta = 1 - \operatorname{cosec}^2 \theta$, $0 \leq \theta \leq 2\pi$
  e $3 \sec \tfrac{1}{2}\theta = 2 \tan^2 \tfrac{1}{2}\theta$, $0 \leq \theta \leq 360°$
  f $(\sec \theta - \cos \theta)^2 = \tan \theta - \sin^2 \theta$, $0 \leq \theta \leq \pi$
  g $\tan^2 2\theta = \sec 2\theta - 1$, $0 \leq \theta \leq 180°$
  h $\sec^2 \theta - (1 + \sqrt{3}) \tan \theta + \sqrt{3} = 1$, $0 \leq \theta \leq 2\pi$

9 Given that $\tan^2 k = 2 \sec k$,
  a find the value of $\sec k$ (4 marks)
  b deduce that $\cos k = \sqrt{2} - 1$. (2 marks)
  c Hence solve, in the interval $0 \leq k \leq 360°$, $\tan^2 k = 2 \sec k$, giving your answers to 1 decimal place. (3 marks)

10 Given that $a = 4 \sec x$, $b = \cos x$ and $c = \cot x$,
  a express b in terms of a (2 marks)
  b show that $c^2 = \dfrac{16}{a^2 - 16}$ (3 marks)

11 Given that $x = \sec \theta + \tan \theta$,
  a show that $\dfrac{1}{x} = \sec \theta - \tan \theta$. (3 marks)
  b Hence express $x^2 + \dfrac{1}{x^2} + 2$ in terms of θ, in its simplest form. (5 marks)

12 Given that $2 \sec^2 \theta - \tan^2 \theta = p$ show that $\operatorname{cosec}^2 \theta = \dfrac{p - 1}{p - 2}$, $p \neq 2$. (5 marks)

Chapter 6

6.5 Inverse trigonometric functions

You need to understand and use the inverse trigonometric functions arcsin x, arccos x and arctan x and their graphs.

- **The inverse function of sin x is called arcsin x.**

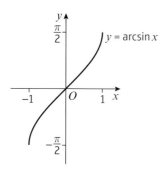

Hint The sin⁻¹ function on your calculator will give principal values in the same range as arcsin.

- **The domain of y = arcsin x is $-1 \leq x \leq 1$.**
- **The range of y = arcsin x is $-\dfrac{\pi}{2} \leq$ arcsin $x \leq \dfrac{\pi}{2}$ or $-90° \leq$ arcsin $x \leq 90°$.**

Example 13

Sketch the graph of y = arcsin x.

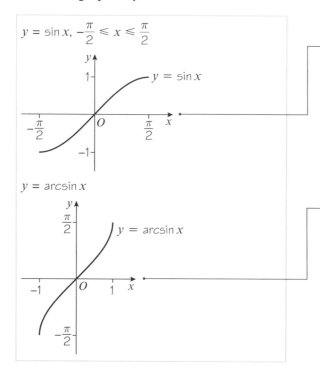

Step 1
Draw the graph of y = sin x, with the restricted domain of $-\dfrac{\pi}{2} \leq x \leq \dfrac{\pi}{2}$

Restricting the domain ensures that the inverse function exists since y = sin x is a **one-to-one** function for the restricted domain. Only one-to-one functions have inverses. ← Section 2.3

Step 2
Reflect in the line $y = x$.
The domain of arcsin x is $-1 \leq x \leq 1$; the range is $-\dfrac{\pi}{2} \leq$ arcsin $x \leq \dfrac{\pi}{2}$

Remember that the x and y coordinates of points interchange when reflecting in $y = x$. For example:
$$\left(\dfrac{\pi}{2}, 1\right) \to \left(1, \dfrac{\pi}{2}\right)$$

Trigonometric functions

- **The inverse function of cos x is called arccos x.**

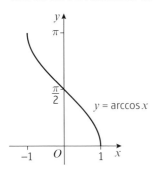

- The domain of $y = \arccos x$ is $-1 \leq x \leq 1$.
- The range of $y = \arccos x$ is $0 \leq \arccos x \leq \pi$ or $0° \leq \arccos x \leq 180°$.

- **The inverse function of tan x is called arctan x.**

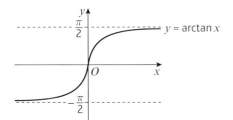

Watch out Unlike arcsin x and arccos x, the function arctan x is defined for all real values of x.

- The domain of $y = \arctan x$ is $x \in \mathbb{R}$.
- The range of $y = \arctan x$ is $-\dfrac{\pi}{2} < \arctan x < \dfrac{\pi}{2}$ or $-90° < \arctan x < 90°$.

Example 14

Work out, in radians, the values of:

a $\arcsin\left(-\dfrac{\sqrt{2}}{2}\right)$ **b** $\arccos(-1)$ **c** $\arctan(\sqrt{3})$

a

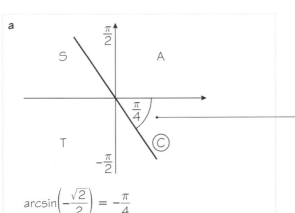

$\arcsin\left(-\dfrac{\sqrt{2}}{2}\right) = -\dfrac{\pi}{4}$

You need to solve, in the interval $-\dfrac{\pi}{2} \leq x \leq \dfrac{\pi}{2}$, the equation $\sin x = -\dfrac{\sqrt{2}}{2}$.

The angle to the horizontal is $\dfrac{\pi}{4}$ and, as sin is −ve, it is in the 4th quadrant.

Online Use your calculator to evaluate inverse trigonometric functions in radians.

b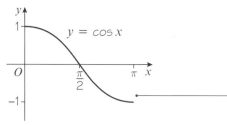

You need to solve, in the interval $0 \leq x \leq \pi$, the equation $\cos x = -1$.

Draw the graph of $y = \cos x$.

arccos(−1) = π

c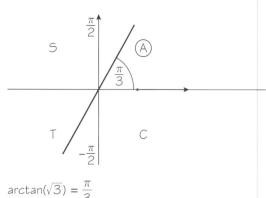

You need to solve, in the interval $-\frac{\pi}{2} < x < \frac{\pi}{2}$, the equation $\tan x = \sqrt{3}$.

The angle to the horizontal is $\frac{\pi}{3}$ and, as tan is +ve, it is in the 1st quadrant.

arctan(√3) = π/3

You can verify these results using the $\sin^{-1}$, $\cos^{-1}$ and $\tan^{-1}$ functions on your calculator.

Exercise 6E

In this exercise, all angles are given in radians.

1 Without using a calculator, work out, giving your answer in terms of π:

   a $\arccos(0)$ **b** $\arcsin(1)$ **c** $\arctan(-1)$ **d** $\arcsin\left(-\frac{1}{2}\right)$

   e $\arccos\left(-\frac{1}{\sqrt{2}}\right)$ **f** $\arctan\left(-\frac{1}{\sqrt{3}}\right)$ **g** $\arcsin\left(\sin\frac{\pi}{3}\right)$ **h** $\arcsin\left(\sin\frac{2\pi}{3}\right)$

2 Find:

   a $\arcsin\left(\frac{1}{2}\right) + \arcsin\left(-\frac{1}{2}\right)$ **b** $\arccos\left(\frac{1}{2}\right) - \arccos\left(-\frac{1}{2}\right)$ **c** $\arctan(1) - \arctan(-1)$

(P) 3 Without using a calculator, work out the values of:

   a $\sin\left(\arcsin\frac{1}{2}\right)$ **b** $\sin\left(\arcsin\left(-\frac{1}{2}\right)\right)$

   c $\tan(\arctan(-1))$ **d** $\cos(\arccos 0)$

(P) 4 Without using a calculator, work out the exact values of:

   a $\sin\left(\arccos\left(\frac{1}{2}\right)\right)$ **b** $\cos\left(\arcsin\left(-\frac{1}{2}\right)\right)$ **c** $\tan\left(\arccos\left(-\frac{\sqrt{2}}{2}\right)\right)$

   d $\sec(\arctan(\sqrt{3}))$ **e** $\csc(\arcsin(-1))$ **f** $\sin\left(2\arcsin\left(\frac{\sqrt{2}}{2}\right)\right)$

Trigonometric functions

5 Given that arcsin $k = \alpha$, where $0 < k < 1$ and α is in radians, write down, in terms of α, the first two positive values of x satisfying the equation $\sin x = k$.

6 Given that x satisfies arcsin $x = k$, where $0 < k < \frac{\pi}{2}$,
  a state the range of possible values of x (1 mark)
  b express, in terms of x,
   i $\cos k$ **ii** $\tan k$ (4 marks)

  Given, instead, that $-\frac{\pi}{2} < k < 0$,

  c how, if at all, are your answers to part **b** affected? (2 marks)

7 Sketch the graphs of:
  a $y = \frac{\pi}{2} + 2\arcsin x$ **b** $y = \pi - \arctan x$
  c $y = \arccos(2x + 1)$ **d** $y = -2\arcsin(-x)$

8 The function f is defined as $f : x \mapsto \arcsin x$, $-1 \leq x \leq 1$, and the function g is such that $g(x) = f(2x)$.
  a Sketch the graph of $y = f(x)$ and state the range of f. (3 marks)
  b Sketch the graph of $y = g(x)$. (2 marks)
  c Define g in the form $g: x \mapsto \ldots$ and give the domain of g. (3 marks)
  d Define g^{-1} in the form $g^{-1}: x \mapsto \ldots$ (2 marks)

9 a Prove that for $0 \leq x \leq 1$, $\arccos x = \arcsin \sqrt{1 - x^2}$ (4 marks)

  b Give a reason why this result is not true for $-1 \leq x \leq 0$. (2 marks)

Challenge

a Sketch the graph of $y = \sec x$, with the restricted domain $0 \leq x \leq \pi, x \neq \frac{\pi}{2}$

b Given that arcsec x is the inverse function of $\sec x$, $0 \leq x \leq \pi, x \neq \frac{\pi}{2}$, sketch the graph of $y = \text{arcsec } x$ and state the range of arcsec x.

Mixed exercise 6

Give any non-exact answers to equations to 1 decimal place.

E/P 1 Solve $\tan x = 2 \cot x$, in the interval $-180° \leq x \leq 90°$. (4 marks)

E/P 2 Given that $p = 2 \sec \theta$ and $q = 4 \cos \theta$, express p in terms of q. (4 marks)

E/P 3 Given that $p = \sin \theta$ and $q = 4 \cot \theta$, show that $p^2 q^2 = 16(1 - p^2)$. (4 marks)

P 4 **a** Solve, in the interval $0 < \theta < 180°$,
   i $\operatorname{cosec} \theta = 2 \cot \theta$ **ii** $2 \cot^2 \theta = 7 \operatorname{cosec} \theta - 8$
   b Solve, in the interval $0 \leq \theta \leq 360°$,
   i $\sec(2\theta - 15°) = \operatorname{cosec} 135°$ **ii** $\sec^2 \theta + \tan \theta = 3$
   c Solve, in the interval $0 \leq x \leq 2\pi$,
   i $\operatorname{cosec}\left(x + \dfrac{\pi}{15}\right) = -\sqrt{2}$ **ii** $\sec^2 x = \dfrac{4}{3}$

E/P 5 Given that $5 \sin x \cos y + 4 \cos x \sin y = 0$, and that $\cot x = 2$, find the value of $\cot y$. (5 marks)

P 6 Prove that:
   a $(\tan \theta + \cot \theta)(\sin \theta + \cos \theta) \equiv \sec \theta + \operatorname{cosec} \theta$
   b $\dfrac{\operatorname{cosec} x}{\operatorname{cosec} x - \sin x} \equiv \sec^2 x$
   c $(1 - \sin x)(1 + \operatorname{cosec} x) \equiv \cos x \cot x$
   d $\dfrac{\cot x}{\operatorname{cosec} x - 1} - \dfrac{\cos x}{1 + \sin x} \equiv 2 \tan x$
   e $\dfrac{1}{\operatorname{cosec} \theta - 1} + \dfrac{1}{\operatorname{cosec} \theta + 1} \equiv 2 \sec \theta \tan \theta$
   f $\dfrac{(\sec \theta - \tan \theta)(\sec \theta + \tan \theta)}{1 + \tan^2 \theta} \equiv \cos^2 \theta$

E/P 7 **a** Prove that $\dfrac{\sin x}{1 + \cos x} + \dfrac{1 + \cos x}{\sin x} \equiv 2 \operatorname{cosec} x$. (4 marks)

   b Hence solve, in the interval $-2\pi \leq x \leq 2\pi$, $\dfrac{\sin x}{1 + \cos x} + \dfrac{1 + \cos x}{\sin x} = -\dfrac{4}{\sqrt{3}}$ (4 marks)

E/P 8 Prove that $\dfrac{1 + \cos \theta}{1 - \cos \theta} \equiv (\operatorname{cosec} \theta + \cot \theta)^2$ (4 marks)

E 9 Given that $\sec A = -3$, where $\dfrac{\pi}{2} < A < \pi$,
   a calculate the exact value of $\tan A$ (3 marks)
   b show that $\operatorname{cosec} A = \dfrac{3\sqrt{2}}{4}$ (3 marks)

10 Given that $\sec \theta = k$, $|k| \geq 1$, and that θ is obtuse, express in terms of k:
   a $\cos \theta$ **b** $\tan^2 \theta$ **c** $\cot \theta$ **d** $\operatorname{cosec} \theta$

Trigonometric functions

(E) **11** Solve, in the interval $0 \leq x \leq 2\pi$, the equation $\sec\left(x + \frac{\pi}{4}\right) = 2$, giving your answers in terms of π. **(5 marks)**

(E/P) **12** Find, in terms of π, the value of $\arcsin\left(\frac{1}{2}\right) - \arcsin\left(-\frac{1}{2}\right)$. **(4 marks)**

(E/P) **13** Solve, in the interval $0 \leq x \leq 2\pi$, the equation $\sec^2 x - \frac{2\sqrt{3}}{3} \tan x - 2 = 0$, giving your answers in terms of π. **(5 marks)**

(E/P) **14 a** Factorise $\sec x \csc x - 2 \sec x - \csc x + 2$. **(2 marks)**
    b Hence solve $\sec x \csc x - 2 \sec x - \csc x + 2 = 0$, in the interval $0 \leq x \leq 360°$. **(4 marks)**

(E/P) **15** Given that $\arctan(x - 2) = -\frac{\pi}{3}$, find the value of x. **(3 marks)**

(E) **16** On the same set of axes sketch the graphs of $y = \cos x$, $0 \leq x \leq \pi$, and $y = \arccos x$, $-1 \leq x \leq 1$, showing the coordinates of points at which the curves meet the axes. **(4 marks)**

(E/P) **17 a** Given that $\sec x + \tan x = -3$, use the identity $1 + \tan^2 x \equiv \sec^2 x$ to find the value of $\sec x - \tan x$. **(3 marks)**
    b Deduce the values of:
        i $\sec x$ **ii** $\tan x$ **(3 marks)**
    c Hence solve, in the interval $-180° \leq x \leq 180°$, $\sec x + \tan x = -3$. **(3 marks)**

(E/P) **18** Given that $p = \sec \theta - \tan \theta$ and $q = \sec \theta + \tan \theta$, show that $p = \frac{1}{q}$. **(4 marks)**

(E/P) **19 a** Prove that $\sec^4 \theta - \tan^4 \theta = \sec^2 \theta + \tan^2 \theta$. **(3 marks)**
    b Hence solve, in the interval $-180° \leq \theta \leq 180°$, $\sec^4 \theta = \tan^4 \theta + 3 \tan \theta$. **(4 marks)**

(P) **20 a** Sketch the graph of $y = \sin x$ and shade in the area representing $\int_0^{\frac{\pi}{2}} \sin x \, dx$.
    b Sketch the graph of $y = \arcsin x$ and shade in the area representing $\int_0^1 \arcsin x \, dx$.
    c By considering the shaded areas explain why $\int_0^{\frac{\pi}{2}} \sin x \, dx + \int_0^1 \arcsin x \, dx = \frac{\pi}{2}$

(P) **21** Show that $\cot 60° \sec 60° = \frac{2\sqrt{3}}{3}$

(E/P) **22 a** Sketch, in the interval $-2\pi \leq x \leq 2\pi$, the graph of $y = 2 - 3 \sec x$. **(3 marks)**
    b Hence deduce the range of values of k for which the equation $2 - 3 \sec x = k$ has no solutions. **(2 marks)**

(P) **23 a** Sketch the graph of $y = 3 \arcsin x - \frac{\pi}{2}$, showing clearly the exact coordinates of the end-points of the curve. **(4 marks)**
    b Find the exact coordinates of the point where the curve crosses the x-axis. **(3 marks)**

24 a Prove that for $0 < x \leq 1$, $\arccos x = \arctan \dfrac{\sqrt{1-x^2}}{x}$

  b Prove that for $-1 \leq x < 0$, $\arccos x = k + \arctan \dfrac{\sqrt{1-x^2}}{x}$, where k is a constant to be found.

Summary of key points

1 • $\sec x = \dfrac{1}{\cos x}$ (undefined for values of x for which $\cos x = 0$)

  • $\operatorname{cosec} x = \dfrac{1}{\sin x}$ (undefined for values of x for which $\sin x = 0$)

  • $\cot x = \dfrac{1}{\tan x}$ (undefined for values of x for which $\tan x = 0$)

  • $\cot x = \dfrac{\cos x}{\sin x}$

2 The graph of $y = \sec x$, $x \in \mathbb{R}$, has symmetry in the y-axis and has period 360° or 2π radians. It has vertical asymptotes at all the values of x for which $\cos x = 0$.

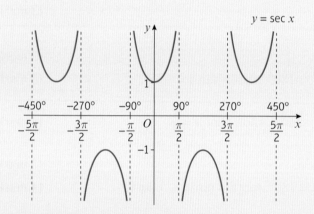

• The domain of $y = \sec x$ is $x \in \mathbb{R}$, $x \neq 90°, 270°, 450°, \ldots$ or any odd multiple of 90°.

• In radians the domain is $x \in \mathbb{R}$, $x \neq \dfrac{\pi}{2}, \dfrac{3\pi}{2}, \dfrac{5\pi}{2}, \ldots$ or any odd multiple of $\dfrac{\pi}{2}$

• The range of $y = \sec x$ is $y \leq -1$ or $y \geq 1$.

3 The graph of $y = \operatorname{cosec} x$, $x \in \mathbb{R}$, has period 360° or 2π radians. It has vertical asymptotes at all the values of x for which $\sin x = 0$.

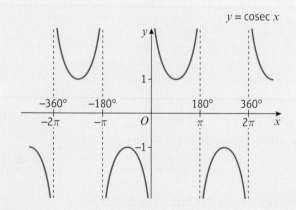

• The domain of $y = \operatorname{cosec} x$ is $x \in \mathbb{R}$, $x \neq 0°, 180°, 360°, \ldots$ or any multiple of 180°.

• In radians the domain is $x \in \mathbb{R}$, $x \neq 0, \pi, 2\pi, \ldots$ or any multiple of π

• The range of $y = \operatorname{cosec} x$ is $y \leq -1$ or $y \geq 1$.

4 The graph of $y = \cot x$, $x \in \mathbb{R}$, has period 180° or π radians. It has vertical asymptotes at all the values of x for which $\tan x = 0$.

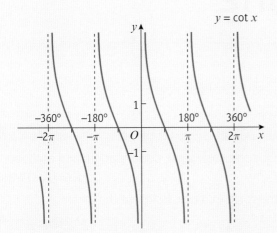

- The domain of $y = \cot x$ is $x \in \mathbb{R}$, $x \neq 0°$, 180°, 360°, ... or any multiple of 180°.
- In radians the domain is $x \in \mathbb{R}$, $x \neq 0, \pi, 2\pi,$ ... or any multiple of π.
- The range of $y = \cot x$ is $y \in \mathbb{R}$.

5 $\sec x = k$ and $\operatorname{cosec} x = k$ have no solutions for $-1 < k < 1$.

6 You can use the identity $\sin^2 x + \cos^2 x \equiv 1$ to prove the following identities:
- $1 + \tan^2 x \equiv \sec^2 x$
- $1 + \cot^2 x \equiv \operatorname{cosec}^2 x$

7 The inverse function of $\sin x$ is called **arcsin x**.
- The domain of $y = \arcsin x$ is $-1 \leq x \leq 1$
- The range of $y = \arcsin x$ is $-\dfrac{\pi}{2} \leq \arcsin x \leq \dfrac{\pi}{2}$ or $-90° \leq \arcsin x \leq 90°$

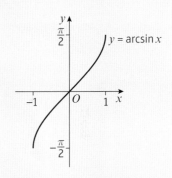

8 The inverse function of $\cos x$ is called **arccos x**.
- The domain of $y = \arccos x$ is $-1 \leq x \leq 1$
- The range of $y = \arccos x$ is $0 \leq \arccos x \leq \pi$ or $0° \leq \arccos x \leq 180°$

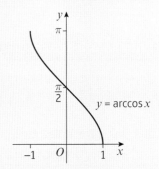

9 The inverse function of $\tan x$ is called **arctan x**.
- The domain of $y = \arctan x$ is $x \in \mathbb{R}$
- The range of $y = \arctan x$ is $-\dfrac{\pi}{2} < \arctan x < \dfrac{\pi}{2}$ or $-90° < \arctan x < 90°$

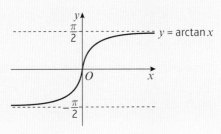

7 Trigonometry and modelling

Objectives

After completing this unit you should be able to:
- Prove and use the addition formulae → **pages 167–173**
- Understand and use the double-angle formulae → **pages 174–177**
- Solve trigonometric equations using the double-angle and addition formulae → **pages 177–181**
- Write expressions of the form $a\cos\theta \pm b\sin\theta$ in the forms $R\cos(\theta \pm \alpha)$ or $R\sin(\theta \pm \alpha)$ → **pages 181–186**
- Prove trigonometric identities using a variety of identities → **pages 186–189**
- Use trigonometric functions to model real-life situations → **pages 189–191**

Prior knowledge check

1. Find the exact values of:
   a $\sin 45°$ b $\cos\dfrac{\pi}{6}$ c $\tan\dfrac{\pi}{3}$ ← Section 5.4

2. Solve the following equations in the interval $0 \leqslant x < 360°$.
   a $\sin(x + 50°) = -0.9$ b $\cos(2x - 30°) = \tfrac{1}{2}$
   c $2\sin^2 x - \sin x - 3 = 0$ ← Year 1, Chapter 10

3. Prove the following:
   a $\cos x + \sin x \tan x \equiv \sec x$ b $\cot x \sec x \sin x \equiv 1$
   c $\dfrac{\cos^2 x + \sin^2 x}{1 + \cot^2 x} \equiv \sin^2 x$ ← Section 6.4

The strength of microwaves at different points within a microwave oven can be modelled using trigonometric functions. → **Exercise 7G Q7**

Trigonometry and modelling

7.1 Addition formulae

The addition formulae for sine, cosine and tangent are defined as follows:

Notation The addition formulae are sometimes called the **compound-angle formulae**.

- $\sin(A + B) \equiv \sin A \cos B + \cos A \sin B$
- $\cos(A + B) \equiv \cos A \cos B - \sin A \sin B$
- $\tan(A + B) \equiv \dfrac{\tan A + \tan B}{1 - \tan A \tan B}$

$\sin(A - B) \equiv \sin A \cos B - \cos A \sin B$

$\cos(A - B) \equiv \cos A \cos B + \sin A \sin B$

$\tan(A - B) \equiv \dfrac{\tan A - \tan B}{1 + \tan A \tan B}$

You can prove these identities using geometric constructions.

Example 1

In the diagram $\angle BAC = \alpha$, $\angle CAE = \beta$ and $AE = 1$. Additionally, lines AB and BC are perpendicular, lines AB and DE are perpendicular, lines AC and EC are perpendicular and lines EF and FC are perpendicular.

Use the diagram, together with known properties of sine and cosine, to prove the following identities:

a $\sin(\alpha + \beta) \equiv \sin \alpha \cos \beta + \cos \alpha \sin \beta$

b $\cos(\alpha + \beta) \equiv \cos \alpha \cos \beta - \sin \alpha \sin \beta$

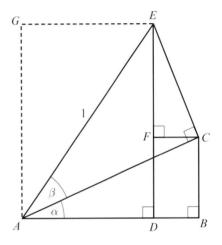

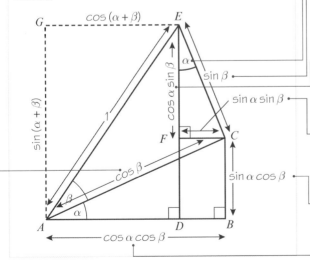

The diagram can be labelled with the following lengths using the properties of sine and cosine.

In triangle ACE, $\cos \beta = \dfrac{AC}{AE} \Rightarrow \cos \beta = \dfrac{AC}{1}$

So $AC = \cos \beta$.

$\angle ACF = \alpha \Rightarrow \angle FCE = 90° - \alpha$. So $\angle FEC = \alpha$.

In triangle ACE, $\sin \beta = \dfrac{EC}{AE} \Rightarrow \sin \beta = \dfrac{EC}{1}$

So $EC = \sin \beta$.

In triangle FEC, $\cos \alpha = \dfrac{FE}{EC} \Rightarrow \cos \alpha = \dfrac{FE}{\sin \beta}$

So $FE = \cos \alpha \sin \beta$.

In triangle FEC, $\sin \alpha = \dfrac{FC}{EC} \Rightarrow \sin \alpha = \dfrac{FC}{\sin \beta}$

So $FC = \sin \alpha \sin \beta$.

In triangle ABC, $\sin \alpha = \dfrac{BC}{AC} \Rightarrow \sin \alpha = \dfrac{BC}{\cos \beta}$

So $BC = \sin \alpha \cos \beta$.

In triangle ABC, $\cos \alpha = \dfrac{AB}{AC} \Rightarrow \cos \alpha = \dfrac{AB}{\cos \beta}$

So $AB = \cos \alpha \cos \beta$.

a Using triangle ADE
$DE = \sin(\alpha + \beta)$
$AD = \cos(\alpha + \beta)$
$DE = DF + FE$
$\Rightarrow \sin(\alpha + \beta) \equiv \sin\alpha \cos\beta + \cos\alpha \sin\beta$
as required

b $AD = AB - DB$
$\Rightarrow \cos(\alpha + \beta) \equiv \cos\alpha \cos\beta - \sin\alpha \sin\beta$
as required

Problem-solving

You are looking for a relationship involving $\sin(\alpha + \beta)$, so consider the right-angled triangle ADE with angle $(\alpha + \beta)$. You can see these relationships more easily on the diagram by looking at $AG = DE$ and $GE = AD$.

Substitute the lengths from the diagram.

Online Explore the proof step-by-step using GeoGebra.

Example 2

Use the results from Example 1 to show that
a $\cos(A - B) \equiv \cos A \cos B + \sin A \sin B$
b $\tan(A + B) \equiv \dfrac{\tan A + \tan B}{1 - \tan A \tan B}$

a Replace B by $-B$ in
$\cos(A + B) \equiv \cos A \cos B - \sin A \sin B$
$\cos(A + (-B)) \equiv \cos A \cos(-B) - \sin A \sin(-B)$
$\cos(A - B) \equiv \cos A \cos B + \sin A \sin B$

$\cos(-B) = \cos B$ and $\sin(-B) = -\sin B$
← Year 1, Chapter 9

b $\tan(A + B) \equiv \dfrac{\sin(A + B)}{\cos(A + B)}$

$\equiv \dfrac{\sin A \cos B + \cos A \sin B}{\cos A \cos B - \sin A \sin B}$

Divide the numerator and denominator by $\cos A \cos B$.

$\equiv \dfrac{\dfrac{\sin A \cos B}{\cos A \cos B} + \dfrac{\cos A \sin B}{\cos A \cos B}}{\dfrac{\cos A \cos B}{\cos A \cos B} - \dfrac{\sin A \sin B}{\cos A \cos B}}$

Cancel where possible.

$\equiv \dfrac{\tan A + \tan B}{1 - \tan A \tan B}$ as required

Example 3

Prove that
$$\dfrac{\cos A}{\sin B} - \dfrac{\sin A}{\cos B} \equiv \dfrac{\cos(A + B)}{\sin B \cos B}$$

$$\text{LHS} \equiv \frac{\cos A}{\sin B} - \frac{\sin A}{\cos B}$$

Write both fractions with a common denominator.

$$\equiv \frac{\cos A \cos B}{\sin B \cos B} - \frac{\sin A \sin B}{\sin B \cos B}$$

Problem-solving

When proving an identity, always keep an eye on the final answer. This can act as a guide as to what to do next.

$$\equiv \frac{\cos A \cos B - \sin A \sin B}{\sin B \cos B}$$

$$\equiv \frac{\cos(A+B)}{\sin B \cos B} \equiv \text{RHS}$$

Use the addition formula in reverse:
$\cos A \cos B - \sin A \sin B \equiv \cos(A+B)$

Example 4

Given that $2\sin(x+y) = 3\cos(x-y)$, express $\tan x$ in terms of $\tan y$.

Expanding $\sin(x+y)$ and $\cos(x-y)$ gives

$2\sin x \cos y + 2\cos x \sin y = 3\cos x \cos y + 3\sin x \sin y$

so $\dfrac{2\sin x \cos y}{\cos x \cos y} + \dfrac{2\cos x \sin y}{\cos x \cos y} = \dfrac{3\cos x \cos y}{\cos x \cos y} + \dfrac{3\sin x \sin y}{\cos x \cos y}$

$2\tan x + 2\tan y = 3 + 3\tan x \tan y$

$2\tan x - 3\tan x \tan y = 3 - 2\tan y$

$\tan x(2 - 3\tan y) = 3 - 2\tan y$

So $\tan x = \dfrac{3 - 2\tan y}{2 - 3\tan y}$

Remember $\tan x = \dfrac{\sin x}{\cos x}$

Dividing each term by $\cos x \cos y$ will produce $\tan x$ and $\tan y$ terms.

Collect all $\tan x$ terms on one side of the equation.

Factorise.

Exercise 7A

1 In the diagram $\angle BAC = \beta$, $\angle CAF = \alpha - \beta$ and $AC = 1$.
   Additionally lines AB and BC are perpendicular.

   a Show each of the following:
   - **i** $\angle FAB = \alpha$
   - **ii** $\angle ABD = \alpha$ and $\angle ECB = \alpha$
   - **iii** $AB = \cos\beta$
   - **iv** $BC = \sin\beta$

   b Use $\triangle ABD$ to write an expression for the lengths
   - **i** AD
   - **ii** BD

   c Use $\triangle BEC$ to write an expression for the lengths
   - **i** CE
   - **ii** BE

   d Use $\triangle FAC$ to write an expression for the lengths
   - **i** FC
   - **ii** FA

   e Use your completed diagram to show that:
   - **i** $\sin(\alpha - \beta) = \sin\alpha\cos\beta - \cos\alpha\sin\beta$
   - **ii** $\cos(\alpha - \beta) = \cos\alpha\cos\beta + \sin\alpha\sin\beta$

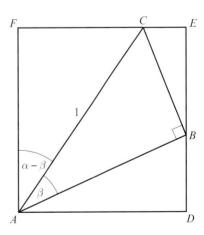

2 Use the formulae for $\sin(A - B)$ and $\cos(A - B)$ to show that
$$\tan(A - B) \equiv \frac{\tan A - \tan B}{1 + \tan A \tan B}$$

3 By substituting $A = P$ and $B = -Q$ into the addition formula for $\sin(A + B)$, show that $\sin(P - Q) \equiv \sin P \cos Q - \cos P \sin Q$.

4 A student makes the mistake of thinking that $\sin(A + B) \equiv \sin A + \sin B$. Choose non-zero values of A and B to show that this identity is not true.

> **Watch out** This is a common mistake. One counter-example is sufficient to disprove the statement.

5 Using the expansion of $\cos(A - B)$ with $A = B = \theta$, show that $\sin^2 \theta + \cos^2 \theta \equiv 1$.

6 a Use the expansion of $\sin(A - B)$ to show that $\sin\left(\frac{\pi}{2} - \theta\right) = \cos \theta$.

b Use the expansion of $\cos(A - B)$ to show that $\cos\left(\frac{\pi}{2} - \theta\right) = \sin \theta$.

7 Write $\sin\left(x + \frac{\pi}{6}\right)$ in the form $p \sin x + q \cos x$ where p and q are constants to be found.

8 Write $\cos\left(x + \frac{\pi}{3}\right)$ in the form $a \cos x + b \sin x$ where a and b are constants to be found.

9 Express the following as a single sine, cosine or tangent:
a $\sin 15° \cos 20° + \cos 15° \sin 20°$
b $\sin 58° \cos 23° - \cos 58° \sin 23°$
c $\cos 130° \cos 80° - \sin 130° \sin 80°$
d $\dfrac{\tan 76° - \tan 45°}{1 + \tan 76° \tan 45°}$
e $\cos 2\theta \cos \theta + \sin 2\theta \sin \theta$
f $\cos 4\theta \cos 3\theta - \sin 4\theta \sin 3\theta$
g $\sin \frac{1}{2}\theta \cos 2\frac{1}{2}\theta + \cos \frac{1}{2}\theta \sin 2\frac{1}{2}\theta$
h $\dfrac{\tan 2\theta + \tan 3\theta}{1 - \tan 2\theta \tan 3\theta}$
i $\sin(A + B) \cos B - \cos(A + B) \sin B$
j $\cos\left(\dfrac{3x + 2y}{2}\right) \cos\left(\dfrac{3x - 2y}{2}\right) - \sin\left(\dfrac{3x + 2y}{2}\right) \sin\left(\dfrac{3x - 2y}{2}\right)$

10 Use the addition formulae for sine or cosine to write each of the following as a single trigonometric function in the form $\sin(x \pm \theta)$ or $\cos(x \pm \theta)$, where $0 < \theta < \dfrac{\pi}{2}$.

a $\dfrac{1}{\sqrt{2}}(\sin x + \cos x)$
b $\dfrac{1}{\sqrt{2}}(\cos x - \sin x)$
c $\dfrac{1}{2}(\sin x + \sqrt{3} \cos x)$
d $\dfrac{1}{\sqrt{2}}(\sin x - \cos x)$

(P) 11 Given that $\cos y = \sin(x + y)$, show that $\tan y = \sec x - \tan x$.

(P) 12 Given that $\tan(x - y) = 3$, express $\tan y$ in terms of $\tan x$.

(P) 13 Given that $\sin x(\cos y + 2\sin y) = \cos x(2\cos y - \sin y)$, find the value of $\tan(x + y)$.

Hint First multiply out the brackets.

(P) 14 In each of the following, calculate the exact value of $\tan x$.

  a $\tan(x - 45°) = \frac{1}{4}$ **b** $\sin(x - 60°) = 3\cos(x + 30°)$ **c** $\tan(x - 60°) = 2$

(E/P) 15 Given that $\tan\left(x + \frac{\pi}{3}\right) = \frac{1}{2}$, show that $\tan x = 8 - 5\sqrt{3}$. **(3 marks)**

(E/P) 16 Prove that
$$\cos\theta + \cos\left(\theta + \frac{2\pi}{3}\right) + \cos\left(\theta + \frac{4\pi}{3}\right) = 0$$

You must show each stage of your working. **(4 marks)**

Challenge

This triangle is constructed from two right-angled triangles T_1 and T_2.

a Find expressions involving x, y, A and B for:
  i the area of T_1
  ii the area of T_2
  iii the area of the large triangle.

b Hence prove that
$\sin(A + B) = \sin A \cos B + \cos A \sin B$

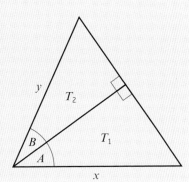

Hint For part **a** your expressions should all involve **all four** variables. You will need to use the formula Area $= \frac{1}{2}ab \sin\theta$ in each case.

7.2 Using the angle addition formulae

The addition formulae can be used to find exact values of trigonometric functions of different angles.

Example 5

Show, using the formula for $\sin(A - B)$, that $\sin 15° = \frac{\sqrt{6} - \sqrt{2}}{4}$

$\sin 15° = \sin(45° - 30°)$
$= \sin 45° \cos 30° - \cos 45° \sin 30°$
$= \left(\frac{1}{2}\sqrt{2}\right)\left(\frac{1}{2}\sqrt{3}\right) - \left(\frac{1}{2}\sqrt{2}\right)\left(\frac{1}{2}\right)$
$= \frac{1}{4}(\sqrt{3}\sqrt{2} - \sqrt{2})$
$= \frac{\sqrt{6} - \sqrt{2}}{4}$

You know the exact values of sin and cos for many angles, e.g. 30°, 45°, 60°, 90°, 180° ..., so write 15° using two of these angles. You could also use $\sin(60° - 45°)$.

Chapter 7

Example 6

Given that $\sin A = -\frac{3}{5}$ and $180° < A < 270°$, and that $\cos B = -\frac{12}{13}$ and B is obtuse, find the value of:

a $\cos(A - B)$ **b** $\tan(A + B)$ **c** $\csc(A - B)$

a $\cos(A - B) \equiv \cos A \cos B + \sin A \sin B$

$\cos^2 A \equiv 1 - \sin^2 A$

$= 1 - \left(-\frac{3}{5}\right)^2$

$= 1 - \frac{9}{25}$

$= \frac{16}{25}$

$\cos A = \pm\frac{4}{5}$

$\sin^2 B \equiv 1 - \cos^2 B$

$= 1 - \left(-\frac{12}{13}\right)^2$

$= 1 - \frac{144}{169}$

$= \frac{25}{169}$

$\sin B = \pm\frac{5}{13}$

$180° < A < 270°$ so $\cos A = -\frac{4}{5}$ B is obtuse so $\sin B = \frac{5}{13}$

$\cos(A - B) = \left(-\frac{4}{5}\right)\left(-\frac{12}{13}\right) + \left(-\frac{3}{5}\right)\left(+\frac{5}{13}\right)$

$= \frac{48}{65} - \frac{15}{65} = \frac{33}{65}$

b $\tan(A + B) \equiv \dfrac{\tan A + \tan B}{1 - \tan A \tan B}$

So $\tan(A + B) = \dfrac{\frac{3}{4} + \left(-\frac{5}{12}\right)}{1 - \left(\frac{3}{4}\right)\left(-\frac{5}{12}\right)}$

$= \dfrac{\frac{1}{3}}{\frac{21}{16}} = \frac{1}{3} \times \frac{16}{21} = \frac{16}{63}$

c $\csc(A - B) \equiv \dfrac{1}{\sin(A - B)}$

$\sin(A - B) \equiv \sin A \cos B - \cos A \sin B$

$\sin(A - B) = \left(\frac{-3}{5}\right)\left(\frac{-12}{13}\right) - \left(\frac{-4}{5}\right)\left(\frac{5}{13}\right) = \frac{56}{65}$

$\csc(A - B) = \dfrac{1}{\left(\frac{56}{65}\right)} = \frac{65}{56}$

You know $\sin A$ and $\cos B$, but need to find $\sin B$ and $\cos A$.

Use $\sin^2 x + \cos^2 x \equiv 1$ to determine $\cos A$ and $\sin B$.

Problem-solving

Remember there are two possible solutions to $\cos^2 A = \frac{16}{25}$. Use a CAST diagram to determine which one to use.

$\cos x$ is negative in the third quadrant, so choose the negative square root $-\frac{4}{5}$. $\sin x$ is positive in the second quadrant (obtuse angle) so choose the positive square root.

Substitute the values for $\sin A$, $\sin B$, $\cos A$ and $\cos B$ into the formula and then simplify.

$\tan A = \dfrac{\sin A}{\cos A} = \dfrac{-\frac{3}{5}}{-\frac{4}{5}} = \frac{3}{4}$

$\tan B = \dfrac{\sin B}{\cos B} = \dfrac{\frac{5}{13}}{-\frac{12}{13}} = -\frac{5}{12}$

Remember $\csc x = \dfrac{1}{\sin x}$

Exercise 7B

1 Without using your calculator, find the exact value of:

a $\cos 15°$ **b** $\sin 75°$ **c** $\sin(120° + 45°)$ **d** $\tan 165°$

Trigonometry and modelling

2 Without using your calculator, find the exact value of:

   a $\sin 30° \cos 60° + \cos 30° \sin 60°$

   b $\cos 110° \cos 20° + \sin 110° \sin 20°$

   c $\sin 33° \cos 27° + \cos 33° \sin 27°$

   d $\cos \dfrac{\pi}{8} \cos \dfrac{\pi}{8} - \sin \dfrac{\pi}{8} \sin \dfrac{\pi}{8}$

   e $\sin 60° \cos 15° - \cos 60° \sin 15°$

   f $\cos 70° (\cos 50° - \tan 70° \sin 50°)$

   g $\dfrac{\tan 45° + \tan 15°}{1 - \tan 45° \tan 15°}$

   h $\dfrac{1 - \tan 15°}{1 + \tan 15°}$

   i $\dfrac{\tan \dfrac{7\pi}{12} - \tan \dfrac{\pi}{3}}{1 + \tan \dfrac{7\pi}{12} \tan \dfrac{\pi}{3}}$

   j $\sqrt{3} \cos 15° - \sin 15°$

(E) 3 a Express $\tan(45° + 30°)$ in terms of $\tan 45°$ and $\tan 30°$. **(2 marks)**

   b Hence show that $\tan 75° = 2 + \sqrt{3}$. **(2 marks)**

(P) 4 Given that $\cot A = \frac{1}{4}$ and $\cot(A + B) = 2$, find the value of $\cot B$.

(E/P) 5 a Using $\cos(\theta + \alpha) \equiv \cos\theta \cos\alpha - \sin\theta \sin\alpha$, or otherwise, show that $\cos 105° = \dfrac{\sqrt{2} - \sqrt{6}}{4}$ **(4 marks)**

   b Hence, or otherwise, show that $\sec 105° = -\sqrt{a}(1 + \sqrt{b})$, where a and b are constants to be found. **(3 marks)**

(P) 6 Given that $\sin A = \frac{4}{5}$ and $\sin B = \frac{1}{2}$, where A and B are both acute angles, calculate the exact value of:

   a $\sin(A + B)$ **b** $\cos(A - B)$ **c** $\sec(A - B)$

(P) 7 Given that $\cos A = -\frac{4}{5}$, and A is an obtuse angle measured in radians, find the exact value of:

   a $\sin A$ **b** $\cos(\pi + A)$ **c** $\sin\left(\dfrac{\pi}{3} + A\right)$ **d** $\tan\left(\dfrac{\pi}{4} + A\right)$

(P) 8 Given that $\sin A = \frac{8}{17}$, where A is acute, and $\cos B = -\frac{4}{5}$, where B is obtuse, calculate the exact value of:

   a $\sin(A - B)$ **b** $\cos(A - B)$ **c** $\cot(A - B)$

(P) 9 Given that $\tan A = \frac{7}{24}$, where A is reflex, and $\sin B = \frac{5}{13}$, where B is obtuse, calculate the exact value of:

   a $\sin(A + B)$ **b** $\tan(A - B)$ **c** $\csc(A + B)$

(P) 10 Given that $\tan A = \frac{1}{5}$ and $\tan B = \frac{2}{3}$, calculate, without using your calculator, the value of $A + B$ in degrees, where:

   a A and B are both acute,

   b A is reflex and B is acute.

7.3 Double-angle formulae

You can use the addition formulae to derive the following double-angle formulae.

- $\sin 2A \equiv 2 \sin A \cos A$
- $\cos 2A \equiv \cos^2 A - \sin^2 A \equiv 2\cos^2 A - 1 \equiv 1 - 2\sin^2 A$
- $\tan 2A \equiv \dfrac{2 \tan A}{1 - \tan^2 A}$

Example 7

Use the double-angle formulae to write each of the following as a single trigonometric ratio.

a $\cos^2 50° - \sin^2 50°$ **b** $\dfrac{2 \tan \frac{\pi}{6}}{1 - \tan^2 \frac{\pi}{6}}$ **c** $\dfrac{4 \sin 70°}{\sec 70°}$

a $\cos^2 50° - \sin^2 50° = \cos(2 \times 50°)$ — Use $\cos 2A \equiv \cos^2 A - \sin^2 A$ in reverse, with $A = 50°$.
$= \cos 100°$

b $\dfrac{2 \tan \frac{\pi}{6}}{1 - \tan^2 \frac{\pi}{6}} = \tan\left(2 \times \dfrac{\pi}{6}\right)$ — Use $\tan 2A \equiv \dfrac{2 \tan A}{1 - \tan^2 A}$ in reverse, with $A = \dfrac{\pi}{6}$.

$= \tan \dfrac{\pi}{3}$

c $\dfrac{4 \sin 70°}{\sec 70°} = 4 \sin 70° \cos 70°$ — $\sec x = \dfrac{1}{\cos x}$ so $\cos x = \dfrac{1}{\sec x}$

$= 2(2 \sin 70° \cos 70°)$ — Recognise this is a multiple of $2 \sin A \cos A$.

$= 2 \sin(2 \times 70°) = 2 \sin 140°$ — Use $\sin 2A \equiv 2 \sin A \cos A$ in reverse with $A = 70°$.

Example 8

Given that $x = 3 \sin \theta$ and $y = 3 - 4 \cos 2\theta$, eliminate θ and express y in terms of x.

The equations can be written as

$\sin \theta = \dfrac{x}{3}$ $\cos 2\theta = \dfrac{3-y}{4}$

Watch out Be careful with this manipulation. Many errors can occur in the early part of a solution.

As $\cos 2\theta \equiv 1 - 2\sin^2 \theta$ for all values of θ,

$\dfrac{3-y}{4} = 1 - 2\left(\dfrac{x}{3}\right)^2$ — θ has been eliminated from this equation. We still need to solve for y.

So $\dfrac{y}{4} = 2\left(\dfrac{x}{3}\right)^2 - \dfrac{1}{4}$

or $y = 8\left(\dfrac{x}{3}\right)^2 - 1$ — The final answer should be in the form $y = \ldots$

Example 9

Given that $\cos x = \frac{3}{4}$, and that $180° < x < 360°$, find the exact value of:

a $\sin 2x$ **b** $\tan 2x$

a $\sin^2 A = 1 - \cos^2 A$ —— Use $\sin^2 A + \cos^2 A = 1$ to determine $\sin A$.

$= 1 - \left(\frac{3}{4}\right)^2$

$= \frac{7}{16}$

$180° < A < 360°$, so $\sin A = -\frac{\sqrt{7}}{4}$ —— $\sin A$ is negative in the third and fourth quadrants, so choose the negative square root.

$\sin 2x = 2 \sin x \cos x$

$= 2\left(-\frac{\sqrt{7}}{4}\right)\left(\frac{3}{4}\right) = -\frac{3\sqrt{7}}{8}$

b $\tan x = \frac{\sin x}{\cos x} = \frac{-\frac{\sqrt{7}}{4}}{\frac{3}{4}}$ —— Find $\tan x$ in simplified surd form, then substitute this value into the double-angle formula for $\tan 2x$.

$= -\frac{\sqrt{7}}{3}$

$\tan 2x = \frac{2 \tan x}{1 - \tan^2 x} = \frac{-\frac{2\sqrt{7}}{3}}{1 - \frac{7}{9}}$ —— Make sure you square all of $\tan x$ when working out $\tan^2 x$:

$\left(-\frac{\sqrt{7}}{3}\right)^2 = \frac{7}{9}$

$= -\frac{2\sqrt{7}}{3} \times \frac{9}{2}$

$= -3\sqrt{7}$

Exercise 7C

P **1** Use the expansion of $\sin(A + B)$ to show that $\sin 2A \equiv 2 \sin A \cos A$.

Hint Set $B = A$.

P **2 a** Using the identity $\cos(A + B) \equiv \cos A \cos B - \sin A \sin B$, show that $\cos 2A \equiv \cos^2 A - \sin^2 A$.

 b Hence show that:
  i $\cos 2A \equiv 2\cos^2 A - 1$
  ii $\cos 2A \equiv 1 - 2\sin^2 A$

Problem-solving

Use $\sin^2 A + \cos^2 A \equiv 1$

P **3** Use the expansion of $\tan(A + B)$ to express $\tan 2A$ in terms of $\tan A$.

Chapter 7

4 Write each of the following expressions as a single trigonometric ratio.

  a $2 \sin 10° \cos 10°$
  b $1 - 2 \sin^2 25°$
  c $\cos^2 40° - \sin^2 40°$

  d $\dfrac{2 \tan 5°}{1 - \tan^2 5°}$
  e $\dfrac{1}{2 \sin (24.5)° \cos (24.5)°}$
  f $6 \cos^2 30° - 3$

  g $\dfrac{\sin 8°}{\sec 8°}$
  h $\cos^2 \dfrac{\pi}{16} - \sin^2 \dfrac{\pi}{16}$

5 Without using your calculator find the exact values of:

  a $2 \sin 22.5° \cos 22.5°$
  b $2 \cos^2 15° - 1$

  c $(\sin 75° - \cos 75°)^2$
  d $\dfrac{2 \tan \dfrac{\pi}{8}}{1 - \tan^2 \dfrac{\pi}{8}}$

6 a Show that $(\sin A + \cos A)^2 \equiv 1 + \sin 2A$. **(3 marks)**

  b Hence find the exact value of $\left(\sin \dfrac{\pi}{8} + \cos \dfrac{\pi}{8}\right)^2$. **(2 marks)**

7 Write the following in their simplest form, involving only one trigonometric function:

  a $\cos^2 3\theta - \sin^2 3\theta$
  b $6 \sin 2\theta \cos 2\theta$
  c $\dfrac{2 \tan \dfrac{\theta}{2}}{1 - \tan^2 \dfrac{\theta}{2}}$

  d $2 - 4 \sin^2 \dfrac{\theta}{2}$
  e $\sqrt{1 + \cos 2\theta}$
  f $\sin^2 \theta \cos^2 \theta$

  g $4 \sin \theta \cos \theta \cos 2\theta$
  h $\dfrac{\tan \theta}{\sec^2 \theta - 2}$
  i $\sin^4 \theta - 2 \sin^2 \theta \cos^2 \theta + \cos^4 \theta$

8 Given that $p = 2 \cos \theta$ and $q = \cos 2\theta$, express q in terms of p.

9 Eliminate θ from the following pairs of equations:

  a $x = \cos^2 \theta,\ y = 1 - \cos 2\theta$
  b $x = \tan \theta,\ y = \cot 2\theta$

  c $x = \sin \theta,\ y = \sin 2\theta$
  d $x = 3 \cos 2\theta + 1,\ y = 2 \sin \theta$

10 Given that $\cos x = \tfrac{1}{4}$, find the exact value of $\cos 2x$.

11 Find the possible values of $\sin \theta$ when $\cos 2\theta = \tfrac{23}{25}$.

12 Given that $\tan \theta = \tfrac{3}{4}$, and that θ is acute,

  a find the exact value of: **i** $\tan 2\theta$ **ii** $\sin 2\theta$ **iii** $\cos 2\theta$

  b deduce the value of $\sin 4\theta$.

Trigonometry and modelling

P 13 Given that $\cos A = -\frac{1}{3}$, and that A is obtuse,

  a find the exact value of: **i** $\cos 2A$ **ii** $\sin A$ **iii** $\csc 2A$

  b show that $\tan 2A = \frac{4\sqrt{2}}{7}$

E/P 14 Given that $\pi < \theta < \frac{3\pi}{2}$, find the value of $\tan\frac{\theta}{2}$ when $\tan\theta = \frac{3}{4}$ **(4 marks)**

E/P 15 Given that $\cos x + \sin x = m$ and $\cos x - \sin x = n$, where m and n are constants, write down, in terms of m and n, the value of $\cos 2x$. **(4 marks)**

E/P 16 In $\triangle PQR$, $PQ = 3$ cm, $PR = 6$ cm, $QR = 5$ cm and $\angle QPR = 2\theta$.

  a Use the cosine rule to show that $\cos 2\theta = \frac{5}{9}$ **(3 marks)**

  b Hence find the exact value of $\sin\theta$. **(2 marks)**

E/P 17 The line l, with equation $y = \frac{3}{4}x$, bisects the angle between the x-axis and the line $y = mx$, $m > 0$. Given that the scales on each axis are the same, and that l makes an angle θ with the x-axis,

  a write down the value of $\tan\theta$ **(1 mark)**

  b show that $m = \frac{24}{7}$ **(3 marks)**

E/P 18 **a** Use the identity $\cos(A + B) \equiv \cos A \cos B - \sin A \sin B$, to show that $\cos 2A \equiv 2\cos^2 A - 1$. **(2 marks)**

  The curves C_1 and C_2 have equations
  $C_1: y = 4\cos 2x$
  $C_2: y = 6\cos^2 x - 3\sin 2x$

  b Show that the x-coordinates of the points where C_1 and C_2 intersect satisfy the equation
  $\cos 2x + 3\sin 2x - 3 = 0$ **(3 marks)**

P 19 Use the fact that $\tan 2A \equiv \frac{\sin 2A}{\cos 2A}$ to derive the formula for $\tan 2A$ in terms of $\tan A$.

> **Hint** Use the identities for $\sin 2A$ and $\cos 2A$ and then divide both the numerator and denominator by $\cos^2 A$.

7.4 Solving trigonometric equations

You can use the addition formulae and the double-angle formulae to help you solve trigonometric equations.

Example 10

Solve $4\cos(\theta - 30°) = 8\sqrt{2}\sin\theta$ in the range $0 \leq \theta \leq 360°$. Round your answer to 1 decimal place.

$4\cos(\theta - 30°) = 8\sqrt{2}\sin\theta$

$4\cos\theta\cos 30° + 4\sin\theta\sin 30° = 8\sqrt{2}\sin\theta$ — Use the formula for $\cos(A - B)$.

$4\cos\theta\left(\dfrac{\sqrt{3}}{2}\right) + 4\sin\theta\left(\dfrac{1}{2}\right) = 8\sqrt{2}\sin\theta$ — Substitute $\cos 30° = \dfrac{\sqrt{3}}{2}$ and $\sin 30° = \dfrac{1}{2}$.

$2\sqrt{3}\cos\theta + 2\sin\theta = 8\sqrt{2}\sin\theta$

$2\sqrt{3}\cos\theta = (8\sqrt{2} - 2)\sin\theta$ — Gather cosine terms on the LHS and sine terms on the RHS of the equation.

$\dfrac{2\sqrt{3}}{8\sqrt{2} - 2} = \tan\theta$ — Divide both sides by $\cos\theta$ and by $(8\sqrt{2} - 2)$.

$\tan\theta = 0.3719\ldots$

$\theta = 20.4°, 200.4°$ — Use a CAST diagram or a sketch graph to find all the solutions in the given range.

Example 11

Solve $3\cos 2x - \cos x + 2 = 0$ for $0 \leq x \leq 360°$.

Using a double angle formula for $\cos 2x$

$3\cos 2x - \cos x + 2 = 0$

becomes

$3(2\cos^2 x - 1) - \cos x + 2 = 0$

$6\cos^2 x - 3 - \cos x + 2 = 0$

$6\cos^2 x - \cos x - 1 = 0$

So $(3\cos x + 1)(2\cos x - 1) = 0$

Solving: $\cos x = -\dfrac{1}{3}$ or $\cos x = \dfrac{1}{2}$

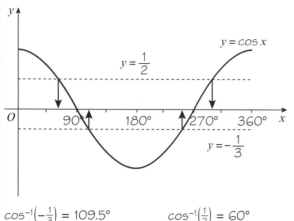

$\cos^{-1}\left(-\dfrac{1}{3}\right) = 109.5°$ $\cos^{-1}\left(\dfrac{1}{2}\right) = 60°$

So $x = 60°, 109.5°, 250.5°, 300°$

Problem-solving

Choose the double angle formula for $\cos 2x$ which only involves $\cos x$:
$\cos 2x \equiv 2\cos^2 x - 1$
This will give you a quadratic equation in $\cos x$.

This quadratic equation factorises:
$6X^2 - X - 1 = (3X + 1)(2X - 1)$

Trigonometry and modelling

Example 12

Solve $2 \tan 2y \tan y = 3$ for $0 \leq y \leq 2\pi$. Give your answers to 2 decimal places.

$2 \tan 2y \tan y = 3$

$2 \left(\dfrac{2 \tan y}{1 - \tan^2 y} \right) \tan y = 3$ ——— Use the double-angle identity for tan.

$\dfrac{4 \tan^2 y}{1 - \tan^2 y} = 3$

$4 \tan^2 y = 3 - 3 \tan^2 y$ ——— This is a quadratic equation in $\tan y$. Because there is a $\tan^2 y$ term but no $\tan y$ term you can solve it directly.

$7 \tan^2 y = 3$

$\tan^2 y = \tfrac{3}{7}$

$\tan y = \pm \sqrt{\tfrac{3}{7}}$

Watch out Remember to include the positive and negative square roots.

$y = 0.58, 2.56, 3.72, 5.70$

Example 13

a By expanding $\sin(2A + A)$ show that $\sin 3A \equiv 3 \sin A - 4 \sin^3 A$.

b Hence, or otherwise, for $0 < \theta < 2\pi$, solve $16 \sin^3 \theta - 12 \sin \theta - 2\sqrt{3} = 0$ giving your answers in terms of π.

a LHS $\equiv \sin 3A \equiv \sin(2A + A)$

$\equiv \sin 2A \cos A + \cos 2A \sin A$ ——— Use the addition formula for $\sin(A + B)$.

$\equiv (2 \sin A \cos A) \cos A$
$\quad + (1 - 2 \sin^2 A) \sin A$

Substitute for $\sin 2A$ and $\cos 2A$. As the answer is in terms of $\sin A$, $\cos 2A \equiv 1 - 2 \sin^2 A$ is the best identity to use.

$\equiv 2 \sin A \cos^2 A + \sin A - 2 \sin^3 A$

$\equiv 2 \sin A(1 - \sin^2 A) + \sin A - 2 \sin^3 A$ ——— Use $\sin^2 A + \cos^2 A \equiv 1$ to substitute for $\cos^2 A$.

$\equiv 2 \sin A - 2 \sin^3 A + \sin A - 2 \sin^3 A$

$\equiv 3 \sin A - 4 \sin^3 A \equiv$ RHS

b $16 \sin^3 \theta - 12 \sin \theta - 2\sqrt{3} = 0$

$16 \sin^3 \theta - 12 \sin \theta = 2\sqrt{3}$

$-4 \sin 3\theta = 2\sqrt{3}$

$\sin 3\theta = -\dfrac{\sqrt{3}}{2}$

Problem-solving

The question says 'hence' so look for an opportunity to use the identity you proved in part **a**. You need to multiply both sides of the identity by -4.

$3\theta = \dfrac{4\pi}{3}, \dfrac{5\pi}{3}, \dfrac{10\pi}{3}, \dfrac{11\pi}{3}, \dfrac{16\pi}{3}, \dfrac{17\pi}{3}$

$\theta = \dfrac{4\pi}{9}, \dfrac{5\pi}{9}, \dfrac{10\pi}{9}, \dfrac{11\pi}{9}, \dfrac{16\pi}{9}, \dfrac{17\pi}{9}$

Use a CAST diagram or a sketch graph to find all answers for 3θ. $0 < \theta < 2\pi$ so $0 < 3\theta < 6\pi$.

Exercise 7D

P 1 Solve, in the interval $0 \leq \theta < 360°$, the following equations. Give your answers to 1 d.p.
  a $3\cos\theta = 2\sin(\theta + 60°)$
  b $\sin(\theta + 30°) + 2\sin\theta = 0$
  c $\cos(\theta + 25°) + \sin(\theta + 65°) = 1$
  d $\cos\theta = \cos(\theta + 60°)$

E/P 2 a Show that $\sin\left(\theta + \dfrac{\pi}{4}\right) \equiv \dfrac{1}{\sqrt{2}}(\sin\theta + \cos\theta)$ **(2 marks)**

  b Hence, or otherwise, solve the equation $\dfrac{1}{\sqrt{2}}(\sin\theta + \cos\theta) = \dfrac{1}{\sqrt{2}}$, $0 \leq \theta \leq 2\pi$. **(4 marks)**

  c Use your answer to part **b** to write down the solutions to $\sin\theta + \cos\theta = 1$ over the same interval. **(2 marks)**

P 3 a Solve the equation $\cos\theta\cos 30° - \sin\theta\sin 30° = 0.5$, for $0 \leq \theta \leq 360°$.

  b Hence write down, in the same interval, the solutions of $\sqrt{3}\cos\theta - \sin\theta = 1$.

P 4 a Given that $3\sin(x - y) - \sin(x + y) = 0$, show that $\tan x = 2\tan y$.

  b Solve $3\sin(x - 45°) - \sin(x + 45°) = 0$, for $0 \leq x \leq 360°$.

P 5 Solve the following equations, in the intervals given.
  a $\sin 2\theta = \sin\theta$, $0 \leq \theta \leq 2\pi$
  b $\cos 2\theta = 1 - \cos\theta$, $-180° < \theta \leq 180°$
  c $3\cos 2\theta = 2\cos^2\theta$, $0 \leq \theta < 360°$
  d $\sin 4\theta = \cos 2\theta$, $0 \leq \theta \leq \pi$
  e $3\cos\theta - \sin\dfrac{\theta}{2} - 1 = 0$, $0 \leq \theta < 720°$
  f $\cos^2\theta - \sin 2\theta = \sin^2\theta$, $0 \leq \theta \leq \pi$
  g $2\sin\theta = \sec\theta$, $0 \leq \theta \leq 2\pi$
  h $2\sin 2\theta = 3\tan\theta$, $0 \leq \theta < 360°$
  i $2\tan\theta = \sqrt{3}(1 - \tan\theta)(1 + \tan\theta)$, $0 \leq \theta \leq 2\pi$
  j $\sin^2\theta = 2\sin 2\theta$, $-180° < \theta < 180°$
  k $4\tan\theta = \tan 2\theta$, $0 \leq \theta \leq 360°$

E/P 6 In $\triangle ABC$, $AB = 4$ cm, $AC = 5$ cm, $\angle ABC = 2\theta$ and $\angle ACB = \theta$. Find the value of θ, giving your answer, in degrees, to 1 decimal place. **(4 marks)**

E/P 7 a Show that $5\sin 2\theta + 4\sin\theta = 0$ can be written in the form $a\sin\theta(b\cos\theta + c) = 0$, stating the values of a, b and c. **(2 marks)**

  b Hence solve, for $0 \leq \theta < 360°$, the equation $5\sin 2\theta + 4\sin\theta = 0$. **(4 marks)**

E/P 8 a Given that $\sin 2\theta + \cos 2\theta = 1$, show that $2\sin\theta(\cos\theta - \sin\theta) = 0$. **(2 marks)**

  b Hence, or otherwise, solve the equation $\sin 2\theta + \cos 2\theta = 1$ for $0 \leq \theta < 360°$. **(4 marks)**

E/P 9 a Prove that $(\cos 2\theta - \sin 2\theta)^2 \equiv 1 - \sin 4\theta$. **(4 marks)**

  b Use the result to solve, for $0 \leq \theta < \pi$, the equation $\cos 2\theta - \sin 2\theta = \dfrac{1}{\sqrt{2}}$.
  Give your answers in terms of π. **(3 marks)**

P) 10 a Show that:

i $\sin\theta \equiv \dfrac{2\tan\dfrac{\theta}{2}}{1+\tan^2\dfrac{\theta}{2}}$

ii $\cos\theta \equiv \dfrac{1-\tan^2\dfrac{\theta}{2}}{1+\tan^2\dfrac{\theta}{2}}$

b By writing the following equations as quadratics in $\tan\dfrac{\theta}{2}$, solve, in the interval $0 \leqslant \theta \leqslant 360°$:

i $\sin\theta + 2\cos\theta = 1$

ii $3\cos\theta - 4\sin\theta = 2$

E/P) 11 a Show that $3\cos^2 x - \sin^2 x \equiv 1 + 2\cos 2x$. **(3 marks)**

b Hence sketch, for $-\pi \leqslant x \leqslant \pi$, the graph of $y = 3\cos^2 x - \sin^2 x$, showing the coordinates of points where the curve meets the axes. **(3 marks)**

E/P) 12 a Express $2\cos^2\dfrac{\theta}{2} - 4\sin^2\dfrac{\theta}{2}$ in the form $a\cos\theta + b$, where a and b are constants. **(4 marks)**

b Hence solve $2\cos^2\dfrac{\theta}{2} - 4\sin^2\dfrac{\theta}{2} = -3$, in the interval $0 \leqslant \theta < 360°$. **(3 marks)**

E/P) 13 a Use the identity $\sin^2 A + \cos^2 A \equiv 1$ to show that $\sin^4 A + \cos^4 A \equiv \tfrac{1}{2}(2 - \sin^2 2A)$. **(5 marks)**

b Deduce that $\sin^4 A + \cos^4 A \equiv \tfrac{1}{4}(3 + \cos 4A)$. **(3 marks)**

c Hence solve $8\sin^4\theta + 8\cos^4\theta = 7$, for $0 < \theta < \pi$. **(3 marks)**

Hint Start by squaring $(\sin^2 A + \cos^2 A)$.

E/P) 14 a By writing 3θ as $2\theta + \theta$, show that $\cos 3\theta \equiv 4\cos^3\theta - 3\cos\theta$. **(4 marks)**

b Hence, or otherwise, for $0 < \theta < \pi$, solve $6\cos\theta - 8\cos^3\theta + 1 = 0$ giving your answer in terms of π. **(5 marks)**

7.5 Simplifying $a\cos x \pm b\sin x$

You can use the addition formulae to simplify some trigonometric expressions:

- **For positive values of a and b,**
  - $a\sin x \pm b\cos x$ can be expressed in the form $R\sin(x \pm \alpha)$
  - $a\cos x \pm b\sin x$ can be expressed in the form $R\cos(x \mp \alpha)$

  with $R > 0$ and $0 < \alpha < 90°$ $\left(\text{or } \dfrac{\pi}{2}\right)$

  where $R\cos\alpha = a$ and $R\sin\alpha = b$ and $R = \sqrt{a^2 + b^2}$.

Use the addition formulae to expand $\sin(x \pm \alpha)$ or $\cos(x \mp \alpha)$, then equate coefficients.

Notation The symbol $\mp$ means that $a\cos x + b\sin x$ will be written in the form $R\cos(x - \alpha)$, and $a\cos x - b\sin x$ will be written in the form $R\cos(x + \alpha)$.

Example 14

Show that you can express $3\sin x + 4\cos x$ in the form:

a $R\sin(x + \alpha)$

b $R\cos(x - \beta)$

where $R > 0$, $0 < \alpha < 90°$, $0 < \beta < 90°$ giving your values of R, α and β to 1 decimal place when appropriate.

a $R\sin(x + \alpha) \equiv R\sin x \cos\alpha + R\cos x \sin\alpha$ — Use $\sin(A + B) \equiv \sin A \cos B + \cos A \sin B$ and multiply through by R.

Let $3\sin x + 4\cos x \equiv R\sin x \cos\alpha + R\cos x \sin\alpha$

So $R\cos\alpha = 3$ and $R\sin\alpha = 4$ — Equate the coefficients of the $\sin x$ and $\cos x$ terms.

$\dfrac{R\sin\alpha}{R\cos\alpha} = \tan\alpha = \dfrac{4}{3}$

$\alpha = \tan^{-1}\left(\dfrac{4}{3}\right)$ — Divide the equations to eliminate R and use $\tan^{-1}$ to find α.

So $\alpha = 53.1°$ (1 d.p.)

$R^2\cos^2\alpha + R^2\sin^2\alpha = 3^2 + 4^2$ — Square and add the equations to eliminate α and find R^2.

$R^2(\cos^2\alpha + \sin^2\alpha) = 25$

$R^2 = 25$, so $R = 5$ — Use $\sin^2\alpha + \cos^2\alpha \equiv 1$.

$3\sin x + 4\cos x \equiv 5\sin(x + 53.1°)$

b $R\cos(x - \beta) \equiv R\cos x \cos\beta + R\sin x \sin\beta$ — Use $\cos(A - B) \equiv \cos A \cos B + \sin A \sin B$ and multiply through by R.

Let $3\sin x + 4\cos x \equiv R\cos x \cos\beta + R\sin x \sin\beta$

So $R\cos\beta = 4$ and $R\sin\beta = 3$ — Equate the coefficients of the $\cos x$ and $\sin x$ terms.

$\dfrac{R\sin\beta}{R\cos\beta} = \tan\beta = \dfrac{3}{4}$ — Divide the equations to eliminate R.

So $\beta = 36.9°$ (1 d.p.)

$R^2\cos^2\beta + R^2\sin^2\beta = 3^2 + 4^2$ — Square and add the equations to eliminate α and find R^2.

$R^2(\cos^2\beta + \sin^2\beta) = 25$

$R^2 = 25$, so $R = 5$ — Remember $\sin^2\alpha + \cos^2\alpha \equiv 1$.

$3\sin x + 4\cos x \equiv 5\cos(x - 36.9°)$

Online Explore how you can transform the graphs of $y = \sin x$ and $y = \cos x$ to obtain the graph of $y = 3\sin x + 4\cos x$ using technology.

Trigonometry and modelling

Example 15

a Show that you can express $\sin x - \sqrt{3} \cos x$ in the form $R \sin(x - \alpha)$, where $R > 0$, $0 < \alpha < \dfrac{\pi}{2}$

b Hence sketch the graph of $y = \sin x - \sqrt{3} \cos x$.

a Set $\sin x - \sqrt{3} \cos x \equiv R \sin(x - \alpha)$
$\sin x - \sqrt{3} \cos x \equiv R \sin x \cos \alpha - R \cos x \sin \alpha$ — Expand $\sin(x - \alpha)$ and multiply by R.

So $R \cos \alpha = 1$ and $R \sin \alpha = \sqrt{3}$ — Equate the coefficients of $\sin x$ and $\cos x$ on both sides of the identity.

Dividing, $\tan \alpha = \sqrt{3}$, so $\alpha = \dfrac{\pi}{3}$

Squaring and adding: $R = 2$

So $\sin x - \sqrt{3} \cos x \equiv 2 \sin\left(x - \dfrac{\pi}{3}\right)$

b $y = \sin x - \sqrt{3} \cos x \equiv 2 \sin\left(x - \dfrac{\pi}{3}\right)$

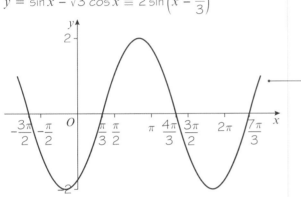

You can sketch $y = 2 \sin\left(x - \dfrac{\pi}{3}\right)$ by translating $y = \sin x$ by $\dfrac{\pi}{3}$ to the right and then stretching by a scale factor of 2 in the y-direction.

Example 16

a Express $2 \cos \theta + 5 \sin \theta$ in the form $R \cos(\theta - \alpha)$, where $R > 0$, $0 < \alpha < 90°$.

b Hence solve, for $0 < \theta < 360°$, the equation $2 \cos \theta + 5 \sin \theta = 3$.

a Set $2 \cos \theta + 5 \sin \theta \equiv R \cos \theta \cos \alpha + R \sin \theta \sin \alpha$

So $R \cos \alpha = 2$ and $R \sin \alpha = 5$ — Equate the coefficients of $\sin x$ and $\cos x$ on both sides of the identity.

Dividing, $\tan \alpha = \dfrac{5}{2}$, so $\alpha = 68.2°$

Squaring and adding: $R = \sqrt{29}$

So $2 \cos \theta + 5 \sin \theta \equiv \sqrt{29} \cos(\theta - 68.2°)$ — Use the result from part **a**: $2 \cos \theta + 5 \sin \theta \equiv \sqrt{29} \cos(\theta - 68.2°)$.

b $\sqrt{29} \cos(\theta - 68.2°) = 3$ — Divide both sides by $\sqrt{29}$.

So $\cos(\theta - 68.2°) = \dfrac{3}{\sqrt{29}}$

$\cos^{-1}\left(\dfrac{3}{\sqrt{29}}\right) = 56.1...°$ — As $0 < \theta < 360°$, the interval for $(\theta - 68.2°)$ is $-68.2° < \theta - 68.2° < 291.8°$.

So $\theta - 68.2° = -56.1...°, 56.1...°$

$\dfrac{3}{\sqrt{29}}$ is positive, so solutions for $\theta - 68.2°$ are in the 1st and 4th quadrants.

$\theta = 12.1°, 124.3°$ (to the nearest 0.1°)

Example 17

$f(\theta) = 12\cos\theta + 5\sin\theta$

a Write $f(\theta)$ in the form $R\cos(\theta - \alpha)$.

b Find the maximum value of $f(\theta)$ and the smallest positive value of θ at which it occurs.

Online Use technology to explore maximums and minimums of curves in the form $R\cos(\theta - \alpha)$.

a Set $12\cos\theta + 5\sin\theta \equiv R\cos(\theta - \alpha)$ — Equate $\sin x$ and $\cos x$ terms and then solve for R and α.

So $12\cos\theta + 5\sin\theta \equiv R\cos\theta\cos\alpha + R\sin\theta\sin\alpha$

So $R\cos\alpha = 12$ and $R\sin\alpha = 5$

$R = 13$ and $\tan\alpha = \frac{5}{12} \Rightarrow \alpha = 22.6°$

So $12\cos\theta + 5\sin\theta \equiv 13\cos(\theta - 22.6°)$

b The maximum value of $13\cos(\theta - 22.6°)$ is 13.

This occurs when $\cos(\theta - 22.6°) = 1$

$\theta - 22.6° = ..., -360°, 0°, 360°, ...$

The smallest positive value of θ is 22.6°.

The maximum value of $\cos x$ is 1 so the maximum value of $\cos(\theta - 22.6°)$ is also 1.

Solve the equation to find the smallest positive value of θ.

Exercise 7E

Unless otherwise stated, give all angles to 1 decimal place and write non-integer values of R in surd form.

1 Given that $5\sin\theta + 12\cos\theta \equiv R\sin(\theta + \alpha)$, find the value of R, $R > 0$, and the value of $\tan\alpha$.

2 Given that $\sqrt{3}\sin\theta + \sqrt{6}\cos\theta \equiv 3\cos(\theta - \alpha)$, where $0 < \alpha < 90°$, find the value of α.

3 Given that $2\sin\theta - \sqrt{5}\cos\theta \equiv -3\cos(\theta + \alpha)$, where $0 < \alpha < 90°$, find the value of α.

4 **a** Show that $\cos\theta - \sqrt{3}\sin\theta$ can be written in the form $R\cos(\theta + \alpha)$, with $R > 0$ and $0 < \alpha < \frac{\pi}{2}$

  b Hence sketch the graph of $y = \cos\theta - \sqrt{3}\sin\theta$, $0 < \theta < \frac{\pi}{2}$, giving the coordinates of points of intersection with the axes.

(P) 5 **a** Express $7\cos\theta - 24\sin\theta$ in the form $R\cos(\theta + \alpha)$, with $R > 0$ and $0 < \alpha < 90°$.

  b The graph of $y = 7\cos\theta - 24\sin\theta$ meets the y-axis at P. State the coordinates of P.

  c Write down the maximum and minimum values of $7\cos\theta - 24\sin\theta$.

  d Deduce the number of solutions, in the interval $0 < \theta < 360°$, of the following equations:

  i $7\cos\theta - 24\sin\theta = 15$ **ii** $7\cos\theta - 24\sin\theta = 26$ **iii** $7\cos\theta - 24\sin\theta = -25$

(E) 6 $f(\theta) = \sin\theta + 3\cos\theta$

  Given $f(\theta) \equiv R\sin(\theta + \alpha)$, where $R > 0$ and $0 < \alpha < 90°$.

  a Find the value of R and the value of α. **(4 marks)**

  b Hence, or otherwise, solve $f(\theta) = 2$ for $0 \leq \theta < 360°$. **(3 marks)**

Trigonometry and modelling

7 **a** Express $\cos 2\theta - 2\sin 2\theta$ in the form $R\cos(2\theta + \alpha)$, where $R > 0$ and $0 < \alpha < \dfrac{\pi}{2}$
Give the value of α to 3 decimal places. **(4 marks)**

b Hence, or otherwise, solve for $0 \leqslant \theta < \pi$, $\cos 2\theta - 2\sin 2\theta = -1.5$, rounding your answers to 2 decimal places. **(4 marks)**

8 Solve the following equations, in the intervals given in brackets.
a $6\sin x + 8\cos x = 5\sqrt{3}$, $[0, 360°]$ **b** $2\cos 3\theta - 3\sin 3\theta = -1$, $[0, 90°]$
c $8\cos \theta + 15\sin \theta = 10$, $[0, 360°]$ **d** $5\sin\dfrac{x}{2} - 12\cos\dfrac{x}{2} = -6.5$, $[-360°, 360°]$

9 **a** Express $3\sin 3\theta - 4\cos 3\theta$ in the form $R\sin(3\theta - \alpha)$, with $R > 0$ and $0 < \alpha < 90°$. **(3 marks)**
b Hence write down the minimum value of $3\sin 3\theta - 4\cos 3\theta$ and the value of θ at which it occurs. **(3 marks)**
c Solve, for $0 \leqslant \theta < 180°$, the equation $3\sin 3\theta - 4\cos 3\theta = 1$. **(3 marks)**

10 **a** Express $5\sin^2\theta - 3\cos^2\theta + 6\sin\theta\cos\theta$ in the form $a\sin 2\theta + b\cos 2\theta + c$, where a, b and c are constants to be found. **(3 marks)**
b Hence find the maximum and minimum values of $5\sin^2\theta - 3\cos^2\theta + 6\sin\theta\cos\theta$. **(4 marks)**
c Solve $5\sin^2\theta - 3\cos^2\theta + 6\sin\theta\cos\theta = -1$ for $0 \leqslant \theta < 180°$, rounding your answers to 1 decimal place. **(4 marks)**

11 A class were asked to solve $3\cos\theta = 2 - \sin\theta$ for $0 \leqslant \theta < 360°$. One student expressed the equation in the form $R\cos(\theta - \alpha) = 2$, with $R > 0$ and $0 < \alpha < 90°$, and correctly solved the equation.
a Find the values of R and α and hence find her solutions.
Another student decided to square both sides of the equation and then form a quadratic equation in $\sin\theta$.
b Show that the correct quadratic equation is $10\sin^2\theta - 4\sin\theta - 5 = 0$.
c Solve this equation, for $0 \leqslant \theta < 360°$.
d Explain why not all of the answers satisfy $3\cos\theta = 2 - \sin\theta$.

12 **a** Given $\cot\theta + 2 = \operatorname{cosec}\theta$, show that $2\sin\theta + \cos\theta = 1$. **(4 marks)**
b Solve $\cot\theta + 2 = \operatorname{cosec}\theta$ for $0 \leqslant \theta < 360°$. **(3 marks)**

13 **a** Given $\sqrt{2}\cos\left(\theta - \dfrac{\pi}{4}\right) + (\sqrt{3} - 1)\sin\theta = 2$, show that $\cos\theta + \sqrt{3}\sin\theta = 2$. **(4 marks)**
b Solve $\sqrt{2}\cos\left(\theta - \dfrac{\pi}{4}\right) + (\sqrt{3} - 1)\sin\theta = 2$ for $0 \leqslant \theta \leqslant 2\pi$. **(2 marks)**

185

Chapter 7

E/P **14** **a** Express $9\cos\theta + 40\sin\theta$ in the form $R\cos(\theta - \alpha)$, where $R > 0$ and $0 < \alpha < 90°$.
Give the value of α to 3 decimal places. **(4 marks)**

b $g(\theta) = \dfrac{18}{50 + 9\cos\theta + 40\sin\theta}$, $0 \leqslant \theta \leqslant 360°$

Calculate:
  i the minimum value of $g(\theta)$ **(2 marks)**
  ii the smallest positive value of θ at which the minimum occurs. **(2 marks)**

E/P **15** $p(\theta) = 12\cos 2\theta - 5\sin 2\theta$
Given that $p(\theta) = R\cos(2\theta + \alpha)$, where $R > 0$ and $0 < \alpha < 90°$,

a find the value of R and the value of α. **(3 marks)**
b Hence solve the equation $12\cos 2\theta - 5\sin 2\theta = -6.5$ for $0 \leqslant \theta < 180°$. **(5 marks)**
c Express $24\cos^2\theta - 10\sin\theta\cos\theta$ in the form $a\cos 2\theta + b\sin 2\theta + c$, where a, b and c are constants to be found. **(3 marks)**
d Hence, or otherwise, find the minimum value of $24\cos^2\theta - 10\sin\theta\cos\theta$. **(2 marks)**

7.6 Proving trigonometric identities

You can use known trigonometric identities to prove other identities.

Example 18

a Show that $2\sin\dfrac{\theta}{2}\cos\dfrac{\theta}{2}\cos\theta \equiv \dfrac{1}{2}\sin 2\theta$.

b Show that $1 + \cos 4\theta \equiv 2\cos^2 2\theta$.

---

a $\sin 2A \equiv 2\sin A \cos A$

$\sin\theta \equiv 2\sin\dfrac{\theta}{2}\cos\dfrac{\theta}{2}$

LHS $\equiv 2\sin\dfrac{\theta}{2}\cos\dfrac{\theta}{2}\cos\theta$

$\equiv \sin\theta\cos\theta$

$\equiv \dfrac{1}{2}\sin 2\theta$

$\equiv$ RHS

b LHS $\equiv 1 + \cos 4\theta$

$\equiv 1 + 2\cos^2 2\theta - 1$

$\equiv 2\cos^2 2\theta$

$\equiv$ RHS

— Substitute $A = \dfrac{\theta}{2}$ into the formula for $\sin 2A$.

Problem-solving

Always be aware that the addition formulae can be altered by making a substitution.

— Use the above result for $2\sin\dfrac{\theta}{2}\cos\dfrac{\theta}{2}$.

— Remember $\sin 2\theta \equiv 2\sin\theta\cos\theta$.

— Use $\cos 2A \equiv 2\cos^2 A - 1$ with $A = 2\theta$.

Trigonometry and modelling

Example 19

Prove the identity $\tan 2\theta \equiv \dfrac{2}{\cot \theta - \tan \theta}$

LHS $\equiv \tan 2\theta \equiv \dfrac{2 \tan \theta}{1 - \tan^2 \theta}$

Divide the numerator and denominator by $\tan \theta$.

So $\tan 2\theta \equiv \dfrac{2}{\dfrac{1}{\tan \theta} - \tan \theta}$

$\equiv \dfrac{2}{\cot \theta - \tan \theta}$

Problem-solving

Dividing the numerator and denominator by a common term can be helpful when trying to rearrange an expression into a required form.

Example 20

Prove that $\sqrt{3} \cos 4\theta + \sin 4\theta \equiv 2 \cos\left(4\theta - \dfrac{\pi}{6}\right)$.

RHS $\equiv 2 \cos\left(4\theta - \dfrac{\pi}{6}\right)$

$\equiv 2 \cos 4\theta \cos \dfrac{\pi}{6} + 2 \sin 4\theta \sin \dfrac{\pi}{6}$

$\equiv 2 \cos 4\theta \left(\dfrac{\sqrt{3}}{2}\right) + 2 \sin 4\theta \left(\dfrac{1}{2}\right)$

$\equiv \sqrt{3} \cos 4\theta + \sin 4\theta \equiv$ LHS

Problem-solving

Sometimes it is easier to begin with the RHS of the identity.

Use the addition formulae.

Write the exact values of $\cos \dfrac{\pi}{6}$ and $\sin \dfrac{\pi}{6}$

Exercise 7F

(P) **1** Prove the following identities.

a $\dfrac{\cos 2A}{\cos A + \sin A} \equiv \cos A - \sin A$

b $\dfrac{\sin B}{\sin A} - \dfrac{\cos B}{\cos A} \equiv 2 \operatorname{cosec} 2A \sin(B - A)$

c $\dfrac{1 - \cos 2\theta}{\sin 2\theta} \equiv \tan \theta$

d $\dfrac{\sec^2 \theta}{1 - \tan^2 \theta} \equiv \sec 2\theta$

e $2(\sin^3 \theta \cos \theta + \cos^3 \theta \sin \theta) \equiv \sin 2\theta$

f $\dfrac{\sin 3\theta}{\sin \theta} - \dfrac{\cos 3\theta}{\cos \theta} \equiv 2$

g $\operatorname{cosec} \theta - 2 \cot 2\theta \cos \theta \equiv 2 \sin \theta$

h $\dfrac{\sec \theta - 1}{\sec \theta + 1} \equiv \tan^2 \dfrac{\theta}{2}$

i $\tan\left(\dfrac{\pi}{4} - x\right) \equiv \dfrac{1 - \sin 2x}{\cos 2x}$

2 Prove the identities:

   a $\sin(A + 60°) + \sin(A - 60°) \equiv \sin A$

   b $\dfrac{\cos A}{\sin B} - \dfrac{\sin A}{\cos B} \equiv \dfrac{\cos(A + B)}{\sin B \cos B}$

   c $\dfrac{\sin(x + y)}{\cos x \cos y} \equiv \tan x + \tan y$

   d $\dfrac{\cos(x + y)}{\sin x \sin y} + 1 \equiv \cot x \cot y$

   e $\cos\left(\theta + \dfrac{\pi}{3}\right) + \sqrt{3}\sin\theta \equiv \sin\left(\theta + \dfrac{\pi}{6}\right)$

   f $\cot(A + B) \equiv \dfrac{\cot A \cot B - 1}{\cot A + \cot B}$

   g $\sin^2(45° + \theta) + \sin^2(45° - \theta) \equiv 1$

   h $\cos(A + B)\cos(A - B) \equiv \cos^2 A - \sin^2 B$

3 a Show that $\tan\theta + \cot\theta \equiv 2\csc 2\theta$. **(3 marks)**
  b Hence find the value of $\tan 75° + \cot 75°$. **(2 marks)**

4 a Show that $\sin 3\theta \equiv 3\sin\theta\cos^2\theta - \sin^3\theta$. **(3 marks)**
  b Show that $\cos 3\theta \equiv \cos^3\theta - 3\sin^2\theta\cos\theta$. **(3 marks)**
  c Hence, or otherwise, show that $\tan 3\theta \equiv \dfrac{3\tan\theta - \tan^3\theta}{1 - 3\tan^2\theta}$ **(4 marks)**
  d Given that θ is acute and that $\cos\theta = \tfrac{1}{3}$, show that $\tan 3\theta = \dfrac{10\sqrt{2}}{23}$ **(3 marks)**

5 a Using $\cos 2A \equiv 2\cos^2 A - 1 \equiv 1 - 2\sin^2 A$, show that:

   i $\cos^2\dfrac{x}{2} \equiv \dfrac{1 + \cos x}{2}$ ii $\sin^2\dfrac{x}{2} \equiv \dfrac{1 - \cos x}{2}$

  b Given that $\cos\theta = 0.6$, and that θ is acute, write down the values of:

   i $\cos\dfrac{\theta}{2}$ ii $\sin\dfrac{\theta}{2}$ iii $\tan\dfrac{\theta}{2}$

  c Show that $\cos^4\dfrac{A}{2} \equiv \dfrac{1}{8}(3 + 4\cos A + \cos 2A)$.

6 Show that $\cos^4\theta \equiv \tfrac{3}{8} + \tfrac{1}{2}\cos 2\theta + \tfrac{1}{8}\cos 4\theta$. You must show each stage of your working. **(6 marks)**

7 Prove that $\sin^2(x + y) - \sin^2(x - y) \equiv \sin 2x \sin 2y$. **(5 marks)**

8 Prove that $\cos 2\theta - \sqrt{3}\sin 2\theta \equiv 2\cos\left(2\theta + \dfrac{\pi}{3}\right)$. **(4 marks)**

9 Prove that $4\cos\left(2\theta - \dfrac{\pi}{6}\right) \equiv 2\sqrt{3} - 4\sqrt{3}\sin^2\theta + 4\sin\theta\cos\theta$. **(4 marks)**

10 Show that:

   a $\cos\theta + \sin\theta \equiv \sqrt{2}\sin\left(\theta + \dfrac{\pi}{4}\right)$

   b $\sqrt{3}\sin 2\theta - \cos 2\theta \equiv 2\sin\left(2\theta - \dfrac{\pi}{6}\right)$

Trigonometry and modelling

Challenge

1 a Show that $\cos(A+B) - \cos(A-B) \equiv -2\sin A \sin B$.

b Hence show that $\cos P - \cos Q \equiv -2\sin\left(\dfrac{P+Q}{2}\right)\sin\left(\dfrac{P-Q}{2}\right)$.

c Express $3\sin x \sin 7x$ as the difference of cosines.

2 a Prove that $\sin P + \sin Q \equiv 2\sin\left(\dfrac{P+Q}{2}\right)\cos\left(\dfrac{P-Q}{2}\right)$.

b Hence, or otherwise, show that $2\sin\dfrac{11\pi}{24}\cos\dfrac{5\pi}{24} = \dfrac{\sqrt{3}+\sqrt{2}}{2}$

7.7 Modelling with trigonometric functions

You can use trigonometric functions to model real-life situations. In trigonometrical modelling questions you will often have to write the model using $R\sin(x \pm \alpha)$ or $R\cos(x \pm \alpha)$ to find maximum or minimum values.

Example 21

The cabin pressure, P, in pounds per square inch (psi) on an aeroplane at cruising altitude can be modelled by the equation $P = 11.5 - 0.5\sin(t-2)$, where t is the time in hours since the cruising altitude was first reached, and angles are measured in radians.

a State the maximum and the minimum cabin pressure.
b Find the time after reaching cruising altitude that the cabin first reaches a maximum pressure.
c Calculate the cabin pressure after 5 hours at a cruising altitude.
d Find all the times during the first 8 hours of cruising that the cabin pressure would be exactly 11.3 psi.

a Maximum pressure = $11.5 - 0.5 \times (-1) = 12$ psi
   Minimum pressure = $11.5 - 0.5 \times 1 = 11$ psi ——— $-1 \leq \sin(t-2) \leq 1$. Use the maximum and minimum values of the sine function to find the maximum and minimum pressure.

b $11.5 - 0.5\sin(t-2) = 12$
   $-0.5\sin(t-2) = 0.5$
   $\sin(t-2) = -1$ ——— Set the model equal to 12, the maximum pressure.

   $t - 2 = \ldots, -\dfrac{\pi}{2}, \dfrac{3\pi}{2}, \ldots$ ——— Remember the model uses radians.

   $t = 0.43$ hours = 26 min ——— Multiply 0.43 by 60 to get the time in minutes.

c $P = 11.5 - 0.5\sin(5-2)$ ——— Substitute $t = 5$.
   $= 11.5 - 0.070\ldots$
   $= 11.43$ psi

Online Explore the solution to this modelling problem graphically using technology.

Chapter 7

d $11.5 - 0.5 \sin(t - 2) = 11.3$ ← Set the model equal to 11.3.
$-0.5 \sin(t - 2) = -0.2$
$\sin(t - 2) = 0.4$
$t - 2 = -3.553..., 0.4115..., 2.73...,$ ← Use $\sin^{-1}(0.4)$ to find the principal solution, then use the properties of the sine function to find other possible solutions in the range $0 \leq t \leq 8$.
$6.6947...$
$t = 2.41$ hours, 4.73 hours.
$t = 2$ h 25 min, 4 h 44 min
$0 \leq t \leq 8$ so $-2 \leq t - 2 \leq 6$. There are two solutions in the required range.

Exercise 7G

P 1 The height, h, of a buoy on a boating lake can be modelled by $h = 0.25 \sin(1800t)°$, where h is the height in metres above the buoy's resting position and t is the time in minutes.
 a State the maximum height the buoy reaches above its resting position according to this model.
 b Calculate the time, to the nearest tenth of a second, at which the buoy is first at a height of 0.1 metres.
 c Calculate the time interval between successive minimum heights of the buoy.

P 2 The angle of displacement of a pendulum, θ, at time t seconds after it is released is modelled as $\theta = 0.03 \cos(25t)$, where all angles are measured in radians.
 a State the maximum displacement of the pendulum according to this model.
 b Calculate the angle of displacement of the pendulum after 0.2 seconds.
 c Find the time taken for the pendulum to return to its starting position.
 d Find all the times in the first half second of motion that the pendulum has a displacement of 0.01 radians.

P 3 The price, P, of stock in pounds during a 9-hour trading window can be modelled by $P = 17.4 + 2 \sin(0.7t - 3)$, where t is the time in hours after the stock market opens, and angles are measured in radians.
 a State the beginning and end price of the stock.
 b Calculate the maximum price of the stock and the time when it occurs.
 c A day trader wants to sell the stock when it firsts shows a profit of £0.40 above the day's starting price. At what time should the trader sell the stock?

P 4 The temperature of an oven can be modelled by the equation $T = 225 - 0.3 \sin(2x - 3)$, where T is the temperate in Celsius and x is the time in minutes after the oven first reaches the desired temperature, and angles are measured in radians.
 a State the minimum temperature of the oven.
 b Find the times during the first 10 minutes when the oven is at a minimum temperature.
 c Calculate the time when the oven first reaches a temperature of 225.2 °C.

E/P 5 a Express $0.3 \sin \theta - 0.4 \cos \theta$ in the form $R \sin(\theta - \alpha)°$, where $R > 0$ and $0 < \alpha < 90°$. Give the value of α to 2 decimal places. **(4 marks)**

b i Find the maximum value of $0.3 \sin \theta - 0.4 \cos \theta$. (2 marks)
  ii Find the value of θ, for $0 < \theta < 180$ at which the maximum occurs. (1 mark)

Jack models the temperature in his house, $T\,°C$, on a particular day by the equation
$$T = 23 + 0.3 \sin (18x)° - 0.4 \cos (18x)°, \; x \geqslant 0$$
where x is the number of minutes since the thermostat was adjusted.

c Calculate the minimum value of T predicted by this model, and the value of x, to 2 decimal places, when this minimum occurs. (3 marks)

d Calculate, to the nearest minute, the times in the first hour when the temperature is predicted, by this model, to be exactly $23\,°C$. (4 marks)

6 a Express $65 \cos \theta - 20 \sin \theta$ in the form $R \cos (\theta + \alpha)$, where $R > 0$ and $0 < \alpha < \dfrac{\pi}{2}$
Give the value of α correct to 4 decimal places. (4 marks)

A city wants to build a large circular wheel as a tourist attraction. The height of a tourist on the circular wheel is modelled by the equation
$$H = 70 - 65 \cos 0.2t + 20 \sin 0.2t$$
where H is the height of the tourist above the ground in metres, t is the number of minutes after boarding and the angles are given in radians. Find:

b the maximum height of the wheel (2 marks)
c the time for one complete revolution (2 marks)
d the number of minutes the tourist will be over 100 m above the ground in each revolution. (4 marks)

7 a Express $200 \sin \theta - 150 \cos \theta$ in the form $R \sin (\theta - \alpha)$, where $R > 0$ and $0 < \alpha < \dfrac{\pi}{2}$
Give the exact value of R and the value of α to 4 decimal places. (4 marks)

The electric field strength, E V/m, in a microwave of width 25 cm can be modelled using the equation
$$E = 1700 + 200 \sin \left(\dfrac{4\pi x}{25}\right) - 150 \cos \left(\dfrac{4\pi x}{25}\right)$$
where x is the distance in cm from the left hand edge of the microwave oven.

b i Calculate the maximum value of E predicted by this model.
  ii Find the values of x, for $0 \leqslant x < 25$, where this maximum occurs. (3 marks)

c Food in the microwave will heat best when the electric field strength at the centre of the food is above 1800 V/m. Find the range of possible locations for the centre of the food. (5 marks)

Challenge

Look at the model for the electric field strength in a microwave oven given in question **7** above. For food of the same type and mass, the energy transferred by the oven is proportional to the square of the electric field strength. Given that a square of chocolate placed at a point of maximum field strength takes 20 seconds to melt,

a estimate the range of locations within the oven that an identical square of chocolate will take longer than 30 seconds to melt.

b State two limitations of the model.

Mixed exercise 7

1. Without using a calculator, find the value of:

   a $\sin 40° \cos 10° - \cos 40° \sin 10°$

   b $\frac{1}{\sqrt{2}} \cos 15° - \frac{1}{\sqrt{2}} \sin 15°$

   c $\dfrac{1 - \tan 15°}{1 + \tan 15°}$

2. Given that $\sin x = \dfrac{1}{\sqrt{5}}$ where x is acute and that $\cos(x - y) = \sin y$, show that $\tan y = \dfrac{\sqrt{5} + 1}{2}$

3. The lines l_1 and l_2, with equations $y = 2x$ and $3y = x - 1$ respectively, are drawn on the same set of axes. Given that the scales are the same on both axes and that the angles l_1 and l_2 make with the positive x-axis are A and B respectively,

   a write down the value of $\tan A$ and the value of $\tan B$;

   b without using your calculator, work out the acute angle between l_1 and l_2.

4. In $\triangle ABC$, $AB = 5$ cm and $AC = 4$ cm, $\angle ABC = (\theta - 30°)$ and $\angle ACB = (\theta + 30°)$. Using the sine rule, show that $\tan \theta = 3\sqrt{3}$.

5. The first three terms of an arithmetic series are $\sqrt{3} \cos \theta$, $\sin(\theta - 30°)$ and $\sin \theta$, where θ is acute. Find the value of θ.

6. Two of the angles, A and B, in $\triangle ABC$ are such that $\tan A = \frac{3}{4}$, $\tan B = \frac{5}{12}$.

   a Find the exact value of: i $\sin(A + B)$ ii $\tan 2B$.

   b By writing C as $180° - (A + B)$, show that $\cos C = -\frac{33}{65}$.

7. The angles x and y are acute angles such that $\sin x = \dfrac{2}{\sqrt{5}}$ and $\cos y = \dfrac{3}{\sqrt{10}}$.

   a Show that $\cos 2x = -\frac{3}{5}$.

   b Find the value of $\cos 2y$.

   c Show without using your calculator, that:

   i $\tan(x + y) = 7$ ii $x - y = \dfrac{\pi}{4}$

8. Given that $\sin x \cos y = \frac{1}{2}$ and $\cos x \sin y = \frac{1}{3}$,

   a show that $\sin(x + y) = 5 \sin(x - y)$.

   Given also that $\tan y = k$, express in terms of k:

   b $\tan x$

   c $\tan 2x$

9. a Given that $\sqrt{3} \sin 2\theta + 2 \sin^2 \theta = 1$, show that $\tan 2\theta = \dfrac{1}{\sqrt{3}}$ (2 marks)

   b Hence solve, for $0 \leq \theta \leq \pi$, the equation $\sqrt{3} \sin 2\theta + 2 \sin^2 \theta = 1$. (4 marks)

10 **a** Show that $\cos 2\theta = 5 \sin \theta$ may be written in the form $a \sin^2 \theta + b \sin \theta + c = 0$, where a, b and c are constants to be found. **(3 marks)**

b Hence solve, for $-\pi \leq \theta \leq \pi$, the equation $\cos 2\theta = 5 \sin \theta$. **(4 marks)**

11 **a** Given that $2 \sin x = \cos(x - 60)°$, show that $\tan x = \dfrac{1}{4 - \sqrt{3}}$ **(4 marks)**

b Hence solve, for $0 \leq x \leq 360°$, $2 \sin x = \cos(x - 60°)$, giving your answers to 1 decimal place. **(2 marks)**

12 **a** Given that $4 \sin(x + 70°) = \cos(x + 20°)$, show that $\tan x = -\tfrac{3}{5} \tan 70°$. **(4 marks)**

b Hence solve, for $0 \leq x \leq 180°$, $4 \sin(x + 70°) = \cos(x + 20°)$, giving your answers to 1 decimal place. **(3 marks)**

13 **a** Given that α is acute and $\tan \alpha = \tfrac{3}{4}$, prove that
$$3 \sin(\theta + \alpha) + 4 \cos(\theta + \alpha) \equiv 5 \cos \theta$$

b Given that $\sin x = 0.6$ and $\cos x = -0.8$, evaluate $\cos(x + 270°)$ and $\cos(x + 540°)$.

14 **a** Prove, by counter-example, that the statement
$$\sec(A + B) \equiv \sec A + \sec B, \text{ for all } A \text{ and } B$$
is false. **(2 marks)**

b Prove that $\tan \theta + \cot \theta \equiv 2 \operatorname{cosec} 2\theta$, $\theta \neq \dfrac{n\pi}{2}$, $n \in \mathbb{Z}$. **(4 marks)**

15 Using $\tan 2\theta \equiv \dfrac{2 \tan \theta}{1 - \tan^2 \theta}$ with an appropriate value of θ,

a show that $\tan \dfrac{\pi}{8} = \sqrt{2} - 1$.

b Use the result in **a** to find the exact value of $\tan \dfrac{3\pi}{8}$

16 **a** Express $\sin x - \sqrt{3} \cos x$ in the form $R \sin(x - \alpha)$, with $R > 0$ and $0 < \alpha < 90°$. **(4 marks)**

b Hence sketch the graph of $y = \sin x - \sqrt{3} \cos x$, for $-360° \leq x \leq 360°$, giving the coordinates of all points of intersection with the axes. **(4 marks)**

17 Given that $7 \cos 2\theta + 24 \sin 2\theta \equiv R \cos(2\theta - \alpha)$, where $R > 0$ and $0 < \alpha < \dfrac{\pi}{2}$, find:

a the value of R and the value of α, to 2 decimal places **(3 marks)**

b the maximum value of $14 \cos^2 \theta + 48 \sin \theta \cos \theta$. **(2 marks)**

c Solve the equation $7 \cos 2\theta + 24 \sin 2\theta = 12.5$, for $0 \leq \theta \leq \pi$, giving your answers to 2 decimal places. **(5 marks)**

193

Chapter 7

E/P 18 a Express $1.5 \sin 2x + 2 \cos 2x$ in the form $R \sin(2x + \alpha)$, where $R > 0$ and $0 < \alpha < \frac{\pi}{2}$, giving your values of R and α to 3 decimal places where appropriate. **(4 marks)**

b Express $3 \sin x \cos x + 4 \cos^2 x$ in the form $a \sin 2x + b \cos 2x + c$, where a, b and c are constants to be found. **(3 marks)**

c Hence, using your answer to part **a**, deduce the maximum value of $3 \sin x \cos x + 4 \cos^2 x$. **(1 mark)**

E/P 19 a Given that $\sin^2 \frac{\theta}{2} = 2 \sin \theta$, show that $\sqrt{17} \sin(\theta + \alpha) = 1$ and state the value of α, where $0 \leq \alpha \leq 90°$. **(3 marks)**

b Hence, or otherwise, solve $\sin^2 \frac{\theta}{2} = 2 \sin \theta$ for $0 \leq \theta \leq 360°$. **(4 marks)**

E/P 20 a Given that $2 \cos \theta = 1 + 3 \sin \theta$, show that $R \cos(\theta + \alpha) = 1$, where R and α are constants to be found, and $0 \leq \alpha \leq 90°$. **(2 marks)**

b Hence, or otherwise, solve $2 \cos \theta = 1 + 3 \sin \theta$ for $0 \leq \theta \leq 360°$. **(4 marks)**

P 21 Using known trigonometric identities, prove the following:

a $\sec \theta \operatorname{cosec} \theta \equiv 2 \operatorname{cosec} 2\theta$

b $\tan\left(\frac{\pi}{4} + x\right) - \tan\left(\frac{\pi}{4} - x\right) \equiv 2 \tan 2x$

c $\sin(x + y) \sin(x - y) \equiv \cos^2 y - \cos^2 x$

d $1 + 2 \cos 2\theta + \cos 4\theta \equiv 4 \cos^2 \theta \cos 2\theta$

E/P 22 a Use the double-angle formulae to prove that $\dfrac{1 - \cos 2x}{1 + \cos 2x} \equiv \tan^2 x$. **(4 marks)**

b Hence find, for $-\pi \leq x \leq \pi$, all the solutions of $\dfrac{1 - \cos 2x}{1 + \cos 2x} = 3$, leaving your answers in terms of π. **(2 marks)**

E/P 23 a Prove that $\cos^4 2\theta - \sin^4 2\theta \equiv \cos 4\theta$. **(4 marks)**

b Hence find, for $0 \leq \theta \leq 180°$, all the solutions of $\cos^4 2\theta - \sin^4 2\theta = \frac{1}{2}$. **(2 marks)**

E/P 24 a Prove that $\dfrac{1 - \cos 2\theta}{\sin 2\theta} \equiv \tan \theta$. **(4 marks)**

b Verify that $\theta = 180°$ is a solution of the equation $\sin 2\theta = 2 - 2 \cos 2\theta$. **(1 mark)**

c Using the result in part **a**, or otherwise, find the two other solutions, $0 < \theta < 360°$, of the equation $\sin 2\theta = 2 - 2 \cos 2\theta$. **(3 marks)**

E/P 25 The curve on an oscilloscope screen satisfies the equation $y = 2 \cos x - \sqrt{5} \sin x$.

a Express the equation of the curve in the form $y = R \cos(x + \alpha)$, where R and α are constants and $R > 0$ and $0 \leq \alpha < \frac{\pi}{2}$. **(4 marks)**

b Find the values of x, $0 \leq x < 2\pi$, for which $y = -1$. **(3 marks)**

Trigonometry and modelling

 26 a Express $1.4 \sin \theta - 5.6 \cos \theta$ in the form $R \sin(\theta - \alpha)$, where R and α are constants, $R > 0$ and $0 < \alpha < 90°$. Round R and α to 3 decimal places. **(4 marks)**

b Hence find the maximum value of $1.4 \sin \theta - 5.6 \cos \theta$ and the smallest positive value of θ for which this maximum occurs. **(3 marks)**

The length of daylight, $d(t)$ at a location in northern Scotland can be modelled using the equation

$$d(t) = 12 - 5.6 \cos\left(\frac{360t}{365}\right)° + 1.4 \sin\left(\frac{360t}{365}\right)°$$

where t is the numbers of days into the year.

c Calculate the minimum number of daylight hours in northern Scotland as given by this model. **(2 marks)**

d Find the value of t when this minimum number of daylight hours occurs. **(1 mark)**

 27 a Express $12 \sin x + 5 \cos x$ in the form $R \sin(x + \alpha)$, where R and α are constants, $R > 0$ and $0 < \alpha < 90°$. Round α to 1 decimal place. **(4 marks)**

A runner's speed, v in m/s, in an endurance race can be modelled by the equation

$$v(x) = \frac{50}{12 \sin\left(\frac{2x}{5}\right)° + 5 \cos\left(\frac{2x}{5}\right)°}, \ 0 \leq x \leq 300$$

where x is the time in minutes since the beginning of the race.

b Find the minimum value of v. **(2 marks)**

c Find the time into the race when this speed occurs. **(1 mark)**

Challenge

1 Prove the identities:

a $\dfrac{\cos 2\theta + \cos 4\theta}{\sin 2\theta - \sin 4\theta} \equiv -\cot \theta$

b $\cos x + 2 \cos 3x + \cos 5x \equiv 4 \cos^2 x \cos 3x$

2 The points A, B and C lie on a circle with centre O and radius 1. AC is a diameter of the circle and point D lies on OC such that $\angle ODB = 90°$.

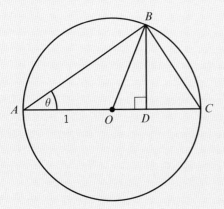

Hint Find expressions for $\angle BOD$ and AB, then consider the lengths OD and DB.

Use this construction to prove:

a $\sin 2\theta \equiv 2 \sin \theta \cos \theta$ **b** $\cos 2\theta \equiv 2 \cos^2 \theta - 1$

Chapter 7

Summary of key points

1. The **addition** (or compound-angle) formulae are:
   - $\sin(A+B) \equiv \sin A \cos B + \cos A \sin B$ $\quad \sin(A-B) \equiv \sin A \cos B - \cos A \sin B$
   - $\cos(A+B) \equiv \cos A \cos B - \sin A \sin B$ $\quad \cos(A-B) \equiv \cos A \cos B + \sin A \sin B$
   - $\tan(A+B) \equiv \dfrac{\tan A + \tan B}{1 - \tan A \tan B}$ $\quad \tan(A-B) \equiv \dfrac{\tan A - \tan B}{1 + \tan A \tan B}$

2. The **double-angle** formulae are:
   - $\sin 2A \equiv 2 \sin A \cos A$
   - $\cos 2A \equiv \cos^2 A - \sin^2 A \equiv 2\cos^2 A - 1 \equiv 1 - 2\sin^2 A$
   - $\tan 2A \equiv \dfrac{2 \tan A}{1 - \tan^2 A}$

3. For positive values of a and b,
   - $a \sin x \pm b \cos x$ can be expressed in the form $R \sin(x \pm \alpha)$
   - $a \cos x \pm b \sin x$ can be expressed in the form $R \cos(x \mp \alpha)$

   with $R > 0$ and $0 < \alpha < 90°$ $\left(\text{or } \dfrac{\pi}{2}\right)$

   where $R \cos \alpha = a$ and $R \sin \alpha = b$ and $R = \sqrt{a^2 + b^2}$.

Parametric equations

8

Objectives

After completing this chapter you should be able to:

- Convert parametric equations into Cartesian form by substitution → pages 198–202
- Convert parametric equations into Cartesian form using trigonometric identities → pages 202–206
- Understand and use parametric equations of curves and sketch parametric curves → pages 206–208
- Solve coordinate geometry problems involving parametric equations → pages 209–213
- Use parametric equations in modelling in a variety of contexts → pages 213–220

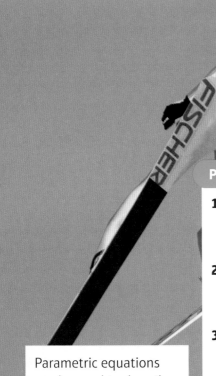

Parametric equations can be used to describe the path of a ski jumper from the point of leaving the ski ramp to the point of landing.
→ Exercise 8E, Q8

Prior knowledge check

1. Rearrange to make t the subject:
   a $x = 4t - kt$ **b** $y = 3t^2$ **c** $y = 2 - 4\ln t$ **d** $x = 1 + 2e^{-3t}$
   ← GCSE Mathematics; Year 1, Chapter 14

2. Write in terms of powers of $\cos x$:
   a $4 + 3\sin^2 x$ **b** $\sin 2x$
   c $\cot x$ **d** $2\cos x + \cos 2x$ ← Section 7.2

3. State the ranges of the following functions.
   a $y = \ln(x+1),\ x > 0$ **b** $y = 2\sin x,\ 0 < x < \pi$
   c $y = x^2 + 4x - 2,\ -4 < x < 1$ **d** $y = \dfrac{1}{2x+5},\ x > -2$
   ← Section 2.2

4. A circle has centre $(0, 4)$ and radius 5. Find the coordinates of the points of intersection of the circle and the line with equation $2y - x - 10 = 0$.
   ← Year 1, Chapter 6

Chapter 8

8.1 Parametric equations

You can write the x- and y-coordinates of each point on a curve as functions of a third variable. This variable is called a parameter and is often represented by the letter t.

- **A curve can be defined using parametric equations $x = p(t)$ and $y = q(t)$. Each value of the parameter, t, defines a point on the curve with coordinates $(p(t), q(t))$.**

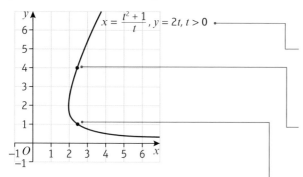

These are the parametric equations of the curve. The domain of the parameter tells you the values of t you would need to substitute to find the coordinates of the points on the curve.

When $t = 2$, $x = \dfrac{2^2 + 1}{2} = 2.5$ and $y = 2 \times 2 = 4$.

This corresponds to the point $(2.5, 4)$.

When $t = 0.5$, $x = \dfrac{0.5^2 + 1}{0.5} = 2.5$ and $y = 2 \times 0.5 = 1$.

This corresponds to the point $(2.5, 1)$.

Watch out The value of the parameter t is generally not equal to either the x- or the y-coordinate, and more than one point on the curve can have the same x-coordinate.

- **You can convert between parametric equations and Cartesian equations by using substitution to eliminate the parameter.**

Notation A Cartesian equation in two dimensions involves the variables x and y only.

You can use the domain and range of the parametric functions to find the domain and range of the resulting Cartesian function.

- **For parametric equations $x = p(t)$ and $y = q(t)$ with Cartesian equation $y = f(x)$:**
  - **the domain of $f(x)$ is the range of $p(t)$**
  - **the range of $f(x)$ is the range of $q(t)$**

Example 1

A curve has parametric equations
$$x = 2t, \quad y = t^2, \quad -3 < t < 3$$

a Find a Cartesian equation of the curve in the form $y = f(x)$.
b State the domain and range of $f(x)$.
c Sketch the curve within the given domain for t.

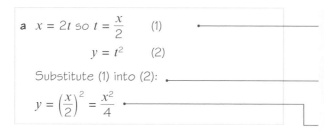

A Cartesian equation only involves the variables x and y, so you need to eliminate t.

Rearrange one equation into the form $t = \ldots$ then substitute into the other equation.

This is a quadratic curve.

198

b $x = 2t$, $-3 < t < 3$
So the domain of f(x) is $-6 < x < 6$.
$y = t^2$, $-3 < t < 3$
So the range of f(x) is $0 \leq y < 9$.

c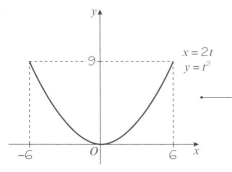

| The domain of f is the range of the parametric function for x. The range of $x = 2t$ over the domain $-3 < t < 3$ is $-6 < x < 6$. ← Section 2.1

The range of f is the range of the parametric function for y. Choose your inequalities carefully. $y = t^2$ can equal 0 in the interval $-3 < t < 3$, so use $\leq$, but it cannot equal 9, so use $<$.

The curve is a graph of $y = \frac{1}{4}x^2$. Use your answers to part **b** to help with your sketch.

Watch out Pay careful attention to the **domain** when sketching parametric curves. The curve is only defined for $-3 < t < 3$, or for $-6 < x < 6$. You should not draw any points on the curve outside that range.

Example 2

A curve has parametric equations
$$x = \ln(t + 3), \quad y = \frac{1}{t + 5}, \quad t > -2$$

a Find a Cartesian equation of the curve of the form $y = f(x)$, $x > k$ where k is a constant to be found.
b Write down the range of f(x).

Online Sketch this parametric curve using technology.

a $x = \ln(t + 3)$
$e^x = t + 3$
So $e^x - 3 = t$
Substitute $t = e^x - 3$ into
$$y = \frac{1}{t + 5} = \frac{1}{e^x - 3 + 5} = \frac{1}{e^x + 2}$$
When $t = -2$: $x = \ln(t + 3) = \ln 1 = 0$
As t increases $\ln(t + 3)$ increases, so the range of the parametric function for x is $x > 0$.
The Cartesian equation is
$$y = \frac{1}{e^x + 2}, \quad x > 0$$

b When $t = -2$: $y = \frac{1}{t + 5} = \frac{1}{3}$
As t increases y decreases, but is always positive, so the range of the parametric function for y is $0 < y < \frac{1}{3}$
The range of f(x) is $0 < y < \frac{1}{3}$

e^x is the inverse function of $\ln x$.

Rearrange the equation for x into the form $t = ...$ then substitute into the equation for y.

To find the domain for f(x), consider the range of values x can take for values of $t > -2$.

You need to consider what value x takes when $t = -2$ **and** what happens when t increases.

The range of f is the range of values y can take within the given range of the parameter.

You could also find the range of f(x) by considering the domain of f(x). $f(0) = \frac{1}{3}$ and f(x) decreases as x increases, so $0 < y < \frac{1}{3}$. ← Section 2.1

Chapter 8

Exercise 8A

1 Find a Cartesian equation for each of these parametric equations, giving your answer in the form $y = f(x)$. In each case find the domain and range of $f(x)$.

  a $x = t - 2, \quad y = t^2 + 1, \quad -4 \leq t \leq 4$

  b $x = 5 - t, \quad y = t^2 - 1, \quad t \in \mathbb{R}$

  c $x = \dfrac{1}{t}, \quad y = 3 - t, \quad t \neq 0$

> **Notation** If the domain of t is given as $t \neq 0$, this implies that t can take any value in $\mathbb{R}$ other than 0.

  d $x = 2t + 1, \quad y = \dfrac{1}{t}, \quad t > 0$

  e $x = \dfrac{1}{t-2}, \quad y = t^2, \quad t > 2$

  f $x = \dfrac{1}{t+1}, \quad y = \dfrac{1}{t-2}, \quad t > 2$

2 For each of these parametric curves:
  i find a Cartesian equation for the curve in the form $y = f(x)$ giving the domain on which the curve is defined
  ii find the range of $f(x)$.

  a $x = 2\ln(5 - t), \quad y = t^2 - 5, \quad t < 4$

  b $x = \ln(t + 3), \quad y = \dfrac{1}{t+5}, \quad t > -2$

  c $x = e^t, \quad y = e^{3t}, \quad t \in \mathbb{R}$

(P) 3 A curve C is defined by the parametric equations $x = \sqrt{t}, \quad y = t(9 - t), \quad 0 \leq t \leq 5$.

  a Find a Cartesian equation of the curve in the form $y = f(x)$, and determine the domain and range of $f(x)$.

  b Sketch C showing clearly the coordinates of any turning points, endpoints and intersections with the coordinate axes.

> **Problem-solving**
> $y = t(9 - t)$ is a quadratic with a negative t^2 term and roots at $t = 0$ and $t = 9$. It will take its **maximum** value when $t = 4.5$.

4 For each of the following parametric curves:
  i find a Cartesian equation for the curve in the form $y = f(x)$
  ii find the domain and range of $f(x)$
  iii sketch the curve within the given domain of t.

  a $x = 2t^2 - 3, \quad y = 9 - t^2, \quad t > 0$

  b $x = 3t - 1, \quad y = (t - 1)(t + 2), \quad -4 < t < 4$

  c $x = t + 1, \quad y = \dfrac{1}{t-1}, \quad t \in \mathbb{R}, \quad t \neq 1$

  d $x = \sqrt{t} - 1, \quad y = 3\sqrt{t}, \quad t > 0$

  e $x = \ln(4 - t), \quad y = t - 2, \quad t < 3$

Parametric equations

P 5 The curves C_1 and C_2 are defined by the following parametric equations.

C_1: $x = 1 + 2t$, $y = 2 + 3t$ $2 < t < 5$ C_2: $x = \dfrac{1}{2t-3}$, $y = \dfrac{t}{2t-3}$ $2 < t < 3$

a Show that both curves are segments of the same straight line.

Notation Straight lines and line segments can be referred to as 'curves' in coordinate geometry.

b Find the length of each line segment.

E/P 6 A curve C has parametric equations

$$x = \dfrac{3}{t} + 2, \quad y = 2t - 3 - t^2, \quad t \in \mathbb{R}, \quad t \neq 0$$

a Determine the ranges of x and y in the given domain of t. **(3 marks)**

b Show that the Cartesian equation of C can be written in the form

$$y = \dfrac{A(x^2 + bx + c)}{(x-2)^2}$$

where A, b and c are integers to be determined. **(3 marks)**

7 A curve has parametric equations

$$x = \ln(t+3), \quad y = \dfrac{1}{t+5}, \quad t > -2$$

a Show that a Cartesian equation of this curve is $y = f(x)$, $x > k$ where k is a constant to be found.

b Write down the range of $f(x)$.

E/P 8 A diagram shows a curve C with parametric equations

$x = 3\sqrt{t}$, $y = t^3 - 2t$, $0 \leq t \leq 2$

a Find a Cartesian equation of the curve in the form $y = f(x)$, and state the domain of $f(x)$. **(3 marks)**

b Show that $\dfrac{dy}{dt} = 0$ when $t = \sqrt{\dfrac{2}{3}}$ **(3 marks)**

c Hence determine the range of $f(x)$. **(2 marks)**

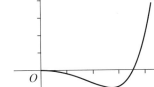

E/P 9 A curve C has parametric equations

$x = t^3 - t$, $y = 4 - t^2$, $t \in \mathbb{R}$

a Show that the Cartesian equation of C can be written in the form

$$x^2 = (a - y)(b - y)^2$$

where a and b are integers to be determined. **(3 marks)**

b Write down the maximum value of the y-coordinate for any point on this curve. **(2 marks)**

201

Chapter 8

Challenge

A curve C has parametric equations

$$x = \frac{1-t^2}{1+t^2}, \quad y = \frac{2t}{1+t^2}, \quad t \in \mathbb{R}$$

a Show that a Cartesian equation for this curve is $x^2 + y^2 = 1$.

b Hence describe C.

8.2 Using trigonometric identities

You can use trigonometric identities to convert trigonometric parametric equations into Cartesian form. In this chapter you will always consider angles measured in radians.

Example 3

A curve has parametric equations $x = \sin t + 2$, $y = \cos t - 3$, $t \in \mathbb{R}$

a Show that a Cartesian equation of the curve is $(x-2)^2 + (y+3)^2 = 1$.

b Hence sketch the curve.

a $x = \sin t + 2$
So $\sin t = x - 2$ (1)
$y = \cos t - 3$
$\cos t = y + 3$ (2)

Substitute (1) and (2) into
$\sin^2 t + \cos^2 t \equiv 1$
$(x - 2)^2 + (y + 3)^2 = 1$

b
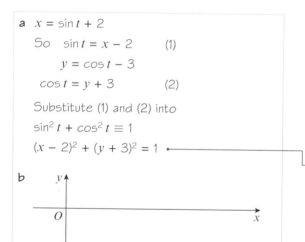

Problem-solving

If you can write expressions for $\sin t$ and $\cos t$ in terms of x and y then you can use the identity $\sin^2 t + \cos^2 t \equiv 1$ to eliminate the parameter, t.
← Year 1, Chapter 10

Your equations in (1) and (2) are in terms of $\sin t$ and $\cos t$ so you need to square them when you substitute. Make sure you square the whole expression.

$(x - a)^2 + (y - b)^2 = r^2$ is the equation of a circle with centre (a, b) and radius r.

So the curve is a circle with centre $(2, -3)$ and radius 1.
← Year 1, Chapter 6

Example 4

A curve is defined by the parametric equations

$$x = \sin t, \quad y = \sin 2t, \quad -\frac{\pi}{2} \leq t \leq \frac{\pi}{2}$$

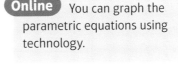

Online You can graph the parametric equations using technology.

a Find a Cartesian equation of the curve in the form

$$y = f(x), \quad -k \leq x \leq k$$

stating the value of the constant k.

b Write down the range of $f(x)$.

a $y = \sin 2t$
$ = 2 \sin t \cos t$
$ = 2x \cos t \qquad (1)$

Use the identity $\sin 2t \equiv 2 \sin t \cos t$, then substitute $x = \sin t$. ← Section 7.2

$\sin^2 t + \cos^2 t \equiv 1$
$\cos^2 t \equiv 1 - \sin^2 t$
$ = 1 - x^2$
$\cos t = \sqrt{1 - x^2} \qquad (2)$

Use the identity $\sin^2 t + \cos^2 t \equiv 1$ together with $x = \sin t$ to find an expression for $\cos t$ in terms of x.

Substitute (2) into (1): $y = 2x\sqrt{1 - x^2}$

When $t = -\frac{\pi}{2}$, $x = \sin\left(-\frac{\pi}{2}\right) = -1$

When $t = \frac{\pi}{2}$, $x = \sin\left(\frac{\pi}{2}\right) = 1$

Watch out Be careful when taking square roots. In this case you don't need to consider the negative square root because $\cos t$ is positive for all values in the domain of the parameter.

The Cartesian equation is $y = 2x\sqrt{1 - x^2}$, $-1 \leq x \leq 1$ so $k = 1$.

To find the domain of $f(x)$, consider the range of $x = \sin t$ for the values of the parameter given.

b $-1 \leq y \leq 1$

Within $-\frac{\pi}{2} \leq t \leq \frac{\pi}{2}$, $y = \sin 2t$ takes a minimum value of -1 and a maximum value of 1.

Example 5

A curve C has parametric equations

$$x = \cot t + 2 \quad y = \csc^2 t - 2, \quad 0 < t < \pi$$

a Find the equation of the curve in the form $y = f(x)$ and state the domain of x for which the curve is defined.

b Hence, sketch the curve.

a $x = \cot t + 2$
$\cot t = x - 2$ (1)
$y = \text{cosec}^2 t - 2$
$\text{cosec}^2 t = y + 2$ (2)

Substitute (1) and (2) into
$1 + \cot^2 t \equiv \text{cosec}^2 t$
$1 + (x - 2)^2 = y + 2$
$1 + x^2 - 4x + 4 = y + 2$
$y = x^2 - 4x + 3$

The range of $x = \cot t + 2$ over the domain $0 < t < \pi$ is all of the real numbers, so the domain of $f(x)$ is $x \in \mathbb{R}$.

b $y = x^2 - 4x + 3 = (x - 3)(x - 1)$ is a quadratic with roots at $x = 3$ and $x = 1$ and y-intercept 3. The minimum point is $(2, -1)$.

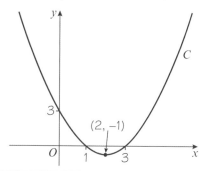

Problem-solving

The parametric equations involve $\cot t$ and $\text{cosec}^2 t$ so you can use the identity $1 + \cot^2 t \equiv \text{cosec}^2 t$.   ← Section 6.4

Rearrange to find expressions for $\cot t$ and $\text{cosec}^2 t$ in terms of x and y.

Expand and rearrange to make y the subject. You could also write the equation as:
$y = (x - 2)^2 - 1$
This is the completed square form which is useful when sketching the curve.

Consider the range of values taken by x over the domain of the parameter. The curve is defined on all of the real numbers, so it is the whole quadratic curve

Online Explore this curve graphically using technology.

Exercise 8B

1 Find the Cartesian equation of the curves given by the following parametric equations:

a $x = 2\sin t - 1$, $y = 5\cos t + 4$, $0 < t < 2\pi$
b $x = \cos t$, $y = \sin 2t$, $0 < t < 2\pi$
c $x = \cos t$, $y = 2\cos 2t$, $0 < t < 2\pi$
d $x = \sin t$, $y = \tan 2t$, $0 < t < \dfrac{\pi}{2}$
e $x = \cos t + 2$, $y = 4\sec t$, $0 < t < \dfrac{\pi}{2}$
f $x = 3\cot t$, $y = \text{cosec}\, t$, $0 < t < \pi$

2 A circle has parametric equations $x = \sin t - 5$, $y = \cos t + 2$
  a Find a Cartesian equation of the circle.
  b Write down the radius and the coordinates of the centre of the circle.
  c Write down a suitable domain of t which defines one full revolution around the circle.

Problem-solving

Think about how x and y change as t varies.

3 A circle has parametric equations $x = 4\sin t + 3$, $y = 4\cos t - 1$. Find the radius and the coordinates of the centre of the circle.

4 A curve is given by the parametric equation $x = \cos t - 2$, $y = \sin t + 3$, $-\pi < t < \pi$.
Sketch the curve.

(P) **5** Find the Cartesian equation of the curves given by the following parametric equations.

a $x = \sin t$, $y = \sin\left(t + \dfrac{\pi}{4}\right)$, $-\dfrac{\pi}{2} < t < \dfrac{\pi}{2}$

b $x = 3\cos t$, $y = 2\cos\left(t + \dfrac{\pi}{6}\right)$, $0 < t < \dfrac{\pi}{3}$

> **Hint** Use the addition formulae and exact values.

c $x = \sin t$, $y = 3\sin(t + \pi)$, $0 < t < 2\pi$

(E) **6** The curve C has parametric equations
$$x = 8\cos t, \quad y = \dfrac{1}{4}\sec^2 t, \quad -\dfrac{\pi}{2} < t < \dfrac{\pi}{2}$$

a Find a Cartesian equation of C. **(4 marks)**

b Sketch the curve C on the appropriate domain. **(3 marks)**

(E) **7** A curve has parametric equations
$$x = 3\cot^2 2t, \quad y = 3\sin^2 2t, \quad 0 < t \leq \dfrac{\pi}{4}$$

Find a Cartesian equation of the curve in the form $y = f(x)$. State the domain on which $f(x)$ is defined. **(6 marks)**

(E/P) **8** A curve C has parametric equations
$$x = \dfrac{1}{3}\sin t, \quad y = \sin 3t, \quad 0 < t < \dfrac{\pi}{2}$$

a Show that the Cartesian equation of the curve is given by
$$y = ax(1 - bx^2)$$
where a and b are integers to be found. **(5 marks)**

b State the domain and range of $y = f(x)$ in the given domain of t. **(2 marks)**

(E/P) **9** Show that the curve with parametric equations
$$x = 2\cos t, \quad y = \sin\left(t - \dfrac{\pi}{6}\right), \quad 0 < t < \pi$$

can be written in the form
$$y = \dfrac{1}{4}\left(\sqrt{12 - 3x^2} - x\right), \quad -2 < x < 2$$ **(6 marks)**

(E/P) **10** A curve has parametric equations
$$x = \tan^2 t + 5, \quad y = 5\sin t, \quad 0 < t < \dfrac{\pi}{2}$$

a Find the Cartesian equation of the curve in the form $y^2 = f(x)$. **(4 marks)**

b Determine the possible values of x and y in the given domain of t. **(2 marks)**

Chapter 8

E/P 11 A curve C has parametric equations

$$x = \tan t, \quad y = 3\sin(t - \pi), \quad 0 < t < \frac{\pi}{2}$$

Find a Cartesian equation of C. **(4 marks)**

Challenge

The curve C is given by the parametric equations:

$$x = \frac{1}{2}\cos 2t, \quad y = \sin\left(t + \frac{\pi}{6}\right), \quad 0 < t < 2\pi$$

Show that a Cartesian equation for C is $(4y^2 - 2 + 2x)^2 + 12x^2 - 3 = 0$.

8.3 Curve sketching

Most parametric curves do not result in curves you will recognise and can sketch easily. You can plot any parametric curve by substituting values of the parameter into each equation.

Example 6

Draw the curve given by parametric equations

$$x = 3\cos t + 4, \quad y = 2\sin t, \quad 0 \leq t \leq 2\pi$$

t	0	$\frac{\pi}{4}$	$\frac{\pi}{2}$	$\frac{3\pi}{4}$	π	$\frac{5\pi}{4}$	$\frac{3\pi}{2}$	$\frac{7\pi}{4}$	2π
$x = 3\cos t + 4$	7	6.12	4	1.88	1	1.88	4	6.12	7
$y = 2\sin t$	0	1.41	2	1.41	0	−1.41	−2	−1.41	0

This parametric curve has Cartesian equation

$$\left(\frac{x-4}{3}\right)^2 + \left(\frac{y}{2}\right)^2 = 1.$$

This isn't a form of curve that you need to be able to recognise, but you can plot the curve using a table of values.

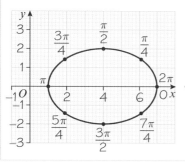

Choose values for t covering the domain of t. For each value of t, substitute to find corresponding values for x and y which will be the coordinates of points on the curve.

Plot the points and draw the curve through the points. The curve is an ellipse.

Example 7

Draw the curve given by the parametric equations $x = 2t$, $y = t^2$, for $-1 \leq t \leq 5$.

t	-1	0	1	2	3	4	5
$x = 2t$	-2	0	2	4	6	8	10
$y = t^2$	1	0	1	4	9	16	25

Online Use technology to graph the parametric equations.

Only calculate values of x and y for values of t in the given domain.

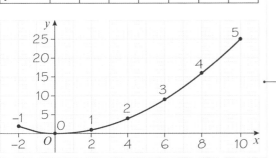

This is a 'partial' graph of the quadratic equation

$$y = \frac{x^2}{4}$$

You could also plot this curve by converting to Cartesian form and considering the domain of and range of the Cartesian function.

The domain is $-2 \leq x \leq 10$ and the range is $0 \leq y \leq 25$.

Exercise 8C

1 A curve is given by the parametric equations

$$x = 2t, \quad y = \frac{5}{t}, \quad t \neq 0$$

Copy and complete the table and draw a graph of the curve for $-5 \leq t \leq 5$.

t	-5	-4	-3	-2	-1	-0.5	0.5	1	2	3	4	5
$x = 2t$	-10	-8				-1						
$y = \frac{5}{t}$	-1	-1.25					10					

2 A curve is given by the parametric equations

$$x = t^2, \quad y = \frac{t^3}{5}$$

Copy and complete the table and draw a graph of the curve for $-4 \leq t \leq 4$.

t	-4	-3	-2	-1	0	1	2	3	4
$x = t^2$	16								
$y = \frac{t^3}{5}$	-12.8								

3 A curve is given by parametric equations

$$x = \tan t + 1, \quad y = \sin t, \quad -\frac{\pi}{4} \leq t \leq \frac{\pi}{3}$$

Copy and complete the table and draw a graph of the curve for the given domain of t.

t	$-\dfrac{\pi}{4}$	$-\dfrac{\pi}{6}$	$-\dfrac{\pi}{12}$	0	$\dfrac{\pi}{12}$	$\dfrac{\pi}{6}$	$\dfrac{\pi}{4}$	$\dfrac{\pi}{3}$
$x = \tan t + 1$	0			1				
$y = \sin t$				0				

4 Sketch the curves given by these parametric equations:

  a $x = t - 2$, $y = t^2 + 1$, $-4 \leqslant t \leqslant 4$

  b $x = 3\sqrt{t}$, $y = t^3 - 2t$, $0 \leqslant t \leqslant 2$

  c $x = t^2$, $y = (2 - t)(t + 3)$, $-5 \leqslant t \leqslant 5$

  d $x = 2\sin t - 1$, $y = 5\cos t + 1$, $-\dfrac{\pi}{4} \leqslant t \leqslant \dfrac{\pi}{4}$

  e $x = \sec^2 t - 3$, $y = 2\sin t + 1$, $-\dfrac{\pi}{4} \leqslant t \leqslant \dfrac{\pi}{2}$

  f $x = t - 3\cos t$, $y = 1 + 2\sin t$, $0 \leqslant t \leqslant 2\pi$

(E) **5** The curve C has parametric equations
$$x = 3 - t,\quad y = t^2 - 2,\quad -2 \leqslant t \leqslant 3$$

  a Find a Cartesian equation of C in the form $y = f(x)$. **(4 marks)**

  b Sketch the curve C on the appropriate domain. **(3 marks)**

(E/P) **6** The curve C has parametric equations
$$x = 9\cos t - 2,\quad y = 9\sin t + 1,\quad -\dfrac{\pi}{6} \leqslant t \leqslant \dfrac{\pi}{2}$$

  a Show that the Cartesian equation of C can be written as
$$(x + a)^2 + (y + b)^2 = c$$
  where a, b and c are integers to be determined. **(4 marks)**

  b Sketch the curve C on the given domain of t. **(3 marks)**

  c Find the length of C. **(2 marks)**

Challenge

Sketch the curve given by the parametric equations on the given domain of t:
$$x = \dfrac{9t}{1 + t^3},\quad y = \dfrac{9t^2}{1 + t^3},\quad t \neq -1$$

Comment on the behaviour of the curve as t approaches -1 from the positive direction and from the negative direction.

8.4 Points of intersection

You need to be able to solve coordinate geometry problems involving parametric equations.

Example 8

The diagram shows a curve C with parametric equations
$x = at^2 + t$, $y = a(t^3 + 8)$, $t \in \mathbb{R}$, where a is a non-zero constant. Given that C passes through the point $(-4, 0)$,

a find the value of a

b find the coordinates of the points A and B where the curve crosses the y-axis.

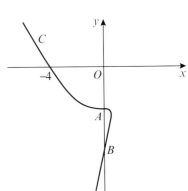

a At point $(-4, 0)$, $x = -4$ and $y = 0$
Hence
$-4 = at^2 + t$ (1)
$0 = a(t^3 + 8)$ (2)

Solving equation (2) for t:
$0 = a(t^3 + 8)$
$0 = t^3 + 8$
$-8 = t^3$
$-2 = t$

So, at the point $(-4, 0)$, $t = -2$.
Since $t = -2$ at $(-4, 0)$, then, from equation (1),
$-4 = a(-2)^2 + (-2)$
$-4 = 4a - 2$
$-2 = 4a$
$-\frac{1}{2} = a$

b At points A and B, the x-coordinate is 0:
$0 = -\frac{1}{2}t^2 + t$
$0 = t\left(-\frac{1}{2}t + 1\right)$
$t = 0$ or $-\frac{1}{2}t + 1 = 0$
                                $t = 2$

At $t = 0$,
$y = -\frac{1}{2}(0^3 + 8)$
   $= -4$

At $t = 2$,
$y = -\frac{1}{2}(2^3 + 8)$
   $= -8$

Therefore,
A is $(0, -4)$ and B is $(0, -8)$.

Online Explore curves with parametric equations of this form using technology.

The point $(-4, 0)$ lies on the curve. You can use this to write two equations for t.

Since a is non-zero, the factor $(t^3 + 8)$ must equal 0.

The value of t is the same in both equations at any given point on the curve.

Substitute $t = -2$ into equation (1).

Substitute $x = 0$ into the parametric equation for x. You now know that $a = -\frac{1}{2}$.

Solve this quadratic equation to find the two values of t corresponding to points a and b.

Substitute each value for the parameter into the parametric equation for y to find the y-coordinates at these points.

You already know that these t-values will give you an x-coordinate of 0. Use the diagram to work out which point is A and which point is B.

Chapter 8

Example 9

A curve is given parametrically by the equations $x = t^2$, $y = 4t$. The line $x + y + 4 = 0$ meets the curve at A. Find the coordinates of A.

$x + y + 4 = 0$

Substitute:

$t^2 + 4t + 4 = 0$ — Find the value of t at A. Solve the equations simultaneously. Substitute $x = t^2$ and $y = 4t$ into $x + y + 4 = 0$.

$(t + 2)^2 = 0$ — Factorise.

$t + 2 = 0$ — Take the square root of each side.

So $t = -2$

Substitute:

$x = t^2$
$= (-2)^2$
$= 4$ — Find the coordinates of A. Substitute $t = -2$ into the parametric equations.

$y = 4t$
$= 4(-2)$
$= -8$

The coordinates of A are $(4, -8)$.

Example 10

The diagram shows a curve C with parametric equations

$$x = \cos t + \sin t, \quad y = \left(t - \frac{\pi}{6}\right)^2, \quad -\frac{\pi}{2} < t < \frac{4\pi}{3}$$

a Find the point where the curve intersects the line $y = \pi^2$.
b Find the coordinates of the points A and B where the curve cuts the y-axis.

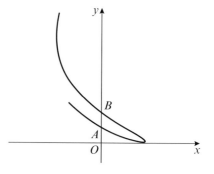

a Curve crosses the line $y = \pi^2$ when

$\left(t - \frac{\pi}{6}\right)^2 = \pi^2$ — For all points on the line $y = \pi^2$. Substitute this into the parametric equation for y and solve to find t.

$t - \frac{\pi}{6} = \pm\pi$ — You are taking square roots of both sides so consider the positive and negative values.

$t = \frac{7\pi}{6}$ or $-\frac{5\pi}{6}$

Reject $t = -\frac{5\pi}{6}$ since this is outside of the domain of t. — $-\frac{\pi}{2} < t < \frac{4\pi}{3}$ so only one of these solutions is a valid value for t.

When $t = \frac{7\pi}{6}$,

210

$$x = \cos\left(\frac{7\pi}{6}\right) + \sin\left(\frac{7\pi}{6}\right) = -\frac{1+\sqrt{3}}{2}$$ ← Substitute into the parametric equation for x. Find an exact value for x.

The point of intersection is $\left(-\frac{1+\sqrt{3}}{2}, \pi^2\right)$

b Curve cuts the y-axis when $x = 0$. So, ← Substitute $x = 0$ into the parametric equation for x and solve the resulting trigonometric equation.

$$\cos t + \sin t = 0$$
$$\sin t = -\cos t$$
$$\tan t = -1$$

Since, $-\frac{\pi}{2} < t < \frac{4\pi}{3}$ ← Consider the range of the parameter. There are two solutions to $\tan t = -1$ in this range. These correspond to the two points of intersection.

$$t = -\frac{\pi}{4} \text{ or } \frac{3\pi}{4}$$

At $t = -\frac{\pi}{4}$, $y = \left(-\frac{\pi}{4} - \frac{\pi}{6}\right)^2 = \frac{25\pi^2}{144}$ ← Substitute each value of t into the equation for y to find the y-coordinates.

At $t = \frac{3\pi}{4}$, $y = \left(\frac{3\pi}{4} - \frac{\pi}{6}\right)^2 = \frac{49\pi^2}{144}$

A is $\left(0, \frac{25\pi^2}{144}\right)$ and B is $\left(0, \frac{49\pi^2}{144}\right)$.

Problem-solving

When you are given a sketch diagram in a question, you can't read off values, but you can check whether your answers have the correct sign. The y-coordinates at both points of intersection should be positive.

Exercise 8D

1 Find the coordinates of the point(s) where the following curves meet the x-axis.

 a $x = 5 + t$, $y = 6 - t$
 b $x = 2t + 1$, $y = 2t - 6$
 c $x = t^2$, $y = (1 - t)(t + 3)$
 d $x = \frac{1}{t}$, $y = (t - 1)(2t - 1)$, $t \neq 0$
 e $x = \frac{2t}{1 + t}$, $y = t - 9$, $t \neq -1$

2 Find the coordinates of the point(s) where the following curves meet the y-axis.

 a $x = 2t$, $y = t^2 - 5$
 b $x = 3t - 4$, $y = \frac{1}{t^2}$, $t \neq 0$
 c $x = t^2 + 2t - 3$, $y = t(t - 1)$
 d $x = 27 - t^3$, $y = \frac{1}{t - 1}$, $t \neq 1$
 e $x = \frac{t - 1}{t + 1}$, $y = \frac{2t}{t^2 + 1}$, $t \neq -1$

(P) 3 A curve has parametric equations $x = 4at^2$, $y = a(2t - 1)$, where a is a constant. The curve passes through the point $(4, 0)$. Find the value of a.

(P) 4 A curve has parametric equations $x = b(2t - 3)$, $y = b(1 - t^2)$, where b is a constant. The curve passes through the point $(0, -5)$. Find the value of b.

5 Find the coordinates of the point of intersection of the line with parametric equations $x = 3t + 2$, $y = 1 - t$ and the line $y + x = 2$.

Chapter 8

6 Find the values of t at the points of intersection of the line $4x - 2y - 15 = 0$ with the parabola $x = t^2$, $y = 2t$ and give the coordinates of these points.

(P) 7 Find the points of intersection of the parabola $x = t^2$, $y = 2t$ with the circle $x^2 + y^2 - 9x + 4 = 0$.

8 Find the coordinates of the point(s) where the following curves meet the x-axis and the y-axis.
   a $x = t^2 - 1$, $y = \cos t$, $0 < t < \pi$
   b $x = \sin 2t$, $y = 2\cos t + 1$, $\pi < t < 2\pi$
   c $x = \tan t$, $y = \sin t - \cos t$, $0 < t < \dfrac{\pi}{2}$

9 Find the coordinates of the point(s) where the following curves meet the x-axis and the y-axis.
   a $x = e^t + 5$, $y = \ln t$, $t > 0$
   b $x = \ln t$, $y = t^2 - 64$, $t > 0$
   c $x = e^{2t} + 1$, $y = 2e^t - 1$, $-1 < t < 1$

10 Find the values of t at the points of intersection of the line $y = -3x + 2$ and the curve with parametric equations $x = t^2$, $y = t$, and give the coordinates of these points.

11 Find the value(s) of t at the point of intersection of the line $y = x - \ln 3$ and the curve with parametric equations $x = \ln(t - 1)$, $y = \ln(2t - 5)$, $t > \dfrac{5}{2}$, and give the exact coordinates of this point.

(E) 12 A curve C has parametric equations
$$x = 6\cos t, \quad y = 4\sin 2t + 2, \quad -\dfrac{\pi}{2} < t < \dfrac{\pi}{2}$$

   a Find the coordinates of the points where the curve intersects the x-axis. **(4 marks)**
   b Show that the curve crosses the line $y = 4$ where
   $t = \dfrac{\pi}{12}$ and $t = \dfrac{5\pi}{12}$ **(3 marks)**
   c Hence determine the coordinates of points where $y = 4$ intersects the curve. **(2 marks)**

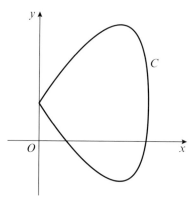

(E/P) 13 Show that the line with equation $y = 2x - 5$ does not intersect the curve with parametric equations $x = 2t$, $y = 4t(t - 1)$. **(4 marks)**

> **Problem-solving**
> Consider the discriminant after substituting.

(E/P) 14 The curve C has parametric equations $x = \sin t$, $y = \cos 2t + 1$, $0 \leq t \leq 2\pi$.
   Given that the line $y = k$, where k is a constant, intersects the curve,
   a show that $0 \leq k \leq 2$ **(3 marks)**
   b show that if the line $y = k$ is a tangent to the curve, then $k = 2$. **(3 marks)**

Parametric equations

E **15** The curve C has parametric equations $x = e^{2t}$, $y = e^t - 1$. The straight line l passes through the points A and B where $t = \ln 2$ and $t = \ln 3$ respectively.

  a Find the points A and B. **(3 marks)**

  b Show that the gradient of the line l is $\frac{1}{5}$. **(2 marks)**

  c Hence, find the equation for line l in the form $ax + by + c = 0$. **(2 marks)**

E/P **16** The curve C has parametric equations $x = \sin t$, $y = \cos t$. The straight line l passes through the points A and B where $t = \dfrac{\pi}{6}$ and $t = \dfrac{\pi}{2}$ respectively. Find an equation for the line l in the form $ax + by + c = 0$. **(7 marks)**

E/P **17** The diagram shows the curve C with parametric equations

$$x = \frac{t-1}{t}, \quad y = t - 4, \quad t \neq 0$$

The curve crosses the y-axis and the x-axis at points A and B respectively.

  a Find the coordinates of A and B. **(4 marks)**

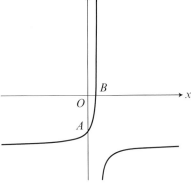

The line l_1 intersects the curve at points A and B. The lines l_2 and l_3 are parallel to l_1 and are distinct tangents to the curve.

  b Show that the two possible equations for l_2 and l_3 are
  $$y = 4x - 4 \text{ and } y = 4x - 12$$ **(6 marks)**

  c Find the coordinates of the point where each tangent meets C. **(4 marks)**

Challenge

The curve C_1 has parametric equations
$$x = e^{2t}, \quad y = 2t + 1$$
The curve C_2 has parametric equations
$$x = e^t, \quad y = 1 + t^2$$
Find the coordinates of the points at which these two curves intersect.

8.5 Modelling with parametric equations

You can use parametric equations to model real-life situations. In mechanics you will use parametric equations with **time** as a parameter to model motion in two dimensions.

Example 11

A plane's position at time t seconds after take-off can be modelled with the following parametric equations:

$$x = (v\cos\theta)t \text{ m}, \quad y = (v\sin\theta)t \text{ m}, \quad t > 0$$

where v is the speed of the plane, θ is the angle of elevation of its path, x is the horizontal distance travelled and y is the vertical distance travelled, relative to a fixed origin.

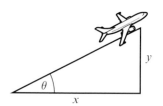

When the plane has travelled 600 m horizontally, it has climbed 120 m
a Find the angle of elevation, θ.

Given that the plane's speed is 50 m s^{-1},
b find the parametric equations for the plane's motion
c find the vertical height of the plane after 10 seconds
d show that the plane's motion is a straight line
e explain why the domain of t, $t > 0$, is not realistic.

a

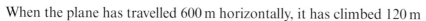

Angle of elevation
$\theta = \tan^{-1}\left(\dfrac{120}{600}\right) = 11.3°$ (1 d.p.)   •——— The model assumes that the angle of elevation will stay constant so the ratio will always be the same regardless of how far along the journey the plane is.

b $x = (v\cos\theta)t$
  $= (50 \times \cos 11.3...)t = 49.0t$ m (3 s.f.)

  $y = (v\sin\theta)t$
  $= (50 \times \sin 11.3...)t = 9.81t$ m (3 s.f.)   •——— Substitute $v = 50$ and $\theta = 11.3$ into the equations for x and y. The units of length, metres, are given with the model.

c At $t = 10$,
  $y = 9.81t = 9.81 \times 10 = 98.1$ m
  So, the plane has climbed 98.1 m after 10 seconds.   •——— Substitute $t = 10$ into y, as y represents the vertical height.

d $x = 49t$
  So, $\dfrac{x}{49} = t$ (1)   •——— Find a Cartesian equation for the plane's path. Rearrange the equation for x to make t the subject.

  $y = 9.81t$ (2)

  So,
  $y = 9.81 \times \dfrac{x}{49} = 0.2x$ •——— Substitute t from (1) into (2).

  Since this is a linear equation, the motion of the plane is a straight line with gradient 0.2   •——— The gradient in this context represents the height gained for every metre travelled horizontally.

e $t > 0$ is not realistic as this would mean the plane would continue climbing forever at the same speed and with the same angle of elevation.

Problem-solving

If you have to comment on a modelling assumption or range of validity, consider whether the assumption is realistic given the context of the question. Make sure you refer to the real-life situation being modelled in your answer.

Example 12

A stone is thrown from the top of a 25 m high cliff with an initial speed of 5 m s^{-1} at an angle of 45°. Its position after t seconds can be described using the following parametric equations

$$x = \frac{5\sqrt{2}}{2} t \text{ m}, \quad y = \left(-4.9t^2 + \frac{5\sqrt{2}}{2} t + 25\right) \text{m}, \quad 0 \leq t \leq k$$

where x is the horizontal distance from the point of projection, y is the vertical distance from the ground and k is a constant.

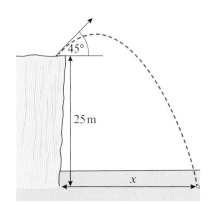

Given that the model is valid from the time the stone is thrown to the time it hits the ground,

a find the value of k

b find the horizontal distance travelled by the stone once it hits the floor.

a The stone hits the ground when $y = 0$:

$$-4.9t^2 + \frac{5\sqrt{2}}{2} t + 25 = 0$$

$$t = \frac{-\frac{5\sqrt{2}}{2} \pm \sqrt{\left(\frac{5\sqrt{2}}{2}\right)^2 - 4(-4.9)(25)}}{2(-4.9)}$$

$t = -1.926...$ or $t = 2.648...$

$t \geq 0$, so the stone hits the ground at

$t = 2.648...$

So $k = 2.65$ (2 d.p.)

b When $t = 2.648...$

$$x = \frac{5\sqrt{2}}{2} t = \frac{5\sqrt{2}}{2} \times 2.648...$$

$= 9.362...$ m

So the horizontal distance travelled by the stone is 9.36 m (2 d.p.).

Online Use the polynomial function on your calculator to solve the quadratic equation.

Use the quadratic formula on your calculator to find two solutions for t.

The model is only valid for $t \geq 0$ so disregard the negative solution.

Substitute this value of t into the parametric equation for x.

Chapter 8

Example 13

The motion of a figure skater relative to a fixed origin, O, at time t minutes is modelled using the parametric equations

$$x = 8\cos 20t, \quad y = 12\sin\left(10t - \frac{\pi}{3}\right), \quad t \geq 0$$

where x and y are measured in metres.

a Find the coordinates of the figure skater at the beginning of his motion.
b Find the coordinates of the point where the figure skater intersects his own path.
c Find the coordinates of the points where the path of the figure skater crosses the y-axis.
d Determine how long it takes the figure skater to complete one complete figure-of-eight motion.

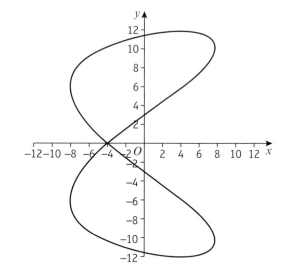

a At $t = 0$,
$x = 8\cos 0 = 8$
$y = 12\sin\left(10 \times 0 - \frac{\pi}{3}\right) = 12\sin\left(-\frac{\pi}{3}\right)$
$= -6\sqrt{3}$

The coordinates of the figure skater at the beginning of his motion are $(8, -6\sqrt{3})$.

— Substitute $t = 0$ into both equations to find the x and y coordinates.

b From the diagram, the figure skater intersects his own path on the x-axis, ie. when $y = 0$.

$12\sin\left(10t - \frac{\pi}{3}\right) = 0$

$\sin\left(10t - \frac{\pi}{3}\right) = 0$

$10t - \frac{\pi}{3} = 0, \pi, 2\pi, \ldots$

$10t = \frac{\pi}{3}, \frac{4\pi}{3}, \frac{7\pi}{3}, \ldots$

$t = \frac{\pi}{30}, \frac{2\pi}{15}, \frac{7\pi}{30}, \ldots$

$x = 8\cos\left(20 \times \frac{\pi}{30}\right) = 8\cos\left(\frac{2\pi}{3}\right) = -4$

So, the figure skater intersects his own path at the point $(-4, 0)$.

— Use the diagram to find information about the point of intersection.

— Substitute $y = 0$ into the equation for y, and solve to find values of t in the domain $t \geq 0$.

There is only one point of intersection so choose any of these values of t. You can use one of the others to check: $8\cos\left(20 \times \frac{2\pi}{15}\right) = 8\cos\left(\frac{8\pi}{3}\right) = -4$

c The figure skater crosses the y-axis when $x = 0$,

$0 = 8\cos 20t$

$0 = \cos 20t$

So, $20t = \dfrac{\pi}{2}, \dfrac{3\pi}{2}, \dfrac{5\pi}{2}, \dfrac{7\pi}{2}, \ldots$

Find solutions to $8\cos 20t = 0$ in the domain $t \geq 0$. There are 4 points of intersection so consider the first 4 solutions, and check that these give different values of y.

Substitute these values into y.

$20t = \dfrac{\pi}{2}$:

$y = 12\sin\left(\dfrac{1}{2} \times \dfrac{\pi}{2} - \dfrac{\pi}{3}\right) = 12\sin\left(-\dfrac{\pi}{12}\right)$

$= -3.11$ (2 d.p.)

Use your calculator to find the corresponding values of y. You can give your answers as decimals or as exact values: $12\sin\left(-\dfrac{\pi}{12}\right) = -3\sqrt{6} + 3\sqrt{2}$

$20t = \dfrac{3\pi}{2}$:

$y = 12\sin\left(\dfrac{1}{2} \times \dfrac{3\pi}{2} - \dfrac{\pi}{3}\right) = 12\sin\left(\dfrac{5\pi}{12}\right)$

$= 11.59$ (2 d.p.)

Online Find points of intersection of this curve with the coordinate axes using technology.

$20t = \dfrac{5\pi}{2}$:

$y = 12\sin\left(\dfrac{1}{2} \times \dfrac{5\pi}{2} - \dfrac{\pi}{3}\right) = 12\sin\left(\dfrac{11\pi}{12}\right)$

$= 3.11$ (2 d.p.)

$20t = \dfrac{7\pi}{2}$:

$y = 12\sin\left(\dfrac{1}{2} \times \dfrac{7\pi}{2} - \dfrac{\pi}{3}\right) = 12\sin\left(\dfrac{17\pi}{12}\right)$

$= -11.59$ (2 d.p.)

So the skater crosses the y-axis at $(0, -3.11)$, $(0, 11.59)$, $(0, 3.11)$, $(0, -11.59)$.

Check that these look sensible from the graph. The motion of the skater appears to be symmetrical about the x-axis so these look right.

d The period of $x = 8\cos 20t$ is $\dfrac{2\pi}{20}$,

so the skater returns to his x-position after $\dfrac{2\pi}{20}$ min, $\dfrac{4\pi}{20}$ min, $\ldots$

The period of $a\cos(bx + c)$ is $\dfrac{2\pi}{b}$ and the period of $a\sin(bx + c)$ is $\dfrac{2\pi}{b}$

The period of $y = 12\sin\left(10t - \dfrac{\pi}{3}\right)$ is $\dfrac{2\pi}{10}$,

so the skater returns to his y-position after $\dfrac{2\pi}{10}$ min, $\dfrac{4\pi}{10}$ min, $\ldots$

Problem-solving

In order for the figure skater to return to his starting position, **both** parametric equations must complete full periods. This occurs at the **least common multiple** of the two periods.

So the skater first completes a full figure-of-eight motion after $\dfrac{2\pi}{10}$ mins $= 0.628\ldots$ mins or 38 seconds (2 s.f.).

Exercise 8E

P 1 A river flows from north to south. The position at time t seconds of a rowing boat crossing the river from west to east is modelled by the parametric equations
$$x = 0.9t \text{ m}, \quad y = -3.2t \text{ m}$$
where x is the distance travelled east and y is the distance travelled north.

Given that the river is 75 m wide,

a find the time taken to get to the other side

b find the distance the boat has been moved off-course due to the current

c show that the motion of the boat is a straight line

d determine the speed of the boat.

P 2 The position of a small plane coming into land at time t seconds after it has started its descent is modelled by the parametric equations
$$x = 80t, \quad y = -9.1t + 3000, \quad 0 \leq t < 330$$
where x is the horizontal distance travelled (in metres) and y is the vertical distance (in metres) of the plane above ground level.

a Find the initial height of the plane.

b Justify the choice of domain, $0 \leq t < 330$, for this model.

c Find the horizontal distance the plane travels between beginning its descent and landing.

P 3 A ball is kicked from the ground with an initial speed of 20 m s^{-1} at an angle of $30°$. Its position after t seconds can be described using the following parametric equations
$$x = 10\sqrt{3}\, t \text{ m}, \quad y = (-4.9t^2 + 10t) \text{ m}, \quad 0 \leq t \leq k$$

a Find the horizontal distance travelled by the ball when it hits the ground.

A player wants to head the ball when it is descending between 1.5 m and 2.5 m off the ground.

b Find the range of time after the ball has been kicked at which the player can head the ball.

c Find the closest horizontal distance from where the ball has been kicked at which the player can head the ball.

P 4 The path of a dolphin leaping out of the water can be modelled with the following parametric equations
$$x = 2t \text{ m}, \quad y = -4.9t^2 + 10t \text{ m}$$
where x is the horizontal distance from the point the dolphin jumps out of the water, y is the height above sea level of the dolphin and t is the time in seconds after the dolphin has started its jump.

a Find the time the dolphin takes to complete a single jump.

b Find the horizontal distance the dolphin travels during a single jump.

c Show that the dolphin's path is modelled by a quadratic curve.

d Find the maximum height of the dolphin.

5 The path of a car on a Ferris wheel at time t minutes is modelled using the parametric equations

$$x = 12\sin t, \quad y = 12 - 12\cos t$$

where x is the horizontal distance in metres of the car from the start of the ride and y is the height in metres above ground level of the car.

a Show that the motion of the car is a circle with radius 12 m.

b Hence, find the maximum height of the car during the journey.

c Find the time taken to complete one revolution of the Ferris wheel and hence calculate the average speed of the car.

6 The cross-section of a bowl design is given by the following parametric equations

$$x = t - 4\sin t, \quad y = 1 - 2\cos t, \quad -\frac{\pi}{2} \leq t \leq \frac{\pi}{2}$$

a Find the length of the opening of the bowl. **(3 marks)**

b Given that the cross-section of the bowl crosses the y-axis at its deepest point, find the depth of the bowl. **(3 marks)**

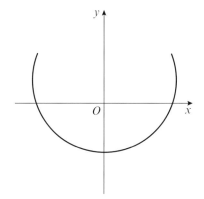

7 A particle is moving in the xy-plane such that its position after time t seconds relative to a fixed origin O is given by the parametric equations

$$x = \frac{t^2 - 3t + 2}{t}, \quad y = 2t, \quad t > 0$$

The diagram shows the path of the particle.

a Find the distance from the origin to the particle at time $t = 0.5$.

b Find the coordinates of the points where the particle crosses the y-axis.

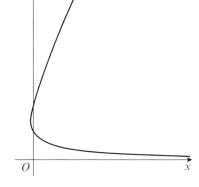

Another particle travels in the same plane with its path given by the equation $y = 2x + 10$.

c Show that the paths of these two particles never intersect.

8 The path of a ski jumper from the point of leaving the ramp to the point of landing is modelled using the parametric equations

$$x = 18t, \quad y = -4.9t^2 + 4t + 10, \quad 0 \leq t \leq k$$

where x is the horizontal distance in metres from the point of leaving the ramp and y is the height in metres above ground level of the ski jumper, after t seconds.

a Find the initial height of the ski jumper. **(1 mark)**

b Find the value of k and hence state the time taken for the ski jumper to complete her jump. **(3 marks)**

c Find the horizontal distance the ski jumper jumps. **(1 mark)**

d Show that the ski jumper's path is a parabola and find the maximum height above ground level of the ski jumper. **(5 marks)**

9 The profile of a hill climb in a bike race is modelled by the following parametric equations

$$x = 50 \tan t \text{ m}, \quad y = 20 \sin 2t \text{ m}, \quad 0 < t \leq \frac{\pi}{2}$$

a Find the value of t at the highest point of the hill climb.

b Hence find the coordinates of the highest point.

c Find the coordinates when $t = 1$ and show that at this point, a cyclist will be descending.

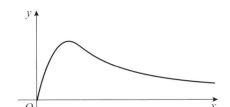

10 A computer model for the shape of the path of a rollercoaster is given by the parametric equations

$$x = 5 + \ln t, \quad y = 5 \sin 2t, \quad 0 < t \leq \frac{\pi}{2}$$

a Find the coordinates of the point where $t = \frac{\pi}{6}$ **(2 marks)**

Given that one unit on the model represents 5 m in real life,

b find the maximum height of the rollercoaster **(1 mark)**

c find the horizontal distance covered during the descent of the rollercoaster. **(4 marks)**

d Hence, find the average gradient of the descent. **(1 mark)**

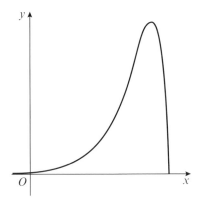

Mixed exercise 8

1 The diagram shows a sketch of the curve with parametric equations

$$x = 4 \cos t, \quad y = 3 \sin t, \quad 0 \leq t < 2\pi$$

a Find the coordinates of the points A and B.

b The point C has parameter $t = \frac{\pi}{6}$. Find the exact coordinates of C.

c Find the Cartesian equation of the curve.

2 The diagram shows a sketch of the curve with parametric equations

$$x = \cos t, \quad y = \tfrac{1}{2} \sin 2t, \quad 0 \leq t < 2\pi$$

The curve is symmetrical about both axes.

Copy the diagram and label the points having parameters $t = 0$, $t = \frac{\pi}{2}$, $t = \pi$ and $t = \frac{3\pi}{2}$.

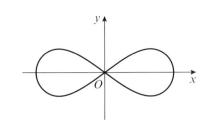

Parametric equations

P **3** A curve has parametric equations

$$x = e^{2t+1} + 1, \quad y = t + \ln 2, \quad t > 1$$

   a Find a Cartesian equation of this curve in the form $y = f(x)$, $x > k$ where k is a constant to be found in exact form.

   b Write down the range of $f(x)$, leaving your answer in exact form.

4 A curve has parametric equations

$$x = \frac{1}{2t+1}, \quad y = 2\ln\left(t + \frac{1}{2}\right), \quad t > \frac{1}{2}$$

Find a Cartesian equation of the curve in the form $y = f(x)$, and state the domain and range of $f(x)$.

P **5** A curve has parametric equations $x = \sin t$, $y = \cos 2t$, $0 \leqslant t < 2\pi$

   a Find a Cartesian equation of the curve.

   The curve cuts the x-axis at $(a, 0)$ and $(b, 0)$.

   b Find the values of a and b.

P **6** A curve has parametric equations $x = \dfrac{1}{1+t}$, $y = \dfrac{1}{(1+t)(1-t)}$, $t > 1$

Express t in terms of x, and hence show that a Cartesian equation of the curve is $y = \dfrac{x^2}{2x-1}$

7 A circle has parametric equations $x = 4\sin t - 3$, $y = 4\cos t + 5$, $0 \leqslant t \leqslant 2\pi$

   a Find a Cartesian equation of the circle.

   b Draw a sketch of the circle.

   c Find the exact coordinates of the points of intersection of the circle with the y-axis.

/P **8** The curve C has parametric equations

$$x = \frac{2-3t}{1+t}, \quad y = \frac{3+2t}{1+t}, \quad 0 \leqslant t \leqslant 4$$

   a Show that the curve C is part of a straight line. (3 marks)

   b Find the length of this line segment. (2 marks)

E **9** A curve C has parametric equations

$$x = t^2 - 2, \quad y = 2t, \quad 0 \leqslant t \leqslant 2$$

   a Find the Cartesian equation of C in the form $y = f(x)$. (3 marks)

   b State the domain and range of $y = f(x)$ in the given domain of t. (3 marks)

   c Sketch the curve in the given domain of t. (2 marks)

Chapter 8

E/P 10 A curve C has parametric equations
$$x = 2\cos t, \quad y = 2\sin t - 5, \quad 0 \leq t \leq \pi$$
a Show that the curve C forms part of a circle. **(3 marks)**
b Sketch the curve in the given domain of t. **(3 marks)**
c Find the length of the curve in the given domain of t. **(3 marks)**

E/P 11 The curve C has parametric equations
$$x = t - 2, \quad y = t^3 - 2t^2, \quad t \in \mathbb{R}$$
a Find a Cartesian equation of C in the form $y = f(x)$. **(3 marks)**
b Sketch the curve C. **(3 marks)**

E/P 12 Show that the line with equation $y = 4x + 20$ is a tangent to the curve with parametric equations $x = t - 3$, $y = 4 - t^2$. **(4 marks)**

E/P 13 The curve C has parametric equations $x = 2\ln t$, $y = t^2 - 1$, $t > 0$
a Find the coordinates of the point where the line $x = 5$ intersects the curve. Give your answer as exact values. **(4 marks)**
b Given that the line $y = k$ intersects the curve, find the range of values for k. **(3 marks)**

E/P 14 The diagram shows the curve C with parametric equations
$$x = 1 + 2t, \quad y = 4^t - 1$$
The curve crosses the y-axis and the x-axis at points A and B respectively.
a Find the coordinates of A and B. **(4 marks)**
The line l intersects the curve at points A and B.
b Find the equation of l in the form $ax + by + c = 0$. **(3 marks)**

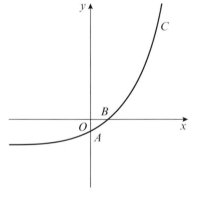

E/P 15 The diagram shows the curve C with parametric equations
$$x = \ln t - \ln\left(\frac{\pi}{2}\right), \quad y = \sin t, \quad 0 < t < 2\pi$$
The curve crosses the y-axis and the x-axis at points A and B respectively. The line l intersects the curve at points A and B. Find the equation of l in the form $ax + by + c = 0$. **(7 marks)**

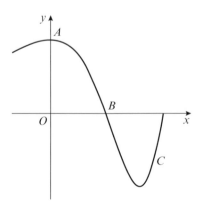

Parametric equations

P) 16 A plane's position at time t seconds during its descent can be modelled with the following parametric equations

$$x = 80t, \quad y = 3000 - 30t, \quad 0 < t < k$$

where x is the horizontal distance travelled in metres and y is the vertical height of the plane in metres.

a Show that the plane's descent is a straight line. **(3 marks)**

Given that the model is valid until the plane is 30 m off the ground,

b find the value of k **(2 marks)**

c determine the distance travelled by the plane in this portion of its descent. **(3 marks)**

P) 17 The path of an arrow path at time t seconds from being fired can be described using the following parametric equations

$$x = 50\sqrt{2}\,t, \quad y = 1.5 - 4.9t^2 + 50\sqrt{2}\,t, \quad 0 \leq t \leq k$$

where x is the horizontal distance from the archer in metres and y is the vertical height of the arrow above level ground.

a Find the furthest horizontal distance the arrow can travel.

A castle is located a horizontal distance of 1000 m from the archer's position. The height of the castle is 10 m.

b Show that the arrow misses the castle.

c Find the distance the archer should step back so that he can hit the top of the castle.

P) 18 The position of a mountain climber t hours after they begin a climb can be modelled with the following parametric equations

$$x = 300\sqrt{t}, \quad y = 244t(4 - t), \quad 0 < t < k$$

where x represents the distance travelled horizontally in metres and y represents the height above sea level in metres.

a Find the highest point above sea level reached by the mountain climber and the time at which she reaches it. **(3 marks)**

Given that the mountain climber completes her climb when she gets back to sea level,

b find the horizontal distance from the beginning of her climb to the end. **(2 marks)**

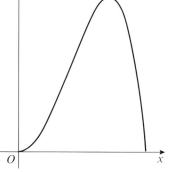

P) 19 A bridge is designed using the following parametric equations:

$$x = \frac{4t}{\pi} - 2\sin t, \quad y = -\cos t, \quad \frac{\pi}{2} < t < \frac{3\pi}{2}$$

Given that 1 unit in the design is 10 m in real life,

a find the highest point of the bridge

b find the width of the widest river this bridge can cross.

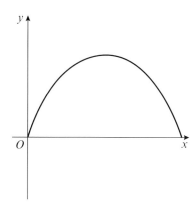

223

20 A BMX cyclist's position on a ramp at time t seconds can be modelled with the parametric equations

$$x = 3(e^t - 1), \quad y = 10(t - 1)^2, \quad 0 \leqslant t \leqslant 1.3$$

where x is the horizontal distance travelled in metres and y is the height above ground level in metres.

a Find the initial height of the cyclist.

b Find the time the cyclist is at her lowest height.

Given that after 1.3 seconds, the cyclist is at the end of the ramp,

c find the height at which the cyclist leaves the ramp.

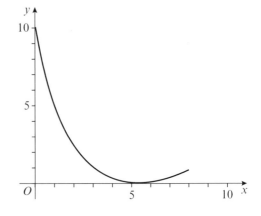

Challenge

Two particles A and B move in the x-y plane, such that their positions relative to a fixed origin at a time t seconds are given, respectively, by the parametric equations:

$$A: x = \frac{2}{t}, \quad y = 3t + 1, \quad t > 0$$

$$B: x = 5 - 2t, \quad y = 2t^2 + 2k - 1, \quad t > 0$$

where k is a non-zero constant.

Given that the particles collide,

a find the value of k

b find the coordinates of the point of collision.

Summary of key points

1 A curve can be defined using parametric equations $x = p(t)$ and $y = q(t)$. Each value of the parameter, t, defines a point on the curve with coordinates $(p(t), q(t))$.

2 You can convert between parametric equations and Cartesian equations by using substitution to eliminate the parameter.

3 For parametric equations $x = p(t)$ and $y = q(t)$ with Cartesian equation $y = f(x)$:
- the domain of $f(x)$ is the range of $p(t)$
- the range of $f(x)$ is the range of $q(t)$

4 You can use parametric equations to model real-life situations. In mechanics you will use parametric equations with time as a parameter to model motion in two dimensions.

Review exercise 2

(E) **1** The diagram shows the curve with equation $y = \sin\left(x + \frac{3\pi}{4}\right)$, $-2\pi \leq x \leq 2\pi$.

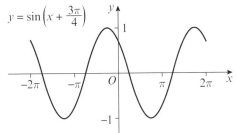

Calculate the coordinates of the points at which the curve meets the coordinate axes. **(3)**

← Section 5.1

(E) **2 a** Sketch, for $0 \leq x \leq 2\pi$, the graph of
$y = \cos\left(x - \frac{\pi}{3}\right)$ **(2)**

b Write down the exact coordinates of the points where the graph meets the coordinate axes. **(3)**

c Solve, for $0 \leq x \leq 2\pi$, the equation $\cos\left(x - \frac{\pi}{3}\right) = -0.27$, giving your answers in radians to 2 decimal places. **(5)**

← Section 5.1

(E) **3** In the diagram, A and B are points on the circumference of a circle centre O and radius 5 cm.
$\angle AOB = \theta$ radians
$AB = 6$ cm

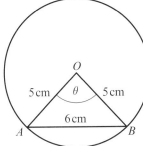

a Find the value of θ. **(2)**

b Calculate the length of the minor arc AB to 3 s.f. **(2)**

← Section 5.2

(E/P) **4** In the diagram, ABC is an equilateral triangle with side 8 cm.
PQ is an arc of a circle centre, A, radius 6 cm.
Find the perimeter of the shaded region in the diagram.

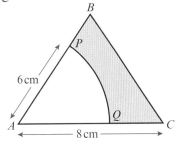

(5)

← Section 5.2

(E/P) **5** In the diagram, AD and BC are arcs of circles with centre O, such that $OA = OD = r$ cm, $AB = DC = 10$ cm and $\angle BOC = \theta$ radians.

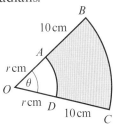

a Given that the area of the shaded region is 40 cm², show that $r = \frac{4}{\theta} - 5$. **(4)**

b Given also that $r = 6\theta$, calculate the perimeter of the shaded region. **(6)**

← Sections 5.2, 5.3

225

Review exercise 2

6 In the diagram,
$AB = 10$ cm, $AC = 13$ cm.
$\angle CAB = 0.6$ radians.
BD is an arc of a circle centre A and radius 10 cm.

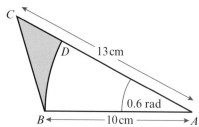

a Calculate the length of the arc BD. **(2)**
b Calculate the shaded area in the diagram to 1 d.p. **(3)**

← Sections 5.2, 5.3

7 The diagram shows the sector OAB of a circle with centre O, radius r cm and angle 1.4 radians.

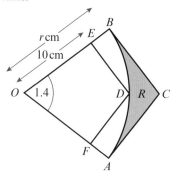

The lines AC and BC are tangent to the circle with centre O. OEB and OFA are straight lines. The line ED is parallel to BC and the line FD is parallel to AC.

a Find the area of sector OAB, giving your answer to 1 decimal place. **(4)**

The region R is bounded by the arc AB and the lines AC and CB.

b Find the perimeter of R, giving your answer to 1 decimal place. **(6)**

← Sections 5.2, 5.3

8 The diagram shows a square, $ABCD$, with side length r, and 2 arcs of circles with centres A and B.

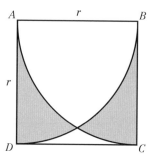

Show that the area of the shaded region is $\dfrac{r^2}{6}(3\sqrt{3} - \pi)$. **(5)**

← Sections 5.2, 5.3

9 a Show that the equation
$3\sin^2 x + 7\cos x + 3 = 0$ can be written as $3\cos^2 x - 7\cos x - 6 = 0$. **(2)**

b Hence solve, for $0 \leqslant x < 2\pi$,
$3\sin^2 x + 7\cos x + 3 = 0$, giving your answers to 2 decimal places. **(3)**

← Section 5.4

10 a Show that, when θ is small,
$\sin 4\theta - \cos 4\theta + \tan 3\theta \approx 8\theta^2 + 7\theta - 1$ **(3)**

b Hence state the approximate value of $\sin 4\theta - \cos 4\theta + \tan 3\theta$ for small values of θ. **(1)**

← Section 5.5

11 a Sketch, in the interval $-2\pi \leqslant x \leqslant 2\pi$, the graph of $y = 4 - 2\cosec x$.
Mark any asymptotes on your graph. **(3)**

b Hence deduce the range of values of k for which the equation $4 - 2\cosec x = k$ has no solutions. **(2)**

← Sections 6.1, 6.2

12 The diagram shows the graph of
$$y = k\sec(\theta - \alpha)$$
The curve crosses the y-axis at the point $(0, 4)$, and the θ-coordinate of its minimum point is $\frac{\pi}{3}$

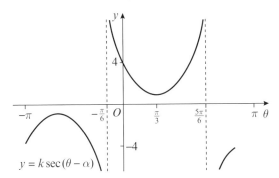

a State, as a multiple of π, the value of α. **(1)**

b Find the value of k. **(2)**

c Find the exact values of θ at the points where the graph crosses the line $y = -2\sqrt{2}$. **(3)**

← Section 6.2

13 a Show that
$$\frac{\cos x}{1 - \sin x} + \frac{1 - \sin x}{\cos x} \equiv 2\sec x$$
(4)

b Hence solve, in the interval $0 \leq x \leq 4\pi$,
$$\frac{\cos x}{1 - \sin x} + \frac{1 - \sin x}{\cos x} = -2\sqrt{2}$$
(4)

← Section 6.3

14 a Prove that
$$\frac{\sin\theta}{\cos\theta} + \frac{\cos\theta}{\sin\theta} = 2\cosec 2\theta, \theta \neq 90n°$$
(3)

b Sketch the graph of $y = 2\cosec 2\theta$ for $0° < \theta < 360°$. **(3)**

c Solve, for $0° < \theta < 360°$, the equation
$$\frac{\sin\theta}{\cos\theta} + \frac{\cos\theta}{\sin\theta} = 3$$, giving your answer to 1 decimal place. **(4)**

← Section 6.3

15

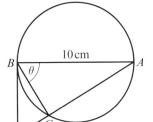

In the diagram, $AB = 10$ cm is the diameter of the circle and BD is the tangent to the circle at B. The chord AC is extended to meet this tangent at D and $\angle ABC = \theta$.

a Show that $BD = 10\cot\theta$. **(4)**

b Given that $BD = \frac{10}{\sqrt{3}}$ cm, calculate the exact length of DC. **(3)**

← Section 6.4

16 a Given that $\sin^2\theta + \cos^2\theta \equiv 1$, show that $1 + \tan^2\theta \equiv \sec^2\theta$. **(2)**

b Solve, for $0° \leq \theta < 360°$, the equation
$$2\tan^2\theta + \sec\theta = 1$$
giving your answers to 1 decimal place. **(6)**

← Section 6.3

17 Given that $a = \cosec x$ and $b = 2\sin x$.

a express a in terms of b

b find the value of $\frac{4 - b^2}{a^2 - 1}$ in terms of b.

← Section 6.4

18 Given that
$$y = \arcsin x, -1 \leq x \leq 1, -\frac{\pi}{2} \leq y \leq \frac{\pi}{2}$$

a express $\arccos x$ in terms of y. **(2)**

b Hence find, in terms of π, the value of $\arcsin x + \arccos x$. **(1)**

← Section 6.5

19 a Prove that for $x \geq 1$,
$$\arccos\frac{1}{x} = \arcsin\frac{\sqrt{x^2 - 1}}{x}$$
(4)

b Explain why this identity is not true for $0 \leq x < 1$. **(2)**

← Section 6.5

Review exercise 2

E) 20 a Sketch the graph of $y = 2\arccos x - \dfrac{\pi}{2}$, showing clearly the exact endpoints of the curve. **(4)**

  b Find the exact coordinates of the point where the curve crosses the x-axis. **(3)**

  ← Section 6.5

E) 21 Given that $\tan\left(x + \dfrac{\pi}{6}\right) = \dfrac{1}{6}$, show that
$$\tan x = \dfrac{72 - 111\sqrt{3}}{321}$$ **(5)**

  ← Section 7.1

E/P) 22 Given that $\sin(x + 30°) = 2\sin(x - 60°)$

  a show that $\tan x = 8 + 5\sqrt{3}$. **(4)**

  b Hence express $\tan(x + 60°)$ in the form $a + b\sqrt{3}$. **(3)**

  ← Section 7.1

E/P) 23 a Use $\sin(\theta + \alpha) = \sin\theta\cos\alpha + \cos\theta\sin\alpha$, or otherwise, to show that
$$\sin 165° = \dfrac{\sqrt{6} - \sqrt{2}}{4}$$ **(4)**

  b Hence, or otherwise, use an algebraic method to show that $\operatorname{cosec} 165° = \sqrt{a} + \sqrt{b}$, where a and b are constants to be found. **(3)**

  ← Sections 7.1, 7.2

E/P) 24 Given that $\cos A = \dfrac{3}{4}$ where $270° < A < 360°$,

  a find the exact value of $\sin 2A$ **(3)**

  b show that $\tan 2A = -3\sqrt{7}$. **(3)**

  ← Section 7.3

E/P) 25 Solve, in the interval $-180° \le x \le 180°$, the equations

  a $\cos 2x + \sin x = 1$ **(3)**

  b $\sin x(\cos x + \operatorname{cosec} x) = 2\cos^2 x$ **(3)**

  giving your answers to 1 decimal place.

  ← Section 7.4

E) 26 $f(x) = 3\sin x + 2\cos x$

  Given $f(x) = R\sin(x + \alpha)$, where $R > 0$ and $0 < \alpha < \dfrac{\pi}{2}$,

  a find the value of R and the value of α. **(4)**

  b Hence find the greatest value of $(3\sin x + 2\cos x)^4$ **(2)**

  c Hence, or otherwise, solve for $0 \le \theta < 2\pi$, $f(x) = 1$, rounding your answers to 3 decimal places. **(3)**

  ← Section 7.5

E) 27 a Prove that
$$\cot\theta - \tan\theta \equiv 2\cot 2\theta,\ \theta \ne \dfrac{n\pi}{2}$$ **(3)**

  b Solve, for $-\pi < \theta < \pi$, the equation
  $$\cot\theta - \tan\theta = 5,$$
  giving your answers to 3 significant figures. **(3)**

  ← Sections 6.3, 7.6

E) 28 a By writing $\cos 3\theta$ as $\cos(2\theta + \theta)$, show that
$$\cos 3\theta \equiv 4\cos^3\theta - 3\cos\theta$$ **(4)**

  b Given that $\cos\theta = \dfrac{\sqrt{2}}{3}$, find the exact value of $\sec 3\theta$. Give your answer in the form $k\sqrt{2}$ where k is a rational constant to be found. **(2)**

  ← Sections 6.3, 7.1

E) 29 Show that $\sin^4\theta \equiv \dfrac{3}{8} - \dfrac{1}{2}\cos 2\theta + \dfrac{1}{8}\cos 4\theta$. You must show each stage of your working. **(6)**

  ← Section 7.6

E/P) 30 a Express $6\sin\theta + 2\cos\theta$ in the form $R\sin(\theta + \alpha)$, where $r < 0$ and $0 < \alpha < \dfrac{\pi}{2}$

  Give the value of α to 2 decimal places. **(4)**

  b i Find the maximum value of $6\sin\theta + 2\cos\theta$ **(2)**

  ii Find the value of θ, for $0 < \theta < \pi$, at which the maximum occurs, giving the value to 2 d.p. **(1)**

  The temperature, in $T\,°C$, on a particular day is modelled by the equation
  $$T = 9 + 6\sin\left(\dfrac{\pi t}{12}\right) + 2\cos\left(\dfrac{\pi t}{12}\right),$$
  $0 \le t \le 24$ where t is the number of hours after 9 a.m.

c Calculate the minimum value of T predicted by this model, and the value of t, to 2 decimal places, when this minimum occurs. **(3)**

d Calculate, to the nearest minute, the times in the first day when the temperature is predicted by this model, to be exactly 14°C. **(4)**

← Section 7.5, 7.7

E 31 A curve C has parametric equations
$$x = 1 - \frac{4}{t}, y = t^2 - 3t + 1, t \in \mathbb{R}, t \neq 0$$

a Determine the ranges of x and y in the given domain of t. **(3)**

b Show that the Cartesian equation of C can be written in the form
$$y = \frac{ax^2 + bx + c}{(1-x)^2},$$ where a, b and c are integers to be found. **(3)**

← Section 8.1

E 32 A curve has parametric equations
$$x = \ln(t+2), y = \frac{3t}{t+3}, t > 4$$

a Find a Cartesian equation of this curve in the form $y = f(x)$, $x > k$, where k is an exact constant to be found. **(4)**

b Write down the range of $f(x)$ in the form $a < y < b$, where a and b are constants to be found. **(2)**

← Section 8.1

E 33 A curve C has parametric equations
$$x = \frac{1}{1+t}, y = \frac{1}{1-t}, -1 < t < 1$$
Show that a Cartesian equation of C is
$$y = \frac{x}{2x-1}$$ **(4)**

← Section 8.1

E/P 34 A curve C has parametric equations
$$x = 2\cos t, y = \cos 3t, 0 \leqslant t \leqslant \frac{\pi}{2}$$

a Find a Cartesian equation of the curve in the form $y = f(x)$, where $f(x)$ is a cubic function. **(5)**

b State the domain and range of $f(x)$ for the given domain of t. **(2)**

← Section 8.2

E/P 35 The curve shown in the figure has parametric equations
$$x = \sin t, y = \sin\left(t + \frac{\pi}{6}\right), -\frac{\pi}{2} \leqslant t \leqslant \frac{\pi}{2}$$

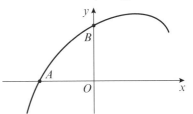

a Show that a Cartesian equation of the curve is
$$y = \frac{\sqrt{3}}{2}x + \frac{1}{2}\sqrt{(1-x^2)}, -1 \leqslant x \leqslant 1$$ **(4)**

b Find the coordinates of the points A and B, where the curve intercepts the x- and y-axes. **(3)**

← Section 8.2

E 36 The curve C has parametric equations
$$x = 3\cos t, y = \cos 2t, 0 \leqslant t \leqslant \pi$$

a Find a Cartesian equation of C. **(4)**

b Sketch the curve C on the appropriate domain, labelling the points where the curve intercepts the x- and y-axes. **(3)**

← Section 8.2, 8.3

E/P 37 The curve C has parametric equations
$$x = 4t, y = 8t(2t - 1), t \in \mathbb{R}.$$
Given that the line with equation $y = 3x + c$, where c is a constant, does not intersect C, find the range of possible values of c. **(5)**

← Section 8.4

Review exercise 2

E) 38 A curve has parametric equations
$x = 3\sin 2t, y = 2\cos t + 1, \frac{\pi}{2} \leq t \leq \frac{3\pi}{2}$

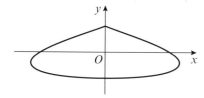

a Find the coordinates of the points where the curve intersects the x-axis. **(4)**

b Show that the curve crosses the line $x = 1.5$ when $t = \frac{13\pi}{12}$ and $t = \frac{17\pi}{12}$ **(3)**

← Section 8.4

E/P) 39 A golf ball is hit from an elevation of 50 m, with an initial speed of 50 m s⁻¹ at an angle of 30° above the horizontal. Its position after t seconds can be described using the following parametric equations:
$x = (25\sqrt{3})t, y = 25t - 4.9t^2 + 50, 0 \leq t \leq k$
where x is the horizontal distance in metres, y is the vertical distance in metres from the ground and k is a constant.

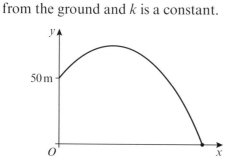

Given that the model is valid from the time the golf ball is hit until the time it hits the ground,

a find the value of k to 2 decimal places. **(3)**

b Find a Cartesian equation for the path of the golf ball in the form $y = f(x)$, and determine the domain of $f(x)$. Give the domain to 1 d.p. **(5)**

← Section 8.5

Challenge

1 A chord of a circle, centre O and radius r, divides the circumference in the ratio $1:3$, as shown in the diagram. Find the ratio of the area of region P to the area of region Q.

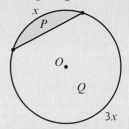

← Section 5.3

2 The diagram shows a circle, centre O. The radius of the circle, OC, is 1, and $\angle CDO = 90°$.

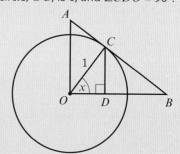

Given that $\angle COD = x$, express the following lengths as single trigonometric functions of x.

a CD **b** OD **c** OA
d AC **e** CB **f** OB

← Section 6.1

3 The curve C has parametric equations
$x = 4\sin t + 3, y = 4\cos t - 1, -\frac{\pi}{2} \leq t \leq \frac{\pi}{4}$

a By finding a Cartesian equation of C in the form $(x - a)^2 + (y - b)^2 = c$, or otherwise, sketch C, labelling the endpoints of the curve with their exact coordinates.

b Find the length of C, giving your answer in terms of π.

← Section 8.3

Differentiation

9

Objectives

After completing this chapter you should be able to:
- Differentiate trigonometric functions → **pages 232–234, 246–251**
- Differentiate exponentials and logarithms → **pages 235–237**
- Differentiate functions using the chain, product and quotient rules → **pages 237–245**
- Differentiate parametric equations → **pages 251–254**
- Differentiate functions which are defined implicitly → **pages 254–257**
- Use the second derivative to describe the behaviour of a function → **pages 257–261**
- Solve problems involving connected rates of change and construct simple differential equations → **pages 261–264**

You can use differentiation to find rates of change in trigonometric and exponential models. The velocity of a wrecking ball could be estimated by modelling its displacement then differentiating.

Prior knowledge check

1 Differentiate:

   a $3x^2 - 5x$ **b** $\dfrac{2}{x} - \sqrt{x}$

   c $4x^2(1 - x^2)$ ← Year 1, Chapter 12

2 Find the equation of the tangent to the curve with equation $y = 8 - x^2$ at the point $(3, -1)$.   ← Year 1, Chapter 12

3 The curve C is defined by the parametric equations
   $$x = 3t^2 - 5t, \quad y = t^3 + 2, \quad t \in \mathbb{R}$$
   Find the coordinates of any points where C intersects the coordinate axes.   ← Section 8.4

4 Solve $2\operatorname{cosec} x - 3\sec x = 0$ in the interval $0 \leq x \leq 2\pi$, giving your answers correct to 3 significant figures.   ← Section 6.3

Chapter 9

9.1 Differentiating sin x and cos x

You need to be able to differentiate sin x and cos x from first principles. You can use the following small angle approximations for sin and cos when the angle is measured in **radians**:

- $\sin x \approx x$
- $\cos x \approx 1 - \frac{1}{2}x^2$

Watch out You will always need to use radians when differentiating trigonometric functions.

This means that $\lim\limits_{h \to 0} \dfrac{\sin h}{h} = \lim\limits_{h \to 0} \dfrac{h}{h} = 1$, and

$$\lim\limits_{h \to 0} \dfrac{\cos h - 1}{h} = \lim\limits_{h \to 0} \dfrac{1 - \frac{1}{2}h^2 - 1}{h} = \lim\limits_{h \to 0} \left(-\tfrac{1}{2}h\right) = 0$$

You will need to use these two limits when you differentiate sin and cos from first principles.

Example 1

Prove, from first principles, that the derivative of sin x is cos x.

You may assume that as $h \to 0$, $\dfrac{\sin h}{h} \to 1$ and $\dfrac{\cos h - 1}{h} \to 0$.

Let $f(x) = \sin x$

$f'(x) = \lim\limits_{h \to 0} \dfrac{f(x + h) - f(x)}{h}$

$= \lim\limits_{h \to 0} \dfrac{\sin(x + h) - \sin x}{h}$

$= \lim\limits_{h \to 0} \dfrac{\sin x \cos h + \cos x \sin h - \sin x}{h}$

$= \lim\limits_{h \to 0} \left(\left(\dfrac{\cos h - 1}{h}\right)\sin x + \left(\dfrac{\sin h}{h}\right)\cos x\right)$

Since $\dfrac{\cos h - 1}{h} \to 0$ and $\dfrac{\sin h}{h} \to 1$ the expression inside the limit tends to $(0 \times \sin x + 1 \times \cos x)$

So $\lim\limits_{h \to 0} \dfrac{\sin(x + h) - \sin x}{h} = \cos x$

Hence the derivative of sin x is cos x.

Problem-solving

Use the rule for differentiating from first principles. This is provided in the formula booklet. If you don't want to use limit notation, you could write an expression for the gradient of the chord joining $(x, \sin x)$ to $(x + h, \sin(x + h))$ and show that as $h \to 0$ the gradient of the chord tends to $\cos x$. ← Year 1, Section 12.2

Use the formula for $\sin(A + B)$ to expand $\sin(x + h)$, then write the resulting expression in terms of $\dfrac{\cos h - 1}{h}$ and $\dfrac{\sin h}{h}$ ← Section 7.1

Make sure you state where you are using the two limits given in the question.

Write down what you have proved.

- If $y = \sin kx$, then $\dfrac{dy}{dx} = k \cos kx$

You can use a similar technique to find the derivative of cos x.

- If $y = \cos kx$, then $\dfrac{dy}{dx} = -k \sin kx$

Online Explore the relationship between sin and cos and their derivatives using technology.

Example 2

Find $\dfrac{dy}{dx}$ given that:

a $y = \sin 2x$ **b** $y = \cos 5x$ **c** $y = 3\cos x + 2\sin 4x$

a $y = \sin 2x$

$\dfrac{dy}{dx} = 2\cos 2x$ —— Use the standard result for $\sin kx$ with $k = 2$.

b $y = \cos 5x$

$\dfrac{dy}{dx} = -5\sin 5x$ —— Use the standard result for $\cos kx$ with $k = 5$.

c $y = 3\cos x + 2\sin 4x$

$\dfrac{dy}{dx} = 3 \times (-\sin x) + 2 \times (4\cos 4x)$ —— Differentiate each term separately.

$= -3\sin x + 8\cos 4x$

Example 3

A curve has equation $y = \tfrac{1}{2}x - \cos 2x$. Find the stationary points on the curve in the interval $0 \leqslant x \leqslant \pi$.

$\dfrac{dy}{dx} = \tfrac{1}{2} - (-2\sin 2x) = \tfrac{1}{2} + 2\sin 2x$ —— Start by differentiating $\tfrac{1}{2}x - \cos 2x$.

Let $\dfrac{dy}{dx} = 0$ and solve for x: —— Stationary points occur when $\dfrac{dy}{dx} = 0$.

$\tfrac{1}{2} + 2\sin 2x = 0$

$2\sin 2x = -\tfrac{1}{2}$

$\sin 2x = -\tfrac{1}{4}$

← Year 1, Chapter 12

$2x = 3.394..., 6.030...$ —— $0 \leqslant x \leqslant \pi$ so the range for $2x$ is $0 \leqslant 2x \leqslant 2\pi$.

$x = 1.70, 3.02$ (3 s.f.)

When $x = 1.70$:

$y = \tfrac{1}{2}(1.70) - \cos(2 \times 1.70) = 1.82$ (3 s.f.)

When $x = 3.02$:

$y = \tfrac{1}{2}(3.02) - \cos(2 \times 3.02) = 0.539$ (3 s.f.)

Watch out Whenever you are using calculus, you must work in **radians**.

Substitute x values into $y = \tfrac{1}{2}x - \cos 2x$ to find the corresponding y values.

The stationary points of $y = \tfrac{1}{2}x - \cos 2x$ in the interval $0 \leqslant x \leqslant \pi$ are $(1.70, 1.82)$ and $(3.02, 0.539)$.

Chapter 9

Exercise 9A

(P) 1 a Given that $f(x) = \cos x$, show that
$$f'(x) = \lim_{h \to 0}\left(\left(\frac{\cos h - 1}{h}\right)\cos x - \frac{\sin h}{h}\sin x\right)$$

> **Problem-solving**
> Use the definition of the derivative and the addition formula for $\cos(A + B)$.

  b Hence prove that $f'(x) = -\sin x$.

2 Differentiate:

  a $y = 2\cos x$ **b** $y = 2\sin\frac{1}{2}x$ **c** $y = \sin 8x$ **d** $y = 6\sin\frac{2}{3}x$

3 Find $f'(x)$ given that:

  a $f(x) = 2\cos x$ **b** $f(x) = 6\cos\frac{5}{6}x$ **c** $f(x) = 4\cos\frac{1}{2}x$ **d** $f(x) = 3\cos 2x$

4 Find $\dfrac{dy}{dx}$ given that:

  a $y = \sin 2x + \cos 3x$

  b $y = 2\cos 4x - 4\cos x + 2\cos 7x$

  c $y = x^2 + 4\cos 3x$

  d $y = \dfrac{1 + 2x\sin 5x}{x}$

5 A curve has equation $y = x - \sin 3x$. Find the stationary points of the curve in the interval $0 \leqslant x \leqslant \pi$.

6 Find the gradient of the curve $y = 2\sin 4x - 4\cos 2x$ at the point where $x = \dfrac{\pi}{2}$.

(P) 7 A curve has the equation $y = 2\sin 2x + \cos 2x$. Find the stationary points of the curve in the interval $0 \leqslant x \leqslant \pi$.

(E/P) 8 A curve has the equation $y = \sin 5x + \cos 3x$. Find the equation of the tangent to the curve at the point $(\pi, -1)$. **(4 marks)**

(E/P) 9 A curve has the equation $y = 2x^2 - \sin x$. Show that the equation of the normal to the curve at the point with x-coordinate π is
$$x + (4\pi + 1)y - \pi(8\pi^2 + 2\pi + 1) = 0$$ **(7 marks)**

(E/P) 10 Prove, from first principles, that the derivative of $\sin x$ is $\cos x$.

You may assume the formula for $\sin(A + B)$ and that as $h \to 0$, $\dfrac{\sin h}{h} \to 1$ and $\dfrac{\cos h - 1}{h} \to 0$.

(5 marks)

Challenge

Prove, from first principles, that the derivative of $\sin(kx)$ is $k\cos(kx)$.

You may assume the formula for $\sin(A + B)$ and that as $h \to 0$, $\dfrac{\sin kh}{h} \to k$ and $\dfrac{\cos kh - 1}{h} \to 0$.

9.2 Differentiating exponentials and logarithms

You need to be able to differentiate expressions involving exponentials and logarithms.

- If $y = e^{kx}$, then $\dfrac{dy}{dx} = ke^{kx}$

- If $y = \ln x$, then $\dfrac{dy}{dx} = \dfrac{1}{x}$

You can use the derivative of e^{kx} to find the derivative of a^{kx} where a is any positive real number.

Watch out For any real constant, k, $\ln kx = \ln k + \ln x$. Since $\ln k$ is also a constant, the derivative of $\ln kx$ is also $\dfrac{1}{x}$.

Example 4

Show that the derivative of a^x is $a^x \ln a$.

Online Explore the function a^x and its derivative using technology.

$$\begin{aligned}\text{Let } y &= a^x \\ &= e^{\ln(a^x)} \\ &= e^{x \ln a} \\ \dfrac{dy}{dx} &= \ln a \, e^{x \ln a} \\ &= \ln a \, e^{\ln(a^x)} \\ &= a^x \ln a\end{aligned}$$

You could also use the laws of logs like this:
$\ln y = \ln a^x = x \ln a \Rightarrow y = e^{x \ln a}$

← Year 1, Chapter 14

$\ln a$ is just a constant so use the standard result for the derivative of e^{kx} with $k = \ln a$.

- If $y = a^{kx}$, where k is a real constant and $a > 0$, then $\dfrac{dy}{dx} = a^{kx} k \ln a$

Example 5

Find $\dfrac{dy}{dx}$ given that:

a $y = e^{3x} + 2^{3x}$ **b** $y = \ln(x^3) + \ln 7x$ **c** $y = \dfrac{2 - 3e^{7x}}{4e^{3x}}$

a $y = e^{3x} + 2^{3x}$
$\dfrac{dy}{dx} = 3e^{3x} + 2^{3x}(3 \ln 2)$

Differentiate each term separately using the standard results for e^{kx} with $k = 3$, and a^{kx} with $a = 2$ and $k = 3$.

b $y = \ln(x^3) + \ln 7x$
$y = 3 \ln x + \ln 7 + \ln x = 4 \ln x + \ln 7$
$\dfrac{dy}{dx} = 4 \times \dfrac{1}{x} + 0 = \dfrac{4}{x}$

Rewrite y using the laws of logs.

Use the standard result for $\ln x$. $\ln 7$ is a constant, so it disappears when you differentiate.

c $y = \dfrac{2 - 3e^{7x}}{4e^{3x}}$
$= \dfrac{1}{2}e^{-3x} - \dfrac{3}{4}e^{4x}$
$\dfrac{dy}{dx} = \dfrac{1}{2} \times (-3e^{-3x}) - \dfrac{3}{4} \times 4e^{4x}$
$= -\dfrac{3}{2}e^{-3x} - 3e^{4x}$

Divide each term in the numerator by the denominator.

Differentiate each term separately using the standard result for e^{kx}.

Chapter 9

Exercise 9B

1 a Find $\dfrac{dy}{dx}$ for each of the following:

    a $y = 4e^{7x}$ **b** $y = 3^x$ **c** $y = \left(\dfrac{1}{2}\right)^x$ **d** $y = \ln 5x$

    e $y = 4\left(\dfrac{1}{3}\right)^x$ **f** $y = \ln(2x^3)$ **g** $y = e^{3x} - e^{-3x}$ **h** $y = \dfrac{(1+e^x)^2}{e^x}$

2 Find $f'(x)$ given that:

    a $f(x) = 3^{4x}$ **b** $f(x) = \left(\dfrac{3}{2}\right)^{2x}$ **c** $f(x) = 2^{4x} + 4^{2x}$ **d** $f(x) = \dfrac{2^{7x} + 8^x}{4^{2x}}$

Hint In parts **c** and **d**, rewrite the terms so that they all have the same base and hence can be simplified.

3 Find the gradient of the curve $y = (e^{2x} - e^{-2x})^2$ at the point where $x = \ln 3$.

(E) **4** Find the equation of the tangent to the curve $y = 2^x + 2^{-x}$ at the point $\left(2, \dfrac{17}{4}\right)$. **(6 marks)**

(E/P) **5** A curve has the equation $y = e^{2x} - \ln x$. Show that the equation of the tangent at the point with x-coordinate 1 is
$$y = (2e^2 - 1)x - e^2 + 1$$
(6 marks)

6 A particular radioactive isotope has an activity, R millicuries at time t days, given by the equation $R = 200 \times 0.9^t$. Find the value of $\dfrac{dR}{dt}$, when $t = 8$.

(P) **7** The population of Cambridge was 37 000 in 1900, and was about 109 000 in 2000. Given that the population, P, at a time t years after 1900 can be modelled using the equation $P = P_0 k^t$,

    a find the values of P_0 and k

    b evaluate $\dfrac{dP}{dt}$ in the year 2000

    c interpret your answer to part **b** in the context of the model.

(P) **8** A student is attempting to differentiate $\ln kx$. The student writes:

> $y = \ln kx$, so $\dfrac{dy}{dx} = k \ln kx$

    Explain the mistake made by the student and state the correct derivative.

(E/P) **9** Prove that the derivative of a^{kx} is $a^{kx} k \ln a$. You may assume that the derivative of e^{kx} is ke^{kx}.

(4 marks)

10 $f(x) = e^{2x} - \ln(x^2) + 4, x > 0$

  a Find $f'(x)$. **(3 marks)**

  The curve with equation $y = f(x)$ has a gradient of 2 at point P. The x-coordinate of P is a.

  b Show that $a(e^{2a} - 1) = 1$. **(2 marks)**

11 A curve C has equation
$$y = 5\sin 3x + 2\cos 3x, \; -\pi \leq x \leq \pi$$

  a Show that the point $P(0, 2)$ lies on C. **(1 mark)**

  b Find an equation of the normal to the curve C at P. **(5 marks)**

12 The point P lies on the curve with equation $y = 2(3^{4x})$. The x-coordinate of P is 1.
Find an equation of the normal to the curve at the point P in the form $y = ax + b$, where a and b are constants to be found in exact form. **(5 marks)**

Challenge

A curve C has the equation $y = e^{4x} - 5x$. Find the equation of the tangent to C that is parallel to the line $y = 3x + 4$.

9.3 The chain rule

You can use the chain rule to differentiate composite functions, or functions of another function.

- **The chain rule is:**

$$\frac{dy}{dx} = \frac{dy}{du} \times \frac{du}{dx}$$

  where y is a function of u and u is another function of x.

Example 6

Given that $y = (3x^4 + x)^5$, find $\frac{dy}{dx}$ using the chain rule.

Let $u = 3x^4 + x$:

$\frac{du}{dx} = 12x^3 + 1$ — Differentiate u with respect to x to get $\frac{du}{dx}$

$y = u^5$

$\frac{dy}{du} = 5u^4$ — Substitute u into the equation for y and differentiate with respect to u to get $\frac{dy}{du}$

Using the chain rule,

$\frac{dy}{dx} = \frac{dy}{du} \times \frac{du}{dx}$

$= 5u^4(12x^3 + 1)$

$\frac{dy}{dx} = 5(3x^4 + x)^4(12x^3 + 1)$ — Use $u = 3x^4 + x$ to write your final answer in terms of x only.

Chapter 9

Example 7

Given that $y = \sin^4 x$, find $\dfrac{dy}{dx}$

$y = \sin^4 x = (\sin x)^4$

Let $u = \sin x$:

$\dfrac{du}{dx} = \cos x$ — Differentiate u with respect to x to get $\dfrac{du}{dx}$

$y = u^4$
$\dfrac{dy}{du} = 4u^3$

Substitute u into the equation for y and differentiate with respect to u to get $\dfrac{dy}{du}$

Using the chain rule,
$$\dfrac{dy}{dx} = \dfrac{dy}{du} \times \dfrac{du}{dx}$$
$$= 4u^3(\cos x)$$

$\dfrac{dy}{dx} = 4\sin^3 x \cos x$ — Substitute $u = \sin x$ back into $\dfrac{dy}{dx}$ to get an answer in terms of x only.

You can write the chain rule using function notation:

- **The chain rule enables you to differentiate a function of a function. In general,**
  - if $y = (f(x))^n$ then $\dfrac{dy}{dx} = n(f(x))^{n-1} f'(x)$
  - if $y = f(g(x))$ then $\dfrac{dy}{dx} = f'(g(x)) g'(x)$

Example 8

Given that $y = \sqrt{5x^2 + 1}$, find $\dfrac{dy}{dx}$ at $(4, 9)$.

$y = \sqrt{5x^2 + 1}$ — This is $y = (f(x))^n$ with $f(x) = 5x^2 + 1$ and $n = \tfrac{1}{2}$
Let $f(x) = 5x^2 + 1$
Then $f'(x) = 10x$

So $\dfrac{dy}{dx} = \tfrac{1}{2}(f(x))^{-\tfrac{1}{2}} f'(x)$.

Using the chain rule:
$\dfrac{dy}{dx} = \tfrac{1}{2}(5x^2 + 1)^{-\tfrac{1}{2}} \times 10x$
$= 5x(5x^2 + 1)^{-\tfrac{1}{2}}$

At $(4, 9)$, $\dfrac{dy}{dx} = 5(4)(5(4)^2 + 1)^{-\tfrac{1}{2}} = \dfrac{20}{9}$ — Substitute $x = 4$ into $\dfrac{dy}{dx}$ to find the required value.

The following particular case of the chain rule is useful for differentiating functions that are not in the form $y = f(x)$.

- $\dfrac{dy}{dx} = \dfrac{1}{\frac{dx}{dy}}$

Hint This is because:
$\dfrac{dy}{dx} \times \dfrac{dx}{dy} = \dfrac{dy}{dy} = 1$

Example 9

Find the value of $\dfrac{dy}{dx}$ at the point $(2, 1)$ on the curve with equation $y^3 + y = x$.

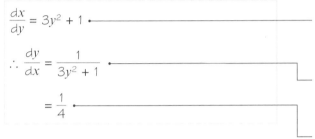

Start with $x = y^3 + y$ and differentiate with respect to y.

$\dfrac{dx}{dy} = 3y^2 + 1$

$\therefore \dfrac{dy}{dx} = \dfrac{1}{3y^2 + 1}$

Use $\dfrac{dy}{dx} = \dfrac{1}{\frac{dx}{dy}}$

$= \dfrac{1}{4}$

Substitute $y = 1$.

Exercise 9C

1 Differentiate:

   a $(1 + 2x)^4$ **b** $(3 - 2x^2)^{-5}$ **c** $(3 + 4x)^{\frac{1}{2}}$ **d** $(6x + x^2)^7$

   e $\dfrac{1}{3 + 2x}$ **f** $\sqrt{7 - x}$ **g** $4(2 + 8x)^4$ **h** $3(8 - x)^{-6}$

2 Differentiate:

   a $e^{\cos x}$ **b** $\cos(2x - 1)$ **c** $\sqrt{\ln x}$ **d** $(\sin x + \cos x)^5$

   e $\sin(3x^2 - 2x + 1)$ **f** $\ln(\sin x)$ **g** $2e^{\cos 4x}$ **h** $\cos(e^{2x} + 3)$

3 Given that $y = \dfrac{1}{(4x + 1)^2}$ find the value of $\dfrac{dy}{dx}$ at $\left(\dfrac{1}{4}, \dfrac{1}{4}\right)$.

(E) **4** A curve C has equation $y = (5 - 2x)^3$. Find the tangent to the curve at the point P with x-coordinate 1. **(7 marks)**

(E) **5** Given that $y = (1 + \ln 4x)^{\frac{3}{2}}$, find the value of $\dfrac{dy}{dx}$ at $x = \dfrac{1}{4}e^3$. **(5 marks)**

(P) **6** Find $\dfrac{dy}{dx}$ for the following curves, giving your answers in terms of y.

   a $x = y^2 + y$ **b** $x = e^y + 4y$ **c** $x = \sin 2y$ **d** $4x = \ln y + y^3$

7 Find the value of $\dfrac{dy}{dx}$ at the point $(8, 2)$ on the curve with equation $3y^2 - 2y = x$.

Problem-solving
Your expression for $\dfrac{dy}{dx}$ will be in terms of y.
Remember to substitute the y-coordinate into the expression to find the gradient.

8 Find the value of $\dfrac{dy}{dx}$ at the point $\left(\dfrac{5}{2}, 4\right)$ on the curve with equation $y^{\frac{1}{2}} + y^{-\frac{1}{2}} = x$.

9 a Differentiate $e^y = x$ with respect to y.

  b Hence, prove that if $y = \ln x$, then $\dfrac{dy}{dx} = \dfrac{1}{x}$.

10 The curve C has equation $x = 4\cos 2y$.

  a Show that the point $Q\left(2, \dfrac{\pi}{6}\right)$ lies on C. **(1 mark)**

  b Show that $\dfrac{dy}{dx} = -\dfrac{1}{4\sqrt{3}}$ at Q. **(4 marks)**

  c Find an equation of the normal to C at Q. Give your answer in the form $ax + by + c = 0$, where a, b and c are exact constants. **(4 marks)**

11 Differentiate:

  a $\sin^2 3x$ **b** $e^{(x+1)^2}$ **c** $\ln(\cos x)^2$

  d $\dfrac{1}{3 + \cos 2x}$ **e** $\sin\left(\dfrac{1}{x}\right)$

12 The curve C has equation $y = \dfrac{4}{(2-4x)^2}$, $x \neq \dfrac{1}{2}$

The point A on C has x-coordinate 3.
Find an equation of the normal to C at A in the form $ax + by + c = 0$, where a, b and c are integers. **(7 marks)**

13 Find the exact value of the gradient of the curve with equation $y = 3^{x^3}$ at the point with coordinates $(1, 3)$. **(4 marks)**

Challenge

Find $\dfrac{dy}{dx}$ given that:

a $y = \sqrt{\sin\sqrt{x}}$ **b** $\ln y = \sin^3(3x + 4)$

9.4 The product rule

You need to be able to differentiate the product of two functions.

- If $y = uv$ then $\dfrac{dy}{dx} = u\dfrac{dv}{dx} + v\dfrac{du}{dx}$,

  where u and v are functions of x.

The product rule in function notation is:

- If $f(x) = g(x)h(x)$ then $f'(x) = g(x)h'(x) + h(x)g'(x)$

Watch out Make sure you can spot the difference between a product of two functions and a function of a function. A product is two separate functions multiplied together.

Example 10

Given that $f(x) = x^2\sqrt{3x-1}$, find $f'(x)$.

Let $u = x^2$ and $v = \sqrt{3x-1} = (3x-1)^{\frac{1}{2}}$

Then $\dfrac{du}{dx} = 2x$ and $\dfrac{dv}{dx} = 3 \times \dfrac{1}{2}(3x-1)^{-\frac{1}{2}}$

Using $\dfrac{dy}{dx} = u\dfrac{dv}{dx} + v\dfrac{du}{dx}$

$f'(x) = x^2 \times \dfrac{3}{2}(3x-1)^{-\frac{1}{2}} + \sqrt{3x-1} \times 2x$

$= \dfrac{3x^2 + 12x^2 - 4x}{2\sqrt{3x-1}}$

$= \dfrac{15x^2 - 4x}{2\sqrt{3x-1}}$

$= \dfrac{x(15x-4)}{2\sqrt{3x-1}}$

Write out your functions u, v, $\dfrac{du}{dx}$ and $\dfrac{dv}{dx}$ before substituting into the product rule. Use the chain rule to differentiate $(3x-1)^{\frac{1}{2}}$

Substitute u, v, $\dfrac{du}{dx}$ and $\dfrac{dv}{dx}$

Example 11

Given that $y = e^{4x}\sin^2 3x$, show that $\dfrac{dy}{dx} = e^{4x}\sin 3x\,(A\cos 3x + B\sin 3x)$, where A and B are constants to be determined.

Let $u = e^{4x}$ and $v = \sin^2 3x = (\sin 3x)^2$

$\dfrac{du}{dx} = 4e^{4x}$ and $\dfrac{dv}{dx} = 2(\sin 3x) \times (3\cos 3x)$

$\dfrac{dy}{dx} = u\dfrac{dv}{dx} + v\dfrac{du}{dx}$

$\dfrac{dy}{dx} = e^{4x} \times (6\sin 3x \cos 3x) + \sin^2 3x \times 4e^{4x}$

$= 6e^{4x}\sin 3x \cos 3x + 4e^{4x}\sin^2 3x$

$= e^{4x}\sin 3x\,(6\cos 3x + 4\sin 3x)$

This is in the required form with $A = 6$ and $B = 4$.

Write out u and v and find $\dfrac{du}{dx}$ and $\dfrac{dv}{dx}$

Use the chain rule to find $\dfrac{dv}{dx}$

Write out the product rule before substituting.

Problem-solving

Write out the value of any constants you have determined at the end of your working. You can use this to check that your answer is in the required form.

Exercise 9D

1 Differentiate:
   a $x(1 + 3x)^5$
   b $2x(1 + 3x^2)^3$
   c $x^3(2x + 6)^4$
   d $3x^2(5x - 1)^{-1}$

2 Differentiate:
   a $e^{-2x}(2x - 1)^5$
   b $\sin 2x \cos 3x$
   c $e^x \sin x$
   d $\sin(5x) \ln(\cos x)$

3 a Find the value of $\dfrac{dy}{dx}$ at the point $(1, 8)$ on the curve with equation $y = x^2(3x - 1)^3$.

   b Find the value of $\dfrac{dy}{dx}$ at the point $(4, 36)$ on the curve with equation $y = 3x(2x + 1)^{\frac{1}{2}}$.

   c Find the value of $\dfrac{dy}{dx}$ at the point $(2, \frac{1}{5})$ on the curve with equation $y = (x - 1)(2x + 1)^{-1}$.

4 Find the stationary points of the curve C with the equation $y = (x - 2)^2(2x + 3)$.

5 A curve C has equation $y = \left(x - \dfrac{\pi}{2}\right)^5 \sin 2x$, $0 < x < \pi$. Find the gradient of the curve at the point with x-coordinate $\dfrac{\pi}{4}$.

(E/P) 6 A curve C has equation $y = x^2 \cos(x^2)$. Find the equation of the tangent to the curve C at the point $P\left(\dfrac{\sqrt{\pi}}{2}, \dfrac{\pi\sqrt{2}}{8}\right)$ in the form $ax + by + c = 0$ where a, b and c are exact constants. **(7 marks)**

(E/P) 7 Given that $y = 3x^2(5x - 3)^3$, show that

$$\dfrac{dy}{dx} = Ax(5x - 3)^n(Bx + C)$$

where n, A, B and C are constants to be determined. **(4 marks)**

(E) 8 A curve C has equation $y = (x + 3)^2 e^{3x}$.

   a Find $\dfrac{dy}{dx}$, using the product rule for differentiation. **(3 marks)**

   b Find the gradient of C at the point where $x = 2$. **(3 marks)**

(E) 9 Differentiate with respect to x:
   a $(2\sin x - 3\cos x) \ln 3x$ **(3 marks)**
   b $x^4 e^{7x-3}$ **(3 marks)**

(E) 10 Find the value of $\dfrac{dy}{dx}$ at the point where $x = 1$ on the curve with equation

$$y = x^5 \sqrt{10x + 6}$$

(6 marks)

Challenge

Find $\dfrac{dy}{dx}$ for the following functions.

a $y = e^x \sin^2 x \cos x$
b $y = x(4x - 3)^6 (1 - 4x)^9$

9.5 The quotient rule

You need to be able to differentiate the quotient of two functions.

- If $y = \dfrac{u}{v}$ then $\dfrac{dy}{dx} = \dfrac{v\dfrac{du}{dx} - u\dfrac{dv}{dx}}{v^2}$ where u and v are functions of x.

The quotient rule in function notation is:

Watch out There is a minus sign in the numerator, so the order of the functions is important.

- If $f(x) = \dfrac{g(x)}{h(x)}$, then $f'(x) = \dfrac{h(x)g'(x) - g(x)h'(x)}{(h(x))^2}$

Example 12

Given that $y = \dfrac{x}{2x + 5}$ find $\dfrac{dy}{dx}$

Let $u = x$ and $v = 2x + 5$: — Let u be the numerator and let v be the denominator.

$\dfrac{du}{dx} = 1$ and $\dfrac{dv}{dx} = 2$

Using $\dfrac{dy}{dx} = \dfrac{v\dfrac{du}{dx} - u\dfrac{dv}{dx}}{v^2}$ — Recognise that y is a quotient and use the quotient rule.

$\dfrac{dy}{dx} = \dfrac{(2x + 5) \times 1 - x \times 2}{(2x + 5)^2}$

$= \dfrac{5}{(2x + 5)^2}$ — Simplify the numerator of the fraction.

Example 13

A curve C with equation $y = \dfrac{\sin x}{e^{2x}}$, $0 < x < \pi$, has a stationary point at P. Find the coordinates of P. Give your answer to 3 significant figures.

Online Explore the graph of this function using technology.

Let $u = \sin x$ and $v = e^{2x}$.

$\dfrac{du}{dx} = \cos x$ and $\dfrac{dv}{dx} = 2e^{2x}$

Write out u and v and find $\dfrac{du}{dx}$ and $\dfrac{dv}{dx}$ before using the quotient rule.

Chapter 9

Using the quotient rule,

$$\frac{dy}{dx} = \frac{v\frac{du}{dx} - u\frac{dv}{dx}}{v^2}$$ — Write out the rule before substituting.

$$\frac{dy}{dx} = \frac{e^{2x}\cos x - \sin x(2e^{2x})}{(e^{2x})^2}$$

$$= \frac{e^{2x}\cos x - 2e^{2x}\sin x}{e^{4x}}$$

$$= \frac{e^{2x}(\cos x - 2\sin x)}{e^{4x}}$$

$$= e^{-2x}(\cos x - 2\sin x)$$ — Simplify your expression for $\frac{dy}{dx}$ as much as possible.

When $\frac{dy}{dx} = 0$: — P is a stationary point so $\frac{dy}{dx} = 0$.

$e^{-2x}(\cos x - 2\sin x) = 0$

$e^{-2x} = 0$ or $\cos x - 2\sin x = 0$

$e^{-2x} = 0$ has no solution.

Problem-solving

If the product of two factors is equal to 0 then one of the factors must be equal to 0.

$\cos x - 2\sin x = 0$

$\cos x = 2\sin x$

$\frac{1}{2} = \tan x$

$x = 0.464$ (3 s.f.) — This is the only solution in the range $0 < x < \pi$.

$y = \frac{\sin x}{e^{2x}}$

$y = \frac{\sin(0.464)}{e^{2 \times 0.464}} = 0.177$ (3 s.f.) — Substitute x into y to find the y-coordinate of the stationary point.

So the coordinates of P are (0.464, 0.177).

Exercise 9E

1 Differentiate:

a $\dfrac{5x}{x+1}$ b $\dfrac{2x}{3x-2}$ c $\dfrac{x+3}{2x+1}$ d $\dfrac{3x^2}{(2x-1)^2}$ e $\dfrac{6x}{(5x+3)^{\frac{1}{2}}}$

2 Differentiate:

a $\dfrac{e^{4x}}{\cos x}$ b $\dfrac{\ln x}{x+1}$ c $\dfrac{e^{-2x}+e^{2x}}{\ln x}$ d $\dfrac{(e^x+3)^3}{\cos x}$ e $\dfrac{\sin^2 x}{\ln x}$

3 Find the value of $\dfrac{dy}{dx}$ at the point $\left(1, \tfrac{1}{4}\right)$ on the curve with equation $y = \dfrac{x}{3x+1}$

4 Find the value of $\dfrac{dy}{dx}$ at the point (12, 3) on the curve with equation $y = \dfrac{x+3}{(2x+1)^{\frac{1}{2}}}$

Differentiation

5 Find the stationary point(s) of the curve C with equation $y = \dfrac{e^{2x+3}}{x}$, $x \neq 0$.

6 Find the equation of the tangent to the curve $y = \dfrac{e^{\frac{1}{3}x}}{x}$ at the point $\left(3, \tfrac{1}{3}e\right)$. **(7 marks)**

7 Find the exact value of $\dfrac{dy}{dx}$ at the point $x = \dfrac{\pi}{9}$ on the curve with equation $y = \dfrac{\ln x}{\sin 3x}$

8 The curve C has equation $x = \dfrac{e^y}{3+2y}$

  a Find the coordinates of the point P where the curve cuts the x-axis. **(1 mark)**

  b Find an equation of the normal to the curve at P, giving your answer in the form $y = mx + c$, where m and c are constants to be found. **(6 marks)**

9 Differentiate $\dfrac{x^4}{\cos 3x}$ with respect to x. **(4 marks)**

10 A curve C has equation $y = \dfrac{e^{2x}}{(x-2)^2}$, $x \neq 2$.

  a Show that

  $$\dfrac{dy}{dx} = \dfrac{Ae^{2x}(Bx - C)}{(x-2)^3}$$

  where A, B and C are integers to be found. **(4 marks)**

  b Find the equation of the tangent of C at the point $x = 1$. **(3 marks)**

11 Given that

$$f(x) = \dfrac{2x}{x+5} + \dfrac{6x}{x^2 + 7x + 10}, \; x > 0$$

  a show that $f(x) = \dfrac{2x}{x+2}$ **(4 marks)**

  b Hence find $f'(3)$. **(3 marks)**

12 The diagram shows a sketch of the curve with equation $y = f(x)$, where

$$f(x) = \dfrac{2\cos 2x}{e^{2-x}}, \; 0 < x < \pi$$

The curve has a maximum turning point at A and a minimum turning point and B as shown in the diagram.

  a Show that the x-coordinates of point A and point B are solutions to the equation $\tan 2x = \tfrac{1}{2}$ **(4 marks)**

  b Find the range of $f(x)$. **(2 marks)**

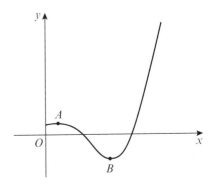

245

9.6 Differentiating trigonometric functions

You can combine all the above rules and apply them to trigonometric functions to obtain standard results.

Example 14

If $y = \tan x$, find $\dfrac{dy}{dx}$

$y = \tan x = \dfrac{\sin x}{\cos x}$

Let $u = \sin x$ and $v = \cos x$

$\dfrac{du}{dx} = \cos x$ and $\dfrac{dv}{dx} = -\sin x$

$\dfrac{dy}{dx} = \dfrac{v\dfrac{du}{dx} - u\dfrac{dv}{dx}}{v^2}$

$\dfrac{dy}{dx} = \dfrac{\cos x \times \cos x - \sin x(-\sin x)}{\cos^2 x}$

$\dfrac{dy}{dx} = \dfrac{\cos^2 x + \sin^2 x}{\cos^2 x}$

$\dfrac{dy}{dx} = \dfrac{1}{\cos^2 x} = \sec^2 x$

You can write $\tan x$ as $\dfrac{\sin x}{\cos x}$ and then use the quotient rule.

Use the identity $\cos^2 x + \sin^2 x \equiv 1$.

You can generalise this method to differentiate $\tan kx$:

- **If $y = \tan kx$, then $\dfrac{dy}{dx} = k \sec^2 kx$**

Example 15

Differentiate **a** $y = x \tan 2x$ **b** $y = \tan^4 x$

a $y = x \tan 2x$

$\dfrac{dy}{dx} = x \times 2\sec^2 2x + \tan 2x$

$= 2x \sec^2 2x + \tan 2x$

b $y = \tan^4 x = (\tan x)^4$

$\dfrac{dy}{dx} = 4(\tan x)^3 (\sec^2 x)$

$= 4 \tan^3 x \sec^2 x$

This is a product. Use $u = x$ and $v = \tan 2x$, together with the product rule.

Use the chain rule with $u = \tan x$.

Differentiation

Example 16

Show that if $y = \text{cosec}\, x$, then $\dfrac{dy}{dx} = -\text{cosec}\, x \cot x$.

$y = \text{cosec}\, x = \dfrac{1}{\sin x}$

Let $u = 1$ and $v = \sin x$

$\dfrac{du}{dx} = 0$ and $\dfrac{dv}{dx} = \cos x$

$\dfrac{dy}{dx} = \dfrac{v\dfrac{du}{dx} - u\dfrac{dv}{dx}}{v^2}$

$\dfrac{dy}{dx} = \dfrac{\sin x \times 0 - 1 \times \cos x}{\sin^2 x}$

$\dfrac{dy}{dx} = -\dfrac{\cos x}{\sin^2 x}$

$\dfrac{dy}{dx} = -\dfrac{1}{\sin x} \times \dfrac{\cos x}{\sin x} = -\text{cosec}\, x \cot x$

Use the quotient rule with $u = 1$ and $v = \sin x$.

$u = 1$ is a constant so $\dfrac{du}{dx} = 0$.

Rearrange your answer into the desired form using the definitions of cosec and cot. ← Section 6.1

You can use similar techniques to differentiate $\sec x$ and $\cot x$ giving you the following general results:

- **If $y = \text{cosec}\, kx$, then $\dfrac{dy}{dx} = -k\,\text{cosec}\, kx \cot kx$**
- **If $y = \sec kx$, then $\dfrac{dy}{dx} = k \sec kx \tan kx$**
- **If $y = \cot kx$, then $\dfrac{dy}{dx} = -k\,\text{cosec}^2 kx$**

Watch out While the standard results for tan, cosec, sec and cot are given in the formulae booklet, learning these results will enable you to differentiate a wide range of functions quickly and confidently.

Example 17

Differentiate: **a** $y = \dfrac{\text{cosec}\, 2x}{x^2}$ **b** $y = \sec^3 x$

a $y = \dfrac{\text{cosec}\, 2x}{x^2}$

So $\dfrac{dy}{dx} = \dfrac{x^2(-2\,\text{cosec}\, 2x \cot 2x) - \text{cosec}\, 2x \times 2x}{x^4}$

$= \dfrac{-2\,\text{cosec}\, 2x(x \cot 2x + 1)}{x^3}$

b $y = \sec^3 x = (\sec x)^3$

$\dfrac{dy}{dx} = 3(\sec x)^2 (\sec x \tan x)$

$= 3 \sec^3 x \tan x$

Use the quotient rule with $u = \text{cosec}\, 2x$ and $v = x^2$.

Use the chain rule with $u = \sec x$.

Chapter 9

When a function is given in the form $x = f(y)$, you can sometimes use the rule $\dfrac{dy}{dx} = \dfrac{1}{\frac{dx}{dy}}$ to find $\dfrac{dy}{dx}$.

Example 18

Given that $x = \sin y$, $-\dfrac{\pi}{2} \leq y \leq \dfrac{\pi}{2}$, show that $\dfrac{dy}{dx} = \dfrac{1}{\sqrt{1-x^2}}$

$x = \sin y$

$\dfrac{dx}{dy} = \cos y$ — Differentiate x with respect to y.

$\dfrac{dy}{dx} = \dfrac{1}{\cos y}$ — Use $\dfrac{dy}{dx} = \dfrac{1}{\frac{dx}{dy}}$. This gives you an expression for $\dfrac{dy}{dx}$ in terms of y.

$\sin^2 y + \cos^2 y \equiv 1$

$\cos y = \sqrt{1 - \sin^2 y} = \sqrt{1 - x^2}$

So $\dfrac{dy}{dx} = \dfrac{1}{\sqrt{1-x^2}}$

Problem-solving

Use the identity $\sin^2 \theta + \cos^2 \theta \equiv 1$ to write $\cos y$ in terms of $\sin y$. This will enable you to find an expression for $\dfrac{dy}{dx}$ in terms of x.

Since $x = \sin y$, $x^2 = \sin^2 y$.

Exercise 9F

1 Differentiate:
   a $y = \tan 3x$
   b $y = 4\tan^3 x$
   c $y = \tan(x - 1)$
   d $y = x^2 \tan \tfrac{1}{2}x + \tan\left(x - \tfrac{1}{2}\right)$

2 Differentiate:
   a $\cot 4x$
   b $\sec 5x$
   c $\csc 4x$
   d $\sec^2 3x$
   e $x \cot 3x$
   f $\dfrac{\sec^2 x}{x}$
   g $\csc^3 2x$
   h $\cot^2(2x - 1)$

3 Find the function $f'(x)$ where $f(x)$ is:
   a $(\sec x)^{\frac{1}{2}}$
   b $\sqrt{\cot x}$
   c $\csc^2 x$
   d $\tan^2 x$
   e $\sec^3 x$
   f $\cot^3 x$

4 Find $f'(x)$ where $f(x)$ is:

a $x^2 \sec 3x$ **b** $\dfrac{\tan 2x}{x}$ **c** $\dfrac{x^2}{\tan x}$ **d** $e^x \sec 3x$

e $\dfrac{\ln x}{\tan x}$ **f** $\dfrac{e^{\tan x}}{\cos x}$

E/P **5** The curve C has equation

$$y = \dfrac{1}{\cos x \sin x}, \ 0 < x \leq \pi$$

    a Find $\dfrac{dy}{dx}$ **(4 marks)**

    b Determine the number of stationary points of the curve C.   **(2 marks)**

    c Find the equation of the tangent at the point where $x = \dfrac{\pi}{3}$, giving your answer in the form $ax + by + c = 0$, where a, b and c are exact constants to be determined.   **(3 marks)**

E/P **6** Show that if $y = \sec x$ then $\dfrac{dy}{dx} = \sec x \tan x$.   **(5 marks)**

E/P **7** Show that if $y = \cot x$ then $\dfrac{dy}{dx} = -\operatorname{cosec}^2 x$.   **(5 marks)**

E/P **8** Given that $x = \cos 2y$, $0 \leq y \leq \dfrac{\pi}{2}$, show that $\dfrac{dy}{dx} = \dfrac{-1}{2\sqrt{1-x^2}}$ **(3 marks)**

E/P **9** Given that $x = \operatorname{cosec} 5y$,

    a find $\dfrac{dy}{dx}$ in terms of y.   **(2 marks)**

    b Hence find $\dfrac{dy}{dx}$ in terms of x.   **(4 marks)**

Challenge

Assuming standard results for $\sin x$ and $\cos x$, prove that:

a the derivative of $\arccos x$ is $-\dfrac{1}{\sqrt{1-x^2}}$

b the derivative of $\arctan x$ is $\dfrac{1}{1+x^2}$

9.7 Parametric differentiation

When functions are defined parametrically, you can find the gradient at a given point without converting into Cartesian form. You can use a variation of the chain rule:

- If x and y are given as functions of a parameter, t: $\dfrac{dy}{dx} = \dfrac{\frac{dy}{dt}}{\frac{dx}{dt}}$

Hint You can obtain this from writing $\dfrac{dy}{dx} \times \dfrac{dx}{dt} = \dfrac{dy}{dt}$

Example 19

Find the gradient at the point P where $t = 2$, on the curve given parametrically by
$$x = t^3 + t, \quad y = t^2 + 1, \quad t \in \mathbb{R}$$

$\dfrac{dx}{dt} = 3t^2 + 1, \quad \dfrac{dy}{dt} = 2t$ ——— First differentiate x and y with respect to the parameter t.

$\dfrac{dy}{dx} = \dfrac{\frac{dy}{dt}}{\frac{dx}{dt}} = \dfrac{2t}{3t^2 + 1}$ ——— This rule will give the gradient function, $\dfrac{dy}{dx}$, in terms of the **parameter**, t.

When $t = 2$, $\dfrac{dy}{dx} = \dfrac{4}{13}$ ——— Substitute $t = 2$ into $\dfrac{2t}{3t^2 + 1}$

So the gradient at P is $\dfrac{4}{13}$

Example 20

Find the equation of the normal at the point P where $\theta = \dfrac{\pi}{6}$, to the curve with parametric equations $x = 3 \sin \theta$, $y = 5 \cos \theta$.

$\dfrac{dx}{d\theta} = 3\cos\theta, \dfrac{dy}{d\theta} = -5\sin\theta \quad 0 \leq \theta < 2\pi$ ——— First differentiate x and y with respect to the parameter θ.

$\therefore \dfrac{dy}{dx} = \dfrac{-5\sin\theta}{3\cos\theta}$ ——— Use the chain rule, $\dfrac{dy}{d\theta} \div \dfrac{dx}{d\theta}$, and substitute $\theta = \dfrac{\pi}{6}$

At point P, where $\theta = \dfrac{\pi}{6}$,

$\dfrac{dy}{dx} = \dfrac{-5 \times \frac{1}{2}}{3 \times \frac{\sqrt{3}}{2}} = \dfrac{-5}{3\sqrt{3}}$

Online Explore the graph of this curve and the normal at this point using technology.

Differentiation

The gradient of the normal at P is $\dfrac{3\sqrt{3}}{5}$,

and at P, $x = \dfrac{3}{2}$, $y = \dfrac{5\sqrt{3}}{2}$

The equation of the normal is

$$y - \dfrac{5\sqrt{3}}{2} = \dfrac{3\sqrt{3}}{5}\left(x - \dfrac{3}{2}\right)$$

$\therefore\ 5y = 3\sqrt{3}x + 8\sqrt{3}$

— The normal is perpendicular to the curve, so its gradient is $-\dfrac{1}{m}$ where m is the gradient of the curve at that point.

— You need to find the coordinates of P. Substitute $\theta = \dfrac{\pi}{6}$ into each of the parametric equations.
 ← Section 8.1

— Use the equation for a line in the form $y - y_1 = m(x - x_1)$

Exercise 9G

1 Find $\dfrac{dy}{dx}$ for each of the following, leaving your answer in terms of the parameter t.

  a $x = 2t$, $y = t^2 - 3t + 2$
  b $x = 3t^2$, $y = 2t^3$
  c $x = t + 3t^2$, $y = 4t$

  d $x = t^2 - 2$, $y = 3t^5$
  e $x = \dfrac{2}{t}$, $y = 3t^2 - 2$
  f $x = \dfrac{1}{2t-1}$, $y = \dfrac{t^2}{2t-1}$

  g $x = \dfrac{2t}{1+t^2}$, $y = \dfrac{1-t^2}{1+t^2}$
  h $x = t^2 e^t$, $y = 2t$
  i $x = 4\sin 3t$, $y = 3\cos 3t$

  j $x = 2 + \sin t$, $y = 3 - 4\cos t$
  k $x = \sec t$, $y = \tan t$
  l $x = 2t - \sin 2t$, $y = 1 - \cos 2t$

  m $x = e^t - 5$, $y = \ln t$, $t > 0$
  n $x = \ln t$, $y = t^2 - 64$, $t > 0$
  o $x = e^{2t} + 1$, $y = 2e^t - 1$, $-1 < t < 1$

(P) 2 a Find the equation of the tangent to the curve with parametric equations $x = 3 - 2\sin t$, $y = t\cos t$, at the point P, where $t = \pi$.

  b Find the equation of the tangent to the curve with parametric equations $x = 9 - t^2$, $y = t^2 + 6t$, at the point P, where $t = 2$.

(P) 3 a Find the equation of the normal to the curve with parametric equations $x = e^t$, $y = e^t + e^{-t}$, at the point P, where $t = 0$.

  b Find the equation of the normal to the curve with parametric equations $x = 1 - \cos 2t$, $y = \sin 2t$, at the point P, where $t = \dfrac{\pi}{6}$

(P) 4 Find the points of zero gradient on the curve with parametric equations

$$x = \dfrac{t}{1-t},\quad y = \dfrac{t^2}{1-t},\quad t \neq 1$$

You do not need to establish whether they are maximum or minimum points.

(P) 5 The curve C has parametric equations $x = e^{2t}$, $y = e^t - 1$, $t \in \mathbb{R}$.

  a Find the equation of the tangent to C at the point A where $t = \ln 2$.

  b Show that the curve C has no stationary points.

Chapter 9

E/P 6 The curve C has parametric equations
$$x = \frac{t^2 - 3t - 4}{t}, \quad y = 2t, \quad t > 0$$
The line l_1 is a tangent to C and is parallel to the line with equation $y = x + 5$.
Find the equation of l_1. (8 marks)

E/P 7 A curve has parametric equations
$$x = 2\sin^2 t, \quad y = 2\cot t, \quad 0 < t < \frac{\pi}{2}$$
   a Find an expression for $\frac{dy}{dx}$ in terms of the parameter t. (4 marks)
   b Find an equation of the tangent to the curve at the point where $t = \frac{\pi}{6}$ (4 marks)

E/P 8 The curve C has parametric equations
$$x = 4\sin t, \quad y = 2\operatorname{cosec} 2t, \quad 0 \leqslant t \leqslant \pi$$
The point A lies on C and has coordinates $\left(2\sqrt{3}, \frac{4\sqrt{3}}{3}\right)$.
   a Find the value of t at the point A. (2 marks)
The line l is a normal to C at A.
   b Show that an equation for l is $9x + 12y - 34\sqrt{3} = 0$. (6 marks)

E/P 9 The curve C has parametric equations
$$x = t^2 + t, \quad y = t^2 - 10t + 5, \quad t \in \mathbb{R}$$
where t is a parameter. Given that at point P, the gradient of C is 2,
   a find the coordinates of P (4 marks)
   b find the equation of the tangent to C at point P (3 marks)
   c show that the tangent to C at point P does not intersect the curve again. (5 marks)

> **Problem-solving**
> Substitute the equations for x and y into the equation of your tangent, and show that the resulting quadratic equation has one unique root.

E/P 10 The curve C has parametric equations
$$x = 2\sin t, \quad y = \sqrt{2}\cos 2t, \quad 0 < t < \pi$$
   a Find an expression for $\frac{dy}{dx}$ in terms of t. (2 marks)

The point A lies on C where $t = \frac{\pi}{3}$. The line l is the normal to C at A.

  b Find an equation for l in the form $ax + by + c = 0$, where a, b and c are exact constants to be found. **(5 marks)**

  c Prove that the line l does not intersect the curve anywhere other than at point A. **(6 marks)**

E/P **11** A curve has parametric equations

$$x = \cos t, \quad y = \tfrac{1}{2}\sin 2t, \quad 0 \leq t < 2\pi$$

  a Find an expression for $\dfrac{dy}{dx}$ in terms of t. **(2 marks)**

  b Find an equation of the tangent to the curve at point A where $t = \dfrac{\pi}{6}$ **(4 marks)**

The lines l_1 and l_2 are two further distinct tangents to the curve. Given that l_1 and l_2 are both parallel to the tangent to the curve at point A,

  c find an equation of l_1 and an equation of l_2. **(6 marks)**

9.8 Implicit differentiation

Some equations are difficult to rearrange into the form $y = f(x)$ or $x = f(y)$. You can sometimes differentiate these equations **implicitly** without rearranging them.

Notation An equation in the form $y = f(x)$ is given **explicitly**.

Equations which involve functions of both x and y such as $x^2 + 2xy = 3$ or $\cos(x + y) = 2x$ are called **implicit** equations.

In general, from the chain rule:

■ $\dfrac{d}{dx}(f(y)) = f'(y)\dfrac{dy}{dx}$

The following two specific results are useful for implicit differentiation:

■ $\dfrac{d}{dx}(y^n) = ny^{n-1}\dfrac{dy}{dx}$

■ $\dfrac{d}{dx}(xy) = x\dfrac{dy}{dx} + y$

When you differentiate implicit equations your expression for $\dfrac{dy}{dx}$ will usually be given in terms of **both x and y**.

Watch out You need to pay careful attention to the variable you are differentiating with respect to.

Example 21

Find $\dfrac{dy}{dx}$ in terms of x and y where $x^3 + x + y^3 + 3y = 6$.

$3x^2 + 1 + 3y^2 \dfrac{dy}{dx} + 3\dfrac{dy}{dx} = 0$ — Differentiate the expression term by term with respect to x.

$\dfrac{dy}{dx}(3y^2 + 3) = -3x^2 - 1$ — Divide both sides by $3y^2 + 3$ and factorise.

$\dfrac{dy}{dx} = -\dfrac{3x^2 + 1}{3(1 + y^2)}$ — Then make $\dfrac{dy}{dx}$ the subject of the formula.

Use $\dfrac{d}{dx}(y^n) = ny^{n-1}\dfrac{dy}{dx}$ with $n = 3$.

Example 22

Given that $4xy^2 + \dfrac{6x^2}{y} = 10$, find the value of $\dfrac{dy}{dx}$ at the point $(1, 1)$.

$\left(4x \times 2y\dfrac{dy}{dx} + 4y^2\right) + \left(\dfrac{12x}{y} - \dfrac{6x^2}{y^2}\dfrac{dy}{dx}\right) = 0$

Substitute $x = 1$, $y = 1$ to give

$\left(8\dfrac{dy}{dx} + 4\right) + \left(12 - 6\dfrac{dy}{dx}\right) = 0$

$16 + 2\dfrac{dy}{dx} = 0$

$\dfrac{dy}{dx} = -8$

Differentiate each term with respect to x.

Use the product rule on each term, expressing $\dfrac{6x^2}{y}$ as $6x^2y^{-1}$.

Find the value of $\dfrac{dy}{dx}$ at $(1, 1)$ by substituting $x = 1$, $y = 1$.

Substitute before rearranging, as this simplifies the working.

Solve to find the value of $\dfrac{dy}{dx}$ at this point.

Differentiation

Example 23

Find the value of $\frac{dy}{dx}$ at the point (1, 1) where $e^{2x} \ln y = x + y - 2$.

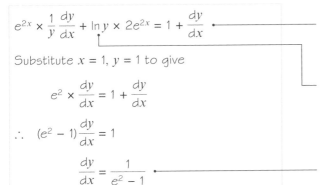

$e^{2x} \times \frac{1}{y}\frac{dy}{dx} + \ln y \times 2e^{2x} = 1 + \frac{dy}{dx}$ — Differentiate each term with respect to x.

Substitute $x = 1$, $y = 1$ to give

$e^2 \times \frac{dy}{dx} = 1 + \frac{dy}{dx}$

Use the product rule applied to the term on the left hand side of the equation, noting that $\ln y$ differentiates to give $\frac{1}{y}\frac{dy}{dx}$

$\therefore (e^2 - 1)\frac{dy}{dx} = 1$

$\frac{dy}{dx} = \frac{1}{e^2 - 1}$ — Rearrange to make $\frac{dy}{dx}$ the subject of the formula.

Exercise 9H

P 1 By writing $u = y^n$, and using the chain rule, show that $\frac{d}{dx}(y^n) = ny^{n-1}\frac{dy}{dx}$

P 2 Use the product rule to show that $\frac{d}{dx}(xy) = x\frac{dy}{dx} + y$.

P 3 Find an expression in terms of x and y for $\frac{dy}{dx}$, given that:

    a $x^2 + y^3 = 2$ **b** $x^2 + 5y^2 = 14$ **c** $x^2 + 6x - 8y + 5y^2 = 13$

    d $y^3 + 3x^2y - 4x = 0$ **e** $3y^2 - 2y + 2xy = x^3$ **f** $x = \frac{2y}{x^2 - y}$

    g $(x - y)^4 = x + y + 5$ **h** $e^x y = xe^y$ **i** $\sqrt{xy} + x + y^2 = 0$

P 4 Find the equation of the tangent to the curve with implicit equation $x^2 + 3xy^2 - y^3 = 9$ at the point (2, 1).

P 5 Find the equation of the normal to the curve with implicit equation $(x + y)^3 = x^2 + y$ at the point (1, 0).

Chapter 9

(P) **6** Find the coordinates of the points of zero gradient on the curve with implicit equation $x^2 + 4y^2 - 6x - 16y + 21 = 0$.

Problem-solving

Find $\dfrac{dy}{dx}$ then set the numerator equal to 0 to find the x-coordinate at the points of 0 gradient. You need to find two corresponding y-coordinates.

(E/P) **7** A curve C is described by the equation
$$2x^2 + 3y^2 - x + 6xy + 5 = 0$$
Find an equation of the tangent to C at the point $(1, -2)$, giving your answer in the form $ax + by + c = 0$, where a, b and c are integers. **(7 marks)**

(E/P) **8** A curve C has equation
$$3^x = y - 2xy$$
Find the exact value of $\dfrac{dy}{dx}$ at the point on C with coordinates $(2, -3)$. **(7 marks)**

(E/P) **9** Find the gradient of the curve with equation
$$\ln(y^2) = \tfrac{1}{2}x \ln(x - 1), \quad x > 1, \quad y > 0$$
at the point on the curve where $x = 4$. Give your answer as an exact value. **(7 marks)**

(E/P) **10** A curve C satisfies $\sin x + \cos y = 0.5$, where $-\pi < x < \pi$ and $-\pi < y < \pi$.

a Find an expression for $\dfrac{dy}{dx}$ **(2 marks)**

b Find the coordinates of the stationary points on C. **(5 marks)**

(E/P) **11** The curve C has the equation $ye^{-3x} - 3x = y^2$.

a Find $\dfrac{dy}{dx}$ in terms of x and y. **(5 marks)**

b Show that the equation of the tangent to C at the origin, O, is $y = 3x$. **(3 marks)**

Challenge

The curve C has implicit equation $6x + y^2 + 2xy = x^2$.

a Show that there are no points on the curve such that $\dfrac{dy}{dx} = 0$.

b Find the coordinates of the two points on C such that $\dfrac{dx}{dy} = 0$.

Differentiation

9.9 Using second derivatives

You can use the second derivative to determine whether a curve is **concave** or **convex** on a given domain.

- **The function f(x) is concave on a given interval if and only if f''(x) ≤ 0 for every value of x in that interval.**
- **The function f(x) is convex on a given interval if and only if f''(x) ≥ 0 for every value of x in that interval.**

Links To find the second derivative, f''(x) or $\frac{d^2y}{dx^2}$, you differentiate **twice** with respect to x.
← Year 1, Chapter 12

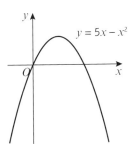

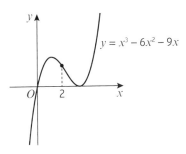

$\frac{d^2y}{dx^2} = -2$ so the curve is concave for all $x \in \mathbb{R}$.

$\frac{d^2y}{dx^2} = e^x$ which is always positive, so the curve is convex for all $x \in \mathbb{R}$.

$\frac{d^2y}{dx^2} = 6x - 12$ so the curve is concave for all $x \leq 2$ and convex for $x \geq 2$.

Example 24

Find the interval on which the function f(x) = $x^3 + 4x + 3$ is concave.

f(x) = $x^3 + 4x + 3$
f'(x) = $3x^2 + 4$
f''(x) = $6x$

For f(x) to be concave, f''(x) ≤ 0
$6x \leq 0$
$x \leq 0$
So f(x) is concave for all $x \leq 0$.

— Differentiate twice to get an expression for f''(x).

Write down the condition for a concave function in your working.

Notation You can also write this interval as $(-\infty, 0]$.

Example 25

Show that the function $f(x) = e^{2x} + x^2$ is convex for all real values of x.

$f(x) = e^{2x} + x^2$
$f'(x) = 2e^{2x} + 2x$
$f''(x) = 4e^{2x} + 2$ — Differentiate twice to get an expression for $f''(x)$.

$e^{2x} > 0$ for all $x \in \mathbb{R}$, so $4e^{2x} + 2 > 2$ for all $x \in \mathbb{R}$

Hence $f''(x) \geq 0$, so f is convex for all $x \in \mathbb{R}$.

Problem-solving

Write down the condition for a convex function and a conclusion.

The point at which a curve changes from being concave to convex (or vice versa) is called a **point of inflection**. The diagram shows the curve with equation $y = x^3 - 2x^2 - 4x + 5$.

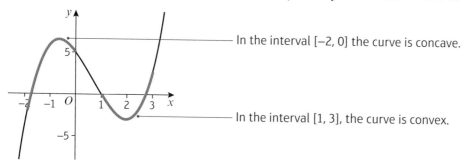

In the interval $[-2, 0]$ the curve is concave.

In the interval $[1, 3]$, the curve is convex.

At some point between 0 and 1 the curve changes from being concave to being convex. This is the point of inflection.

- **A point of inflection is a point at which $f''(x)$ changes sign.**

To find a point of inflection you need to show that $f''(x) = 0$ at that point, and that it has different signs on either side of that point.

Watch out A point of inflection does not have to be a stationary point.

Example 26

The curve C has equation $y = x^3 - 2x^2 - 4x + 5$.

a Show that C is concave on the interval $[-2, 0]$ and convex on the interval $[1, 3]$.
b Find the coordinates of the point of inflection.

a $\dfrac{dy}{dx} = 3x^2 - 4x - 4$

$\dfrac{d^2y}{dx^2} = 6x - 4$

$\dfrac{d^2y}{dx^2} = 6x - 4 \leq 0$ for all $-2 \leq x \leq 0$.

Differentiate $y = x^3 - 2x^2 - 4x + 5$ with respect to x twice.

Consider the value of $6x - 4$ on the interval $[-2, 0]$. $6x - 4$ is a linear function. When $x = -2$, $\dfrac{d^2y}{dx^2} = -16$ and when $x = 0$, $\dfrac{d^2y}{dx^2} = -4$, so $\dfrac{d^2y}{dx^2} \leq 0$ on $[0, 2]$.

Differentiation

Therefore, $y = x^3 - 2x^2 - 4x + 5$ is concave on the interval $[-2, 0]$.

$\dfrac{d^2y}{dx^2} = 6x - 4 \geqslant 0$ for all $1 \leqslant x \leqslant 3$. — Consider the value of $6x - 4$ on the interval $[1, 3]$.

Therefore, $y = x^3 - 2x^2 - 4x + 5$ is convex on the interval $[1, 3]$.

When $x = 1$, $\dfrac{d^2y}{dx^2} = 2$ and when $x = 3$, $\dfrac{d^2y}{dx^2} = 14$.

b $\dfrac{d^2y}{dx^2} = 6x - 4 = 0$ — Find the point where $f''(x) = 0$. You have already determined that $f''(x)$ changes sign on either side of this point.

$6x = 4$

$x = \dfrac{4}{6} = \dfrac{2}{3}$

Substitute x into y gives

$y = \left(\dfrac{2}{3}\right)^3 - 2\left(\dfrac{2}{3}\right)^2 - 4\left(\dfrac{2}{3}\right) + 5 = \dfrac{47}{27}$

So the point of inflection of the curve C is $\left(\dfrac{2}{3}, \dfrac{47}{27}\right)$.

Online Explore the solution to this example graphically using technology.

Exercise 9I

P **1** For each of the following functions, find the interval on which the function is:

i convex **ii** concave

a $f(x) = x^3 - 3x^2 + x - 2$ **b** $f(x) = x^4 - 3x^3 + 2x - 1$ **c** $f(x) = \sin x$, $0 < x < 2\pi$

d $f(x) = -x^2 + 3x - 7$ **e** $f(x) = e^x - x^2$ **f** $f(x) = \ln x$, $x > 0$

P **2** $f(x) = \arcsin x$, $-1 < x < 1$

   a Show that $f'(x) = \dfrac{1}{\sqrt{1 - x^2}}$

   b Hence show that $f(x)$ is concave on the interval $(-1, 0)$.

   c Show that $f(x)$ is convex on the interval $(0, 1)$.

   d Hence deduce the point of inflection of f.

P **3** Find any point(s) of inflection of the following functions.

   a $f(x) = \cos^2 x - 2\sin x$, $0 < x < 2\pi$ **b** $f(x) = -\dfrac{x^3 - 2x^2 + x - 1}{x - 2}$, $x \neq 2$

   c $f(x) = \dfrac{x^3}{x^2 - 4}$, $x \neq \pm 2$

P **4** $f(x) = 2x^2 \ln x$, $x > 0$

   Show that f has exactly one point of inflection and determine the value of x at this point.

P **5** The curve C has equation $y = e^x(x^2 - 2x + 2)$.

   a Find the exact coordinates of the stationary point on C and determine its nature.

   b Find the coordinates of any non-stationary points of inflection on C.

6 The curve C has equation $y = xe^x$.

   a Find the exact coordinates of the stationary point on C and determine its nature.

   b Find the coordinates of any non-stationary points of inflection on C.

   c Hence sketch the graph of $y = xe^x$.

   Problem-solving

   Consider how C behaves for very large positive and negative values of x.

7 For each point on the graph, state whether:

   i $f'(x)$ is positive, negative or zero
   ii $f''(x)$ is positive, negative or zero

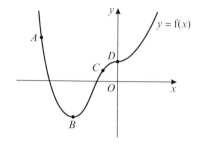

8 $f(x) = \tan x$, $-\frac{\pi}{2} < x < \frac{\pi}{2}$

   Prove that $f(x)$ has exactly one point of inflection, at the origin.

9 Given that $y = x(3x - 1)^5$,

   a find $\frac{dy}{dx}$ and $\frac{d^2y}{dx^2}$ (4 marks)

   b find the points of inflection of y. (4 marks)

10 A student is attempting to find the points of inflection on the curve C with equation

   $y = (x - 5)^4$

   The attempt is shown below:

   > $\frac{dy}{dx} = 4(x - 5)^3$
   >
   > $\frac{d^2y}{dx^2} = 12(x - 5)^2$
   >
   > When $\frac{d^2y}{dx^2} = 0$,
   >
   > $12(x - 5)^2 = 0$
   > $(x - 5)^2 = 0$
   > $x - 5 = 0$
   > $x = 5$
   >
   > Therefore, the curve C has a point of inflection at $x = 5$.

   a Identify the mistake made by the student. (2 marks)

   b Write down the coordinates of the stationary point on C and determine its nature. (2 marks)

Differentiation

E/P **11** A curve C has equation
$$y = \tfrac{1}{3}x^2 \ln x - 2x + 5, \; x > 0$$
Show that the curve C is convex for all $x \geqslant e^{-\frac{3}{2}}$. **(5 marks)**

Challenge

1 Prove that every cubic curve has exactly one point of inflection.

2 The curve C has equation $y = ax^4 + bx^3 + cx^2 + dx + e, a \neq 0$
   a Show that C has at most two points of inflection.
   b Prove that if $3b^2 < 8ac$, then C has no points of inflection.

9.10 Rates of change

- You can use the chain rule to connect rates of change in situations involving more than two variables.

Example 27

Given that the area of a circle A cm² is related to its radius r cm by the formula $A = \pi r^2$, and that the rate of change of its radius in cm s⁻¹ is given by $\dfrac{dr}{dt} = 5$, find $\dfrac{dA}{dt}$ when $r = 3$.

$A = \pi r^2$

$\therefore \dfrac{dA}{dr} = 2\pi r$

Using $\dfrac{dA}{dt} = \dfrac{dA}{dr} \times \dfrac{dr}{dt}$

$\dfrac{dA}{dt} = 2\pi r \times 5$

$= 30\pi$, when $r = 3$.

Problem-solving

In order to be able to apply the chain rule to find $\dfrac{dA}{dt}$ you need to know $\dfrac{dA}{dr}$. You can find it by differentiating $A = \pi r^2$ with respect to r.

You should use the chain rule, giving the derivative which you need to find in terms of known derivatives.

Example 28

The volume of a hemisphere V cm³ is related to its radius r cm by the formula $V = \tfrac{2}{3}\pi r^3$ and the total surface area S cm² is given by the formula $S = \pi r^2 + 2\pi r^2 = 3\pi r^2$. Given that the rate of increase of volume, in cm³ s⁻¹, $\dfrac{dV}{dt} = 6$, find the rate of increase of surface area $\dfrac{dS}{dt}$.

$V = \tfrac{2}{3}\pi r^3$ and $S = 3\pi r^2$

$\dfrac{dV}{dr} = 2\pi r^2$ and $\dfrac{dS}{dr} = 6\pi r$

This is area of circular base plus area of curved surface.

As V and S are functions of r, find $\dfrac{dV}{dr}$ and $\dfrac{dS}{dr}$.

Chapter 9

Now $\dfrac{dS}{dt} = \dfrac{dS}{dr} \times \dfrac{dr}{dV} \times \dfrac{dV}{dt}$

$= 6\pi r \times \dfrac{1}{2\pi r^2} \times 6$

$= \dfrac{18}{r}$

Use the chain rule together with the property that $\dfrac{dr}{dV} = 1 \div \dfrac{dV}{dr}$

An equation which involves a rate of change is called a **differential equation**. You can formulate differential equations from information given in a question.

Links You can use integration to solve differential equations. → Section 11.10

Example 29

In the decay of radioactive particles, the rate at which particles decay is proportional to the number of particles remaining. Write down a differential equation for the rate of change of the number of particles.

Let N be the number of particles and let t be time. The rate of change of the number of particles $\dfrac{dN}{dt}$ is proportional to N.

i.e. $\dfrac{dN}{dt} = -kN$, where k is a positive constant.

The minus sign arises because the number of particles is decreasing.

$\dfrac{dN}{dt} \propto N$ so you can write $\dfrac{dN}{dt} = kN$ where k is the constant of proportion.

Example 30

Newton's law of cooling states that the rate of loss of temperature of a body is proportional to the excess temperature of the body over its surroundings. Write an equation that expresses this law.

Let the temperature of the body be θ degrees and the time be t seconds.

The rate of change of the temperature $\dfrac{d\theta}{dt}$ is proportional to $\theta - \theta_0$, where θ_0 is the temperature of the surroundings.

i.e. $\dfrac{d\theta}{dt} = -k(\theta - \theta_0)$, where k is a positive constant.

$\theta - \theta_0$ is the difference between the temperature of the body and that of its surroundings.

The minus sign arises because the temperature is decreasing. The question mentions loss of temperature.

Differentiation

Example 31

The head of a snowman of radius R cm loses volume by evaporation at a rate proportional to its surface area. Assuming that the head is spherical, that the volume of a sphere is $\frac{4}{3}\pi R^3$ cm^3 and that the surface is $4\pi R^2$ cm^2, write down a differential equation for the rate of change of radius of the snowman's head.

The first sentence tells you that $\frac{dV}{dt} = -kA$, where V cm^3 is the volume, t seconds is time, k is a positive constant and A cm^2 is the surface area of the snowman's head.

Since $V = \frac{4}{3}\pi R^3$ — The question asks for a differential equation in terms of R, so you need to use the expression for V in terms of R.

$$\frac{dV}{dR} = 4\pi R^2$$

$$\therefore \quad \frac{dV}{dt} = \frac{dV}{dR} \times \frac{dR}{dt} = 4\pi R^2 \times \frac{dR}{dt}$$

The chain rule is used here because this is a related rate of change.

But as $\frac{dV}{dt} = -kA$ — Use the expression for A in terms of R.

$$4\pi R^2 \times \frac{dR}{dt} = -k \times 4\pi R^2$$

Divide both sides by the common factor $4\pi R^2$.

$$\therefore \quad \frac{dR}{dt} = -k$$

This gives the rate of change of radius as required.

Exercise 9J

1. Given that $A = \frac{1}{4}\pi r^2$ and that $\frac{dr}{dt} = 6$, find $\frac{dA}{dt}$ when $r = 2$.

2. Given that $y = xe^x$ and that $\frac{dx}{dt} = 5$, find $\frac{dy}{dt}$ when $x = 2$.

3. Given that $r = 1 + 3\cos\theta$ and that $\frac{d\theta}{dt} = 3$, find $\frac{dr}{dt}$ when $\theta = \frac{\pi}{6}$.

4. Given that $V = \frac{1}{3}\pi r^3$ and that $\frac{dV}{dt} = 8$, find $\frac{dr}{dt}$ when $r = 3$.

5. A population is growing at a rate which is proportional to the size of the population. Write down a differential equation for the growth of the population.

6. A curve C has equation $y = f(x)$, $y > 0$. At any point P on the curve, the gradient of C is proportional to the product of the x- and the y-coordinates of P. The point A with coordinates $(4, 2)$ is on C and the gradient of C at A is $\frac{1}{2}$.

    Show that $\frac{dy}{dx} = \frac{xy}{16}$.

7 Liquid is pouring into a container at a constant rate of $30 \, \text{cm}^3 \, \text{s}^{-1}$. At time t seconds liquid is leaking from the container at a rate of $\frac{2}{15} V \, \text{cm}^3 \, \text{s}^{-1}$, where $V \, \text{cm}^3$ is the volume of the liquid in the container at that time.

Show that $-15 \dfrac{dV}{dt} = 2V - 450$.

8 An electrically-charged body loses its charge, Q coulombs, at a rate, measured in coulombs per second, proportional to the charge Q.

Write down a differential equation in terms of Q and t where t is the time in seconds since the body started to lose its charge.

9 The ice on a pond has a thickness x mm at a time t hours after the start of freezing. The rate of increase of x is inversely proportional to the square of x.

Write down a differential equation in terms of x and t.

10 The radius of a circle is increasing at a constant rate of 0.4 cm per second.

a Find $\dfrac{dC}{dt}$, where C is the circumference of the circle, and interpret this value in the context of the model.

b Find the rate at which the area of the circle is increasing when the radius is 10 cm.

c Find the radius of the circle when its area is increasing at the rate of $20 \, \text{cm}^2$ per second.

11 The volume of a cube is decreasing at a constant rate of $4.5 \, \text{cm}^3$ per second. Find:

a the rate at which the length of one side of the cube is decreasing when the volume is $100 \, \text{cm}^3$

b the volume of the cube when the length of one side is decreasing at the rate of 2 mm per second.

12 Fluid flows out of a cylindrical tank with constant cross section. At time t minutes, $t > 0$, the volume of fluid remaining in the tank is $V \, \text{m}^3$. The rate at which the fluid flows in $\text{m}^3 \, \text{min}^{-1}$ is proportional to the square root of V.

Show that the depth, h metres, of fluid in the tank satisfies the differential equation $\dfrac{dh}{dt} = -k\sqrt{h}$, where k is a positive constant.

13 At time, t seconds, the surface area of a cube is $A \, \text{cm}^2$ and the volume is $V \, \text{cm}^3$. The surface area of the cube is expanding at a constant rate of $2 \, \text{cm}^2 \, \text{s}^{-1}$.

a Write an expression for V in terms of A.

b Find an expression for $\dfrac{dV}{dA}$.

c Show that $\dfrac{dV}{dt} = \dfrac{1}{2} V^{\frac{1}{3}}$.

14 An inverted conical funnel is full of salt. The salt is allowed to leave by a small hole in the vertex. It leaves at a constant rate of $6 \, \text{cm}^3 \, \text{s}^{-1}$.

Given that the angle of the cone between the slanting edge and the vertical is $30°$, show that the volume of the salt is $\frac{1}{9} \pi h^3$, where h is the height of salt at time t seconds. Show that the rate of change of the height of the salt in the funnel is inversely proportional to h^2. Write down a differential equation relating h and t.

Mixed Exercise 9

1 Differentiate with respect to x:

   a $\ln x^2$ (3 marks)

   b $x^2 \sin 3x$ (4 marks)

2 a Given that $2y = x - \sin x \cos x$, $0 < x < 2\pi$, show that $\dfrac{dy}{dx} = \sin^2 x$.   (4 marks)

   b Find the coordinates of the points of inflection of the curve.   (4 marks)

3 Differentiate, with respect to x:

   a $\dfrac{\sin x}{x}$, $x > 0$ (4 marks)

   b $\ln \dfrac{1}{x^2 + 9}$ (4 marks)

4 $f(x) = \dfrac{x}{x^2 + 2}$, $x \in \mathbb{R}$

   a Given that $f(x)$ is increasing on the interval $[-k, k]$, find the largest possible value of k.   (4 marks)

   b Find the exact coordinates of the points of inflection of $f(x)$.   (5 marks)

5 The function f is defined for positive real values of x by

$$f(x) = 12 \ln x + x^{\frac{3}{2}}$$

   a Find the set of values of x for which $f(x)$ is an increasing function of x.   (4 marks)

   b Find the coordinates of the point of inflection of the function f.   (4 marks)

6 Given that a curve has equation $y = \cos^2 x + \sin x$, $0 < x < 2\pi$, find the coordinates of the stationary points of the curve.   (6 marks)

7 The maximum point on the curve with equation $y = x\sqrt{\sin x}$, $0 < x < \pi$, is the point A. Show that the x-coordinate of point A satisfies the equation $2 \tan x + x = 0$.   (5 marks)

8 $f(x) = e^{0.5x} - x^2$, $x \in \mathbb{R}$

   a Find $f'(x)$.   (3 marks)

   b By evaluating $f'(6)$ and $f'(7)$, show that the curve with equation $y = f(x)$ has a stationary point at $x = p$, where $6 < p < 7$.   (2 marks)

9 $f(x) = e^{2x} \sin 2x$, $0 < x < \pi$

   a Use calculus to find the coordinates of the turning points on the graph of $y = f(x)$.   (6 marks)

   b Show that $f''(x) = 8e^{2x} \cos 2x$.   (4 marks)

   c Hence, or otherwise, determine which turning point is a maximum and which is a minimum.   (3 marks)

   d Find the points of inflection of $f(x)$.   (2 marks)

10 The curve C has equation $y = 2e^x + 3x^2 + 2$. Find the equation of the normal to C at the point where the curve intercepts the y-axis. Give your answer in the form $ax + by + c = 0$ where a, b and c are integers to be found. **(5 marks)**

11 The curve C has equation $y = f(x)$, where

$$f(x) = 3\ln x + \frac{1}{x}, \quad x > 0$$

The point P is a stationary point on C.
a Calculate the x-coordinate of P. **(4 marks)**

The point Q on C has x-coordinate 1.
b Find an equation for the normal to C at Q. **(4 marks)**

12 The curve C has equation $y = e^{2x} \cos x$.
a Show that the turning points on C occur when $\tan x = 2$. **(4 marks)**
b Find an equation of the tangent to C at the point where $x = 0$. **(4 marks)**

13 Given that $x = y^2 \ln y$, $y > 0$,
a find $\dfrac{dx}{dy}$ **(4 marks)**
b Use your answer to part **a** to find in terms of e, the value of $\dfrac{dy}{dx}$ at $y = e$. **(2 marks)**

14 A curve has equation $f(x) = (x^3 - 2x)e^{-x}$.
a Find $f'(x)$. **(4 marks)**

The normal to C at the origin O intersects C again at P.
b Show that the x-coordinate of P is the solution to the equation $2x^2 = e^x + 4$. **(6 marks)**

15 The diagram shows part of the curve with equation $y = f(x)$ where $f(x) = x(1 + x)\ln x$, $x > 0$
The point A is the minimum point of the curve.
a Find $f'(x)$. **(4 marks)**
b Hence show that the x-coordinate of A is the solution to the equation $x = e^{-\frac{1+x}{1+2x}}$ **(4 marks)**

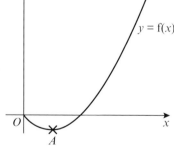

16 The curve C is given by the equations

$$x = 4t - 3, \quad y = \frac{8}{t^2}, \quad t > 0$$

where t is a parameter.
At A, $t = 2$. The line l is the normal to C at A.
a Find $\dfrac{dy}{dx}$ in terms of t. **(4 marks)**
b Hence find an equation of l. **(3 marks)**

Differentiation

17 The curve C is given by the equations $x = 2t$, $y = t^2$, where t is a parameter.
Find an equation of the normal to C at the point P on C where $t = 3$. **(7 marks)**

18 The curve C has parametric equations
$$x = t^3, \quad y = t^2, \quad t > 0$$
Find an equation of the tangent to C at $A\,(1, 1)$. **(7 marks)**

19 A curve C is given by the equations
$$x = 2\cos t + \sin 2t, \quad y = \cos t - 2\sin 2t, \quad 0 < t < \pi$$
where t is a parameter.
a Find $\dfrac{\mathrm{d}x}{\mathrm{d}t}$ and $\dfrac{\mathrm{d}y}{\mathrm{d}t}$ in terms of t. **(3 marks)**
b Find the value of $\dfrac{\mathrm{d}y}{\mathrm{d}x}$ at the point P on C where $t = \dfrac{\pi}{4}$. **(3 marks)**
c Find an equation of the normal to the curve at P. **(3 marks)**

20 A curve is given by $x = 2t + 3$, $y = t^3 - 4t$, where t is a parameter. The point A has parameter $t = -1$ and the line l is the tangent to C at A. The line l also cuts the curve at B.
a Show that an equation for l is $2y + x = 7$. **(6 marks)**
b Find the value of t at B. **(5 marks)**

21 A car has value £V at time t years. A model for V assumes that the rate of decrease of V at time t is proportional to V. Form an appropriate differential equation for V.

22 In a study of the water loss of picked leaves the mass, M grams, of a single leaf was measured at times, t days, after the leaf was picked. It was found that the rate of loss of mass was proportional to the mass M of the leaf.
Write down a differential equation for the rate of change of mass of the leaf.

23 In a pond the amount of pondweed, P, grows at a rate proportional to the amount of pondweed already present in the pond. Pondweed is also removed by fish eating it at a constant rate of Q per unit of time.
Write down a differential equation relating P to t, where t is the time which has elapsed since the start of the observation.

24 A circular patch of oil on the surface of some water has radius r and the radius increases over time at a rate inversely proportional to the radius.
Write down a differential equation relating r and t, where t is the time which has elapsed since the start of the observation.

25 A metal bar is heated to a certain temperature, then allowed to cool down and it is noted that, at time t, the rate of loss of temperature is proportional to the difference between the temperature of the metal bar, θ, and the temperature of its surroundings θ_0.
Write down a differential equation relating θ and t.

Chapter 9

26 The curve C has parametric equations

$$x = 4\cos 2t, \quad y = 3\sin t, \quad -\frac{\pi}{2} < t < \frac{\pi}{2}$$

A is the point $\left(2, \frac{3}{2}\right)$, and lies on C.

a Find the value of t at the point A. **(2 marks)**

b Find $\dfrac{dy}{dx}$ in terms of t. **(3 marks)**

c Show that an equation of the normal to C at A is $6y - 16x + 23 = 0$. **(4 marks)**

The normal at A cuts C again at the point B.

d Find the y-coordinate of the point B. **(6 marks)**

27 The diagram shows the curve C with parametric equations

$$x = a\sin^2 t, \quad y = a\cos t, \quad 0 \leq t \leq \tfrac{1}{2}\pi$$

where a is a positive constant. The point P lies on C and has coordinates $\left(\tfrac{3}{4}a, \tfrac{1}{2}a\right)$.

a Find $\dfrac{dy}{dx}$, giving your answer in terms of t. **(4 marks)**

b Find an equation of the tangent to C at P. **(4 marks)**

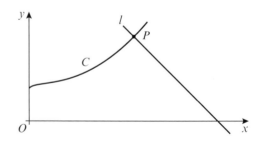

The tangent to C at P cuts the coordinate axes at points A and B.

c Show that the triangle AOB has area ka^2 where k is a constant to be found. **(2 marks)**

28 This graph shows part of the curve C with parametric equations

$$x = (t+1)^2, \quad y = \tfrac{1}{2}t^3 + 3, \quad t > -1$$

P is the point on the curve where $t = 2$.
The line l is the normal to C at P.

Find the equation of l. **(7 marks)**

29 Find the gradient of the curve with equation $5x^2 + 5y^2 - 6xy = 13$ at the point $(1, 2)$. **(7 marks)**

30 Given that $e^{2x} + e^{2y} = xy$, find $\dfrac{dy}{dx}$ in terms of x and y. **(7 marks)**

31 Find the coordinates of the turning points on the curve $y^3 + 3xy^2 - x^3 = 3$. **(7 marks)**

Differentiation

32 a If $(1 + x)(2 + y) = x^2 + y^2$, find $\dfrac{dy}{dx}$ in terms of x and y. (4 marks)

   b Find the gradient of the curve $(1 + x)(2 + y) = x^2 + y^2$ at each of the two points where the curve meets the y-axis. (3 marks)

   c Show also that there are two points at which the tangents to this curve are parallel to the y-axis. (4 marks)

33 A curve has equation $7x^2 + 48xy - 7y^2 + 75 = 0$. A and B are two distinct points on the curve and at each of these points the gradient of the curve is equal to $\dfrac{2}{11}$. Use implicit differentiation to show that the straight line passing through A and B has equation $x + 2y = 0$. (6 marks)

34 Given that $y = x^x$, $x > 0$, $y > 0$, by taking logarithms show that
$$\dfrac{dy}{dx} = x^x(1 + \ln x)$$ (6 marks)

35 a Given that $a^x \equiv e^{kx}$, where a and k are constants, $a > 0$ and $x \in \mathbb{R}$, prove that $k = \ln a$. (2 marks)

   b Hence, using the derivative of e^{kx}, prove that when $y = 2^x$
$$\dfrac{dy}{dx} = 2^x \ln 2$$ (4 marks)

   c Hence deduce that the gradient of the curve with equation $y = 2^x$ at the point $(2, 4)$ is $\ln 16$. (3 marks)

36 A population P is growing at the rate of 9% each year and at time t years may be approximated by the formula
$$P = P_0(1.09)^t, \; t \geq 0$$
where P is regarded as a continuous function of t and P_0 is the population at time $t = 0$.

   a Find an expression for t in terms of P and P_0. (2 marks)

   b Find the time T years when the population has doubled from its value at $t = 0$, giving your answer to 3 significant figures. (4 marks)

   c Find, as a multiple of P_0, the rate of change of population $\dfrac{dP}{dt}$ at time $t = T$. (4 marks)

37 A curve C has equation
$$y = \ln(\sin x), \quad 0 < x < \pi$$

   a Find the stationary point of the curve C. (6 marks)

   b Show that the curve C is concave at all values of x in its given domain. (3 marks)

38 The mass of a radioactive substance t years after first being observed is modelled by the equation
$$m = 40e^{-0.244t}$$

a Find the mass of the substance nine months after it was first observed. **(2 marks)**

b Find $\dfrac{dm}{dt}$ **(2 marks)**

c With reference to the model, interpret the significance of the sign of the value of $\dfrac{dm}{dt}$ found in part **b**. **(1 mark)**

39 The curve C with equation $y = f(x)$ is shown in the diagram, where $f(x) = \dfrac{\cos 2x}{e^x}$, $0 \leqslant x \leqslant \pi$

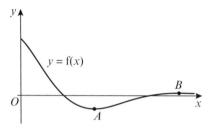

The curve has a local minimum at A and a local maximum at B.

a Show that the x-coordinates of A and B satisfy the equation $\tan 2x = -0.5$ and hence find the coordinates of A and B. **(6 marks)**

b Using your answer to part **a**, find the coordinates of the maximum and minimum turning points on the curve with equation $y = 2 + 4f(x - 4)$. **(3 marks)**

c Determine the values of x for which $f(x)$ is concave. **(5 marks)**

Challenge

The curve C has parametric equations
$$y = 2 \sin 2t, \quad x = 5 \cos\left(t + \dfrac{\pi}{12}\right), \quad 0 \leqslant t \leqslant 2\pi$$

a Find $\dfrac{dy}{dx}$ in terms of t.

b Find the coordinates of the points on C where $\dfrac{dy}{dx} = 0$.

c Find the coordinates of any points where the curve cuts or intersects the coordinate axes, and determine the gradient of the curve at these points.

d Find the coordinates of the points on C where $\dfrac{dx}{dy} = 0$.

e Hence sketch C.

Hint The points on C where $\dfrac{dx}{dy} = 0$ correspond to points where a tangent to the curve would be a vertical line.

Summary of key points

1. For small angles, measured in radians:
   - $\sin x \approx x$
   - $\cos x \approx 1 - \frac{1}{2}x^2$

2. - If $y = \sin kx$, then $\frac{dy}{dx} = k \cos kx$
   - If $y = \cos kx$, then $\frac{dy}{dx} = -k \sin kx$

3. - If $y = e^{kx}$, then $\frac{dy}{dx} = ke^{kx}$
   - If $y = \ln x$, then $\frac{dy}{dx} = \frac{1}{x}$

4. If $y = a^{kx}$, where k is a real constant and $a > 0$, then $\frac{dy}{dx} = a^{kx} k \ln a$

5. The **chain rule** is: $\frac{dy}{dx} = \frac{dy}{du} \times \frac{du}{dx}$
   where y is a function of u and u is another function of x.

6. The chain rule enables you to differentiate a function of a function. In general,
   - if $y = (f(x))^n$ then $\frac{dy}{dx} = n(f(x))^{n-1} f'(x)$
   - if $y = f(g(x))$ then $\frac{dy}{dx} = f'(g(x)) g'(x)$

7. $\frac{dy}{dx} = \frac{1}{\frac{dx}{dy}}$

8. The **product rule**:
   - If $y = uv$ then $\frac{dy}{dx} = u\frac{dv}{dx} + v\frac{du}{dx}$, where u and v are functions of x.
   - If $f(x) = g(x)h(x)$ then $f'(x) = g(x)h'(x) + h(x)g'(x)$

9. The **quotient rule**:
   - If $y = \frac{u}{v}$, then $\frac{dy}{dx} = \frac{v\frac{du}{dx} - u\frac{dv}{dx}}{v^2}$ where u and v are functions of x.
   - If $f(x) = \frac{g(x)}{h(x)}$, then $f'(x) = \frac{h(x)g'(x) - g(x)h'(x)}{(h(x))^2}$

10 • If $y = \tan kx$, then $\dfrac{dy}{dx} = k\sec^2 kx$

• If $y = \operatorname{cosec} kx$, then $\dfrac{dy}{dx} = -k\operatorname{cosec} kx \cot kx$

• If $y = \sec kx$, then $\dfrac{dy}{dx} = k\sec kx \tan kx$

• If $y = \cot kx$, then $\dfrac{dy}{dx} = -k\operatorname{cosec}^2 kx$

11 If x and y are given as functions of a parameter, t: $\dfrac{dy}{dx} = \dfrac{\frac{dy}{dt}}{\frac{dx}{dt}}$

12 • $\dfrac{d}{dx}(f(y)) = f'(y)\dfrac{dy}{dx}$

• $\dfrac{d}{dx}(y^n) = ny^{n-1}\dfrac{dy}{dx}$

• $\dfrac{d}{dx}(xy) = x\dfrac{dy}{dx} + y$

13 • The function f(x) is **concave** on a given interval if and only if f''(x) $\leq$ 0 for every value of x in that interval.

• The function f(x) is **convex** on a given interval if and only if f''(x) $\geq$ 0 for every value of x in that interval.

14 A **point of inflection** is a point at which f''(x) changes sign.

15 You can use the chain rule to connect rates of change in situations involving more than two variables.

Numerical methods

10

Objectives

After completing this chapter you should be able to:
- Locate roots of f(x) = 0 by considering changes of sign → pages 274–277
- Use iteration to find an approximation to the root of the equation f(x) = 0 → pages 278–282
- Use the Newton–Raphson procedure to find approximations to the solutions of equations of the form f(x) = 0 → pages 282–285
- Use numerical methods to solve problems in context → pages 286–289

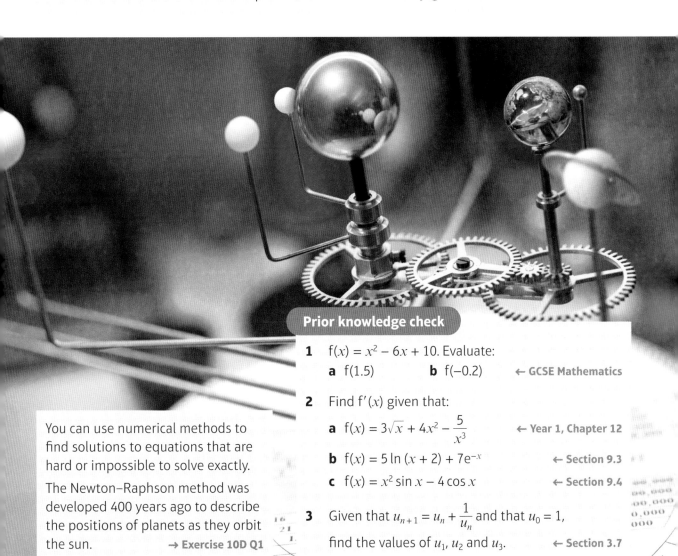

You can use numerical methods to find solutions to equations that are hard or impossible to solve exactly.

The Newton–Raphson method was developed 400 years ago to describe the positions of planets as they orbit the sun. → Exercise 10D Q1

Prior knowledge check

1. $f(x) = x^2 - 6x + 10$. Evaluate:
   a $f(1.5)$ b $f(-0.2)$ ← GCSE Mathematics

2. Find $f'(x)$ given that:
   a $f(x) = 3\sqrt{x} + 4x^2 - \dfrac{5}{x^3}$ ← Year 1, Chapter 12
   b $f(x) = 5\ln(x+2) + 7e^{-x}$ ← Section 9.3
   c $f(x) = x^2 \sin x - 4\cos x$ ← Section 9.4

3. Given that $u_{n+1} = u_n + \dfrac{1}{u_n}$ and that $u_0 = 1$, find the values of u_1, u_2 and u_3.   ← Section 3.7

Chapter 10

10.1 Locating roots

A root of a function is a value of x for which $f(x) = 0$. The graph of $y = f(x)$ will cross the x-axis at points corresponding to the roots of the function.

Links The following two things are identical:
- the roots of the function $f(x)$
- the roots of the equation $f(x) = 0$ ← Year 1, Section 2.3

You can sometimes show that a root exists within a given interval by showing that the function changes sign (from positive to negative, or vice versa) within the interval.

- **If the function $f(x)$ is continuous on the interval $[a, b]$ and $f(a)$ and $f(b)$ have opposite signs, then $f(x)$ has at least one root, x, which satisfies $a < x < b$.**

Notation Continuous means that the function does not 'jump' from one value to another. If the graph of the function has a vertical asymptote between a and b then the function is not continuous on $[a, b]$.

Example 1

The diagram shows a sketch of the curve $y = f(x)$, where $f(x) = x^3 - 4x^2 + 3x + 1$.

a Explain how the graph shows that $f(x)$ has a root between $x = 2$ and $x = 3$.

b Show that $f(x)$ has a root between $x = 1.4$ and $x = 1.5$.

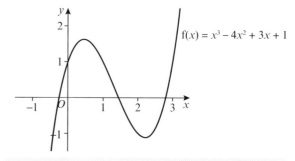

a The graph crosses the x-axis between $x = 2$ and $x = 3$. This means that a root of $f(x)$ lies between $x = 2$ and $x = 3$.

The graph of $y = f(x)$ crosses the x-axis whenever $f(x) = 0$.

b $f(1.4) = (1.4)^3 - 4(1.4)^2 + 3(1.4) + 1 = 0.104$
$f(1.5) = (1.5)^3 - 4(1.5)^2 + 3(1.5) + 1 = -0.125$
There is a change of sign between 1.4 and 1.5, and $f(x)$ is continuous on $[1.4, 1.5]$, so there is at least one root between $x = 1.4$ and $x = 1.5$.

$f(1.4) > 0$ and $f(1.5) < 0$, so there is a change of sign.

$f(x)$ changes sign in the interval $[1.4, 1.5]$, so $f(x)$ must equal zero within this interval.

There are three situations you need to watch out for when using the change of sign rule to locate roots. A change of sign does not necessarily mean there is exactly one root, and the absence of a sign change does not necessarily mean that a root does not exist in the interval.

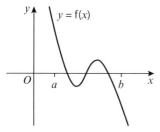

There are multiple roots within the interval $[a, b]$. In this case there is an **odd number** of roots

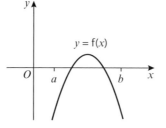

There are multiple roots within the interval $[a, b]$, but a sign change does not occur. In this case there is an **even number** of roots.

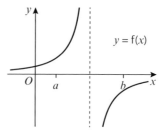

There is a vertical asymptote within interval $[a, b]$. A sign change does occur, but there is no root.

Numerical methods

Example 2

The graph of the function
f(x) = 54x³ − 225x² + 309x − 140 is shown in the diagram.
A student observes that f(1.1) and f(1.6) are both negative and states that f(x) has no roots in the interval (1.1, 1.6).
a Explain by reference to the diagram why the student is incorrect.
b Calculate f(1.3) and f(1.5) and use your answer to explain why there are at least 3 roots in the interval 1.1 < x < 1.7.

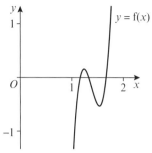

a The diagram shows that there could be two roots in the interval (1.1, 1.6).

> **Notation** The interval (1.1, 1.6) is the set of all real numbers, x, that satisfy $1.1 < x < 1.6$.

b f(1.1) = −0.476 < 0
f(1.3) = 0.088 > 0
f(1.5) = −0.5 < 0
f(1.7) = 0.352 > 0
There is a change of sign between 1.1 and 1.3, between 1.3 and 1.5 and between 1.5 and 1.7, and f(x) is continuous on all three intervals, so there are at least three roots in the interval 1.1 < x < 1.7.

Calculate the values of f(1.1), f(1.3), f(1.5) and f(1.7). Comment on the sign of each answer.

f(x) changes sign at least three times in the interval $1.1 < x < 1.7$ so f(x) must equal zero at least three times within this interval. You need to state that the function is continuous on each interval, but you do not need to show this.

Example 3

a Using the same axes, sketch the graphs of $y = \ln x$ and $y = \frac{1}{x}$. Explain how your diagram shows that the function $f(x) = \ln x - \frac{1}{x}$ has only one root.
b Show that this root lies in the interval $1.7 < x < 1.8$.
c Given that the root of f(x) is α, show that $\alpha = 1.763$ correct to 3 decimal places.

a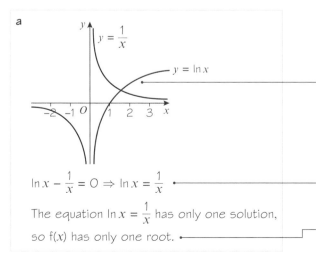

$\ln x - \frac{1}{x} = 0 \Rightarrow \ln x = \frac{1}{x}$

The equation $\ln x = \frac{1}{x}$ has only one solution, so f(x) has only one root.

Sketch $y = \ln x$ and $y = \frac{1}{x}$ on the same axes.
Notice that the curves do intersect.

f(x) has a root where f(x) = 0.

The curves meet at only one point, so there is only one value of x that satisfies the equation $\ln x = \frac{1}{x}$.

b $f(x) = \ln x - \dfrac{1}{x}$

$f(1.7) = \ln 1.7 - \dfrac{1}{1.7} = -0.0576...$

$f(1.8) = \ln 1.8 - \dfrac{1}{1.8} = 0.0322...$

There is a change of sign between 1.7 and 1.8, and $f(x)$ is continuous on [1.7, 1.8], so there is at least one root in the interval $1.7 < x < 1.8$.

c $f(1.7625) = -0.00064... < 0$
$f(1.7635) = 0.00024... > 0$
There is a change of sign in the interval [1.7625, 1.7635] so $1.7625 < \alpha < 1.7635$, so $\alpha = 1.763$ correct to 3 d.p.

Online Locate the root of $f(x) = \ln x - \dfrac{1}{x}$ using technology.

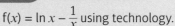

$f(1.7) < 0$ and $f(1.8) > 0$, so there is a change of sign.

You need to state that there is a change of sign in your conclusion, and that the function is continuous.

Problem-solving

To determine a root to a given degree of accuracy you need to show that it lies within a range of values that will all round to the given value.

Numbers in this range will round to 1.763 to 3 d.p.

1.762   1.7625   1.763   1.7635   1.764 x

Exercise 10A

1 Show that each of these functions has at least one root in the given interval.

a $f(x) = x^3 - x + 5, -2 < x < -1$

b $f(x) = x^2 - \sqrt{x} - 10, 3 < x < 4$

c $f(x) = x^3 - \dfrac{1}{x} - 2, -0.5 < x < -0.2$

d $f(x) = e^x - \ln x - 5, 1.65 < x < 1.75$

(E) **2** $f(x) = 3 + x^2 - x^3$

a Show that the equation $f(x) = 0$ has a root, α, in the interval [1.8, 1.9]. **(2 marks)**

b By considering a change of sign of $f(x)$ in a suitable interval, verify that $\alpha = 1.864$ correct to 3 decimal places. **(3 marks)**

(E) **3** $h(x) = \sqrt[3]{x} - \cos x - 1$, where x is in radians.

a Show that the equation $h(x) = 0$ has a root, α, between $x = 1.4$ and $x = 1.5$. **(2 marks)**

b By choosing a suitable interval, show that $\alpha = 1.441$ is correct to 3 decimal places. **(3 marks)**

(E) **4** $f(x) = \sin x - \ln x, x > 0$, where x is in radians.

a Show that $f(x) = 0$ has a root, α, in the interval [2.2, 2.3]. **(2 marks)**

b By considering a change of sign of $f(x)$ in a suitable interval, verify that $\alpha = 2.219$ correct to 3 decimal places. **(3 marks)**

(P) **5** $f(x) = 2 + \tan x, 0 < x < \pi$, where x is in radians.

a Show that $f(x)$ changes sign in the interval [1.5, 1.6].

b State with a reason whether $f(x)$ has a root in the interval [1.5, 1.6].

Numerical methods

6 A student observes that the function $f(x) = \frac{1}{x} + 2$, $x \neq 0$, has a change of sign on the interval $[-1, 1]$. The student writes:

> $y = f(x)$ has a vertical asymptote within this interval so even though there is a change of sign, f(x) has no roots in this interval.

By means of a sketch, or otherwise, explain why the student is incorrect.

7 $f(x) = (105x^3 - 128x^2 + 49x - 6) \cos 2x$, where x is in radians. The diagram shows a sketch of $y = f(x)$.
  a Calculate $f(0.2)$ and $f(0.8)$.
  b Use your answer to part **a** to make a conclusion about the number of roots of $f(x)$ in the interval $0.2 < x < 0.8$.
  c Further calculate $f(0.3)$, $f(0.4)$, $f(0.5)$, $f(0.6)$ and $f(0.7)$.
  d Use your answers to parts **a** and **c** to make an improved conclusion about the number of roots of $f(x)$ in the interval $0.2 < x < 0.8$.

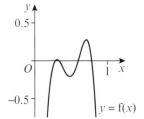

8 a Using the same axes, sketch the graphs of $y = e^{-x}$ and $y = x^2$.
  b Explain why the function $f(x) = e^{-x} - x^2$ has only one root.
  c Show that the function $f(x) = e^{-x} - x^2$ has a root between $x = 0.70$ and $x = 0.71$.

9 a On the same axes, sketch the graphs of $y = \ln x$ and $y = e^x - 4$.
  b Write down the number of roots of the equation $\ln x = e^x - 4$.
  c Show that the equation $\ln x = e^x - 4$ has a root in the interval $(1.4, 1.5)$.

10 $h(x) = \sin 2x + e^{4x}$
  a Show that there is a stationary point, α, of $y = h(x)$ in the interval $-0.9 < x < -0.8$. **(4 marks)**
  b By considering the change of sign of $h'(x)$ in a suitable interval, verify that $\alpha = -0.823$ correct to 3 decimal places. **(2 marks)**

11 a On the same axes, sketch the graphs of $y = \sqrt{x}$ and $y = \frac{2}{x}$ **(2 marks)**
  b With reference to your sketch, explain why the equation $\sqrt{x} = \frac{2}{x}$ has exactly one real root. **(1 mark)**
  c Given that $f(x) = \sqrt{x} - \frac{2}{x}$, show that the equation $f(x) = 0$ has a root r, where $1 < r < 2$. **(2 marks)**
  d Show that the equation $\sqrt{x} = \frac{2}{x}$ may be written in the form $x^p = q$, where p and q are integers to be found. **(2 marks)**
  e Hence write down the exact value of the root of the equation $\sqrt{x} - \frac{2}{x} = 0$. **(1 mark)**

12 $f(x) = x^4 - 21x - 18$
  a Show that there is a root of the equation $f(x) = 0$ in the interval $[-0.9, -0.8]$. **(3 marks)**
  b Find the coordinates of any stationary points on the graph $y = f(x)$. **(3 marks)**
  c Given that $f(x) = (x - 3)(x^3 + ax^2 + bx + c)$, find the values of the constants a, b and c. **(3 marks)**
  d Sketch the graph of $y = f(x)$. **(3 marks)**

10.2 Iteration

An iterative method can be used to find a value of x for which $f(x) = 0$. To perform an iterative procedure, it is usually necessary to manipulate the algebraic function first.

- **To solve an equation of the form $f(x) = 0$ by an iterative method, rearrange $f(x) = 0$ into the form $x = g(x)$ and use the iterative formula $x_{n+1} = g(x_n)$.**

Some iterations will **converge** to a root. This can happen in two ways. One way is that successive iterations get closer and closer to the root from the same direction. Graphically these iterations create a series of steps. The resulting diagram is sometimes referred to as a **staircase diagram**.

$f(x) = x^2 - x - 1$ can produce the iterative formula $x_{n+1} = \sqrt{x_n + 1}$ when $f(x) = 0$. Let $x_0 = 0.5$. Successive iterations produce the following staircase diagram.

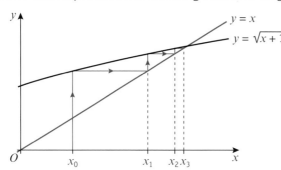

Read up from x_0 on the vertical axis to the curve $y = \sqrt{x + 1}$ to find x_1. You can read across to the line $y = x$ to 'map' this value back onto the x-axis. Repeating the process shows the values of x_n converging to the root of the equation $x = \sqrt{x + 1}$, which is also the root of $f(x)$.

The other way that an iteration converges is that successive iterations alternate being below the root and above the root. These iterations can still converge to the root and the resulting graph is sometimes called a **cobweb diagram**.

$f(x) = x^2 - x - 1$ can produce the iterative formula
$x_{n+1} = \dfrac{1}{x_n - 1}$ when $f(x) = 0$. Let $x_0 = -2$.

Watch out By rearranging the same function in different ways you can find different iterative formulae, which may converge differently.

Successive iterations produce the cobweb diagram, shown on the right.

Not all iterations or starting values converge to a root. When an iteration moves away from a root, often increasingly quickly, you say that it **diverges**.

$f(x) = x^2 - x - 1$ can produce the iterative formula
$x_{n+1} = x_n^2 - 1$ when $f(x) = 0$. Let $x_0 = 2$.

Successive iterations diverge from the root, as shown in the diagram.

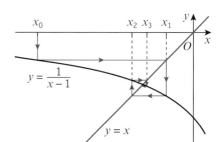

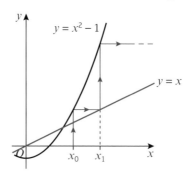

Numerical methods

Example 4

$f(x) = x^2 - 4x + 1$

a Show that the equation $f(x) = 0$ can be written as $x = 4 - \frac{1}{x}$, $x \neq 0$.

$f(x)$ has a root, α, in the interval $3 < x < 4$.

b Use the iterative formula $x_{n+1} = 4 - \frac{1}{x_n}$ with $x_0 = 3$ to find the value of x_1, x_2 and x_3.

a $f(x) = 0$
$x^2 - 4x + 1 = 0$
$x^2 = 4x - 1$ — Add $4x$ to each side and subtract 1 from each side.
$x = 4 - \frac{1}{x}$, $x \neq 0$ — Divide each term by x. This step is only valid if $x \neq 0$.

b $x_1 = 4 - \frac{1}{x_0} = 3.666666...$

$x_2 = 4 - \frac{1}{x_1} = 3.72727...$

$x_3 = 4 - \frac{1}{x_2} = 3.73170...$

Online Use the iterative formula to work out x_1, x_2 and x_3. You can use your calculator to find each value quickly.

Example 5

$f(x) = x^3 - 3x^2 - 2x + 5$

a Show that the equation $f(x) = 0$ has a root in the interval $3 < x < 4$.

b Use the iterative formula $x_{n+1} = \sqrt{\frac{x_n^3 - 2x_n + 5}{3}}$ to calculate the values of x_1, x_2 and x_3, giving your answers to 4 decimal places and taking:

  i $x_0 = 1.5$ **ii** $x_0 = 4$

a $f(3) = (3)^3 - 3(3)^2 - 2(3) + 5 = -1$
$f(4) = (4)^3 - 3(4)^2 - 2(4) + 5 = 13$
There is a change of sign in the interval $3 < x < 4$, and f is continuous, so there is a root of $f(x)$ in this interval. — The graph crosses the x-axis between $x = 3$ and $x = 4$.

b i $x_1 = \sqrt{\frac{x_0^3 - 2x_0 + 5}{3}} = 1.3385...$

$x_2 = \sqrt{\frac{x_1^3 - 2x_1 + 5}{3}} = 1.2544...$

$x_3 = \sqrt{\frac{x_2^3 - 2x_2 + 5}{3}} = 1.2200...$ — Each iteration gets closer to a root, so the sequence $x_0, x_1, x_2, x_3,...$ is **convergent**.

Chapter 10

ii $\quad x_1 = \sqrt{\dfrac{x_0^3 - 2x_0 + 5}{3}} = 4.5092...$

$x_2 = \sqrt{\dfrac{x_1^3 - 2x_1 + 5}{3}} = 5.4058...$

$x_3 = \sqrt{\dfrac{x_2^3 - 2x_2 + 5}{3}} = 7.1219...$

Online Explore the iterations graphically using technology.

Each iteration gets further from a root, so the sequence $x_0, x_1, x_2, x_3,...$ is **divergent**.

Exercise 10B

(P) **1** $f(x) = x^2 - 6x + 2$

    a Show that $f(x) = 0$ can be written as:

        i $x = \dfrac{x^2 + 2}{6}$ **ii** $x = \sqrt{6x - 2}$ **iii** $x = 6 - \dfrac{2}{x}$

    b Starting with $x_0 = 4$, use each iterative formula to find a root of the equation $f(x) = 0$. Round your answers to 3 decimal places.

    c Use the quadratic formula to find the roots to the equation $f(x) = 0$, leaving your answer in the form $a \pm \sqrt{b}$, where a and b are constants to be found.

(P) **2** $f(x) = x^2 - 5x - 3$

    a Show that $f(x) = 0$ can be written as:

        i $x = \sqrt{5x + 3}$ **ii** $x = \dfrac{x^2 - 3}{5}$

    b Let $x_0 = 5$. Show that each of the following iterative formulae gives different roots of $f(x) = 0$.

        i $x_{n+1} = \sqrt{5x_n + 3}$ **ii** $x_{n+1} = \dfrac{x_n^2 - 3}{5}$

(E/P) **3** $f(x) = x^2 - 6x + 1$

    a Show that the equation $f(x) = 0$ can be written as $x = \sqrt{6x - 1}$.   **(1 mark)**

    b Sketch on the same axes the graphs of $y = x$ and $y = \sqrt{6x - 1}$.   **(2 marks)**

    c Write down the number of roots of $f(x)$.   **(1 mark)**

    d Use your diagram to explain why the iterative formula $x_{n+1} = \sqrt{6x_n - 1}$ converges to a root of $f(x)$ when $x_0 = 2$.   **(1 mark)**

$f(x) = 0$ can also be rearranged to form the iterative formula $x_{n+1} = \dfrac{x_n^2 + 1}{6}$

    e By sketching a diagram, explain why the iteration diverges when $x_0 = 10$.   **(2 marks)**

(P) **4** $f(x) = xe^{-x} - x + 2$

    a Show that the equation $f(x) = 0$ can be written as $x = \ln\left|\dfrac{x}{x-2}\right|$, $x \neq 2$.

    $f(x)$ has a root, α, in the interval $-2 < x < -1$.

    b Use the iterative formula $x_{n+1} = \ln\left|\dfrac{x_n}{x_n - 2}\right|$, $x \neq 2$ with $x_0 = -1$ to find, to 2 decimal places, the values of x_1, x_2 and x_3.

Numerical methods

5 $f(x) = x^3 + 5x^2 - 2$

   a Show that $f(x) = 0$ can be written as:

     i $x = \sqrt[3]{2 - 5x^2}$ **ii** $x = \dfrac{2}{x^2} - 5$ **iii** $x = \sqrt{\dfrac{2 - x^3}{5}}$

   b Starting with $x_0 = 10$, use the iterative formula in part **a (ii)** to find a root of the equation $f(x) = 0$. Round your answer to 3 decimal places.

   c Starting with $x_0 = 1$, use the iterative formula in part **a (iii)** to find a different root of the equation $f(x) = 0$. Round your answer to 3 decimal places.

   d Explain why this iterative formulae cannot be used when $x_0 = 2$.

6 $f(x) = x^4 - 3x^3 - 6$

   a Show that the equation $f(x) = 0$ can be written as $x = \sqrt[3]{px^4 + q}$, where p and q are constants to be found. **(2 marks)**

   b Let $x_0 = 0$. Use the iterative formula $x_{n+1} = \sqrt[3]{px_n^4 + q}$, together with your values of p and q from part **a**, to find, to 3 decimal places, the values of x_1, x_2 and x_3. **(3 marks)**

   The root of $f(x) = 0$ is α.

   c By choosing a suitable interval, prove that $\alpha = -1.132$ to 3 decimal places. **(3 marks)**

7 $f(x) = 3\cos(x^2) + x - 2$

   a Show that the equation $f(x) = 0$ can be written as $x = \left(\arccos\left(\dfrac{2-x}{3}\right)\right)^{\frac{1}{2}}$ **(2 marks)**

   b Use the iterative formula $x_{n+1} = \left(\arccos\left(\dfrac{2-x_n}{3}\right)\right)^{\frac{1}{2}}$, $x_0 = 1$ to find, to 3 decimal places, the values of x_1, x_2 and x_3. **(3 marks)**

   c Given that $f(x) = 0$ has only one root, α, show that $\alpha = 1.1298$ correct to 4 decimal places. **(3 marks)**

8 $f(x) = 4\cot x - 8x + 3$, $0 < x < \pi$, where x is in radians.

   a Show that there is a root α of $f(x) = 0$ in the interval $[0.8, 0.9]$. **(2 marks)**

   b Show that the equation $f(x) = 0$ can be written in the form $x = \dfrac{\cos x}{2\sin x} + \dfrac{3}{8}$ **(3 marks)**

   c Use the iterative formula $x_{n+1} = \dfrac{\cos x_n}{2\sin x_n} + \dfrac{3}{8}$, $x_0 = 0.85$ to calculate the values of x_1, x_2 and x_3 giving your answers to 4 decimal places. **(3 marks)**

   d By considering the change of sign of $f(x)$ in a suitable interval, verify that $\alpha = 0.831$ correct to 3 decimal places. **(2 marks)**

9 $g(x) = e^{x-1} + 2x - 15$

   a Show that the equation $g(x) = 0$ can be written as $x = \ln(15 - 2x) + 1$, $x < \dfrac{15}{2}$ **(2 marks)**

The root of $g(x) = 0$ is α.

The iterative formula $x_{n+1} = \ln(15 - 2x_n) + 1$, $x_0 = 3$, is used to find a value for α.

b Calculate the values of x_1, x_2 and x_3 to 4 decimal places. **(3 marks)**

c By choosing a suitable interval, show that $\alpha = 3.16$ correct to 2 decimal places. **(3 marks)**

E/P 10 The diagram shows a sketch of part of the curve with equation $y = f(x)$, where $f(x) = xe^x - 4x$. The curve cuts the x-axis at the points A and B and has a minimum turning point at P, as shown in the diagram.

a Work out the coordinates of A and the coordinates of B. **(3 marks)**

b Find $f'(x)$. **(3 marks)**

c Show that the x-coordinate of P lies between 0.7 and 0.8. **(2 marks)**

d Show that the x-coordinate of P is the solution to the equation $x = \ln\left(\dfrac{4}{x+1}\right)$. **(3 marks)**

To find an approximation for the x-coordinate of P, the iterative formula $x_{n+1} = \ln\left(\dfrac{4}{x_n + 1}\right)$ is used.

e Let $x_0 = 0$. Find the values of x_1, x_2, x_3 and x_4. Give your answers to 3 decimal places. **(3 marks)**

10.3 The Newton–Raphson method

The Newton–Raphson method can be used to find numerical solutions to equations of the form $f(x) = 0$. You need to be able to differentiate $f(x)$ to use this method.

- **The Newton–Raphson formula is**

$$x_{n+1} = x_n - \frac{f(x_n)}{f'(x_n)}$$

Notation The Newton–Raphson method is sometimes called the Newton–Raphson **process** or the Newton–Raphson **procedure**.

The method uses tangent lines to find increasingly accurate approximations of a root. The value of x_{n+1} is the point at which the tangent to the graph at $(x_n, f(x_n))$ intersects the x-axis.

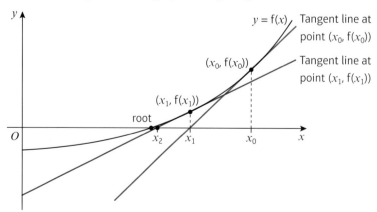

Numerical methods

If the starting value is not chosen carefully, the Newton–Raphson method can converge on a root very slowly, or can fail completely. If the initial value, x_0, is near a turning point or the derivative at this point, $f'(x_0)$, is close to zero, then the tangent at $(x_0, f(x_0))$ will intercept the x-axis a long way from x_0.

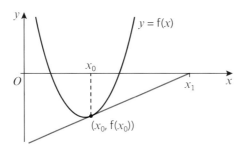

Because x_0 is close to a turning point the gradient of the tangent at $(x_0, f(x_0))$ is small, so it intercepts the x-axis a long way from x_0.

If any value, x_i, in the Newton–Raphson method is **at** a turning point, the method will fail because $f'(x_i) = 0$ and the formula would result in division by zero, which is not valid. Graphically, the tangent line will run parallel to the x-axis, therefore never intersecting.

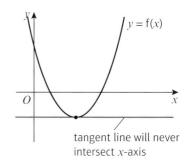

tangent line will never intersect x-axis

Example 6

The diagram shows part of the curve with equation $y = f(x)$, where $f(x) = x^3 + 2x^2 - 5x - 4$.

The point A, with x-coordinate p, is a stationary point on the curve.

The equation $f(x) = 0$ has a root, α, in the interval $1.8 < \alpha < 1.9$.

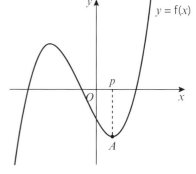

a Explain why $x_0 = p$ is not suitable to use as a first approximation to α when applying the Newton–Raphson method to $f(x)$.

b Using $x_0 = 2$ as a first approximation to α, apply the Newton–Raphson procedure twice to $f(x)$ to find a second approximation to α, giving your answer to 3 decimal places.

c By considering the change of sign in $f(x)$ over an appropriate interval, show that your answer to part **b** is accurate to 3 decimal places.

a It's a turning point, so $f'(p) = 0$, and you cannot divide by zero in the Newton–Raphson formula.

b $f'(x) = 3x^2 + 4x - 5$ — Use $\dfrac{d}{dx}(ax^n) = anx^{n-1}$

Using $x_0 = 2$

$$x_1 = x_0 - \frac{f(x_0)}{f'(x_0)}$$ — Use the Newton–Raphson process twice.

$$x_1 = 2 - \frac{2}{15}$$

$$x_1 = 1.8\dot{6}$$

$$x_2 = x_1 - \frac{f(x_1)}{f'(x_1)}$$

$$x_2 = 1.8\dot{6} - \frac{0.139\,8517}{12.919\,992}$$ — Substitute $x_1 = 1.8\dot{6}$ into the Newton–Raphson formula.

$$x_2 = 1.8558$$

$x_2 = 1.856$ to three decimal places

Use a spreadsheet package to find successive Newton–Raphson approximations.

c $f(1.8555) = -0.00348 < 0$,
$f(1.8565) = 0.00928 > 0$.
Sign change in interval $[1.8555, 1.8565]$ therefore $x = 1.856$ is accurate to 3 decimal places.

Online Explore how the Newton–Raphson method works graphically and algebraically using technology.

Exercise 10C

1 $f(x) = x^3 - 2x - 1$
  a Show that the equation $f(x) = 0$ has a root, α, in the interval $1 < \alpha < 2$.
  b Using $x_0 = 1.5$ as a first approximation to α, apply the Newton–Raphson procedure once to $f(x)$ to find a second approximation to α, giving your answer to 3 decimal places.

(E) 2 $f(x) = x^2 - \dfrac{4}{x} + 6x - 10,\ x \neq 0$.
  a Use differentiation to find $f'(x)$. **(2 marks)**
  The root, α, of the equation $f(x) = 0$ lies in the interval $[-0.4, -0.3]$.
  b Taking -0.4 as a first approximation to α, apply the Newton–Raphson process once to $f(x)$ to obtain a second approximation to α. Give your answer to 3 decimal places. **(4 marks)**

(E/P) 3 The diagram shows part of the curve with equation
$y = f(x)$, where $f(x) = x^{\frac{3}{2}} - e^{-x} + \dfrac{1}{\sqrt{x}} - 2,\ x > 0$.

The point A, with x-coordinate q, is a stationary point on the curve.
The equation $f(x) = 0$ has a root α in the interval $[1.2, 1.3]$.
  a Explain why $x_0 = q$ is not suitable to use as a first approximation when applying the Newton–Raphson method. **(1 mark)**
  b Taking $x_0 = 1.2$ as a first approximation to α, apply the Newton–Raphson process once to $f(x)$ to obtain a second approximation to α. Give your answer to 3 decimal places. **(4 marks)**

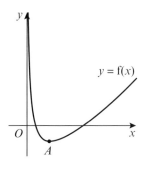

Numerical methods

4 $f(x) = 1 - x - \cos(x^2)$
 a Show that the equation $f(x) = 0$ has a root α in the interval $1.4 < \alpha < 1.5$. **(1 mark)**
 b Using $x_0 = 1.4$ as a first approximation to α, apply the Newton–Raphson procedure once to $f(x)$ to find a second approximation to α, giving your answer to 3 decimal places. **(4 marks)**
 c By considering the change of sign of $f(x)$ over an appropriate interval, show that your answer to part **b** is correct to 3 decimal places. **(2 marks)**

5 $f(x) = x^2 - \dfrac{3}{x^2}, \; x \geqslant 0$
 a Show that a root α of the equation $f(x) = 0$ lies in the interval $[1.3, 1.4]$. **(1 mark)**
 b Differentiate $f(x)$ to find $f'(x)$. **(2 marks)**
 c By taking 1.3 as a first approximation to α, apply the Newton–Raphson process once to $f(x)$ to obtain a second approximation to α. Give your answer to 3 decimal places. **(3 marks)**

6 $y = f(x)$, where $f(x) = x^2 \sin x - 2x + 1$. The points P, Q, and R are roots of the equation. The points A and B are stationary points, with x-coordinates a and b respectively.
 a Show that the curve has a root in each of the following intervals:
  i $[0.6, 0.7]$ **(1 mark)**
  ii $[1.2, 1.3]$ **(1 mark)**
  iii $[2.4, 2.5]$ **(1 mark)**
 b Explain why $x_0 = a$ is not suitable to use as a first approximation to α when applying the Newton–Raphson method to $f(x)$. **(1 mark)**
 c Using $x_0 = 2.4$ as a first approximation, apply the Newton–Raphson method to $f(x)$ to obtain a second approximation. Give your answer to 3 decimal places. **(4 marks)**

7 $f(x) = \ln(3x - 4) - x^2 + 10, \; x > \dfrac{4}{3}$
 a Show that $f(x) = 0$ has a root α in the interval $[3.4, 3.5]$. **(2 marks)**
 b Find $f'(x)$. **(2 marks)**
 c Taking 3.4 as a first approximation to α, apply the Newton–Raphson procedure once to $f(x)$ to obtain a second approximation for α, giving your answer to 3 decimal places. **(3 marks)**

Challenge

$f(x) = \dfrac{1}{5} + xe^{-x^2}$

The diagram shows a sketch of the curve $y = f(x)$. The curve has a horizontal asymptote at $y = \dfrac{1}{5}$.

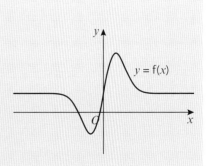

 a Prove that the Newton–Raphson method will fail to converge on a root of $f(x) = 0$ for all values of $x_0 > \dfrac{1}{\sqrt{2}}$

 b Taking -0.5 as a first approximation, use the Newton–Raphson method to find the root of $f(x) = 0$ that lies in the interval $[-1, 0]$, giving your answer to 3 d.p.

Chapter 10

10.4 Applications to modelling

You can use the techniques from this chapter to find solutions to models of real-life situations.

Example 7

The price of a car in £s, x years after purchase, is modelled by the function
$$f(x) = 15\,000\,(0.85)^x - 1000 \sin x, \quad x > 0$$

a Use the model to find the value, to the nearest hundred £s, of the car 10 years after purchase.
b Show that $f(x)$ has a root between 19 and 20.
c Find $f'(x)$.
d Taking 19.5 as a first approximation, apply the Newton–Raphson method once to $f(x)$ to obtain a second approximation for the time when the value of the car is zero. Give your answer to 3 decimal places.
e Criticise this model with respect to the value of the car as it gets older.

a $f(10) = 15\,000\,(0.85)^{10} - 1000 \sin 10$
   $= 3497.13...$

   After 10 years the value of the car is £3500 to the nearest £100.

   Substitute $x = 10$ into the $f(x)$.
   Unless otherwise stated, assume that angles are measured in radians.

b $f(19) = 15\,000\,(0.85)^{19} - 1000 \sin 19$
   $= 534.11... > 0$
   $f(20) = 15\,000\,(0.85)^{20} - 1000 \sin 20$
   $= -331.55... < 0$

   Substitute $x = 19$ and $x = 20$ into $f(x)$.

   There is a change of sign between 19 and 20, and $f(x)$ is continuous on this interval, so there is at least one root in the interval $19 < x < 20$.

   $f(x)$ changes sign in the interval $[19, 20]$, and $f(x)$ is continuous, so $f(x)$ must equal zero within this interval.

c $f'(x) = (15\,000)(0.85)^x (\ln 0.85) - 1000 \cos x$

d $f(19.5) = 15\,000\,(0.85)^{19.5} - 1000 \sin 19.5$
   $= 25.0693...$

   $f'(19.5) = (15\,000)(0.85)^{19.5}(\ln 0.85)$
   $\qquad - 1000 \cos 19.5 = -898.3009...$

   Use the fact that $\dfrac{d}{dx}(a^x) = a^x \ln a$.

   Substitute $x = 19.5$ into $f(x)$ and $f'(x)$.

   $x_{n+1} = x_n - \dfrac{f(x)}{f'(x)}$

   $= 19.5 - \dfrac{25.0693...}{-898.3009...}$

   $= 19.528$

   Apply the Newton–Raphson method once to obtain an improved second estimate.

e In reality, the car can never have a negative value so this model is not reasonable for cars that are approximately 20 or more years old.

Numerical methods

Exercise 10D

1 An astronomer is studying the motion of a planet moving along an elliptical orbit. She formulates the following model relating the angle moved at a given time, E radians, to the angle the planet would have moved if it had been travelling on a circular path, M radians:

$$M = E - 0.1 \sin E, \; E \geq 0$$

In order to predict the position of the planet at a particular time, the astronomer needs to find the value of E when $M = \dfrac{\pi}{6}$

a Show that this value of E is a root of the function $f(x) = x - 0.1 \sin x - k$ where k is a constant to be determined.

b Taking 0.6 as a first approximation, apply the Newton–Raphson procedure once to $f(x)$ to obtain a second approximation for the value of E when $M = \dfrac{\pi}{6}$

c By considering a change of sign on a suitable interval of $f(x)$, show that your answer to part **b** is correct to 3 decimal places.

2 The diagram shows a sketch of part of the curve with equation $v = f(t)$, where $f(t) = \left(10 - \tfrac{1}{2}(t+1)\right)\ln(t+1)$. The function models the velocity in m/s of a skier travelling in a straight line.

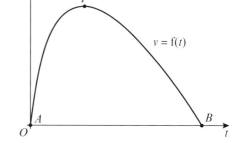

a Find the coordinates of A and B.

b Find $f'(t)$.

c Given that P is a stationary point on the curve, show that the t-coordinate of P lies between 5.8 and 5.9.

d Show that the t-coordinate of P is the solution to

$$t = \dfrac{20}{1 + \ln(t+1)} - 1$$

An approximation for the t-coordinate of P is found using the iterative formula

$$t_{n+1} = \dfrac{20}{1 + \ln(t_n + 1)} - 1$$

e Let $t_0 = 5$. Find the values of t_1, t_2 and t_3. Give your answers to 3 decimal places.

3 The depth of a stream is modelled by the function

$$d(x) = e^{-0.6x}(x^2 - 3x), \; 0 \leq x \leq 3$$

where x is the distance in metres from the left bank of the stream and $d(x)$ is the depth of the stream in metres.

The diagram shows a sketch of $y = d(x)$.

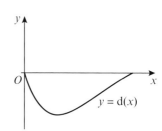

a Explain the condition $0 \leq x \leq 3$.

b Show that $d'(x) = -\frac{1}{5}e^{-0.6x}(ax^2 + bx + c)$, where a, b and c are constants to be found.

c Show that $d'(x) = 0$ can be written in the following ways:

i $x = \sqrt{\dfrac{19x - 15}{3}}$ **ii** $x = \dfrac{3x^2 + 15}{19}$ **iii** $x = \dfrac{19x - 15}{3x}$

d Let $x_0 = 1$. Show that only one of the three iterations converges to a stationary point of $y = d(x)$, and find the x-coordinate at this point correct to 3 decimal places.

e Find the maximum depth of the river in metres to 2 decimal places.

 4 Ed throws a ball for his dog. The vertical height of the ball is modelled by the function

$$h(t) = 40\sin\left(\frac{t}{10}\right) - 9\cos\left(\frac{t}{10}\right) - 0.5t^2 + 9, \ t \geq 0$$

$y = h(t)$ is shown in the diagram.

a Show that the t-coordinate of A is the solution to

$$t = \sqrt{18 + 80\sin\left(\frac{t}{10}\right) - 18\cos\left(\frac{t}{10}\right)}$$

(3 marks)

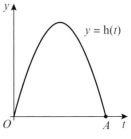

To find an approximation for the t-coordinate of A, the iterative formula

$$t_{n+1} = \sqrt{18 + 80\sin\left(\frac{t_n}{10}\right) - 18\cos\left(\frac{t_n}{10}\right)} \text{ is used.}$$

b Let $t_0 = 8$. Find the values of t_1, t_2, t_3 and t_4. Give your answers to 3 decimal places. (3 marks)

c Find $h'(t)$. (2 marks)

d Taking 8 as a first approximation, apply the Newton–Raphson method once to $h(t)$ to obtain a second approximation for the time when the height of the ball is zero. Give your answer to 3 decimal places. (3 marks)

e Hence suggest an improvement to the range of validity of the model. (2 marks)

 5 The annual number of non-violent crimes, in thousands, in a large town x years after the year 2000 is modelled by the function

$$c(x) = 5e^{-x} + 4\sin\left(\frac{x}{2}\right) + \frac{x}{2}, \ 0 \leq x \leq 10$$

The diagram shows the graph of $y = c(x)$.

a Find $c'(x)$. (2 marks)

b Show that the roots of the following equations correspond to the turning points on the graph of $y = c(x)$.

i $x = 2\arccos\left(\dfrac{5}{2}e^{-x} - \dfrac{1}{4}\right)$ (2 marks)

ii 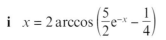 (2 marks)

288

c Let $x_0 = 3$ and $x_{n+1} = 2\arccos\left(\dfrac{5}{2}e^{-x_n} - \dfrac{1}{4}\right)$. Find the values of x_1, x_2, x_3 and x_4. Give your answers to 3 decimal places. **(3 marks)**

d Let $x_0 = 1$ and $x_{n+1} = \ln\left(\dfrac{10}{4\cos\left(\dfrac{x_n}{2}\right) + 1}\right)$. Find the values of x_1, x_2, x_3 and x_4. Give your answers to 3 decimal places. **(3 marks)**

A councillor states that the number of non-violent crimes in the town was increasing between October 2000 and June 2003.

e State, with reasons whether the model supports this claim. **(2 marks)**

Mixed exercise 10

E/P **1** $f(x) = x^3 - 6x - 2$

  a Show that the equation $f(x) = 0$ can be written in the form $x = \pm\sqrt{a + \dfrac{b}{x}}$, and state the values of the integers a and b. **(2 marks)**

  $f(x) = 0$ has one positive root, α.

  The iterative formula $x_{n+1} = \sqrt{a + \dfrac{b}{x_n}}$, $x_0 = 2$ is used to find an approximate value for α.

  b Calculate the values of x_1, x_2, x_3 and x_4 to 4 decimal places. **(3 marks)**

  c By choosing a suitable interval, show that $\alpha = 2.602$ is correct to 3 decimal places. **(3 marks)**

E/P **2** $f(x) = \dfrac{1}{4-x} + 3$

  a Calculate $f(3.9)$ and $f(4.1)$. **(2 marks)**

  b Explain why the equation $f(x) = 0$ does not have a root in the interval $3.9 < x < 4.1$. **(2 marks)**

  The equation $f(x) = 0$ has a single root, α.

  c Use algebra to find the exact value of α. **(2 marks)**

E/P **3** $p(x) = 4 - x^2$ and $q(x) = e^x$.

  a On the same axes, sketch the curves of $y = p(x)$ and $y = q(x)$. **(2 marks)**

  b State the number of positive roots and the number of negative roots of the equation $x^2 + e^x - 4 = 0$. **(1 mark)**

  c Show that the equation $x^2 + e^x - 4 = 0$ can be written in the form $x = \pm(4 - e^x)^{\frac{1}{2}}$ **(2 marks)**

  The iterative formula $x_{n+1} = -(4 - e^{x_n})^{\frac{1}{2}}$, $x_0 = -2$, is used to find an approximate value for the negative root.

  d Calculate the values of x_1, x_2, x_3 and x_4 to 4 decimal places. **(3 marks)**

  e Explain why the starting value $x_0 = 1.4$ will not produce a valid result with this formula. **(2 marks)**

Chapter 10

4 $g(x) = x^5 - 5x - 6$

a Show that $g(x) = 0$ has a root, α, between $x = 1$ and $x = 2$. **(2 marks)**

b Show that the equation $g(x) = 0$ can be written as $x = (px + q)^{\frac{1}{r}}$, where p, q and r are integers to be found. **(2 marks)**

The iterative formula $x_{n+1} = (px + q)^{\frac{1}{r}}$, $x_0 = 1$ is used to find an approximate value for α.

c Calculate the values of x_1, x_2 and x_3 to 4 decimal places. **(3 marks)**

d By choosing a suitable interval, show that $\alpha = 1.708$ is correct to 3 decimal places. **(3 marks)**

5 $g(x) = x^2 - 3x - 5$

a Show that the equation $g(x) = 0$ can be written as $x = \sqrt{3x + 5}$. **(1 mark)**

b Sketch on the same axes the graphs of $y = x$ and $y = \sqrt{3x + 5}$. **(2 marks)**

c Use your diagram to explain why the iterative formula $x_{n+1} = \sqrt{3x_n + 5}$ converges to a root of $g(x)$ when $x_0 = 1$. **(1 mark)**

$g(x) = 0$ can also be rearranged to form the iterative formula $x_{n+1} = \dfrac{x_n^2 - 5}{3}$

d With reference to a diagram, explain why this iterative formula diverges when $x_0 = 7$. **(3 marks)**

6 $f(x) = 5x - 4\sin x - 2$, where x is in radians.

a Show that $f(x) = 0$ has a root, α, between $x = 1.1$ and $x = 1.15$. **(2 marks)**

b Show that $f(x) = 0$ can be written as $x = p \sin x + q$, where p and q are rational numbers to be found. **(2 marks)**

c Starting with $x_0 = 1.1$, use the iterative formula $x_{n+1} = p \sin x_n + q$ with your values of p and q to calculate the values of x_1, x_2, x_3 and x_4 to 3 decimal places. **(3 marks)**

7 a On the same axes, sketch the graphs of $y = \dfrac{1}{x}$ and $y = x + 3$. **(2 marks)**

b Write down the number of roots of the equation $\dfrac{1}{x} = x + 3$. **(1 mark)**

c Show that the positive root of the equation $\dfrac{1}{x} = x + 3$ lies in the interval (0.30, 0.31). **(2 marks)**

d Show that the equation $\dfrac{1}{x} = x + 3$ may be written in the form $x^2 + 3x - 1 = 0$. **(2 marks)**

e Use the quadratic formula to find the positive root of the equation $x^2 + 3x - 1 = 0$ to 3 decimal places. **(2 marks)**

8 $g(x) = x^3 - 7x^2 + 2x + 4$

a Find $g'(x)$. **(2 marks)**

A root α of the equation $g(x) = 0$ lies in the interval $[6.5, 6.7]$.

b Taking 6.6 as a first approximation to α, apply the Newton–Raphson process once to $g(x)$ to obtain a second approximation to α. Give your answer to 3 decimal places. **(4 marks)**

c Given that g(1) = 0, find the exact value of the other two roots of g(x). (3 marks)
d Calculate the percentage error of your answer in part **b**. (2 marks)

E/P 9 $f(x) = 2\sec x + 2x - 3$, $-\dfrac{\pi}{2} < x < \dfrac{\pi}{2}$ where x is in radians.

a Show that $f(x) = 0$ has a solution, α, in the interval $0.4 < x < 0.5$. (2 marks)

b Taking 0.4 as a first approximation to α, apply the Newton–Raphson process once to $f(x)$ to obtain a second approximation to α. Give your answer to 3 decimal places. (4 marks)

c Show that $x = -1.190$ is a different solution, β, of $f(x) = 0$ correct to 3 decimal places. (2 marks)

E/P 10 $f(x) = e^{0.8x} - \dfrac{1}{3-2x}$, $x \neq \dfrac{3}{2}$

a Show that the equation $f(x) = 0$ can be written as $x = 1.5 - 0.5e^{-0.8x}$. (3 marks)

b Use the iterative formula $x_{n+1} = 1.5 - 0.5e^{-0.8x_n}$ with $x_0 = 1.3$ to obtain x_1, x_2 and x_3.
Hence write down one root of $f(x) = 0$ correct to 3 decimal places. (2 marks)

c Show that the equation $f(x) = 0$ can be written in the form $x = p\ln(3 - 2x)$, stating the value of p. (3 marks)

d Use the iterative formula $x_{n+1} = p\ln(3 - 2x_n)$ with $x_0 = -2.6$ and the value of p found in part **c** to obtain x_1, x_2 and x_3. Hence write down a second root of $f(x) = 0$ correct to 2 decimal places. (2 marks)

E/P 11 a By writing $y = x^x$ in the form $\ln y = x\ln x$, show that $\dfrac{dy}{dx} = x^x(\ln x + 1)$. (4 marks)

b Show that the function $f(x) = x^x - 2$ has a root, α, in the interval $[1.4, 1.6]$. (2 marks)

c Taking $x_0 = 1.5$ as a first approximation to α, apply the Newton–Raphson procedure once to obtain a second approximation to α, giving your answer to 4 decimal places. (4 marks)

d By considering a change of sign of $f(x)$ over a suitable interval, show that $\alpha = 1.5596$, correct to 4 decimal places. (3 marks)

E/P 12 The diagram shows part of the curve with equation $y = f(x)$, where $f(x) = \cos(4x) - \frac{1}{2}x$. The curve crosses the x-axis at A, C and D and has a local minimum at B.

a Show that the curve has a root in the interval $[1.3, 1.4]$. (2 marks)

b Use differentiation to find the coordinates of point B. Write each coordinate correct to 3 decimal places. (3 marks)

c Using the iterative formula $x_{n+1} = \dfrac{1}{4}\arccos\left(\dfrac{1}{2}x_n\right)$,

with $x_0 = 0.4$, find the values of x_1, x_2, x_3 and x_4.
Give your answers to 4 decimal places. (3 marks)

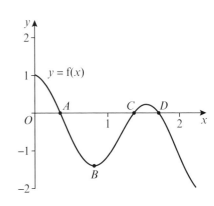

d Using $x_0 = 1.7$ as a first approximation to the root at D, apply the Newton–Raphson procedure once to f(x) to find a second approximation to the root, giving your answer to 3 decimal places. **(4 marks)**

e By considering the change of sign of f(x) over an appropriate interval, show that the answer to part **d** is accurate to 3 decimal places. **(2 marks)**

Challenge

f(x) = $x^6 + x^3 - 7x^2 - x + 3$

The diagram shows a sketch of $y = $ f(x). Points A and B are the points of inflection on the curve.

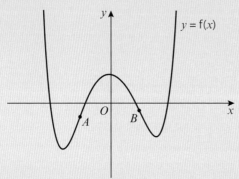

a Show that equation f''(x) = 0 can be written as:

  i $x = \dfrac{7 - 15x^4}{3}$ **ii** $x = \dfrac{7}{15x^3 + 3}$ **iii** $x = \sqrt[4]{\dfrac{7 - 3x}{15}}$

b By choosing a suitable iterative formula and starting value, find an approximation for the x-coordinate of B, correct to 3 decimal places.

c Explain why you cannot use the same iterative formula to find an approximation for the x-coordinate of A.

d Use the Newton–Raphson method to find an estimate for the x-coordinate of A, correct to 3 decimal places.

Summary of key points

1 If the function f(x) is continuous on the interval [a, b] and f(a) and f(b) have opposite signs, then f(x) has at least one root, x, which satisfies $a < x < b$.

2 To solve an equation of the form f(x) = 0 by an iterative method, rearrange f(x) = 0 into the form $x = $ g(x) and use the iterative formula $x_{n+1} = $ g(x_n).

3 The Newton–Raphson formula for approximating the roots of a function f(x) is

$$x_{n+1} = x_n - \dfrac{f(x_n)}{f'(x_n)}$$

Integration

11

Objectives

After completing this chapter you should be able to:

- Integrate standard mathematical functions including trigonometric and exponential functions and use the reverse of the chain rule to integrate functions of the form f($ax + b$) → pages 294–298
- Use trigonometric identities in integration → pages 298–300
- Use the reverse of the chain rule to integrate more complex functions → pages 300–303
- Integrate functions by making a substitution, using integration by parts and using partial fractions → pages 303–313
- Use integration to find the area under a curve → pages 313–317
- Use the trapezium rule to approximate the area under a curve. → pages 317–322
- Solve simple differential equations and model real-life situations with differential equations → pages 322–329

Prior knowledge check

1. Differentiate:

   a $(2x - 7)^6$ b $\sin 5x$

   c $e^{\frac{x}{3}}$ ← Sections 9.1, 9.2, 9.3

2. Given $f(x) = 8x^{\frac{1}{2}} - 6x^{-\frac{1}{2}}$

   a find $\int f(x)\, dx$

   b find $\int_4^9 f(x)\, dx$ ← Year 1, Chapter 13

3. Write $\dfrac{3x + 22}{(4x - 1)(x + 3)}$ as partial fractions.   ← Section 1.3

4. Find the area of the region R bounded by the curve $y = x^2 + 1$, the x-axis and the lines $x = -1$ and $x = 2$.

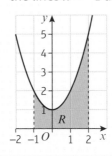

← Year 1, Chapter 13

Integration can be used to solve differential equations. Archaeologists use differential equations to estimate the age of fossilised plants and animals.   → Exercise 11K Q9

Chapter 11

11.1 Integrating standard functions

Integration is the inverse of differentiation. You can use your knowledge of derivatives to integrate familiar functions.

① $\int x^n \, dx = \dfrac{x^{n+1}}{n+1} + c$

Watch out This is true for all values of n except -1.

② $\int e^x \, dx = e^x + c$

③ $\int \dfrac{1}{x} \, dx = \ln|x| + c$

Notation When finding $\int \dfrac{1}{x} \, dx$ it is usual to write the answer as $\ln|x| + c$. The modulus sign removes difficulties that could arise when evaluating the integral for negative values of x.

④ $\int \cos x \, dx = \sin x + c$

⑤ $\int \sin x \, dx = -\cos x + c$

Links For example, if $y = \cos x$ then $\dfrac{dy}{dx} = -\sin x$.
This means that $\int (-\sin x) \, dx = \cos x + c$ and hence $\int \sin x \, dx = -\cos x + c$. ← Section 9.1

⑥ $\int \sec^2 x \, dx = \tan x + c$

⑦ $\int \cosec x \cot x \, dx = -\cosec x + c$

⑧ $\int \cosec^2 x \, dx = -\cot x + c$

⑨ $\int \sec x \tan x \, dx = \sec x + c$

Example 1

Find the following integrals.

a $\int \left(2\cos x + \dfrac{3}{x} - \sqrt{x}\right) dx$ **b** $\int \left(\dfrac{\cos x}{\sin^2 x} - 2e^x\right) dx$

a $\int 2\cos x \, dx = 2\sin x + c$ — Integrate each term separately.

$\int \dfrac{3}{x} \, dx = 3\ln|x| + c$ — Use ④.

$\int \sqrt{x} \, dx = \int x^{\frac{1}{2}} \, dx = \dfrac{2}{3} x^{\frac{3}{2}} + c$ — Use ③.

So $\int \left(2\cos x + \dfrac{3}{x} - \sqrt{x}\right) dx$ — Use ①.

$= 2\sin x + 3\ln|x| - \dfrac{2}{3} x^{\frac{3}{2}} + c$

This is an indefinite integral so don't forget the $+c$.

b $\dfrac{\cos x}{\sin^2 x} = \dfrac{\cos x}{\sin x} \times \dfrac{1}{\sin x} = \cot x \cosec x$

Look at the list of integrals of standard functions and express the integrand in terms of these standard functions.

$\int (\cot x \cosec x) \, dx = -\cosec x + c$

$\int 2e^x \, dx = 2e^x + c$

Remember the minus sign.

So $\int \left(\dfrac{\cos x}{\sin^2 x} - 2e^x \, dx\right)$

$= -\cosec x - 2e^x + c$

Integration

Example 2

Given that a is a positive constant and
$\int_a^{3a} \frac{2x+1}{x}\,dx = \ln 12$, find the exact value of a.

Problem-solving

Integrate as normal and write the limits as a and $3a$. Substitute these limits into your integral to get an expression in a and set this equal to $\ln 12$. Solve the resulting equation to find the value of a.

$$\int_a^{3a} \frac{2x+1}{x}\,dx$$
$$= \int_a^{3a}\left(2 + \frac{1}{x}\right)dx$$

Separate the terms by dividing by x, then integrate term by term.

$$= [2x + \ln x]_a^{3a}$$
$$= (6a + \ln 3a) - (2a + \ln a)$$

Remember the limits are a and $3a$.

$$= 4a + \ln\left(\frac{3a}{a}\right)$$

Substitute $3a$ and a into the integrated expression.

$$= 4a + \ln 3$$

So, $4a + \ln 3 = \ln 12$

$4a = \ln 12 - \ln 3$

Use the laws of logarithms: $\ln a - \ln b = \ln\left(\frac{a}{b}\right)$

$4a = \ln 4$

$a = \frac{1}{4}\ln 4$

$\ln 12 - \ln 3 = \ln\left(\frac{12}{3}\right) = \ln 4$

Exercise 11A

Online Use your calculator to check your value of a using numerical integration.

1 Integrate the following with respect to x.

  a $3\sec^2 x + \frac{5}{x} + \frac{2}{x^2}$
  b $5e^x - 4\sin x + 2x^3$
  c $2(\sin x - \cos x + x)$
  d $3\sec x \tan x - \frac{2}{x}$
  e $5e^x + 4\cos x - \frac{2}{x^2}$
  f $\frac{1}{2x} + 2\operatorname{cosec}^2 x$
  g $\frac{1}{x} + \frac{1}{x^2} + \frac{1}{x^3}$
  h $e^x + \sin x + \cos x$
  i $2\operatorname{cosec} x \cot x - \sec^2 x$
  j $e^x + \frac{1}{x} - \operatorname{cosec}^2 x$

2 Find the following integrals.

  a $\int\left(\frac{1}{\cos^2 x} + \frac{1}{x^2}\right)dx$
  b $\int\left(\frac{\sin x}{\cos^2 x} + 2e^x\right)dx$
  c $\int\left(\frac{1 + \cos x}{\sin^2 x} + \frac{1+x}{x^2}\right)dx$
  d $\int\left(\frac{1}{\sin^2 x} + \frac{1}{x}\right)dx$
  e $\int \sin x(1 + \sec^2 x)\,dx$
  f $\int \cos x(1 + \operatorname{cosec}^2 x)\,dx$
  g $\int \operatorname{cosec}^2 x(1 + \tan^2 x)\,dx$
  h $\int \sec^2 x(1 - \cot^2 x)\,dx$
  i $\int \sec^2 x(1 + e^x \cos^2 x)\,dx$
  j $\int\left(\frac{1 + \sin x}{\cos^2 x} + \cos^2 x \sec x\right)dx$

3 Evaluate the following. Give your answers as exact values.

  a $\int_3^7 2e^x\,dx$
  b $\int_1^6 \frac{1+x}{x^3}\,dx$
  c $\int_{\frac{\pi}{2}}^{\pi} -5\sin x\,dx$
  d $\int_{-\frac{\pi}{4}}^{0} \sec x(\sec x + \tan x)\,dx$

Watch out When applying limits to integrated trigonometric functions, always work in radians.

Chapter 11

4 Given that a is a positive constant and $\int_a^{2a} \dfrac{3x-1}{x}\,dx = 6 + \ln\left(\dfrac{1}{2}\right)$, find the exact value of a. **(4 marks)**

5 Given that a is a positive constant and $\int_{\ln 1}^{\ln a} e^x + e^{-x}\,dx = \dfrac{48}{7}$, find the exact value of a. **(4 marks)**

6 Given $\int_2^b (3e^x + 6e^{-2x})\,dx = 0$, find the value of b. **(4 marks)**

7 $f(x) = \dfrac{1}{8}x^{\frac{3}{2}} - \dfrac{4}{x},\ x > 0$

  a Solve the equation $f(x) = 0$. **(2 marks)**

  b Find $\int f(x)\,dx$. **(2 marks)**

  c Evaluate $\int_1^4 f(x)\,dx$, giving your answer in the form $p + q\ln r$, where p, q and r are rational numbers. **(3 marks)**

11.2 Integrating f(ax + b)

If you know the integral of a function $f(x)$ you can integrate a function of the form $f(ax + b)$ using the reverse of the chain rule for differentiation.

Example 3

Find the following integrals.

a $\int \cos(2x + 3)\,dx$ **b** $\int e^{4x+1}\,dx$ **c** $\int \sec^2 3x\,dx$

---

a Consider $y = \sin(2x + 3)$:

$\dfrac{dy}{dx} = \cos(2x + 3) \times 2$

So $\int \cos(2x + 3)\,dx = \dfrac{1}{2}\sin(2x + 3) + c$

Integrating $\cos x$ gives $\sin x$, so try $\sin(2x + 3)$.

Use the chain rule. Remember to multiply by the derivative of $2x + 3$ which is 2.

This is 2 times the required expression so you need to divide $\sin(2x + 3)$ by 2.

b Consider $y = e^{4x+1}$:

$\dfrac{dy}{dx} = e^{4x+1} \times 4$

So $\int e^{4x+1}\,dx = \dfrac{1}{4}e^{4x+1} + c$

The integral of e^x is e^x, so try e^{4x+1}.

This is 4 times the required expression so you divide by 4.

c Consider $y = \tan 3x$:

$\dfrac{dy}{dx} = \sec^2 3x \times 3$

So $\int \sec^2 3x\,dx = \dfrac{1}{3}\tan 3x + c$

Recall ⑥. Let $y = \tan 3x$ and differentiate using the chain rule. This is 3 times the required expression so you divide by 3.

Integration

In general:

- $\int f'(ax+b)\,dx = \dfrac{1}{a}f(ax+b) + c$

Watch out You cannot use this method to integrate an expression such as $\cos(2x^2+3)$ since it is not in the form $f(ax+b)$.

Example 4

Find the following integrals:

a $\displaystyle\int \dfrac{dx}{3x+2}$
b $\displaystyle\int (2x+3)^4\,dx$

Notation $\displaystyle\int \dfrac{dx}{3x+2}$ is another way of writing $\displaystyle\int \dfrac{1}{3x+2}\,dx$

a Consider $y = \ln(3x+2)$ — Integrating $\dfrac{1}{x}$ gives $\ln|x|$ so try $\ln(3x+2)$.

So $\dfrac{dy}{dx} = \dfrac{1}{3x+2} \times 3$

So $\displaystyle\int \dfrac{1}{3x+2}\,dx = \dfrac{1}{3}\ln|3x+2| + c$

The 3 comes from the chain rule. It is 3 times the required expression, so divide by 3.

b Consider $y = (2x+3)^5$ — To integrate $(ax+b)^n$ try $(ax+b)^{n+1}$.

So $\dfrac{dy}{dx} = 5 \times (2x+3)^4 \times 2$

$= 10 \times (2x+3)^4$

So $\displaystyle\int (2x+3)^4\,dx = \dfrac{1}{10}(2x+3)^5 + c$

The 5 comes from the exponent and the 2 comes from the chain rule.

This answer is 10 times the required expression, so divide by 10.

Exercise 11B

1 Integrate the following:

a $\sin(2x+1)$
b $3e^{2x}$

c $4e^{x+5}$
d $\cos(1-2x)$

e $\operatorname{cosec}^2 3x$
f $\sec 4x \tan 4x$

g $3\sin\left(\tfrac{1}{2}x+1\right)$
h $\sec^2(2-x)$
i $\operatorname{cosec} 2x \cot 2x$
j $\cos 3x - \sin 3x$

Hint For part **a** consider $y = \cos(2x+1)$. You do not need to write out this step once you are confident with using this method.

2 Find the following integrals.

a $\displaystyle\int \left(e^{2x} - \tfrac{1}{2}\sin(2x-1)\right)dx$
b $\displaystyle\int (e^x+1)^2\,dx$

c $\displaystyle\int \sec^2 2x(1+\sin 2x)\,dx$
d $\displaystyle\int \dfrac{3-2\cos\tfrac{1}{2}x}{\sin^2 \tfrac{1}{2}x}\,dx$

e $\displaystyle\int \left(e^{3-x} + \sin(3-x) + \cos(3-x)\right)dx$

3 Integrate the following:

a $\dfrac{1}{2x+1}$
b $\dfrac{1}{(2x+1)^2}$
c $(2x+1)^2$
d $\dfrac{3}{4x-1}$

e $\dfrac{3}{1-4x}$
f $\dfrac{3}{(1-4x)^2}$
g $(3x+2)^5$
h $\dfrac{3}{(1-2x)^3}$

4 Find the following integrals.

a $\int \left(3\sin(2x+1) + \dfrac{4}{2x+1}\right) dx$

b $\int (e^{5x} + (1-x)^5) \, dx$

c $\int \left(\dfrac{1}{\sin^2 2x} + \dfrac{1}{1+2x} + \dfrac{1}{(1+2x)^2}\right) dx$

d $\int \left((3x+2)^2 + \dfrac{1}{(3x+2)^2}\right) dx$

5 Evaluate:

a $\int_{\frac{\pi}{4}}^{\frac{3\pi}{4}} \cos(\pi - 2x) \, dx$

b $\int_{\frac{1}{2}}^{1} \dfrac{12}{(3-2x)^4} \, dx$

c $\int_{\frac{2\pi}{9}}^{\frac{5\pi}{18}} \sec^2(\pi - 3x) \, dx$

d $\int_{2}^{3} \dfrac{5}{7-2x} \, dx$

E/P **6** Given $\int_{3}^{b} (2x-6)^2 \, dx = 36$, find the value of b. **(4 marks)**

E/P **7** Given $\int_{e^2}^{e^8} \dfrac{dx}{kx} = \dfrac{1}{4}$, find the value of k. **(4 marks)**

E/P **8** Given $\int_{\frac{\pi}{4k}}^{\frac{\pi}{3k}} (1 - \pi \sin kx) \, dx = \pi(7 - 6\sqrt{2})$, find the exact value of k. **(7 marks)**

Problem-solving

Calculate the value of the indefinite integral in terms of k and solve the resulting equation.

Challenge

Given $\int_{5}^{11} \dfrac{1}{ax+b} \, dx = \dfrac{1}{a} \ln\left(\dfrac{41}{17}\right)$, and that a and b are integers with $0 < a < 10$, find two different pairs of values for a and b.

11.3 Using trigonometric identities

- **Trigonometric identities can be used to integrate expressions. This allows an expression that cannot be integrated to be replaced by an identical expression that can be integrated.**

Links Make sure you are familiar with the standard trigonometric identities. The list of identities in the summary of Chapter 7 will be useful. ← page 196

Example 5

Find $\int \tan^2 x \, dx$

Since $\sec^2 x \equiv 1 + \tan^2 x$

$\tan^2 x \equiv \sec^2 x - 1$

So $\int \tan^2 x \, dx = \int (\sec^2 x - 1) \, dx$

$= \int \sec^2 x \, dx - \int 1 \, dx$

$= \tan x - x + c$

You cannot integrate $\tan^2 x$ but you can integrate $\sec^2 x$ directly.

Using ⑥.

Example 6

Show that $\int_{\pi/12}^{\pi/8} \sin^2 x \, dx = \frac{\pi}{48} + \frac{1-\sqrt{2}}{8}$

You cannot integrate $\sin^2 x$ directly. Use the trigonometric identity to write it in terms of $\cos 2x$.

Recall $\cos 2x \equiv 1 - 2\sin^2 x$

So $\sin^2 x \equiv \frac{1}{2}(1 - \cos 2x)$

So $\int_{\pi/12}^{\pi/8} \sin^2 x \, dx = \int_{\pi/12}^{\pi/8} \left(\frac{1}{2} - \frac{1}{2}\cos 2x\right) dx$

$= \left[\frac{1}{2}x - \frac{1}{4}\sin 2x\right]_{\pi/12}^{\pi/8}$

$= \left(\frac{\pi}{16} - \frac{1}{4}\sin\left(\frac{\pi}{4}\right)\right) - \left(\frac{\pi}{24} - \frac{1}{4}\sin\left(\frac{\pi}{6}\right)\right)$

$= \left(\frac{\pi}{16} - \frac{1}{4}\left(\frac{\sqrt{2}}{2}\right)\right) - \left(\frac{\pi}{24} - \frac{1}{4}\left(\frac{1}{2}\right)\right)$

$= \left(\frac{\pi}{16} - \frac{\pi}{24}\right) + \frac{1}{4}\left(\frac{1}{2} - \frac{\sqrt{2}}{2}\right)$

$= \left(\frac{3\pi}{48} - \frac{2\pi}{48}\right) + \frac{1-\sqrt{2}}{8}$

$= \frac{\pi}{48} + \frac{1-\sqrt{2}}{8}$

Use the reverse chain rule. If $y = \sin 2x$, $\frac{dy}{dx} = 2\cos 2x$. Adjust for the constant.

Substitute the limits into the integrated expression.

Problem-solving

You will save lots of time in your exam if you are familiar with the exact values for trigonometric functions given in radians.

Write $\sin\left(\frac{\pi}{4}\right)$ in its rationalised denominator form, as $\frac{\sqrt{2}}{2}$ rather than $\frac{1}{\sqrt{2}}$. This will make it easier to simplify your fractions.

Watch out This is a 'show that' question so don't use your calculator to simplify the fractions. Show each line of your working carefully.

Example 7

Find:

a $\int \sin 3x \cos 3x \, dx$

b $\int (\sec x + \tan x)^2 \, dx$

a $\int \sin 3x \cos 3x \, dx = \int \frac{1}{2}\sin 6x \, dx$

$= -\frac{1}{2} \times \frac{1}{6}\cos 6x + c$

$= -\frac{1}{12}\cos 6x + c$

Remember $\sin 2A \equiv 2\sin A \cos A$, so $\sin 6x \equiv 2\sin 3x \cos 3x$.

Use the reverse chain rule.

Simplify $\frac{1}{2} \times \frac{1}{6}$ to $\frac{1}{12}$

b $(\sec x + \tan x)^2$

$\equiv \sec^2 x + 2\sec x \tan x + \tan^2 x$

$\equiv \sec^2 x + 2\sec x \tan x + (\sec^2 x - 1)$

$\equiv 2\sec^2 x + 2\sec x \tan x - 1$

So $\int (\sec x + \tan x)^2 \, dx$

$= \int (2\sec^2 x + 2\sec x \tan x - 1) \, dx$

$= 2\tan x + 2\sec x - x + c$

Multiply out the bracket.

Write $\tan^2 x$ as $\sec^2 x - 1$. Then all the terms are standard integrals.

Integrate each term using ⑥ and ⑨.

Exercise 11C

1 Integrate the following:

a $\cot^2 x$
b $\cos^2 x$
c $\sin 2x \cos 2x$
d $(1 + \sin x)^2$
e $\tan^2 3x$
f $(\cot x - \text{cosec } x)^2$
g $(\sin x + \cos x)^2$
h $\sin^2 x \cos^2 x$
i $\dfrac{1}{\sin^2 x \cos^2 x}$
j $(\cos 2x - 1)^2$

Hint For part **a**, use $1 + \cot^2 x \equiv \text{cosec}^2 x$.
For part **c**, use $\sin 2A \equiv 2 \sin A \cos A$, making a suitable substitution for A.

2 Find the following integrals.

a $\displaystyle\int \dfrac{1 - \sin x}{\cos^2 x} \, dx$
b $\displaystyle\int \dfrac{1 + \cos x}{\sin^2 x} \, dx$
c $\displaystyle\int \dfrac{\cos 2x}{\cos^2 x} \, dx$
d $\displaystyle\int \dfrac{\cos^2 x}{\sin^2 x} \, dx$
e $\displaystyle\int \dfrac{(1 + \cos x)^2}{\sin^2 x} \, dx$
f $\displaystyle\int (\cot x - \tan x)^2 \, dx$
g $\displaystyle\int (\cos x - \sin x)^2 \, dx$
h $\displaystyle\int (\cos x - \sec x)^2 \, dx$
i $\displaystyle\int \dfrac{\cos 2x}{1 - \cos^2 2x} \, dx$

(E/P) 3 Show that $\displaystyle\int_{\frac{\pi}{4}}^{\frac{\pi}{2}} \sin^2 x \, dx = \dfrac{2 + \pi}{8}$ **(4 marks)**

4 Find the exact value of each of the following:

a $\displaystyle\int_{\frac{\pi}{6}}^{\frac{\pi}{3}} \dfrac{1}{\sin^2 x \cos^2 x} \, dx$
b $\displaystyle\int_{\frac{\pi}{6}}^{\frac{\pi}{4}} (\sin x - \text{cosec } x)^2 \, dx$
c $\displaystyle\int_{0}^{\frac{\pi}{4}} \dfrac{(1 + \sin x)^2}{\cos^2 x} \, dx$
d $\displaystyle\int_{\frac{3\pi}{8}}^{\frac{\pi}{2}} \dfrac{\sin 2x}{1 - \sin^2 2x} \, dx$

(E/P) 5 a By expanding $\sin(3x + 2x)$ and $\sin(3x - 2x)$ using the double-angle formulae, or otherwise, show that $\sin 5x + \sin x \equiv 2 \sin 3x \cos 2x$. **(4 marks)**

b Hence find $\displaystyle\int \sin 3x \cos 2x \, dx$ **(3 marks)**

(E/P) 6 $f(x) = 5 \sin^2 x + 7 \cos^2 x$

a Show that $f(x) = \cos 2x + 6$. **(3 marks)**
b Hence, find the exact value of $\displaystyle\int_0^{\frac{\pi}{4}} f(x) \, dx$. **(4 marks)**

(E/P) 7 a Show that $\cos^4 x \equiv \tfrac{1}{8} \cos 4x + \tfrac{1}{2} \cos 2x + \tfrac{3}{8}$ **(4 marks)**
b Hence find $\displaystyle\int \cos^4 x \, dx$. **(4 marks)**

11.4 Reverse chain rule

If a function can be written in the form $k \dfrac{f'(x)}{f(x)}$, you can integrate it using the reverse of the chain rule for differentiation.

Example 8

Find

a $\displaystyle\int \dfrac{2x}{x^2 + 1} \, dx$
b $\displaystyle\int \dfrac{\cos x}{3 + 2 \sin x} \, dx$

Problem-solving

If $f(x) = 3 + 2 \sin x$, then $f'(x) = 2 \cos x$.
By adjusting for the constant, the numerator is the derivative of the denominator.

Integration

a Let $I = \int \dfrac{2x}{x^2 + 1} dx$

Consider $y = \ln|x^2 + 1|$ — This is equal to the original integrand, so you don't need to adjust it.

Then $\dfrac{dy}{dx} = \dfrac{1}{x^2 + 1} \times 2x$

So $I = \ln|x^2 + 1| + c$ — Since integration is the reverse of differentiation.

b Let $I = \int \dfrac{\cos x}{3 + 2\sin x} dx$

Let $y = \ln|3 + 2\sin x|$ — Try differentiating $y = \ln|3 + 2\sin x|$.

$\dfrac{dy}{dx} = \dfrac{1}{3 + 2\sin x} \times 2\cos x$ — The derivative of $\ln|3 + 2\sin x|$ is twice the original integrand, so you need to divide it by 2.

So $I = \dfrac{1}{2}\ln|3 + 2\sin x| + c$

- **To integrate expressions of the form $\int k \dfrac{f'(x)}{f(x)} dx$, try $\ln|f(x)|$ and differentiate to check, and then adjust any constant.**

Watch out You can't use this method to integrate a function such as $\dfrac{1}{x^2 + 3}$ because the derivative of $x^2 + 3$ is $2x$, and the top of the fraction does not contain an x term.

You can use a similar method with functions of the form $kf'(x)(f(x))^n$.

Example 9

Find:

a $\int 3\cos x \sin^2 x \, dx$ **b** $\int x(x^2 + 5)^3 \, dx$

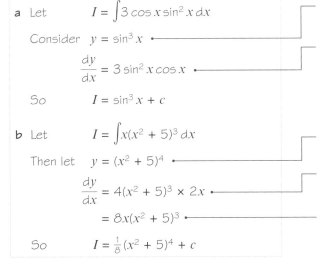

a Let $I = \int 3\cos x \sin^2 x \, dx$ — Try differentiating $\sin^3 x$.

Consider $y = \sin^3 x$ — This is equal to the original integrand, so you don't need to adjust it.

$\dfrac{dy}{dx} = 3\sin^2 x \cos x$

So $I = \sin^3 x + c$

b Let $I = \int x(x^2 + 5)^3 \, dx$ — Try differentiating $(x^2 + 5)^4$.

Then let $y = (x^2 + 5)^4$

$\dfrac{dy}{dx} = 4(x^2 + 5)^3 \times 2x$ — The $2x$ comes from differentiating $x^2 + 5$.

$= 8x(x^2 + 5)^3$ — This is 8 times the required expression so you divide by 8.

So $I = \dfrac{1}{8}(x^2 + 5)^4 + c$

- **To integrate an expression of the form $\int k f'(x)(f(x))^n \, dx$, try $(f(x))^{n+1}$ and differentiate to check, and then adjust any constant.**

Example 10

Use integration to find $\int \dfrac{\operatorname{cosec}^2 x}{(2 + \cot x)^3} dx$

Let $I = \int \dfrac{\operatorname{cosec}^2 x}{(2 + \cot x)^3} dx$ — This is in the form $\int k f'(x)(f(x))^n dx$ with $f(x) = 2 + \cot x$ and $n = -3$.

Let $y = (2 + \cot x)^{-2}$

$\dfrac{dy}{dx} = -2(2 + \cot x)^{-3} \times (-\operatorname{cosec}^2 x)$ — Use the chain rule.

$= 2(2 + \cot x)^{-3} \operatorname{cosec}^2 x$ — This is 2 times the required answer so you need to divide by 2.

So $I = \tfrac{1}{2}(2 + \cot x)^{-2} + c$

Example 11

Given that $\int_0^\theta 5 \tan x \sec^4 x \, dx = \dfrac{15}{4}$ where $0 < \theta < \dfrac{\pi}{2}$, find the exact value of θ.

Let $I = \int_0^\theta 5 \tan x \sec^4 x \, dx$ — This is in the form $\int k f'(x)(f(x))^n dx$ with $f(x) = \sec x$ and $n = 4$.

Let $y = \sec^4 x$

$\dfrac{dy}{dx} = 4 \sec^3 x \times \sec x \tan x$

$= 4 \sec^4 x \tan x$ — This is $\tfrac{4}{5}$ times the required answer so you need to divide by $\tfrac{4}{5}$.

So $I = \left[\dfrac{5}{4} \sec^4 x\right]_0^\theta = \dfrac{15}{4}$

$\left(\dfrac{5}{4} \sec^4 \theta\right) - \left(\dfrac{5}{4} \sec^4 0\right) = \dfrac{15}{4}$ — Substitute the limits into the integrated expression.

$\dfrac{5}{4} \sec^4 \theta - \dfrac{5}{4} = \dfrac{15}{4}$ — $\sec 0 = \dfrac{1}{\cos 0} = \dfrac{1}{1} = 1$

$\dfrac{5}{4} \sec^4 \theta = \dfrac{20}{4}$

$\sec^4 \theta = 4$ — Take the 4th root of both sides.

$\sec \theta = \pm \sqrt{2}$ — The solutions to $\cos \theta = \pm \dfrac{1}{\sqrt{2}}$ are $\theta = -\dfrac{\pi}{4}, \dfrac{\pi}{4}, \dfrac{3\pi}{4}, \dfrac{5\pi}{4} \ldots$

$\theta = \dfrac{\pi}{4}$ — The only solution within the given range for θ is $\dfrac{\pi}{4}$

Online Check your solution by using your calculator.

Exercise 11D

Hint Decide carefully whether each expression is in the form $k \dfrac{f'(x)}{f(x)}$ or $k f'(x)(f(x))^n$.

1 Integrate the following functions.

a $\dfrac{x}{x^2 + 4}$

b $\dfrac{e^{2x}}{e^{2x} + 1}$

c $\dfrac{x}{(x^2 + 4)^3}$

d $\dfrac{e^{2x}}{(e^{2x} + 1)^3}$

e $\dfrac{\cos 2x}{3 + \sin 2x}$

f $\dfrac{\sin 2x}{(3 + \cos 2x)^3}$

g $x e^{-x^2}$

h $\cos 2x (1 + \sin 2x)^4$

i $\sec^2 x \tan^2 x$

j $\sec^2 x (1 + \tan^2 x)$

Integration

2 Find the following integrals.

a $\int (x+1)(x^2+2x+3)^4 \, dx$

b $\int \operatorname{cosec}^2 2x \cot 2x \, dx$

c $\int \sin^5 3x \cos 3x \, dx$

d $\int \cos x \, e^{\sin x} \, dx$

e $\int \dfrac{e^{2x}}{e^{2x}+3} \, dx$

f $\int x(x^2+1)^{\frac{3}{2}} \, dx$

g $\int (2x+1)\sqrt{x^2+x+5} \, dx$

h $\int \dfrac{2x+1}{\sqrt{x^2+x+5}} \, dx$

i $\int \dfrac{\sin x \cos x}{\sqrt{\cos 2x + 3}} \, dx$

j $\int \dfrac{\sin x \cos x}{\cos 2x + 3} \, dx$

3 Find the exact value of each of the following:

a $\int_0^3 (3x^2+10x)\sqrt{x^3+5x^2+9} \, dx$

b $\int_{\frac{\pi}{9}}^{\frac{2\pi}{9}} \dfrac{6 \sin 3x}{1 - \cos 3x} \, dx$

c $\int_4^7 \dfrac{x}{x^2-1} \, dx$

d $\int_0^{\frac{\pi}{4}} \sec^2 x \, e^{4 \tan x} \, dx$

E/P **4** Given that $\int_0^k kx^2 e^{x^3} dx = \tfrac{2}{3}(e^8 - 1)$, find the value of k. **(3 marks)**

P **5** Given that $\int_0^\theta 4 \sin 2x \cos^4 2x \, dx = \tfrac{4}{5}$ where $0 < \theta < \pi$, find the exact value of θ.

E/P **6 a** By writing $\cot x = \dfrac{\cos x}{\sin x}$, find $\int \cot x \, dx$. **(2 marks)**

b Show that $\int \tan x \, dx \equiv \ln|\sec x| + c$. **(3 marks)**

11.5 Integration by substitution

- **Sometimes you can simplify an integral by changing the variable. The process is similar to using the chain rule in differentiation and is called integration by substitution.**

In your exam you will often be told which substitution to use.

Example 12

Find $\int x\sqrt{2x+5} \, dx$ using the substitutions:

a $u = 2x + 5$

b $u^2 = 2x + 5$

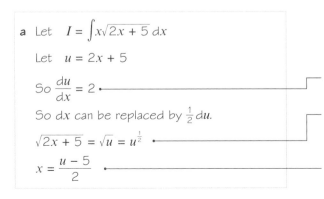

a Let $I = \int x\sqrt{2x+5} \, dx$

Let $u = 2x + 5$

So $\dfrac{du}{dx} = 2$

So dx can be replaced by $\tfrac{1}{2} du$.

$\sqrt{2x+5} = \sqrt{u} = u^{\frac{1}{2}}$

$x = \dfrac{u-5}{2}$

You need to replace each 'x' term with a corresponding 'u' term. Start by finding the relationship between dx and du.

So $dx = \tfrac{1}{2} du$.

Next rewrite the function in terms of $u = 2x + 5$.

Rearrange $u = 2x + 5$ to get $2x = u - 5$ and hence $x = \dfrac{u-5}{2}$

So $I = \int \left(\dfrac{u-5}{2}\right) u^{\frac{1}{2}} \times \dfrac{1}{2} du$ — Rewrite I in terms of u and simplify.

$= \int \dfrac{1}{4}(u-5) u^{\frac{1}{2}} du$

$= \int \dfrac{1}{4}(u^{\frac{3}{2}} - 5u^{\frac{1}{2}}) du$

$= \dfrac{1}{4} \times \dfrac{u^{\frac{5}{2}}}{\frac{5}{2}} - \dfrac{5u^{\frac{3}{2}}}{4 \times \frac{3}{2}} + c$ — Multiply out the brackets and integrate using rules from your Year 1 book. ← Year 1, Chapter 13

$= \dfrac{u^{\frac{5}{2}}}{10} - \dfrac{5u^{\frac{3}{2}}}{6} + c$ — Simplify.

So $I = \dfrac{(2x+5)^{\frac{5}{2}}}{10} - \dfrac{5(2x+5)^{\frac{3}{2}}}{6} + c$ — Finally rewrite the answer in terms of x.

b Let $I = \int x\sqrt{2x+5}\, dx$ — First find the relationship between dx and du.

$u^2 = 2x + 5$

$2u \dfrac{du}{dx} = 2$ — Using implicit differentiation, cancel 2 and rearrange to get $dx = u\,du$.

So replace dx with $u\,du$.

$\sqrt{2x+5} = u$

and $x = \dfrac{u^2 - 5}{2}$ — Rewrite the integrand in terms of u. You will need to make x the subject of $u^2 = 2x + 5$.

So $I = \int \left(\dfrac{u^2-5}{2}\right) u \times u\,du$

$= \int \left(\dfrac{u^4}{2} - \dfrac{5u^2}{2}\right) du$ — Multiply out the brackets and integrate.

$= \dfrac{u^5}{10} - \dfrac{5u^3}{6} + c$

So $I = \dfrac{(2x+5)^{\frac{5}{2}}}{10} - \dfrac{5(2x+5)^{\frac{3}{2}}}{6} + c$ — Rewrite answer in terms of x.

Example 13

Use the substitution $u = \sin x + 1$ to find
$\int \cos x \sin x (1 + \sin x)^3\, dx$

Let $I = \int \cos x \sin x (1 + \sin x)^3\, dx$ — First replace the dx.

Let $u = \sin x + 1$

$\dfrac{du}{dx} = \cos x$ — $\cos x$ appears in the integrand, so you can write this as $du = \cos x\, dx$ and substitute.

So substitute $\cos x\, dx$ with du.

Integration

$(\sin x + 1)^3 = u^3$
$\sin x = u - 1$

So $\quad I = \int (u-1)u^3\, du$

$\quad = \int (u^4 - u^3)\, du$

$\quad = \dfrac{u^5}{5} - \dfrac{u^4}{4} + c$

So $\quad I = \dfrac{(\sin x + 1)^5}{5} - \dfrac{(\sin x + 1)^4}{4} + c$

— Use $u = \sin x + 1$ to substitute for the remaining terms, rearranging where required to get $\sin x = u - 1$.

— Rewrite I in terms of u.

— Multiply out the brackets and integrate in the usual way.

Problem-solving

Although it looks different, $\int \sin 2x(1 + \sin x)^3\, dx$ can be integrated in exactly the same way. Remember $\sin 2x \equiv 2\sin x \cos x$, so the above integral would just need adjusting by a factor of 2.

Example 14

Prove that $\int \dfrac{1}{\sqrt{1-x^2}}\, dx = \arcsin x + c$.

Let $\quad I = \int \dfrac{1}{\sqrt{1-x^2}}\, dx$

Let $\quad x = \sin\theta$

$\dfrac{dx}{d\theta} = \cos\theta$

So replace dx with $\cos\theta\, d\theta$.

$I = \int \dfrac{1}{\sqrt{1-\sin^2\theta}} \cos\theta\, d\theta$

$\quad = \int \dfrac{1}{\sqrt{\cos^2\theta}} \cos\theta\, d\theta$

$\quad = \int \dfrac{1}{\cos\theta} \cos\theta\, d\theta$

$\quad = \int 1\, d\theta = \theta + c$

$x = \sin\theta \Rightarrow \theta = \arcsin x$

So $\quad I = \arcsin x + c$

— This substitution is not obvious at first. Think how the integral will be transformed by using trigonometric identities. ← **Chapter 7**

— The substitution is in the form $x = f(\theta)$, so find $\dfrac{dx}{d\theta}$ to work out the relationship between dx and $d\theta$.

— Make the substitution, and replace dx with $\cos\theta\, d\theta$.

— Remember $\sin^2 A + \cos^2 A \equiv 1$

— Remember to use your substitution to write the final answer in terms of x, not θ.

Example 15

Use integration by substitution to evaluate:

a $\int_0^2 x(x+1)^3\, dx$ $\quad$ **b** $\int_0^{\frac{\pi}{2}} \cos x\sqrt{1 + \sin x}\, dx$

Watch out If you use integration by substitution to evaluate a definite integral, you have to be careful of whether your limits are x values or u values. You can use a table to keep track.

a Let $\quad I = \int_0^2 x(x+1)^3\, dx$

Let $\quad u = x + 1$

$\dfrac{du}{dx} = 1$

305

so replace dx with du and replace $(x + 1)^3$ with u^3, and x with $u - 1$.

x	u
2	3
0	1

Replace each term in x with a term in u in the usual way.

Change the limits. When $x = 2$, $u = 2 + 1 = 3$ and when $x = 0$, $u = 1$.

So $I = \int_1^3 (u - 1)u^3 \, du$

$= \int_1^3 (u^4 - u^3) \, du$

$= \left[\dfrac{u^5}{5} - \dfrac{u^4}{4} \right]_1^3$

$= \left(\dfrac{243}{5} - \dfrac{81}{4} \right) - \left(\dfrac{1}{5} - \dfrac{1}{4} \right)$

$= 48.4 - 20 = 28.4$

Note that the new u limits replace their corresponding x limits.

Multiply out and integrate. Remember there is no need for a constant of integration with definite integrals.

The integral can now be evaluated using the limits for u without having to change back into x.

b $\int_0^{\frac{\pi}{2}} \cos x \sqrt{1 + \sin x} \, dx$

$u = 1 + \sin x \Rightarrow \dfrac{du}{dx} = \cos x$, so replace $\cos x \, dx$ with du and replace $\sqrt{1 + \sin x}$ with $u^{\frac{1}{2}}$.

Use $u = 1 + \sin x$.

x	u
$\dfrac{\pi}{2}$	2
0	1

Remember that limits for integrals involving trigonometric functions will always be in radians. $x = \dfrac{\pi}{2}$ means $u = 1 + 1 = 2$ and $x = 0$ means $u = 1 + 0 = 1$.

So $I = \int_1^2 u^{\frac{1}{2}} \, du$

$= \left[\dfrac{2}{3} u^{\frac{3}{2}} \right]_1^2$

$= \left(\dfrac{2}{3} 2^{\frac{3}{2}} \right) - \left(\dfrac{2}{3} \right)$

So $I = \dfrac{2}{3}(2\sqrt{2} - 1)$

Rewrite the integral in terms of u.

Remember that $2^{\frac{3}{2}} = \sqrt{8} = 2\sqrt{2}$.

Problem-solving

You could also convert the integral back into a function of x and use the original limits.

Exercise 11E

1 Use the substitutions given to find:

a $\int x\sqrt{1 + x} \, dx$; $u = 1 + x$

b $\int \dfrac{1 + \sin x}{\cos x} \, dx$; $u = \sin x$

c $\int \sin^3 x \, dx$; $u = \cos x$

d $\int \dfrac{2}{\sqrt{x}(x - 4)} \, dx$; $u = \sqrt{x}$

e $\int \sec^2 x \tan x \sqrt{1 + \tan x} \, dx$; $u^2 = 1 + \tan x$

f $\int \sec^4 x \, dx$; $u = \tan x$

2 Use the substitutions given to find the exact values of:

a $\int_0^5 x\sqrt{x + 4} \, dx$; $u = x + 4$

b $\int_0^2 x(2 + x)^3 \, dx$; $u = 2 + x$

c $\int_0^{\frac{\pi}{2}} \sin x \sqrt{3 \cos x + 1} \, dx$; $u = \cos x$

d $\int_0^{\frac{\pi}{3}} \sec x \tan x \sqrt{\sec x + 2} \, dx$; $u = \sec x$

e $\int_1^4 \dfrac{dx}{\sqrt{x}(4x - 1)}$; $u = \sqrt{x}$

(P) 3 By choosing a suitable substitution, find:

a $\int x(3 + 2x)^5 \, dx$ b $\int \dfrac{x}{\sqrt{1+x}} \, dx$ c $\int \dfrac{\sqrt{x^2+4}}{x} \, dx$

(P) 4 By choosing a suitable substitution, find the exact values of:

a $\int_2^7 x\sqrt{2+x} \, dx$ b $\int_2^5 \dfrac{1}{1+\sqrt{x-1}} \, dx$ c $\int_0^{\frac{\pi}{2}} \dfrac{\sin 2\theta}{1+\cos\theta} \, d\theta$

(E) 5 Using the substitution $u^2 = 4x + 1$, or otherwise, find the exact value of $\int_6^{20} \dfrac{8x}{\sqrt{4x+1}} \, dx$. **(8 marks)**

(E/P) 6 Use the substitution $u^2 = e^x - 2$ to show that $\int_{\ln 3}^{\ln 4} \dfrac{e^{4x}}{e^x - 2} \, dx = \dfrac{a}{b} + c \ln d$, where a, b, c and d are integers to be found. **(7 marks)**

(E/P) 7 Prove that $-\int \dfrac{1}{\sqrt{1-x^2}} \, dx = \arccos x + c$. **(5 marks)**

(E/P) 8 Use the substitution $u = \cos x$ to show

$\int_0^{\frac{\pi}{3}} \sin^3 x \cos^2 x \, dx = \dfrac{47}{480}$ **(7 marks)**

Hint Use exact trigonometric values to change the limits in x to limits in u.

(E/P) 9 Using a suitable trigonometric substitution for x, find $\int_{\frac{1}{2}}^{\frac{\sqrt{3}}{2}} x^2\sqrt{1-x^2} \, dx$. **(8 marks)**

Challenge

By using a substitution of the form $x = k \sin u$, show that

$\int \dfrac{1}{x^2\sqrt{9-x^2}} \, dx = -\dfrac{\sqrt{9-x^2}}{9x} + c$

11.6 Integration by parts

You can rearrange the product rule for differentiation:

$\dfrac{d}{dx}(uv) = u\dfrac{dv}{dx} + v\dfrac{du}{dx}$

Links u and v are both functions of x. ← Section 9.4

$u\dfrac{dv}{dx} = \dfrac{d}{dx}(uv) - v\dfrac{du}{dx}$

$\int u\dfrac{dv}{dx} \, dx = \int \dfrac{d}{dx}(uv) \, dx - \int v\dfrac{du}{dx} \, dx$

Differentiating a function and then integrating it leaves the original function unchanged.

So, $\int \dfrac{d}{dx}(uv) \, dx = uv$.

■ **This method is called integration by parts.** $\int u\dfrac{dv}{dx} \, dx = uv - \int v\dfrac{du}{dx} \, dx$

To use integration by parts you need to write the function you are integrating in the form $u\dfrac{dv}{dx}$. You will have to choose what to set as u and what to set as $\dfrac{dv}{dx}$.

Chapter 11

Example 16

Find $\int x \cos x \, dx$

Let $I = \int x \cos x \, dx$

$u = x \Rightarrow \dfrac{du}{dx} = 1$

$\dfrac{dv}{dx} = \cos x \Rightarrow v = \sin x$

Using the integration by parts formula:

$I = x \sin x - \int \sin x \times 1 \, dx$

$= x \sin x + \cos x + c$

Problem-solving

For expressions like $x \cos x$, $x^2 \sin x$ and $x^3 e^x$ let u equal the x^n term. When the expression involves $\ln x$, for example $x^2 \ln x$, let u equal the $\ln x$ term.

Let $u = x$ and $\dfrac{dv}{dx} = \cos x$.

Find expressions for u, v, $\dfrac{du}{dx}$ and $\dfrac{dv}{dx}$

Take care to differentiate u but integrate $\dfrac{dv}{dx}$

Notice that $\int v \dfrac{du}{dx} dx$ is a simpler integral than $\int u \dfrac{dv}{dx} dx$.

Example 17

Find $\int x^2 \ln x \, dx$

Let $I = \int x^2 \ln x \, dx$

$u = \ln x \Rightarrow \dfrac{du}{dx} = \dfrac{1}{x}$

$\dfrac{dv}{dx} = x^2 \Rightarrow v = \dfrac{x^3}{3}$

$I = \dfrac{x^3}{3} \ln x - \int \dfrac{x^3}{3} \times \dfrac{1}{x} dx$

$= \dfrac{x^3}{3} \ln x - \int \dfrac{x^2}{3} dx$

$= \dfrac{x^3}{3} \ln x - \dfrac{x^3}{9} + c$

Since there is a $\ln x$ term, let $u = \ln x$ and $\dfrac{dv}{dx} = x^2$.

Find expressions for u, v, $\dfrac{du}{dx}$ and $\dfrac{dv}{dx}$

Take care to differentiate u but integrate $\dfrac{dv}{dx}$

Apply the integration by parts formula.

Simplify the $v \dfrac{du}{dx}$ term.

It is sometimes necessary to use integration by parts twice, as shown in the following example.

Example 18

Find $\int x^2 e^x \, dx$

Let $I = \int x^2 e^x \, dx$

$u = x^2 \Rightarrow \dfrac{du}{dx} = 2x$

$\dfrac{dv}{dx} = e^x \Rightarrow v = e^x$

So $I = x^2 e^x - \int 2x e^x \, dx$

$u = 2x \Rightarrow \dfrac{du}{dx} = 2$

$\dfrac{dv}{dx} = e^x \Rightarrow v = e^x$

There is no $\ln x$ term, so let $u = x^2$ and $\dfrac{dv}{dx} = e^x$.

Find expressions for u, v, $\dfrac{du}{dx}$ and $\dfrac{dv}{dx}$

Take care to differentiate u but integrate $\dfrac{dv}{dx}$

Apply the integration by parts formula.

Notice that this integral is simpler than I but still not one you can write down. It has a similar structure to I and so you can use integration by parts again with $u = 2x$ and $\dfrac{dv}{dx} = e^x$.

Integration

So $I = x^2e^x - \left(2xe^x - \int 2e^x dx\right)$ ← Apply the integration by parts formula for a second time.

$= x^2e^x - 2xe^x + \int 2e^x dx$

$= x^2e^x - 2xe^x + 2e^x + c$

Example 19

Evaluate $\int_1^2 \ln x \, dx$, leaving your answer in terms of natural logarithms.

Let $I = \int_1^2 \ln x \, dx = \int_1^2 \ln x \times 1 \, dx$

$u = \ln x \Rightarrow \dfrac{du}{dx} = \dfrac{1}{x}$

$\dfrac{dv}{dx} = 1 \Rightarrow v = x$

$I = [x \ln x]_1^2 - \int_1^2 x \times \dfrac{1}{x} dx$

$= (2 \ln 2) - (1 \ln 1) - \int_1^2 1 \, dx$

$= 2\ln 2 - [x]_1^2$

$= 2 \ln 2 - (2 - 1)$

$= 2 \ln 2 - 1$

Write the expression to be integrated as $\ln x \times 1$, then $u = \ln x$ and $\dfrac{dv}{dx} = 1$.
Remember if an expression involves $\ln x$ you should always set $u = \ln x$.

Problem-solving

Apply limits to the uv term and the $\int v \dfrac{du}{dx} dx$ term separately.

Evaluate the limits on uv and remember $\ln 1 = 0$.

Exercise 11F

1 Find the following integrals.

a $\int x \sin x \, dx$ **b** $\int xe^x \, dx$ **c** $\int x \sec^2 x \, dx$

d $\int x \sec x \tan x \, dx$ **e** $\int \dfrac{x}{\sin^2 x} dx$

Hint You will need to use these standard results. In your exam they will be given in the formulae booklet:
- $\int \tan x \, dx = \ln|\sec x| + c$
- $\int \sec x \, dx = \ln|\sec x + \tan x| + c$
- $\int \cot x \, dx = \ln|\sin x| + c$
- $\int \cosec x \, dx = -\ln|\cosec x + \cot x| + c$

2 Find the following integrals.

a $\int 3 \ln x \, dx$ **b** $\int x \ln x \, dx$ **c** $\int \dfrac{\ln x}{x^3} dx$

d $\int (\ln x)^2 \, dx$ **e** $\int (x^2 + 1) \ln x \, dx$

3 Find the following integrals.

a $\int x^2 e^{-x} \, dx$ **b** $\int x^2 \cos x \, dx$ **c** $\int 12x^2 (3 + 2x)^5 \, dx$ **d** $\int 2x^2 \sin 2x \, dx$ **e** $\int 2x^2 \sec^2 x \tan x \, dx$

4 Evaluate the following:

a $\int_0^{\ln 2} xe^{2x} \, dx$ **b** $\int_0^{\frac{\pi}{2}} x \sin x \, dx$ **c** $\int_0^{\frac{\pi}{2}} x \cos x \, dx$ **d** $\int_1^2 \dfrac{\ln x}{x^2} dx$

e $\int_0^1 4x(1 + x)^3 \, dx$ **f** $\int_0^{\pi} x \cos \tfrac{1}{4} x \, dx$ **g** $\int_0^{\frac{\pi}{3}} \sin x \ln(\sec x) \, dx$

5 a Use integration by parts to find $\int x \cos 4x \, dx$. (3 marks)

   b Use your answer to part **a** to find $\int x^2 \sin 4x \, dx$. (3 marks)

6 a Find $\int \sqrt{8 - x} \, dx$. (2 marks)

   b Using integration by parts, or otherwise, show that
   $$\int (x - 2)\sqrt{8 - x} \, dx = -\tfrac{2}{5}(8 - x)^{\frac{3}{2}}(x + 2) + c$$ (6 marks)

   c Hence find $\int_4^7 (x - 2)\sqrt{8 - x} \, dx$. (2 marks)

7 a Find $\int \sec^2 3x \, dx$. (3 marks)

   b Using integration by parts, or otherwise, find $\int x \sec^2 3x \, dx$. (6 marks)

   c Hence show that $\int_{\frac{\pi}{18}}^{\frac{\pi}{9}} x \sec^2 3x \, dx = p\pi - q \ln 3$, finding the exact values of the constants p and q. (4 marks)

11.7 Partial fractions

- **Partial fractions can be used to integrate algebraic fractions.**

Using partial fractions enables an expression that looks hard to integrate to be transformed into two or more expressions that are easier to integrate.

Links Make sure you are confident expressing algebraic fractions as partial fractions ← Chapter 1

Example 20

Use partial fractions to find the following integrals.

a $\int \dfrac{x - 5}{(x + 1)(x - 2)} \, dx$

b $\int \dfrac{8x^2 - 19x + 1}{(2x + 1)(x - 2)^2} \, dx$

c $\int \dfrac{2}{1 - x^2} \, dx$

a $\dfrac{x - 5}{(x + 1)(x - 2)} \equiv \dfrac{A}{x + 1} + \dfrac{B}{x - 2}$

Split the expression to be integrated into partial fractions.

So $x - 5 \equiv A(x - 2) + B(x + 1)$

Let $x = -1$: $-6 = A(-3)$ so $A = 2$

Let $x = 2$: $-3 = B(3)$ so $B = -1$

Let $x = -1$ and 2.

So $\int \dfrac{x - 5}{(x + 1)(x - 2)} \, dx$

Rewrite the integral and integrate each term as in ← Section 11.2

$= \int \left(\dfrac{2}{x + 1} - \dfrac{1}{x - 2} \right) dx$

$= 2 \ln|x + 1| - \ln|x - 2| + c$

Remember to use the modulus when using ln in integration.

$= \ln \left| \dfrac{(x + 1)^2}{x - 2} \right| + c$

The answer could be left in this form, but sometimes you may be asked to combine the ln terms using the rules of logarithms.

← Year 1, Chapter 14

b Let $I = \int \dfrac{8x^2 - 19x + 1}{(2x+1)(x-2)^2} dx$ — It is sometimes useful to label the integral as I.

$\dfrac{8x^2 - 19x + 1}{(2x+1)(x-2)^2} \equiv \dfrac{A}{2x+1} + \dfrac{B}{(x-2)^2} + \dfrac{C}{x-2}$ — Remember the partial fraction form for a repeated factor in the denominator.

$8x^2 - 19x + 1 \equiv A(x-2)^2 + B(2x+1) + C(2x+1)(x-2)$

Let $x = 2$: $-5 = 0 + 5B + 0$ so $B = -1$

Let $x = -\tfrac{1}{2}$: $12\tfrac{1}{2} = \tfrac{25}{4}A + 0 + 0$ so $A = 2$

Let $x = 0$: Then $1 = 4A + B - 2C$

So $1 = 8 - 1 - 2C$ so $C = 3$

$I = \int \left(\dfrac{2}{2x+1} - \dfrac{1}{(x-2)^2} + \dfrac{3}{x-2} \right) dx$ — Rewrite the integral using the partial fractions. Note that using I saves copying the question again.

$= \tfrac{2}{2} \ln|2x+1| + \dfrac{1}{x-2} + 3\ln|x-2| + c$ — Don't forget to divide by 2 when integrating $\dfrac{1}{2x+1}$ and remember that the integral of $\dfrac{1}{(x-2)^2}$ does not involve ln.

$= \ln|2x+1| + \dfrac{1}{x-2} + \ln|x-2|^3 + c$

$= \ln|(2x+1)(x-2)^3| + \dfrac{1}{x-2} + c$ — Simplify using the laws of logarithms.

c Let $I = \int \dfrac{2}{1-x^2} dx$

$\dfrac{2}{1-x^2} = \dfrac{2}{(1-x)(1+x)} = \dfrac{A}{1-x} + \dfrac{B}{1+x}$ — Remember that $1 - x^2$ can be factorised using the difference of two squares.

$2 = A(1+x) + B(1-x)$

Let $x = -1$ then $2 = 2B$ so $B = 1$

Let $x = 1$ then $2 = 2A$ so $A = 1$

So $I = \int \left(\dfrac{1}{1+x} + \dfrac{1}{1-x} \right) dx$ — Rewrite the integral using the partial fractions.

$= \ln|1+x| - \ln|1-x| + c$

$= \ln\left|\dfrac{1+x}{1-x}\right| + c$ — Notice the minus sign that comes from integrating $\dfrac{1}{1-x}$

When the degree of the polynomial in the numerator is greater than or equal to the degree of the denominator, it is necessary to first divide the numerator by the denominator.

Example 21

Find $\int \dfrac{9x^2 - 3x + 2}{9x^2 - 4} dx$

Let $I = \int \dfrac{9x^2 - 3x + 2}{9x^2 - 4} dx$ — First divide the numerator by $9x^2 - 4$.

$$\begin{array}{r} 1 \\ 9x^2 - 4 \overline{)9x^2 - 3x + 2} \\ \underline{9x^2 -4} \\ -3x + 6 \end{array}$$

$9x^2 \div 9x^2$ gives 1, so put this on top and subtract $1 \times (9x^2 - 4)$. This leaves a remainder of $-3x + 6$.

311

so $I = \int \left(1 + \dfrac{6 - 3x}{9x^2 - 4}\right) dx$

$\dfrac{6 - 3x}{9x^2 - 4} \equiv \dfrac{A}{3x - 2} + \dfrac{B}{3x + 2}$ ← Factorise $9x^2 - 4$ and then split into partial fractions.

Let $x = -\dfrac{2}{3}$ then $8 = -4B$ so $B = -2$

Let $x = \dfrac{2}{3}$ then $4 = 4A$ so $A = 1$

So $I = \int \left(1 + \dfrac{1}{3x - 2} - \dfrac{2}{3x + 2}\right) dx$ ← Rewrite the integral using the partial fractions.

$= x + \dfrac{1}{3}\ln|3x - 2| - \dfrac{2}{3}\ln|3x + 2| + c$ ← Integrate and don't forget the $\dfrac{1}{3}$

$= x + \dfrac{1}{3}\ln\left|\dfrac{3x - 2}{(3x + 2)^2}\right| + c$ ← Simplify using the laws of logarithms.

Exercise 11G

1 Use partial fractions to integrate the following:

a $\dfrac{3x + 5}{(x + 1)(x + 2)}$ **b** $\dfrac{3x - 1}{(2x + 1)(x - 2)}$ **c** $\dfrac{2x - 6}{(x + 3)(x - 1)}$ **d** $\dfrac{3}{(2 + x)(1 - x)}$

2 Find the following integrals.

a $\displaystyle\int \dfrac{2(x^2 + 3x - 1)}{(x + 1)(2x - 1)} dx$ **b** $\displaystyle\int \dfrac{x^3 + 2x^2 + 2}{x(x + 1)} dx$ **c** $\displaystyle\int \dfrac{x^2}{x^2 - 4} dx$ **d** $\displaystyle\int \dfrac{x^2 + x + 2}{3 - 2x - x^2} dx$

(E/P) 3 $f(x) = \dfrac{4}{(2x + 1)(1 - 2x)}$, $x \neq \pm\dfrac{1}{2}$

  a Given that $f(x) = \dfrac{A}{2x + 1} + \dfrac{B}{1 - 2x}$, find the value of the constants A and B. **(3 marks)**

  b Hence find $\int f(x) dx$, writing your answer as a single logarithm. **(4 marks)**

  c Find $\displaystyle\int_1^2 f(x) dx$, giving your answer in the form $\ln k$ where k is a rational constant. **(2 marks)**

(E/P) 4 $f(x) = \dfrac{17 - 5x}{(3 + 2x)(2 - x)^2}$, $-\dfrac{3}{2} < x < 2$.

  a Express $f(x)$ in partial fractions. **(4 marks)**

  b Hence find the exact value of $\displaystyle\int_0^1 \dfrac{17 - 5x}{(3 + 2x)(2 - x)^2} dx$, writing your answer in the form $a + \ln b$, where a and b are constants to be found. **(5 marks)**

(E/P) 5 $f(x) = \dfrac{9x^2 + 4}{9x^2 - 4}$, $x \neq \pm\dfrac{2}{3}$

  a Given that $f(x) = A + \dfrac{B}{3x - 2} + \dfrac{C}{3x + 2}$, find the values of the constants A, B and C. **(4 marks)**

  b Hence find the exact value of
  $\displaystyle\int_{-\frac{1}{3}}^{\frac{1}{3}} \dfrac{9x^2 + 4}{9x^2 - 4} dx$, writing your answer in the form $a + b \ln c$, where a, b and c are rational numbers to be found.

  Problem-solving
  Simplify the integral as much as possible before substituting your limits.

  (5 marks)

Integration

E/P 6 $f(x) = \dfrac{6 + 3x - x^2}{x^3 + 2x^2}$, $x > 0$

    a Express $f(x)$ in partial fractions. **(4 marks)**

    b Hence find the exact value of $\displaystyle\int_2^4 \dfrac{6 + 3x - x^2}{x^3 + 2x^2}\,dx$, writing your answer in the form $a + \ln b$, where a and b are rational numbers to be found. **(5 marks)**

E/P 7 $\dfrac{32x^2 + 4}{(4x + 1)(4x - 1)} \equiv A + \dfrac{B}{4x + 1} + \dfrac{C}{4x - 1}$

    a Find the value of the constants A, B and C. **(4 marks)**

    b Hence find the exact value of $\displaystyle\int_1^2 \dfrac{32x^2 + 4}{(4x + 1)(4x - 1)}\,dx$ writing your answer in the form $2 + k \ln m$, giving the values of the rational constants k and m. **(5 marks)**

11.8 Finding areas

You need to be able to use the integration techniques from this chapter to find areas under curves.

Example 22

The diagram shows part of the curve $y = \dfrac{9}{\sqrt{4 + 3x}}$

The region R is bounded by the curve, the x-axis and the lines $x = 0$ and $x = 4$, as shown in the diagram. Use integration to find the area of R.

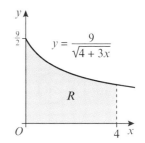

Remember $\dfrac{1}{\sqrt{x}} = \dfrac{1}{x^{\frac{1}{2}}} = x^{-\frac{1}{2}}$

$\text{Area} = \displaystyle\int_0^4 \dfrac{9}{\sqrt{4 + 3x}}\,dx$

$= 9\displaystyle\int_0^4 (4 + 3x)^{-\frac{1}{2}}\,dx$ ← Use the chain rule in reverse. If $y = (4 + 3x)^{\frac{1}{2}}$, $\dfrac{dy}{dx} = \dfrac{3}{2}(4 + 3x)^{-\frac{1}{2}}$. Adjust for the constant.

$= 6\left[(4 + 3x)^{\frac{1}{2}}\right]_0^4$

$= 6\left((4 + 3 \times 4)^{\frac{1}{2}} - (4 + 3 \times 0)^{\frac{1}{2}}\right)$ ← Substitute the limits.

$= 6(\sqrt{16} - \sqrt{4})$

$= 12$ ← You don't need to give units when finding areas under graphs in pure maths.

- **The area bounded by two curves can be found using integration:**

$$\text{Area of } R = \int_a^b (f(x) - g(x))\,dx = \int_a^b f(x)\,dx - \int_a^b g(x)\,dx$$

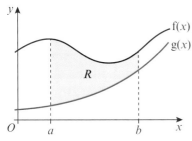

Watch out You can only use this formula if the two curves do not intersect between a and b.

313

Example 23

The diagram shows part of the curves $y = f(x)$ and $y = g(x)$, where $f(x) = \sin 2x$ and $g(x) = \sin x \cos^2 x$, $0 \leq x \leq \dfrac{\pi}{2}$

The region R is bounded by the two curves. Use integration to find the area of R.

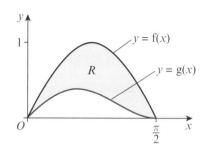

$$\text{Area} = \int_a^b (f(x) - g(x))\, dx$$

— The region R is bounded by two curves.

$$= \int_0^{\frac{\pi}{2}} (\sin 2x - \sin x \cos^2 x)\, dx$$

— Substitute the limits and functions given in the question.

$$= \left[-\tfrac{1}{2}\cos 2x + \tfrac{1}{3}\cos^3 x \right]_0^{\frac{\pi}{2}}$$

$$= \left(-\tfrac{1}{2}(-1) + \tfrac{1}{3}(0) \right) - \left(-\tfrac{1}{2} + \tfrac{1}{3} \right) = \tfrac{2}{3}$$

Online Explore the area between two curves using technology.

You can use integration to find the area under a curve defined by parametric equations. It is often easier to integrate with respect to the parameter.

Links For a parametric curve, x and y are given as functions of a parameter, t. ← **Chapter 8**

Example 24

The curve C has parametric equations
$x = t(1 + t)$, $y = \dfrac{1}{1 + t}$, $t \geq 0$

Find the exact area of the region R, bounded by C, the x-axis and the lines $x = 0$ and $x = 2$.

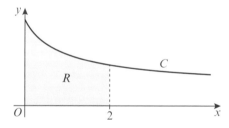

$$\text{Area} = \int y\, dx = \int y \dfrac{dx}{dt}\, dt$$

$x = t(1 + t)$, so $\dfrac{dx}{dt} = 1 + 2t$

When $x = 0$, $t(1 + t) = 0$, so $\underline{t = 0}$ or $t = -1$
When $x = 2$, $t(1 + t) = 2$
$\quad t^2 + t - 2 = 0$
$\quad (t + 2)(t - 1) = 0$, so $t = -2$ or $\underline{t = 1}$

So Area $= \int_0^1 y \dfrac{dx}{dt}\, dt = \int_0^1 \dfrac{1}{1+t}(1 + 2t)\, dt$

$$= \int_0^1 \left(2 - \dfrac{1}{1 + t} \right) dt$$

$$= [2t - \ln|1 + t|]_0^1$$

$$= (2 - \ln 2) - (0 - \ln 1)$$

$$= 2 - \ln 2$$

— Use a change of variable to write the integral in terms of the parameter, t. Using the chain rule, you can replace dx with $\dfrac{dx}{dt}\, dt$.

— $x = t + t^2$

Watch out You will be integrating **with respect to t** so you need to convert the limits from values of x to values of t. Use the parametric equation for x, and choose solutions that are within the domain of the parameter, $t \geq 0$.

— Write $\dfrac{1 + 2t}{1 + t}$ in the form $A + \dfrac{B}{1 + t}$
Dividing $2t + 1$ by $t + 1$ gives 2 with remainder -1.

Integration

Exercise 11H

1 Find the area of the finite region R bounded by the curve with equation $y = f(x)$, the x-axis and the lines $x = a$ and $x = b$.

a $f(x) = \dfrac{2}{1+x}$; $a = 0, b = 1$
b $f(x) = \sec x$; $a = 0, b = \dfrac{\pi}{3}$
c $f(x) = \ln x$; $a = 1, b = 2$

d $f(x) = \sec x \tan x$; $a = 0, b = \dfrac{\pi}{4}$
e $f(x) = x\sqrt{4 - x^2}$; $a = 0, b = 2$

2 Find the exact area of the finite region bounded by the curve $y = f(x)$, the x-axis and the lines $x = a$ and $x = b$ where:

a $f(x) = \dfrac{4x - 1}{(x+2)(2x+1)}$; $a = 0, b = 2$
b $f(x) = \dfrac{x}{(x+1)^2}$; $a = 0, b = 2$

c $f(x) = x \sin x$; $a = 0, b = \dfrac{\pi}{2}$
d $f(x) = \cos x \sqrt{2 \sin x + 1}$; $a = 0, b = \dfrac{\pi}{6}$

e $f(x) = xe^{-x}$; $a = 0, b = \ln 2$

(E) 3 The diagram shows a sketch of the curve with equation, $y = f(x)$, where $f(x) = \dfrac{4x + 3}{(x+2)(2x-1)}$, $x > \dfrac{1}{2}$.

Find the area of the shaded region bounded by the curve, the x-axis and the lines $x = 1$ and $x = 2$. **(7 marks)**

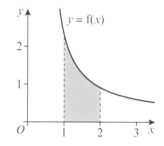

(E) 4 The diagram shows a sketch of the curve with equation $y = f(x)$, where $f(x) = e^{0.5x} + \dfrac{1}{x}$, $x > 0$.

Find the area of the shaded region bounded by the curve, the x-axis and the lines $x = 2$ and $x = 4$. **(7 marks)**

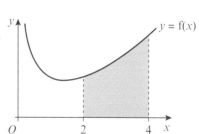

(P) 5 The diagram shows a sketch of the curve with equation $y = g(x)$, where $g(x) = x \sin x$.

a Write down the coordinates of points A, B and C.

b Find the area of the shaded region.

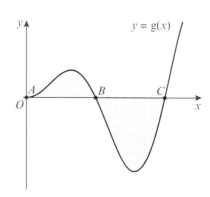

> **Watch out** Find the area of each region separately and then add the answers. Remember areas cannot be negative, so take the absolute value of any negative area.

Chapter 11

E/P 6 The diagram shows a sketch of the curve with equation $y = x^2 \ln x$. The shaded region is bounded by the curve, the x-axis and the line $x = 2$.

a Use integration by parts to find $\int x^2 \ln x \, dx$. **(3 marks)**

b Hence find the exact area of the shaded region, giving your answer in the form $\frac{2}{3}(a \ln 2 + b)$, where a and b are integers. **(5 marks)**

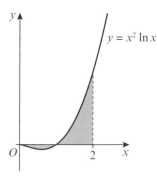

E/P 7 The diagram shows a sketch of the curve with equation $y = 3\cos x \sqrt{\sin x + 1}$.

a Find the coordinates of the points A, B, C and D. **(3 marks)**

b Use a suitable substitution to find
$$\int 3\cos x \sqrt{\sin x + 1} \, dx$$ **(5 marks)**

c Show that the regions R_1 and R_2 have the same area, and find the exact value of this area in the form $\sqrt{a}$, where a is a positive integer to be found. **(3 marks)**

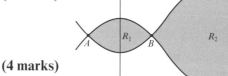

P 8 $f(x) = x^2$ and $g(x) = 3x - x^2$

a On the same axes, sketch the graphs of $y = f(x)$ and $y = g(x)$, and find the coordinates of any points of intersection of the two curves.

b Find the area of the finite region bounded by the two curves.

E/P 9 The diagram shows a sketch of part of the curves with equations $y = 2\cos x + 2$ and $y = -2\cos x + 4$.

a Find the coordinates of the points A, B and C. **(2 marks)**

b Find the area of region R_1 in the form $a\sqrt{3} + \frac{b\pi}{c}$, where a, b and c are integers to be found. **(4 marks)**

c Show that the ratio of $R_2 : R_1$ can be expressed as $(3\sqrt{3} + 2\pi) : (3\sqrt{3} - \pi)$. **(5 marks)**

P 10 The diagrams show the curves $y = \sin \theta$, $0 \le \theta \le 2\pi$ and $y = \sin 2\theta$, $0 \le \theta \le 2\pi$.

By choosing suitable limits, show that the total shaded area in the first diagram is equal to the total shaded area in the second diagram, and state the exact value of this shaded area.

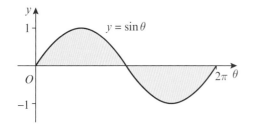 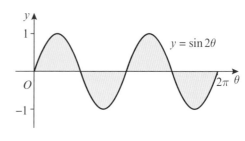

316

Integration

P **11** The diagram shows parts of the graphs of $y = \sin x$ and $y = \cos x$.

  a Find the coordinates of point A.
  b Find the areas of:
   i R_1 **ii** R_2 **iii** R_3
  c Show that the ratio of areas $R_1 : R_2$ can be written as $\sqrt{2} : 2$.

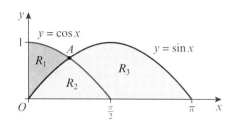

P **12** The curve C has parametric equations $x = t^3$, $y = t^2$, $t \geqslant 0$. Show that the exact area of the region bounded by the curve, the x-axis and the lines $x = 0$ and $x = 4$ is $k\sqrt[3]{2}$, where k is a rational constant to be found.

E/P **13** The curve C has parametric equations

$$x = \sin t, \; y = \sin 2t, \; 0 \leqslant t \leqslant \frac{\pi}{2}$$

The finite region R is bounded by the curve and the x-axis. Find the exact area of R. **(6 marks)**

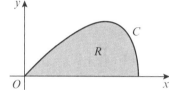

E/P **14** This graph shows part of the curve C with parametric equations $x = (t + 1)^2$, $y = \frac{1}{2}t^3 + 3$, $t \geqslant -1$

P is the point on the curve where $t = 2$.
The line S is the normal to C at P.

  a Find an equation of S. **(5 marks)**

The shaded region R is bounded by C, S, the x-axis and the line with equation $x = 1$.

  b Using integration, find the area of R. **(5 marks)**

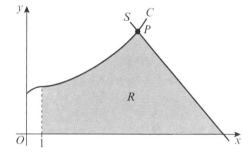

Challenge

The diagram shows the curves $y = \sin 2x$ and $y = \cos x$, $0 \leqslant x \leqslant \frac{\pi}{4}$

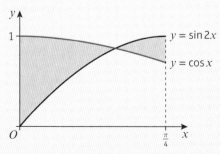

Find the exact value of the total shaded area on the diagram.

11.9 The trapezium rule

If you cannot integrate a function algebraically, you can use a numerical method to approximate the area beneath a curve.

Consider the curve $y = f(x)$:

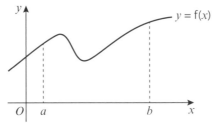

To approximate the area given by $\int_a^b y \, dx$, you can divide the area up into n equal strips. Each strip will be of width h, where $h = \dfrac{b-a}{n}$

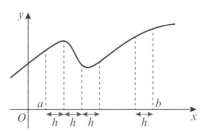

Next you calculate the value of y for each value of x that forms a boundary of one of the strips. So you find y for $x = a$, $x = a + h$, $x = a + 2h$, $x = a + 3h$ and so on up to $x = b$.
You can label these values $y_0, y_1, y_2, y_3, \ldots, y_n$.

Hint Notice that for n strips there will be $n + 1$ values of x and $n + 1$ values of y.

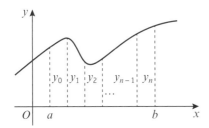

Finally you join adjacent points to form n trapezia and approximate the original area by the sum of the areas of these n trapeziums.

You may recall from GCSE maths that the area of a trapezium like this:

is given by $\frac{1}{2}(y_0 + y_1)h$. The required area under the curve is therefore given by:

$$\int_a^b y \, dx \approx \tfrac{1}{2}h(y_0 + y_1) + \tfrac{1}{2}h(y_1 + y_2) + \ldots + \tfrac{1}{2}h(y_{n-1} + y_n)$$

Factorising gives:

$$\int_a^b y \, dx \approx \tfrac{1}{2}h(y_0 + y_1 + y_1 + y_2 + y_2 \ldots + y_{n-1} + y_{n-1} + y_n)$$

or $\int_a^b y \, dx \approx \tfrac{1}{2}h(y_0 + 2(y_1 + y_2 \ldots + y_{n-1}) + y_n)$

This formula is given in the formula booklet but you will need to know how to use it.

Integration

- **The trapezium rule:**

$$\int_a^b y\,dx \approx \tfrac{1}{2}h(y_0 + 2(y_1 + y_2 \ldots + y_{n-1}) + y_n)$$

where $h = \dfrac{b-a}{n}$ and $y_i = f(a + ih)$

Example 25

The diagram shows a sketch of the curve $y = \sec x$. The finite region R is bounded by the curve, the x-axis, the y-axis and the line $x = \dfrac{\pi}{3}$.
The table shows the corresponding values of x and y for $y = \sec x$.

x	0	$\dfrac{\pi}{12}$	$\dfrac{\pi}{6}$	$\dfrac{\pi}{4}$	$\dfrac{\pi}{3}$
y	1	1.035			2

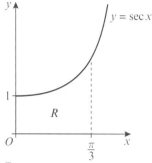

a Complete the table with the values of y corresponding to $x = \dfrac{\pi}{6}$ and $x = \dfrac{\pi}{4}$, giving your answers to 3 decimal places.

b Use the trapezium rule, with all the values of y in the completed table, to obtain an estimate for the area of R, giving your answer to 2 decimal places.

c Explain with a reason whether your estimate in part **b** will be an underestimate or an overestimate.

a $\sec\dfrac{\pi}{6} = \dfrac{1}{\cos\dfrac{\pi}{6}} \approx 1.155$

$\sec\dfrac{\pi}{4} = \dfrac{1}{\cos\dfrac{\pi}{4}} \approx 1.414$

x	0	$\dfrac{\pi}{12}$	$\dfrac{\pi}{6}$	$\dfrac{\pi}{4}$	$\dfrac{\pi}{3}$
y	1	1.035	1.155	1.414	2

Substitute $h = \dfrac{\pi}{12}$ and the five y-values into the formula.

b $I = \int_a^b y\,dx \approx \tfrac{1}{2}h(y_0 + 2(y_1 + y_2 \ldots + y_{n-1}) + y_n)$

$I \approx \tfrac{1}{2}\left(\dfrac{\pi}{12}\right)(1 + 2(1.035 + 1.155 + 1.414) + 2)$

$= \dfrac{\pi}{24} \times 10.208$

$= 1.336\,224\,075\ldots = 1.34$ (2 d.p.)

Online Explore under- and overestimation when using the trapezium rule, using GeoGebra.

c The answer would be an overestimate. The graph is convex so the lines connecting two endpoints would be above the curve, giving a greater answer than the real answer.

Problem-solving

If $f(x)$ is convex on the interval $[a, b]$ then the trapezium rule will give an overestimate for $\int_a^b f(x)\,dx$. If it is concave then it will give an underestimate.

← Section 9.9

Exercise 11I

1 The diagram shows a sketch of the curve with equation $y = \sqrt{1 + \tan x}$, $0 \leq x \leq \dfrac{\pi}{3}$

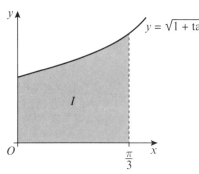

a Complete the table with the values for y corresponding to $x = \dfrac{\pi}{12}$ and $x = \dfrac{\pi}{4}$ **(1 mark)**

x	0	$\dfrac{\pi}{12}$	$\dfrac{\pi}{6}$	$\dfrac{\pi}{4}$	$\dfrac{\pi}{3}$
y	1		1.2559		1.6529

Given that $I = \displaystyle\int_0^{\frac{\pi}{3}} \sqrt{1 + \tan x}\, dx$,

b use the trapezium rule:

  i with the values of y at $x = 0$, $\dfrac{\pi}{6}$ and $\dfrac{\pi}{3}$ to find an approximate value for I, giving your answer to 4 significant figures; **(3 marks)**

  ii with the values of y at $x = 0$, $\dfrac{\pi}{12}, \dfrac{\pi}{6}, \dfrac{\pi}{4}$ and $\dfrac{\pi}{3}$ to find an approximate value for I, giving your answer to 4 significant figures. **(3 marks)**

2 The diagram shows the region R bounded by the x-axis and the curve with equation $y = \cos\dfrac{5\theta}{2}$, $-\dfrac{\pi}{5} \leq \theta \leq \dfrac{\pi}{5}$

The table shows corresponding values of θ and y for $y = \cos\dfrac{5\theta}{2}$

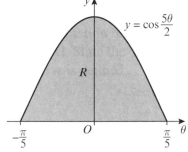

θ	$-\dfrac{\pi}{5}$	$-\dfrac{\pi}{10}$	0	$\dfrac{\pi}{10}$	$\dfrac{\pi}{5}$
y	0		1		0

a Complete the table giving the missing values for y to 4 decimal places. **(1 mark)**

b Using the trapezium rule, with all the values for y in the completed table, find an approximation for the area of R, giving your answer to 3 decimal places. **(4 marks)**

c State, with a reason, whether your approximation in part **b** is an underestimate or an overestimate. **(1 mark)**

d Use integration to find the exact area of R. **(3 marks)**

e Calculate the percentage error in your answer in part **b**. **(2 marks)**

Integration

E 3 The diagram shows a sketch of the curve with equation $y = \dfrac{1}{\sqrt{e^x + 1}}$

The shaded region R is bounded by the curve, the x-axis, the y-axis and the line $x = 2$.

a Complete the table giving values of y to 3 decimal places. **(2 marks)**

x	0	0.5	1	1.5	2
y	0.707	0.614	0.519		0.345

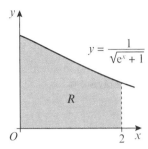

b Use the trapezium rule, with all the values from your table, to estimate the area of the region R, giving your answer to 2 decimal places. **(4 marks)**

E/P 4 The diagram shows the curve with equation $y = (x - 2)\ln x + 1$, $x > 0$.

a Complete the table with the values of y corresponding to $x = 2$ and $x = 2.5$. **(1 mark)**

x	1	1.5	2	2.5	3
y	1	0.7973			2.0986

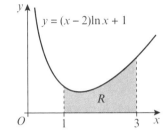

Given that $I = \displaystyle\int_1^3 ((x - 2)\ln x + 1)\,dx$,

b use the trapezium rule

   i with values of y at $x = 1, 2$ and 3 to find an approximate value for I, giving your answer to 4 significant figures. **(3 marks)**

   ii with values of y at $x = 1, 1.5, 2, 2.5$ and 3 to find another approximate value for I, giving your answer to 4 significant figures. **(3 marks)**

c Use the diagram to explain why an increase in the number of values improves the accuracy of the approximation. **(1 mark)**

d Show by integration, that the exact value of $\displaystyle\int_1^3 ((x - 2)\ln x + 1)\,dx$ is $-\frac{3}{2}\ln 3 + 4$. **(6 marks)**

E/P 5 The diagram shows the curve with equation $y = x\sqrt{2 - x}$, $0 \leqslant x \leqslant 2$.

a Complete the table with the value of y corresponding to $x = 1.5$. **(1 mark)**

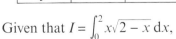

x	0	0.5	1	1.5	2
y	0	0.6124	1		0

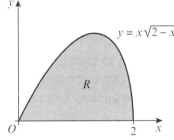

Given that $I = \displaystyle\int_0^2 x\sqrt{2 - x}\,dx$,

b use the trapezium rule with four strips to find an approximate value for I, giving your answer to 4 significant figures. **(5 marks)**

c By using an appropriate substitution, or otherwise, find the exact value of $\displaystyle\int_0^2 x\sqrt{2 - x}\,dx$, leaving your answer in the form $2^q p$, where p and q are rational constants. **(4 marks)**

d Calculate the percentage error of the approximation in part **b**. **(2 marks)**

321

E/P 6 The diagram shows part of the curve with equation
$$y = \frac{4x - 5}{(x - 3)(2x + 1)}$$

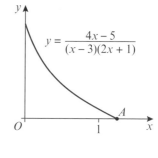

a Show that the coordinates of point A are $\left(\frac{5}{4}, 0\right)$. **(1 mark)**

b Complete the table with the value of y corresponding to $x = 0.75$. Give your answer to 4 decimal places. **(1 mark)**

x	0	0.25	0.5	0.75	1	1.25
y	1.6667	0.9697	0.6		0.1667	0

Given that $I = \int_0^{\frac{5}{4}} \frac{4x - 5}{(x - 3)(2x + 1)} \, dx$,

c use the trapezium rule with values of y at $x = 0, 0.25, 0.5, 0.75, 1$ and 1.25 to find an approximate value for I, giving your answer to 4 significant figures. **(3 marks)**

d Find the exact value of $\int_0^{\frac{5}{4}} \frac{4x - 5}{(x - 3)(2x + 1)} \, dx$, giving your answer in the form $\ln\left(\frac{a}{b}\right)$. **(4 marks)**

e Calculate the percentage error of the approximation in part **c**. **(2 marks)**

E/P 7 $I = \int_0^3 e^{\sqrt{2x+1}} \, dx$

a Given that $y = e^{\sqrt{2x+1}}$, complete the table of values of y corresponding to $x = 0.5$, 1 and 1.5. **(2 marks)**

x	0	0.5	1	1.5	2	2.5	3
y	2.7183				9.3565	11.5824	14.0940

b Use the trapezium rule, with all the values of y in the completed table, to obtain an estimate for the original integral, I, giving your answer to 4 significant figures. **(3 marks)**

c Use the substitution $t = \sqrt{2x + 1}$ to show that I may be expressed as $\int_a^b kte^t \, dt$, giving the values of the constants a, b and k. **(5 marks)**

d Use integration by parts to evaluate this integral, and hence find the value of I correct to 4 significant figures. **(4 marks)**

11.10 Solving differential equations

Integration can be used to solve differential equations. In this chapter you will solve first order differential equations by **separating the variables**.

- When $\dfrac{dy}{dx} = f(x)g(y)$ you can write

$$\int \frac{1}{g(y)} \, dy = \int f(x) \, dx$$

Notation A first order differential equation contains nothing higher than a first order derivative, for example $\dfrac{dy}{dx}$. A second order differential equation would have a term that contains a second order derivative, for example $\dfrac{d^2y}{dx^2}$

The solution to a differential equation will be a function.
When you integrate to solve a differential equation you still need to include a constant of integration. This gives the **general solution** to the differential equation. It represents a **family** of solutions, all with different constants. Each of these solutions satisfies the original differential equation.

Integration

For the first order differential equation $\frac{dy}{dx} = 12x^2 - 1$, the general solution is $y = 4x^3 - x + c$, or $y = x(2x - 1)(2x + 1) + c$.

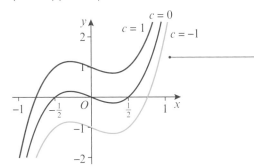

Each of these curves represents a **particular solution** of the differential equation, for different values of the constant c. Together, the curves form a **family of solutions**.

Online Explore families of solutions using technology.

Example 26

Find a general solution to the differential equation $(1 + x^2)\frac{dy}{dx} = x \tan y$.

$$\frac{dy}{dx} = \frac{x}{1 + x^2} \tan y$$

Write the equation in the form $\frac{dy}{dx} = f(x)g(y)$.

$$\int \frac{1}{\tan y} dy = \int \frac{x}{1 + x^2} dx$$

Now **separate the variables**:
$\frac{1}{g(y)} dy = f(x) dx$

$$\int \cot y \, dy = \int \frac{x}{1 + x^2} dx$$

Use $\cot y = \frac{1}{\tan y}$

$$\ln|\sin y| = \tfrac{1}{2}\ln|1 + x^2| + c$$

$\int \cot x \, dx = \ln|\sin x| + c$

or $\quad \ln|\sin y| = \tfrac{1}{2}\ln|1 + x^2| + \ln k$

Don't forget the $+c$ which can be written as $\ln k$.

$\ln|\sin y| = \ln|k\sqrt{1 + x^2}|$

Combining logs.

so $\quad \sin y = k\sqrt{1 + x^2}$

Finally remove the ln. Sometimes you might be asked to give your answer in the form $y = f(x)$. This question did not specify that so it is acceptable to give the answer in this form.

Sometimes you are interested in one specific solution to a differential equation. You can find a **particular solution** to a first-order differential equation if you know **one point** on the curve. This is sometimes called a **boundary condition**.

Example 27

Find the particular solution to the differential equation

$$\frac{dy}{dx} = \frac{-3(y - 2)}{(2x + 1)(x + 2)}$$

given that $x = 1$ when $y = 4$. Leave your answer in the form $y = f(x)$.

Hint The boundary condition in this question is that $x = 1$ when $y = 4$.

Chapter 11

$$\int \frac{1}{y-2} \, dy = \int \frac{-3}{(2x+1)(x+2)} \, dx$$ — First separate the variables. Make sure the function on the left-hand side is in terms of y only, and the function on the right-hand side is in terms of x only.

$$\frac{-3}{(2x+1)(x+2)} \equiv \frac{A}{(2x+1)} + \frac{B}{(x+2)}$$

$$-3 = A(x+2) + B(2x+1)$$

Let $x = -2$: $\quad -3 = -3B \text{ so } B = 1$ — Convert the fraction on the RHS to partial fractions.

Let $x = -\frac{1}{2}$: $\quad -3 = \frac{3}{2}A \text{ so } A = -2$

So

$$\int \frac{1}{y-2} \, dy = \int \left(\frac{1}{x+2} - \frac{2}{2x+1}\right) dx$$ — Rewrite the integral using the partial fractions.

$\ln|y-2| = \ln|x+2| - \ln|2x+1| + \ln k$ — Integrate and use $+\ln k$ instead of $+c$.

$\ln|y-2| = \ln\left|\frac{k(x+2)}{2x+1}\right|$ — Combine ln terms.

$y - 2 = k\left(\frac{x+2}{2x+1}\right)$ — Remove ln.

$4 - 2 = k\left(\frac{1+2}{2+1}\right) \Rightarrow k = 2$ — Use the condition $x = 1$ when $y = 4$ by substituting these values into the general solution and solving to find k.

So $\quad y = 2 + 2\left(\frac{x+2}{2x+1}\right)$

$y = 3 + \dfrac{3}{2x+1}$ — Substitute $k = 2$ and write the answer in the form $y = f(x)$ as requested.

Exercise 11J

1 Find general solutions to the following differential equations. Give your answers in the form $y = f(x)$.

a $\dfrac{dy}{dx} = (1+y)(1-2x)$

b $\dfrac{dy}{dx} = y \tan x$

c $\cos^2 x \dfrac{dy}{dx} = y^2 \sin^2 x$

d $\dfrac{dy}{dx} = 2e^{x-y}$

2 Find particular solutions to the following differential equations using the given boundary conditions.

a $\dfrac{dy}{dx} = \sin x \cos^2 x; \ y = 0, \ x = \dfrac{\pi}{3}$

b $\dfrac{dy}{dx} = \sec^2 x \sec^2 y; \ y = 0, \ x = \dfrac{\pi}{4}$

c $\dfrac{dy}{dx} = 2\cos^2 y \cos^2 x; \ y = \dfrac{\pi}{4}, \ x = 0$

d $\sin y \cos x \dfrac{dy}{dx} = \dfrac{\cos y}{\cos x}, \ y = 0, \ x = 0$

3 a Find the general solution to the differential equation

$x^2 \dfrac{dy}{dx} = y + xy$, giving your answer in the form $y = g(x)$.

Hint Begin by factorising the right-hand side of the equation.

b Find the particular solution to the differential equation that satisfies the boundary condition $y = e^4$ at $x = -1$.

(E) **4** Given that $x = 0$ when $y = 0$, find the particular solution to the differential equation
$(2y + 2yx)\frac{dy}{dx} = 1 - y^2$, giving your answer in the form $y = g(x)$. **(6 marks)**

(E/P) **5** Find the general solution to the differential equation $e^{x+y}\frac{dy}{dx} = 2x + xe^y$, giving your answer in the form $\ln|g(y)| = f(x)$. **(6 marks)**

(E) **6** Find the particular solution to the differential equation $(1 - x^2)\frac{dy}{dx} = xy + y$, with boundary condition $y = 6$ at $x = 0.5$. Give your answer in the form $y = f(x)$. **(8 marks)**

(E) **7** Find the particular solution to the differential equation $(1 + x^2)\frac{dy}{dx} = x - xy^2$, with boundary condition $y = 2$ at $x = 0$. Give your answer in the form $y = f(x)$. **(8 marks)**

(E) **8** Find the particular solution to the differential equation $\frac{dy}{dx} = xe^{-y}$, with boundary condition $y = \ln 2$ at $x = 4$. Give your answer in the form $y = f(x)$. **(8 marks)**

(E/P) **9** Find the particular solution to the differential equation $\frac{dy}{dx} = \cos^2 y + \cos 2x \cos^2 y$, with boundary condition $y = \frac{\pi}{4}$ at $x = \frac{\pi}{4}$. Give your answer in the form $\tan y = f(x)$. **(8 marks)**

(E) **10** Given that $y = 1$ at $x = \frac{\pi}{2}$, solve the differential equation $\frac{dy}{dx} = xy \sin x$. **(6 marks)**

(E) **11 a** Find $\int \frac{3x + 4}{x} dx$, $x > 0$. **(2 marks)**
    b Given that $y = 16$ at $x = 1$, solve the differential equation $\frac{dy}{dx} = \frac{3x\sqrt{y} + 4\sqrt{y}}{x}$ giving your answer in the form $y = f(x)$. **(6 marks)**

(E) **12 a** Express $\frac{8x - 18}{(3x - 8)(x - 2)}$ in partial fractions. **(3 marks)**
    b Given that $x \geq 3$, find the general solution to the differential equation
$(x - 2)(3x - 8)\frac{dy}{dx} = (8x - 18)y$ **(5 marks)**
    c Hence find the particular solution to this differential equation that satisfies $y = 8$ at $x = 3$, giving your answer in the form $y = f(x)$. **(4 marks)**

(P) **13 a** Find the general solution of $\frac{dy}{dx} = 2x - 4$.
    b On the same axes, sketch three different particular solutions to this differential equation.

(E/P) **14 a** Find the general solution to the differential equation $\frac{dy}{dx} = -\frac{1}{(x + 2)^2}$ **(3 marks)**
    b On the same axes, sketch three different particular solutions to this differential equation. **(3 marks)**
    c Write down the particular solution that passes through the point $(8, 3.1)$. **(1 mark)**

Chapter 11

E/P **15 a** Show that the general solution to the differential equation $\dfrac{dy}{dx} = -\dfrac{x}{y}$ can be written in the form $x^2 + y^2 = c$. **(3 marks)**

b On the same axes, sketch three different particular solutions to this differential equation. **(3 marks)**

c Write down the particular solution that passes through the point $(0, 7)$. **(1 mark)**

11.11 Modelling with differential equations

Differential equations can be used to model real-life situations.

Example 28

The rate of increase of a population P of microorganisms at time t, in hours, is given by
$$\frac{dP}{dt} = 3P$$
Initially the population was of size 8.

a Find a model for P in the form $P = Ae^{3t}$, stating the value of A.
b Find, to the nearest hundred, the size of the population at time $t = 2$.
c Find the time at which the population will be 1000 times its starting value.
d State one limitation of this model for large values of t.

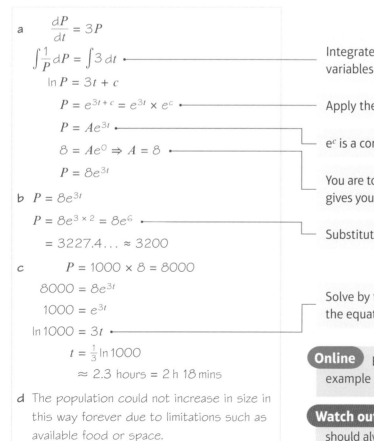

a $\dfrac{dP}{dt} = 3P$

$\int \dfrac{1}{P} dP = \int 3\, dt$ — Integrate this function by separating the variables.

$\ln P = 3t + c$

$P = e^{3t + c} = e^{3t} \times e^{c}$ — Apply the laws of indices.

$P = Ae^{3t}$ — e^c is a constant so write it as A.

$8 = Ae^0 \Rightarrow A = 8$

$P = 8e^{3t}$ — You are told that the initial population was 8. This gives you the boundary condition $P = 8$ when $t = 0$.

b $P = 8e^{3t}$

$P = 8e^{3 \times 2} = 8e^6$ — Substitute $t = 2$.

$= 3227.4\ldots \approx 3200$

c $P = 1000 \times 8 = 8000$

$8000 = 8e^{3t}$

$1000 = e^{3t}$ — Solve by taking the natural log of both sides of the equation.

$\ln 1000 = 3t$

$t = \tfrac{1}{3} \ln 1000$

≈ 2.3 hours $= 2$ h 18 mins

Online Explore the solution to this example graphically using technology.

d The population could not increase in size in this way forever due to limitations such as available food or space.

Watch out When commenting on a model you should always refer to the context of the question.

Integration

Example 29

Water in a manufacturing plant is held in a large cylindrical tank of diameter 20 m. Water flows out of the bottom of the tank through a tap at a rate proportional to the cube root of the volume.

a Show that t minutes after the tap is opened, $\dfrac{dh}{dt} = -k\sqrt[3]{h}$ for some constant k.

b Show that the general solution to this differential equation may be written as $h = (P - Qt)^{\frac{3}{2}}$, where P and Q are constants.

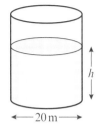

Initially the height of the water is 27 m. 10 minutes later, the height is 8 m.

c Find the values of the constants P and Q.

d Find the time in minutes when the water is at a depth of 1 m.

a $V = \pi r^2 h = 100\pi h$

Use the formula for the volume of a cylinder. The diameter is 20, so the radius is 10.

$\dfrac{dV}{dh} = 100\pi$

$\dfrac{dV}{dt} = -c\sqrt[3]{V}$

$= -c\sqrt[3]{100\pi h}$

Problem-solving

You need to use the information given in the question to construct a mathematical model. Water flows out at a rate proportional to the cube root of the volume.

$\dfrac{dV}{dt}$ is negative as the water is flowing out of the tank, so the volume is decreasing.

$\dfrac{dh}{dt} = \dfrac{dh}{dV} \times \dfrac{dV}{dt}$

$\dfrac{dh}{dt} = \dfrac{1}{100\pi} \times (-c\sqrt[3]{100\pi h})$

$= \left(\dfrac{-c\sqrt[3]{100\pi}}{100\pi}\right)\sqrt[3]{h}$

So $\dfrac{dh}{dt} = -k\sqrt[3]{h}$, where $k = \dfrac{c\sqrt[3]{100\pi}}{100\pi}$

Use the chain rule to find $\dfrac{dh}{dt}$

Substitute for $\dfrac{dh}{dV}$ and $\dfrac{dV}{dt}$

$\dfrac{dh}{dV} = \dfrac{1}{\frac{dV}{dh}} = \dfrac{1}{100\pi}$

b $\int h^{-\frac{1}{3}} dh = -\int k\, dt$

$\dfrac{3}{2}h^{\frac{2}{3}} = -kt + c$

$h^{\frac{2}{3}} = -\dfrac{2}{3}kt + \dfrac{2}{3}c$

$h^{\frac{2}{3}} = -Qt + P$

$h = (P - Qt)^{\frac{3}{2}}$

c was the constant of proportionality and π is constant so $\dfrac{c \times \sqrt[3]{100\pi}}{100\pi} = k$ is a constant.

Integrate this function by separating the variables.

Let $Q = \frac{2}{3}k$ and $P = \frac{2}{3}c$

c $t = 0, h = 27$

$27 = P^{\frac{3}{2}} \Rightarrow P = 9$

$t = 10, h = 8$

$8 = (9 - 10Q)^{\frac{3}{2}}$

$4 = 9 - 10Q$

$Q = \dfrac{1}{2}$

Use the boundary conditions to find the values of P and Q. If there are two boundary conditions then you should consider the initial condition (when $t = 0$) first.

Chapter 11

d $h = (9 - \frac{1}{2}t)^{\frac{3}{2}}$

$1 = (9 - \frac{1}{2}t)^{\frac{3}{2}}$

$1 = 9 - \frac{1}{2}t$

$t = 16$ minutes

⟵ Set $h = 1$ and solve the resulting equation to find the corresponding value of t.

Exercise 11K

1 The rate of increase of a population P of rabbits at time t, in years, is given by $\frac{dP}{dt} = kP, k > 0$. Initially the population was of size 200.
  a Solve the differential equations giving P in terms of k and t. **(3 marks)**
  b Given that $k = 3$, find the time taken for the population to reach 4000. **(4 marks)**
  c State a limitation of this model for large values of t. **(1 mark)**

2 The mass M at time t of the leaves of a certain plant varies according to the differential equation
$$\frac{dM}{dt} = M - M^2$$
  a Given that at time $t = 0$, $M = 0.5$, find an expression for M in terms of t. **(5 marks)**
  b Find a value of M when $t = \ln 2$. **(2 marks)**
  c Explain what happens to the value of M as t increases. **(1 mark)**

3 The thickness of ice x, in cm, on a pond is increasing at a rate that is inversely proportional to the square of the existing thickness of ice. Initially, the thickness is 1 cm. After 20 days, the thickness is 2 cm.
  a Show that the thickness of ice can be modelled by the equation $x = \sqrt[3]{\frac{7}{20}t + 1}$. **(7 marks)**
  b Find the time taken for the ice to increase in thickness from 2 cm to 3 cm. **(2 marks)**

4 A mug of tea, with a temperature $T\,°C$ is made and left to cool in a room with a temperature of 25 °C. The rate at which the tea cools is proportional to the difference in temperature between the tea and the room.
  a Show that this process can be described by the differential equation $\frac{dT}{dt} = -k(T - 25)$, explaining why k is a positive constant. **(3 marks)**
  Initially the tea is at a temperature of 85 °C. 10 minutes later the tea is at 55 °C.
  b Find the temperature, to 1 decimal place, of the tea after 15 minutes. **(7 marks)**

5 The rate of change of the surface area of a drop of oil, A mm², at time t minutes can be modelled by the equation $\frac{dA}{dt} = \frac{A^{\frac{3}{2}}}{10t^2}$
  Given that the surface area of the drop is 1 mm² at $t = 1$,
  a find an expression for A in terms of t **(7 marks)**
  b show that the surface area of the drop cannot exceed $\frac{400}{361}$ mm². **(2 marks)**

6 A bath tub is modelled as a cuboid with a base area of 6000 cm². Water flows into the bath tub from a tap at a rate of 12 000 cm³/min. At time t minutes, the depth of water in the bath tub is h cm. Water leaves the bottom of the bath through an open plughole at a rate of $500h$ cm³/min.

  a Show that t minutes after the tap has been opened, $60\dfrac{dh}{dt} = 120 - 5h$. (3 marks)

  When $t = 0$, $h = 6$ cm.

  b Find the value of t when $h = 10$ cm. (5 marks)

7 a Express $\dfrac{1}{P(10\,000 - P)}$ using partial fractions. (3 marks)

  The deer population, P, in a reservation can be modelled by the differential equation
  $$\dfrac{dP}{dt} = \dfrac{1}{20\,000} P(10\,000 - P)$$
  where t is the time in years since the study began. Given that the initial deer population is 2500,

  b solve the differential equation giving your answer in the form $P = \dfrac{a}{b + ce^{-0.5t}}$ (6 marks)

  c Find the maximum deer population according to the model. (2 marks)

8 Liquid is pouring into a container at a constant rate of 40 cm³ s⁻¹ and is leaking from the container at a rate of $\tfrac{1}{4}V$ cm³ s⁻¹, where V cm³ is the volume of liquid in the container.

  a Show that $-4\dfrac{dV}{dt} = V - 160$. (2 marks)

  Given that $V = 5000$ when $t = 0$,

  b find the solution to the differential equation in the form $V = a + be^{-\frac{1}{4}t}$, where a and b are constants to be found (7 marks)

  c write down the limiting value of V as $t \to \infty$. (1 mark)

9 Fossils are aged using a process called carbon dating. The amount of carbon remaining in a fossil, R, decreases over time, t, measured in years. The rate of decrease of carbon is proportional to the remaining carbon.

  a Given that initially the amount of carbon is R_0, show that $R = R_0 e^{-kt}$ (4 marks)

  It is known that the half-life of carbon is 5730 years. This means that after 5730 years the amount of carbon remaining has reduced by half.

  b Find the exact value of k. (3 marks)

  c A fossil is found with 10% of its expected carbon remaining. Determine the age of the fossil to the nearest year. (3 marks)

11.12 Integration as the limit of a sum

You can approximate the area under a curve as the sum of the areas of a number of thin rectangular strips. As you make these strips thinner and thinner, the approximation becomes more accurate. You can use limit notation to formalise the idea of a definite integral as the limit of a sum of the areas of these rectangular strips.

- $\displaystyle\int_a^b f(x)\,dx = \lim_{\delta x \to 0} \sum_{x=a}^{b} f(x)\,\delta x$

Chapter 11

Example 30

The diagram shows a sketch of the curve with equation $y = \sin x$.

The area under the curve between $x = 1$ and $x = 2$ can be thought of a series of thin strips of height y and width δx.

Calculate $\lim\limits_{\delta x \to 0} \sum\limits_{x=1}^{2} \sin x \, \delta x$, giving your answer correct to 4 significant figures.

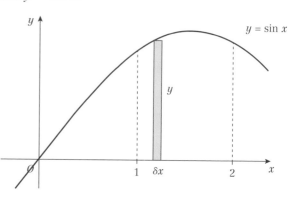

$$\lim_{\delta x \to 0} \sum_{x=1}^{2} \sin x \, \delta x = \int_{1}^{2} \sin x \, dx$$
$$= [-\cos x]_{1}^{2}$$
$$= -\cos 2 - (-\cos 1)$$
$$= 0.9564 \ (4 \text{ s.f.})$$

This is the limit as the width of the strip tends towards 0. Use the formula from the previous page to convert this limit into a definite integral.

Exercise 11L

1 Find the exact value of:

   a $\lim\limits_{\delta x \to 0} \sum\limits_{x=3}^{12} x^2 \, \delta x$ **b** $\lim\limits_{\delta x \to 0} \sum\limits_{x=9}^{25} \sqrt{x} \, \delta x$ **c** $\lim\limits_{\delta x \to 0} \sum\limits_{x=5}^{10} \left(x\sqrt{x-1}\right) \delta x$

(E/P) **2** Calculate $\lim\limits_{\delta x \to 0} \sum\limits_{x=2}^{3} \ln x \, \delta x$, giving your answer in the form $p + \ln q$, where p and q are rational numbers to be found. **(5 marks)**

(E) **3** The diagram shows a sketch of the curve with equation $y = \sqrt[3]{x}$, $x > 0$.

The area under the curve between $x = 2$ and $x = 5$ can be thought of as a series of thin strips of height y and width δx.

Calculate $\lim\limits_{\delta x \to 0} \sum\limits_{x=2}^{5} \sqrt[3]{x} \, \delta x$, giving your answer correct to 4 significant figures. **(3 marks)**

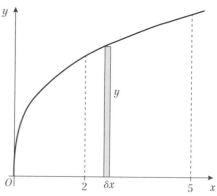

Mixed exercise 11

1 By choosing a suitable method of integration, find:

   a $\int (2x - 3)^7 \, dx$ **b** $\int x\sqrt{4x - 1} \, dx$ **c** $\int \sin^2 x \cos x \, dx$

   d $\int x \ln x \, dx$ **e** $\int \dfrac{4 \sin x \cos x}{4 - 8 \sin^2 x} \, dx$ **f** $\int \dfrac{1}{3 - 4x} \, dx$

Integration

2 By choosing a suitable method, evaluate the following definite integrals. Write your answers as exact values.

a $\displaystyle\int_{-3}^{0} x(x^2 + 3)^5 \, dx$

b $\displaystyle\int_{0}^{\frac{\pi}{4}} x \sec^2 x \, dx$

c $\displaystyle\int_{1}^{4} \left(16x^{\frac{3}{2}} - \frac{2}{x}\right) dx$

d $\displaystyle\int_{\frac{\pi}{12}}^{\frac{\pi}{3}} (\cos x + \sin x)(\cos x - \sin x) \, dx$

e $\displaystyle\int_{1}^{4} \frac{4}{16x^2 + 8x - 3} \, dx$

f $\displaystyle\int_{0}^{\ln 2} \frac{1}{1 + e^x} \, dx$

E/P 3 a Show that $\displaystyle\int_{1}^{e} \frac{1}{x^2} \ln x \, dx = 1 - \frac{2}{e}$ **(5 marks)**

b Given that $p > 1$, show that $\displaystyle\int_{1}^{p} \frac{1}{(x+1)(2x-1)} \, dx = \frac{1}{3} \ln \frac{4p - 2}{p + 1}$ **(5 marks)**

E/P 4 Given $\displaystyle\int_{\frac{1}{2}}^{b} \left(\frac{2}{x^3} - \frac{1}{x^2}\right) dx = \frac{9}{4}$, find the value of b. **(4 marks)**

E/P 5 Given $\displaystyle\int_{0}^{\theta} \cos x \sin^3 x \, dx = \frac{9}{64}$, where θ is a positive constant, find the smallest possible value of θ. Use algebraic integration and show your method clearly. **(4 marks)**

E 6 Using the substitution $t^2 = x + 1$, where $x > -1$,

a find $\displaystyle\int \frac{x}{\sqrt{x+1}} \, dx$. **(5 marks)**

b Hence evaluate $\displaystyle\int_{0}^{3} \frac{x}{\sqrt{x+1}} \, dx$. **(2 marks)**

E 7 a Use integration by parts to find $\displaystyle\int x \sin 8x \, dx$. **(4 marks)**

b Use your answer to part **a** to find $\displaystyle\int x^2 \cos 8x \, dx$. **(4 marks)**

E/P 8 $f(x) = \dfrac{5x^2 - 8x + 1}{2x(x-1)^2}$

a Given that $f(x) = \dfrac{A}{x} + \dfrac{B}{x-1} + \dfrac{C}{(x-1)^2}$, find the values of the constants A, B and C. **(4 marks)**

b Hence find $\displaystyle\int f(x) \, dx$. **(4 marks)**

c Hence show that $\displaystyle\int_{4}^{9} f(x) \, dx = \ln\left(\frac{32}{3}\right) - \frac{5}{24}$ **(4 marks)**

E/P 9 Given that $y = x^{\frac{3}{2}} + \dfrac{48}{x}$, $x > 0$,

a find the value of x and the value of y when $\dfrac{dy}{dx} = 0$. **(3 marks)**

b Show that the value of y which you found is a minimum. **(2 marks)**

The finite region R is bounded by the curve with equation $y = x^{\frac{3}{2}} + \dfrac{48}{x}$, the lines $x = 1$, $x = 4$ and the x-axis.

c Find, by integration, the area of R giving your answer in the form $p + q \ln r$, where the numbers p, q and r are constants to be found. **(4 marks)**

E/P 10 a Find $\displaystyle\int x^2 \ln 2x \, dx$. **(6 marks)**

b Hence show that the exact value of $\displaystyle\int_{\frac{1}{2}}^{3} x^2 \ln 2x \, dx$ is $9 \ln 6 - \dfrac{215}{72}$. **(4 marks)**

331

11 The diagram shows the graph of $y = (1 + \sin 2x)^2$, $0 \leq x \leq \frac{3\pi}{4}$

a Show that $(1 + \sin 2x)^2 \equiv \frac{1}{2}(3 + 4\sin 2x - \cos 4x)$. **(4 marks)**

b Hence find the area of the shaded region R. **(4 marks)**

c Find the coordinates of A, the turning point on the graph. **(3 marks)**

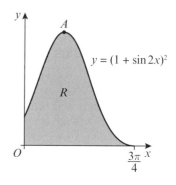

12 a Find $\int xe^{-x}\,dx$. **(4 marks)**

b Given that $y = \frac{\pi}{4}$ at $x = 0$, solve the differential equation

$$e^x \frac{dy}{dx} = \frac{x}{\sin 2y}$$ **(4 marks)**

13 a Find $\int x \sin 2x\, dx$. **(5 marks)**

b Given that $y = 0$ at $x = \frac{\pi}{4}$, solve the differential equation $\frac{dy}{dx} = x \sin 2x \cos^2 y$. **(5 marks)**

14 a Obtain the general solution to the differential equation

$$\frac{dy}{dx} = xy^2,\ y > 0$$ **(3 marks)**

b Given also that $y = 1$ at $x = 1$, show that

$$y = \frac{2}{3 - x^2},\ -\sqrt{3} < x < \sqrt{3}$$

is a particular solution to the differential equation. **(3 marks)**

The curve C has equation $y = \frac{2}{3 - x^2}$, $x \neq \pm\sqrt{3}$

c Write down the gradient of C at the point $(1, 1)$. **(1 mark)**

d Hence write down an equation of the tangent to C at the points $(1, 1)$, and find the coordinates of the point where it again meets the curve. **(4 marks)**

15 a Using the substitution $u = 1 + 2x$, or otherwise, find

$$\int \frac{4x}{(1 + 2x)^2}\,dx,\ x \neq -\frac{1}{2}$$ **(5 marks)**

b Given that $y = \frac{\pi}{4}$ when $x = 0$, solve the differential equation

$$(1 + 2x)^2 \frac{dy}{dx} = \frac{x}{\sin^2 y}$$ **(5 marks)**

Integration

16 The diagram shows the curve with equation $y = xe^{2x}$, $-\frac{1}{2} \leq x \leq \frac{1}{2}$.

The finite region R_1 bounded by the curve, the x-axis and the line $x = -\frac{1}{2}$ has area A_1.

The finite region R_2 bounded by the curve, the x-axis and the line $x = \frac{1}{2}$ has area A_2.

a Find the exact values of A_1 and A_2 by integration. **(6 marks)**

b Show that $A_1 : A_2 = (e - 2) : e$. **(4 marks)**

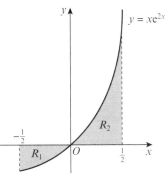

17 a Find $\int x^2 e^{-x} \, dx$. **(5 marks)**

b Use your answer to part **a** to find the solution to the differential equation $\dfrac{dy}{dx} = x^2 e^{3y - x}$, given that $y = 0$ when $x = 0$. Express your answer in the form $y = f(x)$. **(7 marks)**

18 The diagram shows part of the curve $y = e^{3x} + 1$ and the line $y = 8$.

The curve and the line intersect at the point $(h, 8)$.

a Find h, giving your answer in terms of natural logarithms. **(3 marks)**

The region R is bounded by the curve, the x-axis, the y-axis and the line $x = h$.

b Use integration to show the area of R is $2 + \frac{1}{3} \ln 7$. **(5 marks)**

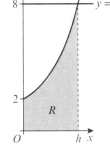

19 a Given that
$$\frac{x^2}{x^2 - 1} \equiv A + \frac{B}{x - 1} + \frac{C}{x + 1}$$
find the values of the constants A, B and C. **(4 marks)**

b Given that $x = 2$ at $t = 1$, solve the differential equation
$$\frac{dx}{dt} = 2 - \frac{2}{x^2}, \quad x > 1$$
You do not need to simplify your final answer. **(7 marks)**

20 The curve with equation $y = e^{2x} - e^{-x}$, $0 \leq x \leq 1$, is shown in the diagram. The finite region enclosed by the curve, the x-axis and the line $x = 1$ is shaded.

The table below shows the corresponding values of x and y with the y values given to 5 decimal places as appropriate.

x	0	0.25	0.5	0.75	1
y	0	0.86992	2.11175		7.02118

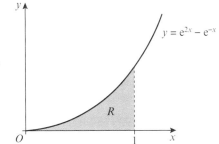

a Complete the table with the missing value for y. Give your answer to 5 decimal places. **(1 mark)**

b Use the trapezium rule, with all the values of y in the table, to obtain an estimate for the area of R, giving your answer to 4 decimal places. **(3 marks)**

c State, with a reason, whether your answer to part **b** is an overestimate or an underestimate. **(1 mark)**

d Use integration to find the exact value of R. Write your answer in the form $\dfrac{e^3 + Pe + Q}{2e}$ where P and Q are constants to be found. **(6 marks)**

e Find the percentage error in the answer to part **b**. **(2 marks)**

E/P 21 The rate, in $\text{cm}^3\,\text{s}^{-1}$, at which oil is leaking from an engine sump at any time t seconds is proportional to the volume of oil, $V\,\text{cm}^3$, in the sump at that instant. At time $t = 0$, $V = A$.

 a By forming and integrating a differential equation, show that
 $$V = Ae^{-kt}$$
 where k is a positive constant. **(5 marks)**

 b Sketch a graph to show the relation between V and t. **(2 marks)**

 Given further that $V = \tfrac{1}{2}A$ at $t = T$,

 c show that $kT = \ln 2$. **(3 marks)**

E/P 22 a Show that the general solution to the differential equation $\dfrac{dy}{dx} = \dfrac{x}{k - y}$ can be written in the form $x^2 + (y - k)^2 = c$. **(4 marks)**

 b Describe the family of curves that satisfy this differential equation when $k = 2$. **(2 marks)**

E/P 23 The diagram shows a sketch of the curve $y = f(x)$, where $f(x) = \tfrac{1}{5}x^2 \ln x - x + 2$, $x > 0$.

The region R, shown in the diagram, is bounded by the curve, the x-axis and the lines with equations $x = 1$ and $x = 4$.

The table below shows the corresponding values of x and y with the y values given to 4 decimal places as appropriate.

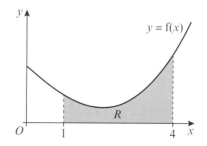

x	1	1.5	2	2.5	3	3.5	4
y	1	0.6825	0.5545	0.6454		1.5693	2.4361

 a Complete the table with the missing value of y. **(1 mark)**

 b Use the trapezium rule, with all the values of y in the table, to obtain an estimate for the area of R, giving your answer to 3 decimal places. **(3 marks)**

 c Explain how the trapezium rule could be used to obtain a more accurate estimate for the area of R. **(1 mark)**

 d Show that the exact area of R can be written in the form $\dfrac{a}{b} + \dfrac{c}{d} \ln e$, where a, b, c, d and e are integers. **(6 marks)**

 e Find the percentage error in the answer in part **b**. **(2 marks)**

E 24 a Find $\int x(1 + 2x^2)^5 \, dx$. **(3 marks)**

 b Given that $y = \dfrac{\pi}{8}$ at $x = 0$, solve the differential equation
 $$\dfrac{dy}{dx} = x(1 + 2x^2)^5 \cos^2 2y$$
 (5 marks)

334

25 By using an appropriate trigonometric substitution, find $\int \frac{1}{1+x^2} dx$. **(5 marks)**

26 Obtain the solution to
$$x(x+2)\frac{dy}{dx} = y, \ y > 0, \ x > 0$$
for which $y = 2$ at $x = 2$, giving your answer in the form $y^2 = f(x)$. **(7 marks)**

27 An oil spill is modelled as a circular disc with radius r km and area A km². The rate of increase of the area of the oil spill, in km²/day at time t days after it occurs is modelled as:
$$\frac{dA}{dt} = k \sin\left(\frac{t}{3\pi}\right), \ 0 \leqslant t \leqslant 12$$
a Show that $\frac{dr}{dt} = \frac{k}{2\pi r} \sin\left(\frac{t}{3\pi}\right)$ **(2 marks)**

Given that the radius of the spill at time $t = 0$ is 1 km, and the radius of the spill at time $t = \pi^2$ is 2 km:

b find an expression for r^2 in terms of t **(7 marks)**

c find the time, in days and hours to the nearest hour, after which the radius of the spill is 1.5 km. **(3 marks)**

28 The diagram shows the curve C with parametric equations
$x = 3t^2, \ y = \sin 2t, \ t \geqslant 0$.

a Write down the value of t at the point A where the curve crosses the x-axis. **(1 mark)**

b Find, in terms of π, the exact area of the shaded region bounded by C and the x-axis. **(6 marks)**

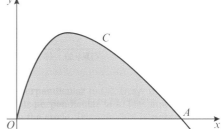

29 The curve shown has parametric equations
$x = 5\cos\theta, \ y = 4\sin\theta, \ 0 \leqslant \theta \leqslant 2\pi$

a Find the gradient of the curve at the point P at which $\theta = \frac{\pi}{4}$ **(3 marks)**

b Find an equation of the tangent to the curve at the point P. **(3 marks)**

c Find the exact area of the shaded region bounded by the tangent PR, the curve and the x-axis. **(6 marks)**

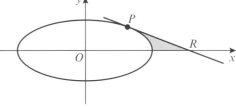

30 The curve C has parametric equations
$x = 1 - t^2, \ y = 2t - t^3, \ t \in \mathbb{R}$

The line L is a normal to the curve at the point P where the curve intersects the positive y-axis. Find the exact area of the region R bounded by the curve C, the line L and the x-axis, as shown on the diagram. **(7 marks)**

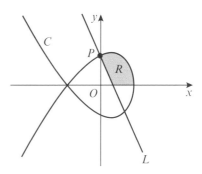

Challenge

Given $f(x) = x^2 - x - 2$, find:

a $\int_{-3}^{3} |f(x)| \, dx$ **b** $\int_{-3}^{3} f(|x|) \, dx$

Hint Draw a sketch of each function.

Summary of key points

1. $\int x^n \, dx = \dfrac{x^{n+1}}{n+1} + c$ $\int e^x \, dx = e^x + c$ $\int \dfrac{1}{x} \, dx = \ln|x| + c$

   $\int \cos x \, dx = \sin x + c$ $\int \sin x \, dx = -\cos x + c$ $\int \sec^2 x = \tan x + c$

   $\int \cosec x \cot x \, dx = -\cosec x + c$ $\int \cosec^2 x \, dx = -\cot x + c$ $\int \sec x \tan x \, dx = \sec x + c$

2. $\int f'(ax + b) \, dx = \dfrac{1}{a} f(ax + b) + c$

3. Trigonometric identities can be used to integrate expressions. This allows an expression that cannot be integrated to be replaced by an identical expression that can be integrated.

4. To integrate expressions of the form $\int k \dfrac{f'(x)}{f(x)} \, dx$, try $\ln|f(x)|$ and differentiate to check, and then adjust any constant.

5. To integrate an expression of the form $\int k f'(x)(f(x))^n \, dx$, try $(f(x))^{n+1}$ and differentiate to check, and then adjust any constant.

6. Sometimes you can simplify an integral by changing the variable. This process is similar to using the chain rule in differentiation and is called **integration by substitution**.

7. The **integration by parts** formula is given by: $\int u \dfrac{dv}{dx} \, dx = uv - \int v \dfrac{du}{dx} \, dx$

8. Partial fractions can be used to integrate algebraic fractions.

9. The area bounded by two curves can be found using integration:
   Area of $R = \int_a^b (f(x) - g(x)) \, dx = \int_a^b f(x) \, dx - \int_a^b g(x) \, dx$

10. The **trapezium rule** is:
    $\int_a^b y \, dx \approx \tfrac{1}{2} h(y_0 + 2(y_1 + y_2 \ldots + y_{n-1}) + y_n)$
    where $h = \dfrac{b-a}{n}$ and $y_i = f(a + ih)$.

11. When $\dfrac{dy}{dx} = f(x)g(y)$ you can write
    $\int \dfrac{1}{g(y)} \, dy = \int f(x) \, dx$

12. You can use limit notation to formalise the idea of a definite integral as the limit of a sum of areas of rectangular strips.
    $\int_a^b f(x) \, dx = \lim_{\delta x \to 0} \sum_{x=a}^{b} f(x) \delta x$

12 Vectors

Objectives
After completing this chapter you should be able to:
- Understand 3D Cartesian coordinates → **pages 338–339**
- Use vectors in three dimensions → **pages 340–344**
- Use vectors to solve geometric problems → **pages 344–348**
- Model 3D motion in mechanics with vectors → **pages 348–350**

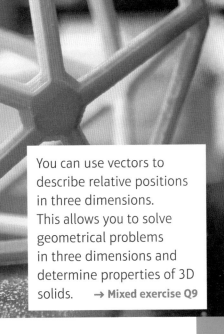

Prior knowledge check

1. Given that $\mathbf{p} = 3\mathbf{i} - \mathbf{j}$ and $\mathbf{q} = -\mathbf{i} + 2\mathbf{j}$, calculate:
   a $2\mathbf{p} + \mathbf{q}$ b $-3\mathbf{p} + 4\mathbf{q}$
   ← Year 1, Section 11.2

2. Given that $\mathbf{a} = 5\mathbf{i} - 3\mathbf{j}$, work out:
   a the magnitude of $\mathbf{a}$
   b the unit vector that is parallel to $\mathbf{a}$.
   ← Year 1, Section 11.3

3. M is the midpoint of the line segment AB. Given that $\overrightarrow{AB} = 4\mathbf{i} + \mathbf{j}$,
   a find $\overrightarrow{BM}$ in terms of $\mathbf{i}$ and $\mathbf{j}$.
   The point P lies on AB such that $AP:PB = 3:1$.
   b Find $\overrightarrow{AP}$ in terms of $\mathbf{i}$ and $\mathbf{j}$.
   ← Year 1, Section 11.5

You can use vectors to describe relative positions in three dimensions. This allows you to solve geometrical problems in three dimensions and determine properties of 3D solids. → **Mixed exercise Q9**

12.1 3D coordinates

Cartesian coordinate axes in three dimensions are usually called x-, y- and z-axes, each being at right angles to each of the others.

The coordinates of a point in three dimensions are written as (x, y, z).

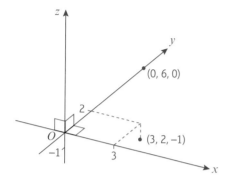

> **Hint** To visualise this, think of the x- and y-axes being drawn on a flat surface and the z-axis sticking up from the surface.

You can use Pythagoras' theorem in 3D to find distances on a 3D coordinate grid.

- **The distance from the origin to the point (x, y, z) is $\sqrt{x^2 + y^2 + z^2}$.**

Example 1

Find the distance from the origin to the point $P(4, -7, -1)$.

$$OP = \sqrt{4^2 + (-7)^2 + (-1)^2}$$
$$= \sqrt{16 + 49 + 1}$$
$$= \sqrt{66}$$

Substitute the values of x, y and z into the formula. You don't need to give units with distances on a coordinate grid.

You can also use Pythagoras' theorem to find the distance between two points.

- **The distance between the points (x_1, y_1, z_1) and (x_2, y_2, z_2) is**
$$\sqrt{(x_1 - x_2)^2 + (y_1 - y_2)^2 + (z_1 - z_2)^2}$$

Example 2

Find the distance between the points $A(1, 3, 4)$ and $B(8, 6, -5)$, giving your answer to 1 d.p.

$$AB = \sqrt{(1-8)^2 + (3-6)^2 + (4-(-5))^2}$$
$$= \sqrt{(-7)^2 + (-3)^2 + 9^2}$$
$$= \sqrt{49 + 9 + 81}$$
$$= \sqrt{139} = 11.8 \text{ (1 d.p.)}$$

Be careful with the signs – use brackets when you substitute.

Example 3

The coordinates of A and B are $(5, 0, 3)$ and $(4, 2, k)$ respectively.

Given that the distance from A to B is 3 units, find the possible values of k.

$AB = \sqrt{(5-4)^2 + (0-2)^2 + (3-k)^2} = 3$

$\sqrt{1 + 4 + (9 - 6k + k^2)} = 3$

$1 + 4 + 9 - 6k + k^2 = 9$

$k^2 - 6k + 5 = 0$

$(k - 5)(k - 1) = 0$

$k = 1$ or $k = 5$

Problem-solving

Use Pythagoras' theorem to form a quadratic equation in k.

Square both sides of the equation.

Solve to find the two possible values of k.

Online Explore the solution to this example visually in 3D using GeoGebra.

Exercise 12A

1 Find the distance from the origin to the point $P(2, 8, -4)$.

2 Find the distance from the origin to the point $P(7, 7, 7)$.

3 Find the distance between A and B when they have the following coordinates:
   a $A(3, 0, 5)$ and $B(1, -1, 8)$
   b $A(8, 11, 8)$ and $B(-3, 1, 6)$
   c $A(3, 5, -2)$ and $B(3, 10, 3)$
   d $A(-1, -2, 5)$ and $B(4, -1, 3)$

(P) 4 The coordinates of A and B are $(7, -1, 2)$ and $(k, 0, 4)$ respectively.
   Given that the distance from A to B is 3 units, find the possible values of k.

(P) 5 The coordinates of A and B are $(5, 3, -8)$ and $(1, k, -3)$ respectively.
   Given that the distance from A to B is $3\sqrt{10}$ units, find the possible values of k.

Challenge

a The points $A(1, 3, -2)$, $B(1, 3, 4)$ and $C(7, -3, 4)$ are three vertices of a solid cube. Write down the coordinates of the remaining five vertices.

An ant walks from A to C along the surface of the cube.

b Determine the length of the shortest possible route the ant can take.

12.2 Vectors in 3D

You can use 3D vectors to describe position and displacement relative to the x-, y- and z-axes. You can represent 3D vectors as column vectors or using the unit vectors **i**, **j** and **k**.

- **The unit vectors along the x-, y- and z-axes are denoted by i, j and k respectively.**

$$\mathbf{i} = \begin{pmatrix} 1 \\ 0 \\ 0 \end{pmatrix} \quad \mathbf{j} = \begin{pmatrix} 0 \\ 1 \\ 0 \end{pmatrix} \quad \mathbf{k} = \begin{pmatrix} 0 \\ 0 \\ 1 \end{pmatrix}$$

- **For any 3D vector $p\mathbf{i} + q\mathbf{j} + r\mathbf{k} = \begin{pmatrix} p \\ q \\ r \end{pmatrix}$**

Links 3D vectors obey all the same addition and scalar multiplication rules as 2D vectors.
← Year 1, Chapter 11

Example 4

Consider the points $A(1, 5, -2)$ and $B(0, -3, 7)$.

a Find the position vectors of A and B in **ijk** notation.
b Find the vector $\overrightarrow{AB}$ as a column vector.

a $\overrightarrow{OA} = \mathbf{i} + 5\mathbf{j} - 2\mathbf{k}$, $\overrightarrow{OB} = -3\mathbf{j} + 7\mathbf{k}$

b $\overrightarrow{AB} = \overrightarrow{OB} - \overrightarrow{OA}$

$$= \begin{pmatrix} 0 \\ -3 \\ 7 \end{pmatrix} - \begin{pmatrix} 1 \\ 5 \\ -2 \end{pmatrix} = \begin{pmatrix} -1 \\ -8 \\ 9 \end{pmatrix}$$

The position vector of a point is the vector from the origin to that point.

$\overrightarrow{OB}$ has no component in the **i** direction. You could write it as $0\mathbf{i} - 3\mathbf{j} + 7\mathbf{k}$.

When adding and subtracting vectors it is often easier to write them as column vectors.

Example 5

The vectors **a** and **b** are given as $\mathbf{a} = \begin{pmatrix} 2 \\ -3 \\ 5 \end{pmatrix}$ and $\mathbf{b} = \begin{pmatrix} 4 \\ -2 \\ 0 \end{pmatrix}$.

a Find:
  i $4\mathbf{a} + \mathbf{b}$ **ii** $2\mathbf{a} - 3\mathbf{b}$

b State with a reason whether each of these vectors is parallel to $4\mathbf{i} - 5\mathbf{k}$.

Online Perform calculations on 3D vectors using your calculator.

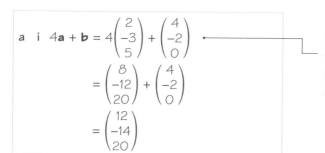

Use the rules for scalar multiplication and addition of vectors:

$$\lambda \begin{pmatrix} p \\ q \\ r \end{pmatrix} = \begin{pmatrix} \lambda p \\ \lambda q \\ \lambda r \end{pmatrix} \text{ and } \begin{pmatrix} p \\ q \\ r \end{pmatrix} + \begin{pmatrix} u \\ v \\ w \end{pmatrix} = \begin{pmatrix} p + u \\ q + v \\ r + w \end{pmatrix}$$

ii $\quad 2\mathbf{a} - 3\mathbf{b} = 2\begin{pmatrix} 2 \\ -3 \\ 5 \end{pmatrix} - 3\begin{pmatrix} 4 \\ -2 \\ 0 \end{pmatrix}$

$= \begin{pmatrix} 4 \\ -6 \\ 10 \end{pmatrix} - \begin{pmatrix} 12 \\ -6 \\ 0 \end{pmatrix}$

$= \begin{pmatrix} -8 \\ 0 \\ 10 \end{pmatrix}$

b i $\quad 4\mathbf{a} + \mathbf{b} = \begin{pmatrix} 12 \\ -14 \\ 20 \end{pmatrix} = 3\begin{pmatrix} 4 \\ -\frac{14}{3} \\ \frac{20}{3} \end{pmatrix}$

which is not a multiple of $\begin{pmatrix} 4 \\ 0 \\ -5 \end{pmatrix}$

Two vectors are parallel if one is a multiple of the other. Make the x-components the same and compare the y- and z-components with $4\mathbf{i} - 5\mathbf{k}$.

$4\mathbf{a} + \mathbf{b}$ is not parallel to $4\mathbf{i} - 5\mathbf{k}$

ii $\quad 2\mathbf{a} - 3\mathbf{b} = \begin{pmatrix} -8 \\ 0 \\ 10 \end{pmatrix} = -2\begin{pmatrix} 4 \\ 0 \\ -5 \end{pmatrix}$

which is a multiple of $\begin{pmatrix} 4 \\ 0 \\ -5 \end{pmatrix}$

Watch out $4\mathbf{i} - 5\mathbf{k} = 4\mathbf{i} + 0\mathbf{j} - 5\mathbf{k}$. Make sure you include a 0 in the $\mathbf{j}$-component of the column vector.

$2\mathbf{a} - 3\mathbf{b}$ is parallel to $4\mathbf{i} - 5\mathbf{k}$

Example 6

Find the magnitude of $\mathbf{a} = 2\mathbf{i} - \mathbf{j} + 4\mathbf{k}$ and hence find $\hat{\mathbf{a}}$, the unit vector in the direction of $\mathbf{a}$.

The magnitude of **a** is given by
$|\mathbf{a}| = \sqrt{2^2 + (-1)^2 + 4^2}$ ——— Use Pythagoras' theorem.

$= \sqrt{21}$

$\hat{\mathbf{a}} = \dfrac{\mathbf{a}}{|\mathbf{a}|} = \dfrac{1}{\sqrt{21}}(2\mathbf{i} - \mathbf{j} + 4\mathbf{k})$ ——— You could also write this as $\dfrac{2}{\sqrt{21}}\mathbf{i} - \dfrac{1}{\sqrt{21}}\mathbf{j} + \dfrac{4}{\sqrt{21}}\mathbf{k}$

Online Check your answer using the vector functions on your calculator.

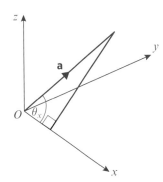

You can find the angle between a given vector and any of the coordinate axes by considering the appropriate right-angled triangle.

- If the vector $\mathbf{a} = x\mathbf{i} + y\mathbf{j} + z\mathbf{k}$ makes an angle θ_x with the positive x-axis then $\cos \theta_x = \dfrac{x}{|\mathbf{a}|}$ and similarly for the angles θ_y and θ_z.

Hint This rule also works with vectors in two dimensions.

341

Example 7

Find the angles that the vector $\mathbf{a} = 2\mathbf{i} - 3\mathbf{j} - \mathbf{k}$ makes with each of the positive coordinate axes to 1 d.p.

$|\mathbf{a}| = \sqrt{2^2 + (-3)^3 + (-1)^2} = \sqrt{4 + 9 + 1} = \sqrt{14}$ — First find $|\mathbf{a}|$ since you will be using it three times.

$\cos\theta_x = \dfrac{x}{|\mathbf{a}|} = \dfrac{2}{\sqrt{14}} = 0.5345...$

$\theta_x = 57.7°$ (1 d.p.)

Write down at least 4 d.p., or use the answer button on your calculator to enter the exact value.

$\cos\theta_y = \dfrac{y}{|\mathbf{a}|} = \dfrac{-3}{\sqrt{14}} = -0.8017...$

$\theta_y = 143.3°$ (1 d.p.)

The formula also works with negative components. The y-component is negative, so the vector makes an obtuse angle with the positive y-axis.

$\cos\theta_z = \dfrac{z}{|\mathbf{a}|} = \dfrac{-1}{\sqrt{14}} = -0.2672...$

$\theta_z = 105.5°$ (1 d.p.)

Example 8

The points A and B have position vectors $4\mathbf{i} + 2\mathbf{j} + 7\mathbf{k}$ and $3\mathbf{i} + 4\mathbf{j} - \mathbf{k}$ relative to a fixed origin, O. Find $\overrightarrow{AB}$ and show that $\triangle OAB$ is isosceles.

$\overrightarrow{OA} = \mathbf{a} = \begin{pmatrix} 4 \\ 2 \\ 7 \end{pmatrix}$, $\overrightarrow{OB} = \mathbf{b} = \begin{pmatrix} 3 \\ 4 \\ -1 \end{pmatrix}$ — Write down the position vectors of A and B.

$\overrightarrow{AB} = \mathbf{b} - \mathbf{a} = \begin{pmatrix} 3 \\ 4 \\ -1 \end{pmatrix} - \begin{pmatrix} 4 \\ 2 \\ 7 \end{pmatrix} = \begin{pmatrix} -1 \\ 2 \\ -8 \end{pmatrix}$ — Use $\overrightarrow{AB} = \mathbf{b} - \mathbf{a}$.

$|\overrightarrow{AB}| = \sqrt{(-1)^2 + 2^2 + (-8)^2} = \sqrt{69}$ — This is the length of the line segment AB.

$|\overrightarrow{OA}| = \sqrt{4^2 + 2^2 + 7^2} = \sqrt{69}$

$|\overrightarrow{OB}| = \sqrt{3^2 + 4^2 + (-1)^2} = \sqrt{26}$

Find the lengths of the other sides OA and OB of $\triangle OAB$.

So $\triangle OAB$ is isosceles, with $AB = OA$.

Online Explore the solution to this example visually in 3D using GeoGebra.

Exercise 12B

1 The vectors $\mathbf{a}$ and $\mathbf{b}$ are defined by $\mathbf{a} = \begin{pmatrix} 1 \\ 2 \\ -4 \end{pmatrix}$ and $\mathbf{b} = \begin{pmatrix} 4 \\ -3 \\ 5 \end{pmatrix}$.

   a Find:

      i $\mathbf{a} - \mathbf{b}$ ii $-\mathbf{a} + 3\mathbf{b}$

   b State with a reason whether each of these vectors is parallel to $6\mathbf{i} - 10\mathbf{j} + 18\mathbf{k}$.

2 The vectors $\mathbf{a}$ and $\mathbf{b}$ are defined by $\mathbf{a} = \begin{pmatrix} 3 \\ 2 \\ -1 \end{pmatrix}$ and $\mathbf{b} = \begin{pmatrix} -3 \\ -2 \\ 4 \end{pmatrix}$.

   Show that the vector $3\mathbf{a} + 2\mathbf{b}$ is parallel to $6\mathbf{i} + 4\mathbf{j} + 10\mathbf{k}$.

3 The vectors **a** and **b** are defined by $\mathbf{a} = \begin{pmatrix} 1 \\ 2 \\ -4 \end{pmatrix}$ and $\mathbf{b} = \begin{pmatrix} p \\ q \\ r \end{pmatrix}$.

   Given that $\mathbf{a} + 2\mathbf{b} = 5\mathbf{i} + 4\mathbf{j}$, find the values of p, q and r.

4 Find the magnitude of:
   a $3\mathbf{i} + 5\mathbf{j} + \mathbf{k}$
   b $4\mathbf{i} - 2\mathbf{k}$
   c $\mathbf{i} + \mathbf{j} - \mathbf{k}$
   d $5\mathbf{i} - 9\mathbf{j} - 8\mathbf{k}$
   e $\mathbf{i} + 5\mathbf{j} - 7\mathbf{k}$

5 Given that $\mathbf{p} = \begin{pmatrix} 5 \\ 0 \\ 2 \end{pmatrix}$, $\mathbf{q} = \begin{pmatrix} 2 \\ 1 \\ -3 \end{pmatrix}$ and $\mathbf{r} = \begin{pmatrix} 7 \\ -4 \\ 2 \end{pmatrix}$, find in column vector form:

   a $\mathbf{p} + \mathbf{q}$
   b $\mathbf{q} - \mathbf{r}$
   c $\mathbf{p} + \mathbf{q} + \mathbf{r}$
   d $3\mathbf{p} - \mathbf{r}$
   e $\mathbf{p} - 2\mathbf{q} + \mathbf{r}$

6 The position vector of the point A is $2\mathbf{i} - 7\mathbf{j} + 3\mathbf{k}$ and $\overrightarrow{AB} = 5\mathbf{i} + 4\mathbf{j} - \mathbf{k}$. Find the position vector of the point B.

7 Given that $\mathbf{a} = t\mathbf{i} + 2\mathbf{j} + 3\mathbf{k}$, and that $|\mathbf{a}| = 7$, find the possible values of t.

8 Given that $\mathbf{a} = 5t\mathbf{i} + 2t\mathbf{j} + t\mathbf{k}$, and that $|\mathbf{a}| = 3\sqrt{10}$, find the possible values of t.

9 The points A, B and C have coordinates $(2, 1, 4)$, $(3, -2, 4)$ and $(-1, 2, 2)$.
   a Find, in terms of **i**, **j** and **k**:
      i the position vectors of A, B and C
      ii $\overrightarrow{AC}$
   b Find the exact value of:
      i $|\overrightarrow{AC}|$
      ii $|\overrightarrow{OC}|$

10 P is the point $(3, 0, 7)$ and Q is the point $(-1, 3, -5)$. Find:
   a the vector $\overrightarrow{PQ}$
   b the distance between P and Q
   c the unit vector in the direction of $\overrightarrow{PQ}$.

11 $\overrightarrow{OA}$ is the vector $4\mathbf{i} - \mathbf{j} - 2\mathbf{k}$ and $\overrightarrow{OB}$ is the vector $-2\mathbf{i} + 3\mathbf{j} + \mathbf{k}$. Find:
   a the vector $\overrightarrow{AB}$
   b the distance between A and B
   c the unit vector in the direction of $\overrightarrow{AB}$.

Chapter 12

12 Find the unit vector in the direction of each of the following vectors.

a $\mathbf{p} = \begin{pmatrix} 3 \\ -4 \\ -2 \end{pmatrix}$
b $\mathbf{q} = \begin{pmatrix} \sqrt{2} \\ -4 \\ -\sqrt{7} \end{pmatrix}$
c $\mathbf{r} = \begin{pmatrix} \sqrt{5} \\ -2\sqrt{2} \\ -\sqrt{3} \end{pmatrix}$

(E/P) **13** The points A, B and C have position vectors $\begin{pmatrix} 8 \\ -7 \\ 4 \end{pmatrix}$, $\begin{pmatrix} 8 \\ -3 \\ 3 \end{pmatrix}$ and $\begin{pmatrix} 12 \\ -6 \\ 3 \end{pmatrix}$ respectively.

a Find the vectors $\overrightarrow{AB}$, $\overrightarrow{AC}$ and $\overrightarrow{BC}$. **(3 marks)**
b Find $|\overrightarrow{AB}|$, $|\overrightarrow{AC}|$ and $|\overrightarrow{BC}|$ giving your answers in exact form. **(6 marks)**
c Describe triangle ABC. **(1 mark)**

(E) **14** A is the point $(3, 4, 8)$, B is the point $(1, -2, 5)$ and C is the point $(7, -5, 7)$.

a Find the vectors $\overrightarrow{AB}$, $\overrightarrow{AC}$ and $\overrightarrow{BC}$. **(3 marks)**
b Hence find the lengths of the sides of triangle ABC. **(6 marks)**
c Given that angle $ABC = 90°$ find the size of angle BAC. **(2 marks)**

15 For each of the given vectors, find the angle made by the vector with:
  i the positive x-axis
  ii the positive y-axis
  iii the positive z-axis

a $-\mathbf{i} + 7\mathbf{j} + \mathbf{k}$
b $\begin{pmatrix} 3 \\ 4 \\ 7 \end{pmatrix}$
c $\begin{pmatrix} 2 \\ 0 \\ -3 \end{pmatrix}$

(P) **16** A scalene triangle has the coordinates $(2, 0, 0)$, $(5, 0, 0)$ and $(4, 2, 3)$. Work out the area of the triangle.

(E/P) **17** The diagram shows the triangle PQR.
Given that $\overrightarrow{PQ} = 3\mathbf{i} - \mathbf{j} + 2\mathbf{k}$ and
$\overrightarrow{QR} = -2\mathbf{i} + 4\mathbf{j} + 3\mathbf{k}$, show that
$\angle PQR = 78.5°$ to 1 d.p. **(5 marks)**

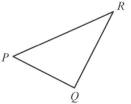

Challenge

Find the acute angle that the vector $\mathbf{a} = -2\mathbf{i} + 6\mathbf{j} - 3\mathbf{k}$ makes with the xy-plane. Give your answer to 1 d.p.

12.3 Solving geometric problems

You need to be able to solve geometric problems involving vectors in three dimensions.

Example 9

A, B, C and D are the points $(2, -5, -8)$, $(1, -7, -3)$, $(0, 15, -10)$ and $(2, 19, -20)$ respectively.

a Find $\overrightarrow{AB}$ and $\overrightarrow{DC}$, giving your answers in the form $p\mathbf{i} + q\mathbf{j} + r\mathbf{k}$.
b Show that the lines AB and DC are parallel and that $\overrightarrow{DC} = 2\overrightarrow{AB}$.
c Hence describe the quadrilateral $ABCD$.

Vectors

a $\vec{AB} = \vec{OB} - \vec{OA}$
$= (\mathbf{i} - 7\mathbf{j} - 3\mathbf{k}) - (2\mathbf{i} - 5\mathbf{j} - 8\mathbf{k})$
$= -\mathbf{i} - 2\mathbf{j} + 5\mathbf{k}$
$\vec{DC} = \vec{OC} - \vec{OD}$
$= (15\mathbf{j} - 10\mathbf{k}) - (2\mathbf{i} + 19\mathbf{j} - 20\mathbf{k})$
$= -2\mathbf{i} - 4\mathbf{j} + 10\mathbf{k}$

b $2\vec{AB} = 2(-\mathbf{i} - 2\mathbf{j} + 5\mathbf{k})$
$= -2\mathbf{i} - 4\mathbf{j} + 10\mathbf{k} = \vec{DC}$
So AB is parallel to DC and half as long.

c There are two unequal parallel sides, so $ABCD$ is a trapezium.

Watch out AB refers to the **line segment** between A and B (or its length), whereas $\vec{AB}$ refers to the **vector** from A to B. Note that $AB = BA$ but $\vec{AB} \neq \vec{BA}$.

Problem-solving
If you can't work out what shape it is, draw a sketch showing AB and DC.

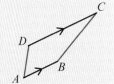

Online Explore the solution to this example visually in 3D using GeoGebra.

Example 10

P, Q and R are the points $(4, -9, -3)$, $(7, -7, -7)$ and $(8, -2, -0)$ respectively. Find the coordinates of the point S so that $PQRS$ forms a parallelogram.

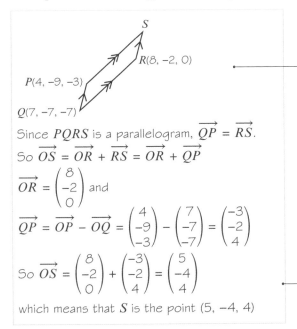

Since $PQRS$ is a parallelogram, $\vec{QP} = \vec{RS}$.
So $\vec{OS} = \vec{OR} + \vec{RS} = \vec{OR} + \vec{QP}$

$\vec{OR} = \begin{pmatrix} 8 \\ -2 \\ 0 \end{pmatrix}$ and

$\vec{QP} = \vec{OP} - \vec{OQ} = \begin{pmatrix} 4 \\ -9 \\ -3 \end{pmatrix} - \begin{pmatrix} 7 \\ -7 \\ -7 \end{pmatrix} = \begin{pmatrix} -3 \\ -2 \\ 4 \end{pmatrix}$

So $\vec{OS} = \begin{pmatrix} 8 \\ -2 \\ 0 \end{pmatrix} + \begin{pmatrix} -3 \\ -2 \\ 4 \end{pmatrix} = \begin{pmatrix} 5 \\ -4 \\ 4 \end{pmatrix}$

which means that S is the point $(5, -4, 4)$.

Draw a sketch. The vertices in a 2D shape are given in order, either clockwise or anticlockwise. It doesn't matter if the positions on the sketch don't correspond to the real positions in 3D – it is still a helpful way to visualise the problem.

You could also go from O to S via P:
$\vec{OS} = \vec{OP} + \vec{PS}$
$= \vec{OP} + \vec{QR}$

In two dimensions you saw that if **a** and **b** are two non-parallel vectors and $p\mathbf{a} + q\mathbf{b} = r\mathbf{a} + s\mathbf{b}$ then $p = r$ and $q = s$. In other words, in two dimensions with two vectors you can **compare coefficients** on both sides of an equation. In three dimensions you have to extend this rule:

Notation **Coplanar vectors** are vectors which are in the same plane.
Non-coplanar vectors are vectors which are **not** in the same plane.

- If **a**, **b** and **c** are vectors in three dimensions which do not all lie on the same plane then you can compare their coefficients on both sides of an equation.

In particular, since the vectors **i**, **j** and **k** are non-coplanar, if $p\mathbf{i} + q\mathbf{j} + r\mathbf{k} = u\mathbf{i} + v\mathbf{j} + w\mathbf{k}$ then $p = u$, $q = v$ and $r = w$.

Example 11

Given that $3\mathbf{i} + (p + 2)\mathbf{j} + 120\mathbf{k} = p\mathbf{i} - q\mathbf{j} + 4pqr\mathbf{k}$, find the values of p, q and r.

> Comparing coefficients of $\mathbf{i}$ gives $p = 3$.
>
> Comparing coefficients of $\mathbf{j}$ gives $p + 2 = -q$
> so $q = -(3 + 2) = -5$.
>
> Comparing coefficients of $\mathbf{k}$ gives
> $120 = 4pqr$ so $r = \dfrac{120}{4 \times 3 \times (-5)} = -2$

Since $\mathbf{i}$, $\mathbf{j}$ and $\mathbf{k}$ do not lie in the same plane you can compare coefficients.

When comparing coefficients like this just write the coefficients. For example, write $3 = p$, not $3\mathbf{i} = p\mathbf{i}$.

Example 12

The diagram shows a cuboid whose vertices are O, A, B, C, D, E, F and G. Vectors $\mathbf{a}$, $\mathbf{b}$ and $\mathbf{c}$ are the position vectors of the vertices A, B and C respectively. Prove that the diagonals OE and BG bisect each other.

Hint Bisect means 'cut into two equal parts'. In this case you need to prove that **both** diagonals are bisected.

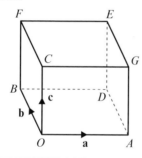

Problem-solving

If there is a point of intersection, H, it must lie on both diagonals. You can reach H directly from O (travelling along OE), or by first travelling to B then travelling along BG. Use this to write two expressions for $\overrightarrow{OH}$.

> Suppose there is a point of intersection, H, of OE and BG.
> $\overrightarrow{OH} = r\overrightarrow{OE}$ for some scalar r.
> But $\overrightarrow{OH} = \overrightarrow{OB} + \overrightarrow{BH}$ and $\overrightarrow{BH} = s\overrightarrow{BG}$
> for some scalar s, so $\overrightarrow{OH} = \overrightarrow{OB} + s\overrightarrow{BG}$.
> So $r\overrightarrow{OE} = \overrightarrow{OB} + s\overrightarrow{BG}$ (1)
> Now $\overrightarrow{OE} = \overrightarrow{OA} + \overrightarrow{AD} + \overrightarrow{DE} = \mathbf{a} + \mathbf{b} + \mathbf{c}$,
> $\overrightarrow{OB} = \mathbf{b}$ and $\overrightarrow{BG} = \overrightarrow{OG} - \overrightarrow{OB} = \mathbf{a} + \mathbf{c} - \mathbf{b}$
> So (1) becomes $r(\mathbf{a} + \mathbf{b} + \mathbf{c}) = \mathbf{b} + s(\mathbf{a} + \mathbf{c} - \mathbf{b})$
> Comparing coefficients in $\mathbf{a}$ and $\mathbf{b}$ gives $r = s$
> and $r = 1 - s$
> Solving simultaneously gives $r = s = \frac{1}{2}$
> These solutions also satisfy the coefficients of $\mathbf{c}$ so the lines do intersect at H.
> $OH = \frac{1}{2}OE$ so H bisects OE.
> $BH = \frac{1}{2}BG$ so H bisects BG, as required.

H lies on the line OE, so $\overrightarrow{OH}$ must be some scalar multiple of $\overrightarrow{OE}$.

Use the fact that H lies on both diagonals to find two different expression for $\overrightarrow{OH}$. You can equate these expressions and compare coefficients.

$\mathbf{a}$, $\mathbf{b}$ and $\mathbf{c}$ are three non-coplanar vectors so you can compare coefficients.

In order for the lines to intersect, the values of r and s must satisfy equation (1) completely:
$\frac{1}{2}(\mathbf{a} + \mathbf{b} + \mathbf{c}) = \mathbf{b} + \frac{1}{2}(\mathbf{a} + \mathbf{c} - \mathbf{b})$
The coefficients of $\mathbf{a}$, $\mathbf{b}$ and $\mathbf{c}$ all match so both ways of writing the vector OH are identical.

Vector proofs such as this one often avoid any coordinate geometry, which tends to be messy and complicated, especially in three dimensions.

Exercise 12C

P 1 The points A, B and C have position vectors $\begin{pmatrix} 1 \\ -4 \\ 8 \end{pmatrix}$, $\begin{pmatrix} 4 \\ 4 \\ 7 \end{pmatrix}$ and $\begin{pmatrix} 10 \\ 0 \\ 30 \end{pmatrix}$ relative to a fixed origin, O.

  a Show that:

  i $|\overrightarrow{OA}| = |\overrightarrow{OB}|$ **ii** $|\overrightarrow{AC}| = |\overrightarrow{BC}|$

  b Hence describe the quadrilateral OACB.

P 2 The points A, B and C have coordinates (2, 1, 5), (4, 4, 3) and (2, 7, 5) respectively.

  a Show that triangle ABC is isosceles.
  b Find the area of triangle ABC.
  c Find a point D such that ABCD is a parallelogram.

P 3 The points A, B, C and D have coordinates (7, 12, −1), (11, 2, −9), (14, −14, 3) and (8, 1, 15) respectively.

  a Show that AB and CD are parallel, and find the ratio AB : CD in its simplest form.
  b Hence describe the quadrilateral ABCD.

P 4 Given that $(3a + b)\mathbf{i} + \mathbf{j} + ac\mathbf{k} = 7\mathbf{i} - b\mathbf{j} + 4\mathbf{k}$, find the values of a, b and c.

P 5 The points A and B have position vectors $10\mathbf{i} - 23\mathbf{j} + 10\mathbf{k}$ and $p\mathbf{i} + 14\mathbf{j} - 22\mathbf{k}$ respectively, relative to a fixed origin O, where p is a constant.

Given that △OAB is isosceles, find **three** possible positions of point B.

E/P 6 The diagram shows a triangle ABC.
Given that $\overrightarrow{AB} = 7\mathbf{i} - \mathbf{j} + 2\mathbf{k}$ and $\overrightarrow{BC} = -\mathbf{i} + 5\mathbf{k}$

  a find the area of triangle ABC. **(7 marks)**

The point D is such that $\overrightarrow{AD} = 3\overrightarrow{AB}$, and the point E is such that $\overrightarrow{AE} = 3\overrightarrow{AC}$.

  b Find the area of triangle ADE. **(2 marks)**

P 7 A parallelepiped is a three-dimensional figure formed by six parallelograms. The diagram shows a parallelepiped with vertices O, A, B, C, D, E, F, and G.
a, **b** and **c** are the vectors $\overrightarrow{OA}$, $\overrightarrow{OB}$ and $\overrightarrow{OC}$ respectively.
Prove that the diagonals OF and AG bisect each other.

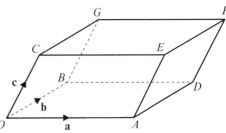

P 8 The diagram shows a cuboid whose vertices are O, A, B, C, D, E, F and G. **a**, **b** and **c** are the position vectors of the vertices A, B and C respectively. The point M lies on OE such that OM : ME = 3 : 1. The straight line AP passes through point M. Given that AM : MP = 3 : 1, prove that P lies on the line EF and find the ratio FP : PE.

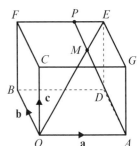

347

Chapter 12

> **Challenge**
>
> 1 **a**, **b** and **c** are the vectors $\begin{pmatrix} 1 \\ 0 \\ 4 \end{pmatrix}$, $\begin{pmatrix} 2 \\ 0 \\ -3 \end{pmatrix}$ and $\begin{pmatrix} -5 \\ 3 \\ 1 \end{pmatrix}$ respectively. Find scalars p, q and r such that $p\mathbf{a} + q\mathbf{b} + r\mathbf{c} = \begin{pmatrix} 28 \\ -12 \\ -4 \end{pmatrix}$
>
> 2 The diagram shows a cuboid with vertices O, A, B, C, D, E, F and G. M is the midpoint of FE and N is the midpoint of AG.
>
>   **a**, **b** and **c** are the position vectors of the vertices A, B and C respectively.
>
>   Prove that the lines OM and BN trisect the diagonal AF.
>
>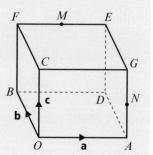
>
> **Hint** Trisect means divide into three equal parts.

12.4 Application to mechanics

3D vectors can be used to model problems in mechanics in the same way as you have previously used 2D vectors.

Example 13

A particle of mass 0.5 kg is acted on by three forces:
$$\mathbf{F_1} = (2\mathbf{i} - \mathbf{j} + 2\mathbf{k})\,\text{N}$$
$$\mathbf{F_2} = (-\mathbf{i} + 3\mathbf{j} - 3\mathbf{k})\,\text{N}$$
$$\mathbf{F_3} = (4\mathbf{i} - 3\mathbf{j} - 2\mathbf{k})\,\text{N}$$

a Find the resultant force **R** acting on the particle.
b Find the acceleration of the particle, giving your answer in the form $(p\mathbf{i} + q\mathbf{j} + r\mathbf{k})\,\text{m s}^{-2}$.
c Find the magnitude of the acceleration.

Given that the particle starts at rest,

d find the distance travelled by the particle in the first 6 seconds of its motion.

a $\begin{pmatrix} 2 \\ -1 \\ 2 \end{pmatrix} + \begin{pmatrix} -1 \\ 3 \\ -3 \end{pmatrix} + \begin{pmatrix} 4 \\ -3 \\ -2 \end{pmatrix} = \begin{pmatrix} 5 \\ -1 \\ -3 \end{pmatrix}$ — $\mathbf{R} = \mathbf{F_1} + \mathbf{F_2} + \mathbf{F_3}$
It is easier to add vectors in column vector form.

$\mathbf{R} = (5\mathbf{i} - \mathbf{j} - 3\mathbf{k})\,\text{N}$

b $\mathbf{F} = m\mathbf{a}$ — Use the vector form of $\mathbf{F} = m\mathbf{a}$. This is Newton's second law of motion.
$(5\mathbf{i} - \mathbf{j} - 3\mathbf{k}) = 0.5\,\mathbf{a}$
$\mathbf{a} = (10\mathbf{i} - 2\mathbf{j} - 6\mathbf{k})\,\text{m s}^{-2}$
← Statistics and Mechanics Year 1, Chapter 10

c $|\mathbf{a}| = \sqrt{10^2 + (-2)^2 + (-6)^2} = \sqrt{140}\,\text{m s}^{-2}$

d $u = 0$, $a = \sqrt{140}\,\text{m s}^{-2}$, $t = 6\text{s}$, $s = ?$
$s = ut + \tfrac{1}{2}at^2$
$= \tfrac{1}{2} \times \sqrt{140} \times 6^2$
$= 36\sqrt{35}\,\text{m}$

Because the particle starts from rest the acceleration acts in the same direction as the motion of the particle. So it is moving in a straight line with constant acceleration. You can use the *suvat* formulae to find the distance travelled.
← Statistics and Mechanics Year 1, Chapter 9

Exercise 12D

1. A particle is acted upon by forces of $(3\mathbf{i} - 2\mathbf{j} + \mathbf{k})$ N, $(7\mathbf{i} + 4\mathbf{j} + 3\mathbf{k})$ N and $(-5\mathbf{i} - 3\mathbf{j})$ N.
   a Work out the resultant force **R**.
   b Find the exact magnitude of the resultant force.

(P) 2. A particle, initially at rest, is acted upon by a force that causes the particle to accelerate at $(4\mathbf{i} - 2\mathbf{j} + 3\mathbf{k})$ m s^{-2} for 2 seconds. Work out the distance travelled by the particle.

3. A body of mass 4 kg is moving with a constant velocity when it is acted upon by a force of $(2\mathbf{i} - 5\mathbf{j} + 3\mathbf{k})$ N.
   a Find the acceleration of the body while the force acts.
   b Find the magnitude of this acceleration to 3 s.f.

(P) 4. A particle of mass 6 kg is acted on by two forces, $\mathbf{F}_1$ and $\mathbf{F}_2$. Given that $\mathbf{F}_1 = (7\mathbf{i} + 3\mathbf{j} + \mathbf{k})$ N, and that the particle is accelerating at $(2\mathbf{i} - \mathbf{k})$ m s^{-2}, find $\mathbf{F}_2$, giving your answer in the form $(p\mathbf{i} + q\mathbf{j} + r\mathbf{k})$ N.

(P) 5. A particle of mass 2 kg is in static equilibrium and is acted upon by three forces:
   $\mathbf{F}_1 = (\mathbf{i} - \mathbf{j} - 2\mathbf{k})$ N
   $\mathbf{F}_2 = (-\mathbf{i} + 3\mathbf{j} + b\mathbf{k})$ N
   $\mathbf{F}_3 = (a\mathbf{j} - 2\mathbf{k})$ N
   a Find the values of the constants a and b.
   $\mathbf{F}_2$ is removed. Work out:
   b the resultant force **R**
   c the acceleration of the particle, giving your answer in the form $(p\mathbf{i} + q\mathbf{j} + r\mathbf{k})$ m s^{-2}
   d the magnitude of this acceleration
   e the angle the acceleration vector makes with the unit vector **j**.

Mixed exercise 12

(P) 1. The points $A(2, 7, 3)$ and $B(4, 3, 5)$ are joined to form the line segment AB. The point M is the midpoint of AB. Find the distance from M to the point $C(5, 8, 7)$.

(P) 2. The coordinates of P and Q are $(2, 3, a)$ and $(a - 2, 6, 7)$. Given that the distance from P to Q is $\sqrt{14}$, find the possible values of a.

(P) 3. $\overrightarrow{AB}$ is the vector $-3\mathbf{i} + t\mathbf{j} + 5\mathbf{k}$, where $t > 0$. Given that $|\overrightarrow{AB}| = 5\sqrt{2}$, show that $\overrightarrow{AB}$ is parallel to $6\mathbf{i} - 8\mathbf{j} - \tfrac{5}{2}t\mathbf{k}$.

(P) 4. P is the point $(5, 6, -2)$, Q is the point $(2, -2, 1)$ and R is the point $(2, -3, 6)$.
   a Find the vectors $\overrightarrow{PQ}$, $\overrightarrow{PR}$ and $\overrightarrow{QR}$.
   b Hence, or otherwise, find the area of triangle PQR.

Chapter 12

E/P 5 The points D, E and F have position vectors $\begin{pmatrix} 1 \\ 0 \\ 0 \end{pmatrix}$, $\begin{pmatrix} 5 \\ 3 \\ 4 \end{pmatrix}$ and $\begin{pmatrix} 2 \\ -1 \\ 8 \end{pmatrix}$ respectively.

  a Find the vectors $\overrightarrow{DE}$, $\overrightarrow{EF}$ and $\overrightarrow{FD}$. (3 marks)

  b Find $|\overrightarrow{DE}|$, $|\overrightarrow{EF}|$ and $|\overrightarrow{FD}|$ giving your answers in exact form. (6 marks)

  c Describe triangle DEF. (1 mark)

E 6 P is the point $(-6, 2, 1)$, Q is the point $(3, -2, 1)$ and R is the point $(1, 3, -2)$.

  a Find the vectors $\overrightarrow{PQ}$, $\overrightarrow{PR}$ and $\overrightarrow{QR}$. (3 marks)

  b Hence find the lengths of the sides of triangle PQR. (6 marks)

  c Given that angle $QRP = 90°$ find the size of angle PQR. (2 marks)

E/P 7 The diagram shows the triangle ABC.

Given that $\overrightarrow{AB} = -\mathbf{i} + \mathbf{j}$ and $\overrightarrow{BC} = \mathbf{i} - 3\mathbf{j} + \mathbf{k}$,
find $\angle ABC$ to 1 d.p.

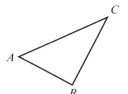

(5 marks)

E/P 8 The diagram shows the quadrilateral $ABCD$.

Given that $\overrightarrow{AB} = \begin{pmatrix} 6 \\ -2 \\ 11 \end{pmatrix}$ and $\overrightarrow{AC} = \begin{pmatrix} 15 \\ 8 \\ 5 \end{pmatrix}$, find the area of the quadrilateral.

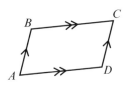

(7 marks)

P 9 A is the point $(2, 3, -2)$, B is the point $(0, -2, 1)$ and C is the point $(4, -2, -5)$. When A is reflected in the line BC it is mapped to the point D.

  a Work out the coordinates of the point D.

  b Give the mathematical name for the shape $ABCD$.

  c Work out the area of $ABCD$.

P 10 The diagram shows a tetrahedron $OABC$. $\mathbf{a}$, $\mathbf{b}$ and $\mathbf{c}$ are the position vectors of A, B and C respectively.

P, Q, R, S, T and U are the midpoints of OC, AB, OA, BC, OB and AC respectively.

Prove that the line segments PQ, RS and TU meet at a point and bisect each other.

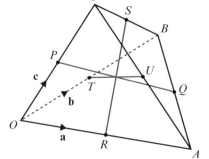

P 11 A particle of mass 2 kg is acted upon by three forces:

$\mathbf{F}_1 = (b\mathbf{i} + 2\mathbf{j} + \mathbf{k})$ N
$\mathbf{F}_2 = (3\mathbf{i} - b\mathbf{j} + 2\mathbf{k})$ N
$\mathbf{F}_3 = (-2\mathbf{i} + 2\mathbf{j} + (4 - b)\mathbf{k})$ N

Given that the particle accelerates at 3.5 m s^{-2}, work out the possible values of b.

350

Vectors

P **12** In this question **i** and **j** are the unit vectors due east and due north respectively, and **k** is the unit vector acting vertically upwards.

A BASE jumper descending with a parachute is modelled as a particle of mass 50 kg subject to forces describing the wind, **W**, and air resistance, **F**, where:

$\mathbf{W} = (20\mathbf{i} + 16\mathbf{j})$ N
$\mathbf{F} = (-4\mathbf{i} - 3\mathbf{j} + 450\mathbf{k})$ N

a With reference to the model, suggest a reason why the **k** component of **F** is greater than the other components.

b Taking $g = 9.8$ m s^{-2}, find the resultant force acting on the BASE jumper.

c Given that the BASE jumper starts from rest and travels a distance of 180 m before landing, find the total time of the descent.

Challenge

A student writes the following hypothesis:

If **a**, **b** and **c** are three non-parallel vectors in three dimensions, then
$p\mathbf{a} + q\mathbf{b} + r\mathbf{c} = s\mathbf{a} + t\mathbf{b} + u\mathbf{c} \Rightarrow p = s, q = t$ and $r = u$

Show, by means of a counter-example, that this hypothesis is not true.

Summary of key points

1 The distance from the origin to the point (x, y, z) is $\sqrt{x^2 + y^2 + z^2}$

2 The distance between the points (x_1, y_1, z_1) and (x_2, y_2, z_2) is $\sqrt{(x_1 - x_2)^2 + (y_1 - y_2)^2 + (z_1 - z_2)^2}$

3 The unit vectors along the x-, y- and z-axes are denoted by **i**, **j** and **k** respectively.

$\mathbf{i} = \begin{pmatrix} 1 \\ 0 \\ 0 \end{pmatrix} \qquad \mathbf{j} = \begin{pmatrix} 0 \\ 1 \\ 0 \end{pmatrix} \qquad \mathbf{k} = \begin{pmatrix} 0 \\ 0 \\ 1 \end{pmatrix}$

Any 3D vector can be written in column form as $p\mathbf{i} + q\mathbf{j} + r\mathbf{k} = \begin{pmatrix} p \\ q \\ r \end{pmatrix}$

4 If the vector $\mathbf{a} = x\mathbf{i} + y\mathbf{j} + z\mathbf{k}$ makes an angle θ_x with the positive x-axis then $\cos\theta_x = \dfrac{x}{|\mathbf{a}|}$ and similarly for the angles θ_y and θ_z.

5 If **a**, **b** and **c** are vectors in three dimensions which do not all lie in the same plane then you can compare their coefficients on both sides of an equation.

Review exercise 3

E/P 1. A curve has equation $y = \frac{1}{2}x^2 + 4\cos x$. Show that an equation of the normal to the curve at $x = \frac{\pi}{2}$ is
$$8y(8 - \pi) - 16x + \pi(\pi^2 - 8\pi + 8) = 0 \quad (7)$$
← Section 9.1

E/P 2. A curve has equation $y = e^{3x} - \ln(x^2)$. Show that an equation of the tangent at $x = 2$ is $y - (3e^6 - 1)x - 2 + \ln 4 + 5e^6 = 0$ (6)
← Section 9.2

E/P 3. A curve has equation
$$y = -\frac{3}{(4 - 6x)^2}, \quad x \neq \frac{2}{3}$$
Find an equation of the normal to the curve at $x = 1$ in the form $ax + by + c = 0$, where a, b and c are integers. (7)
← Section 9.3

E 4. A curve C has equation $y = (2x - 3)^2 e^{2x}$.
  a. Use the product rule to find $\frac{dy}{dx}$ (3)
  b. Hence find the coordinates of the stationary points of C. (3)
← Section 9.4

E 5. The curve C has equation $y = \frac{(x - 1)^2}{\sin x}$
  a. Use the quotient rule to find $\frac{dy}{dx}$ (3)
  b. Show that the equation of the tangent to the curve at $x = \frac{\pi}{2}$ is
  $$y = (\pi - 2)x + \left(1 - \frac{\pi^2}{4}\right) \quad (4)$$
← Section 9.5

E/P 6. a. Show that if $y = \operatorname{cosec} x$ then
$$\frac{dy}{dx} = -\operatorname{cosec} x \cot x \quad (4)$$
  b. Given $x = \operatorname{cosec} 6y$, find $\frac{dy}{dx}$ in terms of x. (6)
← Section 9.6

E/P 7. Assuming standard results for $\sin x$ and $\cos x$, prove that the derivative of $\arcsin x$ is $\frac{1}{\sqrt{1 - x^2}}$ (5)
← Section 9.6

E 8. A curve has parametric equations
$$x = 2\cot t, \quad y = 2\sin^2 t, \quad 0 < t \leq \frac{\pi}{2}$$
  a. Find $\frac{dy}{dx}$ in terms of t. (3)
  b. Find an equation of the tangent to the curve at the point where $t = \frac{\pi}{4}$ (3)
  c. Find a Cartesian equation of the curve in the form $y = f(x)$. State the domain on which the curve is defined. (3)
← Section 9.7

E/P 9. The curve C has parametric equations
$$x = \frac{1}{1 + t}, \quad y = \frac{1}{1 - t}, \quad -1 < t < 1$$
The line l is a tangent to C at the point where $t = \frac{1}{2}$
  a. Find an equation for the line l. (5)
  b. Show that a Cartesian equation for the curve C is $y = \frac{x}{2x - 1}$ (3)
← Section 9.7

Review exercise 3

10 A curve C is described by the equation
$$3x^2 - 2y^2 + 2x - 3y + 5 = 0$$
Find an equation of the normal to C at the point $(0, 1)$, giving your answer in the form $ax + by + c = 0$, where a, b and c are integers. **(7)**

← Section 9.8

11 A set of curves is given by the equation
$$\sin x + \cos y = 0.5$$
a Use implicit differentiation to find an expression for $\dfrac{dy}{dx}$ **(4)**

For $-\pi < x < \pi$ and $-\pi < y < \pi$

b find the coordinates of the points where $\dfrac{dy}{dx} = 0$. **(3)**

← Section 9.8

12 A curve C has equation
$$y = x^2 e^{-x}, \quad x < 0$$
Show that C is convex for all $x < 0$. **(5)**

← Sections 9.4, 9.9

13 The volume of a spherical balloon of radius r cm is V cm^3, where $V = \frac{4}{3}\pi r^3$.

a Find $\dfrac{dV}{dr}$ **(1)**

The volume of the balloon increases with time t seconds according to the formula
$$\dfrac{dV}{dt} = \dfrac{1000}{(2t+1)^2}, \quad t \geq 0.$$

b Find an expression in terms of r and t for $\dfrac{dr}{dt}$ **(3)**

← Section 9.10

14 $g(x) = x^3 - x^2 - 1$

a Show that there is a root α of $g(x) = 0$ in the interval $[1.4, 1.5]$. **(2)**

b By considering a change of sign of $g(x)$ in a suitable interval, verify that $\alpha = 1.466$ correct to 3 decimal places. **(3)**

← Section 10.1

15 $p(x) = \cos x + e^{-x}$

a Show that there is a root α of $p(x) = 0$ in the interval $[1.7, 1.8]$. **(2)**

b By considering a change of sign of $p(x)$ in a suitable interval, verify that $\alpha = 1.746$ correct to 3 decimal places. **(3)**

← Section 10.1

16 $f(x) = e^{x-2} - 3x + 5$

a Show that the equation $f(x) = 0$ can be written as
$$x = \ln(3x - 5) + 2, x > \tfrac{5}{3}$$ **(2)**

The root of $f(x) = 0$ is α.

The iterative formula
$x_{n+1} = \ln(3x_n - 5) + 2, x_0 = 4$ is used to find a value for α.

b Calculate the values of x_1, x_2 and x_3 to 4 decimal places. **(3)**

← Section 10.2

17 $f(x) = \dfrac{1}{(x-2)^3} + 4x^2, \quad x \neq 2$

a Show that there is a root α of $f(x) = 0$ in the interval $[0.2, 0.3]$. **(2)**

b Show that the equation $f(x) = 0$ can be written in the form $x = \sqrt[3]{\dfrac{-1}{4x^2}} + 2$. **(3)**

c Use the iterative formula
$$x_{n+1} = \sqrt[3]{\dfrac{-1}{4x_n^2}} + 2, \ x_0 = 1$$ to calculate the values of x_1, x_2, x_3 and x_4 giving your answers to 4 decimal places. **(3)**

d By considering the change of sign of $f(x)$ in a suitable interval, verify that $\alpha = 1.524$ correct to 3 decimal places. **(2)**

← Section 10.2

Review exercise 3

18 The diagram shows part of the curve with equation $y = f(x)$, where $f(x) = \frac{1}{10}x^2 e^x - 2x - 10$. The point A, with x-coordinate a, is a stationary point on the curve. The equation $f(x) = 0$ has a root α in the interval $[2.9, 3.0]$.

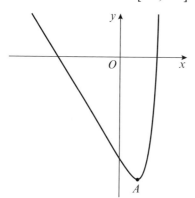

a Explain why $x_0 = a$ is not suitable to use as a first approximation if using the Newton–Raphson process to find an approximation for α. **(1)**

b Taking $x_0 = 2.9$ as a first approximation to α, apply the Newton–Raphson process once to find f(x) to obtain a second approximation to α. Give your answer to 3 decimal places. **(4)**

← Section 10.3

19 $f(x) = \frac{3}{10}x^3 - x^{\frac{2}{3}} + \frac{1}{x} - 4$, $x \neq 0$

a Show that there is a root α of $f(x) = 0$ in the intervals
  i $[0.2, 0.3]$ **(1)**
  ii $[2.6, 2.7]$ **(1)**

b Show that the equation $f(x) = 0$ can be written in the form
$$x = \sqrt[3]{\frac{10}{3}\left(4 + x^{\frac{2}{3}} - \frac{1}{x}\right)}$$ **(3)**

c Use the iterative formula,
$$x_{n+1} = \sqrt[3]{\frac{10}{3}\left(4 + x_n^{\frac{2}{3}} - \frac{1}{x_n}\right)}, x_0 = 2.5$$ to calculate the values of x_1, x_2, x_3 and x_4 giving your answers to 4 decimal places. **(3)**

d Taking $x_0 = 0.3$ as a first approximation to α, apply the Newton–Raphson process once to find f(x) to obtain a second approximation to α. Give your answer to 3 decimal places. **(4)**

← Sections 10.2, 10.3

20 The value in dollars of a stock x hours into a 14-hour trading window can be modelled by the function
$$v(x) = 0.12 \cos\left(\frac{2x}{5}\right) - 0.35 \sin\left(\frac{2x}{5}\right) + 120$$
where $0 \leq x \leq 14$.

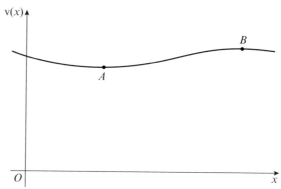

Given that v(x) can be written in the form $R\cos\left(\frac{2x}{5} + \alpha\right) + 120$ where $R > 0$ and $0 \leq \alpha \leq \frac{\pi}{2}$,

a find the value of R and the value of α, correct to 4 decimal places. **(3)**

b Use your answer to part **a** to find v′(x). **(2)**

c Show that the curve has a turning point in the interval $[4.7, 4.8]$. **(2)**

d Taking $x = 12.6$ as a first approximation, apply the Newton–Raphson method once to v′(x) to obtain a second approximation for the time when the value of the stock is a maximum. Give your answer to 3 decimal places. **(3)**

e By considering the change of sign of v′(x) in a suitable interval, verify that the x-coordinate at point B is 12.6067, correct to 4 decimal places. **(2)**

← Sections 7.5, 9.3, 10.4

21 Given $\int_a^3 (12 - 3x)^2 \, dx = 78$, find the value of a. **(4)**

← Section 11.2

22 a By expanding $\cos(5x + 2x)$ and $\cos(5x - 2x)$ using the double-angle formulae, or otherwise, show that
$\cos 7x + \cos 3x \equiv 2 \cos 5x \cos 2x$. **(4)**

b Hence find $\int 6 \cos 5x \cos 2x \, dx$ **(3)**

← Sections 7.1, 11.3

23 Given that $\int_0^m mx^3 e^{x^4} dx = \frac{3}{4}(e^{81} - 1)$, find the value of m. **(3)**

← Section 11.4

24 Using the substitution $u^2 = 2x - 1$, or otherwise, find the exact value of
$$\int_1^5 \frac{3x}{\sqrt{2x-1}} dx$$ **(6)**

← Section 11.5

25 Use the substitution $u = 1 - x^2$ to find the exact value of
$$\int_0^{\frac{1}{2}} \frac{x^3}{(1-x^2)^{\frac{1}{2}}} dx$$ **(6)**

← Section 11.5

26 $f(x) = (x^2 + 1) \ln x$

Find the exact value of $\int_1^e f(x) \, dx$. **(7)**

← Section 11.6

27 a Express $\dfrac{5x+3}{(2x-3)(x+2)}$ in partial fractions. **(3)**

b Hence find the exact value of
$$\int_2^6 \frac{5x+3}{(2x-3)(x+2)} dx,$$ giving your answer as a single logarithm. **(4)**

← Section 11.7

28 The curve shown in the diagram has parametric equations
$x = t - 2\sin t, \ y = 1 - 2\cos t, \ 0 \leqslant t \leqslant 2\pi$

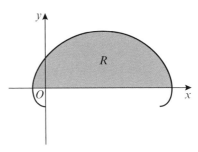

a Show that the curve crosses the x-axis where $t = \dfrac{\pi}{3}$ and $t = \dfrac{5\pi}{3}$ **(3)**

The finite region R is enclosed by the curve and the x-axis, as shown shaded in the diagram.

b Show that the area R is given by
$$\int_{\frac{\pi}{3}}^{\frac{5\pi}{3}} (1 - 2\cos t)^2 \, dt$$ **(3)**

c Use this integral to find the exact value of the shaded area. **(4)**

29 The curve shown in the diagram has parametric equations
$x = a \cos 3t, \ y = a \sin t, \ -\dfrac{\pi}{6} \leqslant t \leqslant \dfrac{\pi}{6}$.

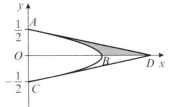

The curve meets the axes at points A, B and C, as shown.

The straight lines shown are tangents to the curve at the points A and C and meet the x-axis at point D.

a Find, in terms of a, the area of the finite region between the curve, the tangent at A and the x-axis, shown shaded in the diagram.

Given that the total area of the finite region between the two tangents and the curve is 10 cm²

b find the value of a.

Review exercise 3

30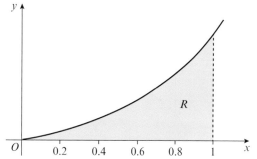

The diagram shows the graph of the curve with equation
$$y = xe^{2x}, \; x \geq 0.$$
The finite region R bounded by the lines $x = 1$, the x-axis and the curve is shown shaded in the diagram.

a Use integration to find the exact area of R. **(4)**

The table shows values of x and y between 0 and 1.

x	0	0.2	0.4	0.6	0.8	1
$y = xe^{2x}$	0	0.29836		1.99207		7.38906

b Find the missing values in the table. **(1)**

c Using the trapezium rule, with all the values for y in the completed table, find an approximation for the area of R, giving your answer to 4 significant figures. **(4)**

d Calculate the percentage error in your answer in part **c**. **(2)**

← Sections, 11.6, 11.9

31 a Express $\dfrac{2x - 1}{(x - 1)(2x - 3)}$ in partial fractions. **(4)**

b Given that $x \geq 2$, find the general solution of the differential equation
$$(2x - 3)(x - 1)\dfrac{dy}{dx} = (2x - 1)y \quad \textbf{(4)}$$

c Hence find the particular solution of this differential equation that satisfies $y = 10$ at $x = 2$, giving your answer in the form $y = f(x)$. **(2)**

← Sections 11.7, 11.10

32 A spherical balloon is being inflated in such a way that the rate of increase of its volume, $V \text{ cm}^3$, with respect to time t seconds is given by $\dfrac{dV}{dt} = \dfrac{k}{V}$, where k is a positive constant.

Given that the radius of the balloon is r cm, and that $V = \frac{4}{3}\pi r^3$,

a prove that r satisfies the differential equation
$$\dfrac{dr}{dt} = \dfrac{B}{r^5}$$
where B is a constant. **(4)**

b Find a general solution of the differential equation obtained in part **a**. **(4)**

← Sections 9.10, 11.10, 11.11

33 Liquid is pouring into a container at a constant rate of $20 \text{ cm}^3 \text{ s}^{-1}$ and is leaking out at a rate proportional to the volume of the liquid already in the container.

a Explain why, at time t seconds, the volume, $V \text{ cm}^3$, of liquid in the container satisfies the differential equation
$$\dfrac{dV}{dt} = 20 - kV$$
where k is a positive constant. **(2)**

The container is initially empty.

b By solving the differential equation, show that
$$V = A + Be^{-kt}$$
giving the values of A and B in terms of k. **(5)**

Given also that $\dfrac{dV}{dt} = 10$ when $t = 5$,

c find the volume of liquid in the container at 10 s after the start. **(3)**

← Sections 11.10, 11.11

34 The rate of decrease of the concentration of a drug in the blood stream is proportional to the concentration C of that drug which is present at that time. The time t is measured in hours from the administration of the drug and C is measured in micrograms per litre.

a Show that this process is described by the differential equation $\frac{dC}{dt} = -kC$, explaining why k is a positive constant. (2)

b Find the general solution of the differential equation, in the form $C = f(t)$. (4)

After 4 hours, the concentration of the drug in the bloodstream is reduced to 10% of its starting value C_0.

c Find the exact value of k. (3)

← Sections 11.10, 11.11

35 The coordinates of P and Q are $(-1, 4, 6)$ and $(8, -4, k)$ respectively. Given that the distance from P to Q is $7\sqrt{5}$ units, find the possible values of k. (3)

← Section 12.1

36 The diagram shows the triangle ABC.

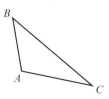

Given that $\overrightarrow{AB} = -\mathbf{i} + 6\mathbf{j} + 4\mathbf{k}$ and $\overrightarrow{AC} = 5\mathbf{i} - 2\mathbf{j} - 3\mathbf{k}$, find the size of $\angle BAC$ to one decimal place. (5)

← Section 12.2

37 P is the point $(-6, 3, 2)$ and Q is the point $(4, -2, 0)$. Find:

a the vector $\overrightarrow{PQ}$ (1)

b the unit vector in the direction of $\overrightarrow{PQ}$ (2)

c the angle $\overrightarrow{PQ}$ makes with the positive z-axis. (2)

The vector $\overrightarrow{AB} = 30\mathbf{i} - 15\mathbf{j} + 6\mathbf{k}$.

d Explain, with a reason, whether the vectors $\overrightarrow{AB}$ and $\overrightarrow{PQ}$ are parallel. (2)

← Section 12.2

38 The vertices of triangle MNP have coordinates $M(-2, 0, 5)$, $N(8, -5, 1)$ and $P(k, -2, -6)$. Given that triangle MNP is isosceles and k is a positive integer, find the value of k. (4)

← Section 12.3

39 Given that
$$-6\mathbf{i} + 40\mathbf{j} + 16\mathbf{k} = 3p\mathbf{i} + (8 + qr)\mathbf{j} + 2pr\mathbf{k}$$
find the values of p, q and r. (3)

← Section 12.3

Challenge

1 The curve C has implicit equation
$$ay + x^2 + 4xy = y^2$$

a Find, in terms of a where necessary, the coordinates of the points such that $\frac{dy}{dx} = 0$.

b Given that $a \neq 0$, show that there does not exist a point where $\frac{dx}{dy} = 0$. ← Section 9.8

2 The diagram shows the curves $y = \sin x + 2$ and $y = \cos 2x + 2$, $0 \leq x \leq \frac{3\pi}{2}$

Find the exact value of the total shaded area on the diagram.

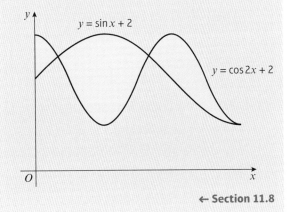

← Section 11.8

Exam-style practice

Mathematics
A Level
Paper 1: Pure Mathematics

Time: 2 hours
You must have: Mathematical Formulae and Statistical Tables, Calculator

1 A curve C has parametric equations $x = \sin^2 t$, $y = 2 \tan t$, $0 \leq t < \dfrac{\pi}{2}$

   Find $\dfrac{dy}{dx}$ in terms of t. (3)

2 Find the set of values of x for which
   a $2(7x - 5) - 6x < 10x - 7$ (2)
   b $|2x + 5| - 3 > 0$ (4)
   c **both** $2(7x - 5) - 6x < 10x - 7$ **and** $|2x + 5| - 3 > 0$. (1)

3 The line with equation $2x + y - 3 = 0$ does not intersect the circle with equation $x^2 + kx + y^2 + 4y = 4$
   a Show that $5x^2 + (k - 20)x + 17 > 0$. (4)
   b Find the range of possible values of k. Write your answer in exact form. (3)

4 Prove, for an angle θ measured in radians, that the derivative of $\cos \theta$ is $-\sin \theta$.
   You may assume the compound angle formula for $\cos(A \pm B)$, and that
   $$\lim_{h \to 0}\left(\frac{\sin h}{h}\right) = 1 \text{ and } \lim_{h \to 0}\left(\frac{\cos h - 1}{h}\right) = 0.$$ (5)

5 $f(x) = (3 + px)^6$, $x \in \mathbb{R}$
   Given that the coefficient of x^2 is 19 440,
   a find two possible values of p. (4)
   Given further that the coefficient of x^5 is negative,
   b find the coefficient of x^5. (2)

Exam-style practice

6 The point R with x-coordinate 2 lies on the curve with equation $y = x^2 + 4x - 2$. The normal to the curve at R intersects the curve again at a point T. Find the coordinates of T, giving your answers in their simplest form. **(6)**

7 A geometric series has first term a and common ratio r. The second term of the series is 96 and the sum to infinity of the series is 600.
   a Show that $25r^2 - 25r + 4 = 0$. **(4)**
   b Find the two possible values of r. **(2)**
   For the larger value of r:
   c find the corresponding value of a. **(1)**
   d find the smallest value of n for which S_n exceeds 599.9. **(3)**

8 The diagram shows part of the graph of $y = f(x)$, $x \in \mathbb{R}$. The points B and D are stationary points of the graph.

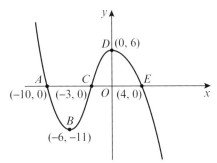

Sketch, on separate diagrams, the graphs of:
   a $y = |f(x)|$ **(3)**
   b $y = -f(x) + 5$ **(3)**
   c $y = 2f(x - 3)$ **(3)**

9 Find all the solutions, in the interval $0 \leq x \leq 2\pi$, to the equation $31 - 25\cos x = 19 - 12\sin^2 x$, giving each solution to 2 decimal places. **(5)**

10 The value, V, of a car decreases over time, t, measured in years. The rate of decrease in value of the car is proportional to the value of the car at that time.
   a Given that the initial value of the car is V_0, show that $V = V_0 e^{-kt}$ **(4)**
   The value of the car after 2 years is £25 000 and after 5 years is £15 000.
   b Find the exact value of k and the value of V_0 to the nearest hundred pounds. **(3)**
   c Find the age of the car when its value is £5000. **(3)**
   A different model of car has the same initial value but is known to depreciate in value more slowly.
   d State which constant in the original equation should be changed to model a slower depreciation in value, and give a possible new value for this constant. **(1)**

Exam-style practice

11 The diagram shows the positions of 4 cities: A, B, C and D. The distances, in miles, between each pair of cities, as measured in a straight-line, are labelled on the diagram. A new road is to be built between cities B and D.

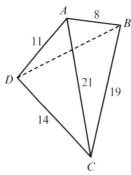

   a What is the minimum possible length of this road? Give your answer to 1 decimal place. (7)
   b Explain why your answer to part a is a minimum. (1)

12 A footballer takes a free-kick. The path of the ball towards the goal can be modelled by the equation $y = -0.01x^2 + 0.22x + 1.58$, $x \geq 0$, where x is the horizontal distance from the goal in metres and y is the height of the ball in metres. The goal is 2.44 m high.
   a Rewrite y in the form $A - B(x + C)^2$, where A, B and C are constants to be found. (3)
   b Using your answer to part a, state the distance from goal at which the ball is at the greatest height and its height at this point. (2)
   c How far from the goal is the football when it is kicked? (2)
   d The football is headed towards the goal. The keeper can save any ball that would cross the goal line at a height of up to 1.5 m. Explain with a reason whether the free kick will result in a goal. (2)

13 A box in the shape of a rectangular prism has a lid that overlaps the box by 3 cm, as shown. The width of the box is x cm, and the length of the box is double the width. The height of the box is h cm. The box and lid can be created exactly from a piece of cardboard of area 5356 cm². The box has volume, V cm³.

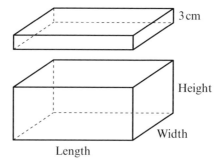

   a Show that $V = \frac{2}{3}(2678x - 9x^2 - 2x^3)$ (5)
   Given that x can vary
   b use differentiation to find the positive value of x, to 2 decimal places, for which V is stationary. (4)
   c Prove that this value of x gives a maximum value of V. (2)
   d Find this maximum value of V. (1)
   Given that V takes its maximum value,
   e determine the percentage of the area of cardboard that is used in the lid. (2)

Exam-style practice

Mathematics
A Level
Paper 2: Pure Mathematics

Time: 2 hours
You must have: Mathematical Formulae and Statistical Tables, Calculator

1. The graph of $y = ax^2 + bx + c$ has a maximum at $(-2, 8)$ and passes through $(-4, 4)$. Find the values of a, b and c. **(3)**

2. The points $P(6, 4)$ and $Q(0, 28)$ lie on the straight line l_1 as shown.

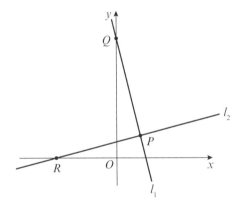

   a Work out an equation for the straight line l_1. **(2)**

   The straight line l_2 is perpendicular to l_1 and passes through the point P.

   b Work out an equation for the straight line l_2. **(2)**

   c Work out the coordinates of R. **(2)**

   d Work out the area of $\triangle PQR$. **(3)**

3. The function f is defined by $f: x \rightarrow e^{3x} - 1, \quad x \in \mathbb{R}$.
   Find $f^{-1}(x)$ and state its domain. **(4)**

4. A student is asked to solve the equation $\log_4(x + 3) + \log_4(x + 4) = \frac{1}{2}$
   The student's attempt is shown.

   ```
   log₄(x + 3) + log₄(x + 4) = ½
           (x + 3) + (x + 4) = 2
                      2x + 7 = 2
                          2x = -5
                           x = -5/2
   ```

Exam-style practice

    a Identify the error made by the student. **(1)**

    b Solve the equation correctly. **(4)**

5 The function p has domain $-14 \leq x \leq 10$, and is linear from $(-14, 18)$ to $(-6, -6)$ and from $(-6, -6)$ to $(10, 2)$.

    a Sketch $y = p(x)$. **(2)**

    b Write down the range of $p(x)$. **(1)**

    c Find the values of a, such that $p(a) = -3$. **(2)**

6 $f(x) = x^3 - kx^2 - 10x + k$

    a Given that $(x + 2)$ is a factor of $f(x)$, find the value of k. **(2)**

    b Hence, or otherwise, find all the solutions to the equation $f(x) = 0$, leaving your answers in the form $p \pm \sqrt{q}$ when necessary. **(4)**

7 In $\triangle DEF$, $DE = x - 3$ cm, $DF = x - 10$ cm and $\angle EDF = 30°$. Given that the area of the triangle is 11 cm²,

    a show that x satisfies the equation $x^2 - 13x - 14 = 0$ **(3)**

    b calculate the value of x. **(2)**

8 The curve C has parametric equations $x = 6 \sin t + 5$, $y = 6 \cos t - 2$, $-\dfrac{\pi}{3} \leq t \leq \dfrac{3\pi}{4}$

    a Show that the Cartesian equation of C can be written as $(x + h)^2 + (y + k)^2 = c$, where h, k and c are integers to be determined. **(4)**

    b Find the length of C. Write your answer in the form $p\pi$, where p is a rational number to be found. **(3)**

9 $\dfrac{4x^2 + 7x}{(x - 2)(x + 4)} \equiv A + \dfrac{B}{x - 2} + \dfrac{C}{x + 4}$

    a Find the values of the constants A, B and C. **(4)**

    b Hence, or otherwise, expand $\dfrac{4x^2 + 7x}{(x - 2)(x + 4)}$ in ascending powers of x, as far as the term in x^2. Give each coefficient as a simplified fraction. **(6)**

10 OAB is a triangle. $\overrightarrow{OA} = \mathbf{a}$ and $\overrightarrow{OB} = \mathbf{b}$. The points M and N are midpoints of OB and BA respectively.

The triangle midsegment theorem states that 'In a triangle, the line joining the midpoints of any two sides will be parallel to the third side and half its length.'

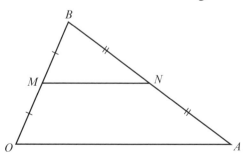

Use vectors to prove the triangle midsegment theorem. **(4)**

11 The diagram shows the region R bounded by the x-axis and the curve with equation $y = x^2(\sin x + \cos x)$, $0 \leqslant x \leqslant \dfrac{3\pi}{4}$

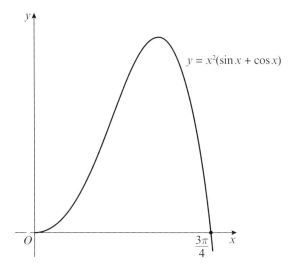

The table shows corresponding values of x and y for $y = x^2(\sin x + \cos x)$.

x	0	$\dfrac{\pi}{8}$	$\dfrac{\pi}{4}$	$\dfrac{3\pi}{8}$	$\dfrac{\pi}{2}$	$\dfrac{5\pi}{8}$	$\dfrac{3\pi}{4}$
y	0	0.20149	0.87239	1.81340		2.08648	0

a Copy and complete the table giving the missing value for y to 5 decimal places. (1)

b Using the trapezium rule, with all the values for y in the completed table, find an approximation for the area of R, giving your answer to 3 decimal places. (4)

c Use integration to find the exact area of R, giving your answer to 3 decimal places. (6)

d Calculate, to one decimal place, the percentage error in your approximation in part b. (1)

12 Ruth wants to save money for her newborn daughter to pay for university costs. In the first year she saves £1000. Each year she plans to save £150 more, so that she will save £1150 in the second year, £1300 in the third year, and so on.

a Find the amount Ruth will save in the 18th year. (2)

b Find the total amount that Ruth will have saved over the 18 years. (3)

Ruth decides instead to increase the amount she saves by 10% each year.

c Calculate the total amount Ruth will have saved after 18 years under this scheme. (4)

13 a Express $0.09 \cos x + 0.4 \sin x$ in the form $R\cos(x - \alpha)$, where $R > 0$ and $0 < \alpha < \dfrac{\pi}{2}$

Give the value of α to 4 decimal places. (4)

The height of a swing above the ground can be modelled using the equation

$$h = \dfrac{16.4}{0.09 \cos\left(\dfrac{t}{2}\right) + 0.4 \sin\left(\dfrac{t}{2}\right)}, \ 0 \leqslant t \leqslant 5.4,$$ where h is the height of the swing, in cm, and t is the time, in seconds, since the swing was initially at its greatest height.

b Calculate the minimum value of h predicted by this model, and the value of t, to 2 decimal places, when this minimum value occurs. (3)

c Calculate, to the nearest hundredth of a second, the times when the swing is at a height of exactly 100 cm. (4)

14 The diagram shows the height, h, in metres of a rollercoaster during the first few seconds of the ride. The graph is $y = h(t)$, where $h(t) = -10e^{-0.3(t-6.4)} - 10e^{0.8(t-6.4)} + 70$.

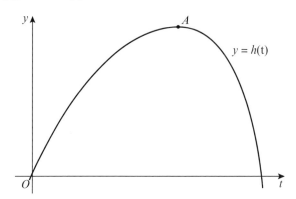

a Find $h'(t)$. (3)

b Show that when $h'(t) = 0$, $t = \dfrac{5}{4} \ln\left(\dfrac{3e^{-0.3(t-6.4)}}{8}\right) + 6.4$ (2)

To find an approximation for the t-coordinate of A, the iterative formula

$$t_{n+1} = \dfrac{5}{4} \ln\left(\dfrac{3e^{-0.3(t_n - 6.4)}}{8}\right) + 6.4 \text{ is used.}$$

c Let $t_0 = 5$. Find the values of t_1, t_2, t_3 and t_4. Give your answers to 4 decimal places. (3)

d By choosing a suitable interval, show that the t-coordinate of A is 5.508, correct to 3 decimal places. (2)

Answers

CHAPTER 1

Prior knowledge 1
1. **a** $(x-1)(x-5)$ **b** $(x+4)(x-4)$ **c** $(3x-5)(3x+5)$
2. **a** $\dfrac{x-3}{x+6}$ **b** $\dfrac{x+4}{3x+1}$ **c** $-\dfrac{x+5}{x+3}$
3. **a** even **b** either **c** either **d** odd

Exercise 1A
1. B At least one multiple of three is odd.
2. **a** At least one rich person is not happy.
   b There is at least one prime number between 10 million and 11 million.
   c If p and q are prime numbers there exists a number of the form $(pq + 1)$ that is not prime.
   d There is a number of the form $2^n - 1$ that is neither a prime nor a multiple of 3.
   e None of the above statements are true.
3. **a** There exists a number n such that n^2 is odd but n is even.
   b n is even so write $n = 2k$
   $n^2 = (2k)^2 = 4k^2 = 2(2k^2) \Rightarrow n^2$ is even.
   This contradicts the assumption that n^2 is odd.
   Therefore if n^2 is odd then n must be odd.
4. **a** Assumption: there is a greatest even integer $2n$.
   $2(n+1)$ is also an integer and $2(n+1) > 2n$
   $2n + 2 = $ even $+$ even $=$ even
   So there exists an even integer greater that $2n$.
   This contradicts the assumption.
   Therefore there is no greatest even integer.
   b Assumption: there exists a number n such that n^3 is even but n is odd.
   n is odd so write $n = 2k + 1$
   $n^3 = (2k+1)^3 = 8k^3 + 12k^2 + 6k + 1$
   $= 2(4k^3 + 6k^2 + 3k) + 1 \Rightarrow n^3$ is odd.
   This contradicts the assumption that n^3 is even.
   Therefore, if n^3 is even then n must be even.
   c Assumption: if pq is even then neither p nor q is even.
   p is odd, $p = 2k + 1$
   q is odd, $q = 2m + 1$
   $pq = (2k+1)(2m+1) = 4km + 2k + 2m + 1$
   $= 2(2km + k + m) + 1 \Rightarrow pq$ is odd.
   This contradicts the assumption that pq is even.
   Therefore, if pq is even then at least one of p and q is even.
   d Assumption: if $p + q$ is odd then neither p nor q is odd
   p is even, $p = 2k$
   q is even, $q = 2m$
   $p + q = 2k + 2m = 2(k + m) \Rightarrow p + q$ is even
   This contradicts the assumption then $p + q$ is odd.
   Therefore, if $p + q$ is odd then at least one of p and q is odd.
5. **a** Assumption: if ab is an irrational number then neither a nor b is irrational.
   a is rational, $a = \dfrac{c}{d}$ where c and d are integers.
   b is rational, $b = \dfrac{e}{f}$ where e and f are integers.
   $ab = \dfrac{ce}{df}$, ce is an integer, df is an integer.
   Therefore ab is a rational number.
   This contradicts assumption then ab is irrational.
   Therefore if ab is an irrational number then at least one of a and b is an irrational number.
   b Assumption: neither a nor b is irrational.
   a is rational, $a = \dfrac{c}{d}$ where c and d are integers.
   b is rational, $b = \dfrac{e}{f}$ where e and f are integers.
   $a + b = \dfrac{cf + de}{df}$
   cf, de and df are integers.
   So $a + b$ is rational. This contradicts the assumption that $a + b$ is irrational.
   Therefore if $a + b$ is irrational then at least one of a and b is irrational.
   c Many possible answers e.g. $a = 2 - \sqrt{2}$, $b = \sqrt{2}$.
6. Assumption: there exists integers a and b such that $21a + 14b = 1$.
   Since 21 and 14 are multiples of 7, divide both sides by 7.
   So now $3a + 2b = \tfrac{1}{7}$
   $3a$ is also an integer. $2b$ is also an integer.
   The sum of two integers will always be an integer, so $3a + 2b = $ 'an integer'.
   This contradicts the statement that $3a + 2b = \tfrac{1}{7}$.
   Therefore there exists no integers a and b for which $21a + 14b = 1$.
7. **a** Assumption: There exists a number n such that n^2 is a multiple of 3, but n is not a multiple of 3.
   We know that all multiples of 3 can be written in the form $n = 3k$, therefore $3k + 1$ and $3k + 2$ are not multiples of 3.
   Let $n = 3k + 1$
   $n^2 = (3k + 1)^2 = 9k^2 + 6k + 1 = 3(3k^2 + 2k) + 1$
   In this case n^2 is *not* a multiple of 3.
   Let $m = 3k + 2$
   $m^2 = (3k + 2)^2 = 9k^2 + 12k + 4 = 3(3k^2 + 4k + 1) + 1$
   In this case m^2 is also not a multiple of 3.
   This contradicts the assumption that n^2 is a multiple of 3.
   Therefore if n^2 is a multiple of 3, n is a multiple of 3.
   b Assumption: $\sqrt{3}$ is a rational number.
   Then $\sqrt{3} = \dfrac{a}{b}$ for some integers a and b.
   Further assume that this fraction is in its simplest terms: there are no common factors between a and b.
   So $3 = \dfrac{a^2}{b^2}$ or $a^2 = 3b^2$.
   Therefore a^2 must be a multiple of 3.
   We know from part **a** that this means a must also be a multiple of 3.
   Write $a = 3c$, which means $a^2 = (3c)^2 = 9c^2$.
   Now $9c^2 = 3b^2$, or $3c^2 = b^2$.
   Therefore b^2 must be a multiple of 3, which implies b is also a multiple of 3.
   If a and b are both multiples of 3, this contradicts the statement that there are no common factors between a and b.
   Therefore, $\sqrt{3}$ is an irrational number.

Answers

8 Assumption: there is an integer solution to the equation $x^2 - y^2 = 2$.
Remember that $x^2 - y^2 = (x - y)(x + y) = 2$
To make a product of 2 using integers, the possible pairs are: $(2, 1), (1, 2), (-2, -1)$ and $(-1, -2)$.
Consider each possibility in turn.
$x - y = 2$ and $x + y = 1 \Rightarrow x = \frac{3}{2}, y = -\frac{1}{2}$.
$x - y = 1$ and $x + y = 2 \Rightarrow x = \frac{3}{2}, y = \frac{1}{2}$.
$x - y = -2$ and $x + y = -1 \Rightarrow x = -\frac{3}{2}, y = \frac{1}{2}$.
$x - y = -1$ and $x + y = -2 \Rightarrow x = -\frac{3}{2}, y = -\frac{1}{2}$.
This contradicts the statement that there is an integer solution to the equation $x^2 - y^2 = 2$.
Therefore the original statement must be true: There are no integer solutions to the equation $x^2 - y^2 = 2$.

9 Assumption: $\sqrt[3]{2}$ is rational and can be written in the form $\sqrt[3]{2} = \frac{a}{b}$ and there are no common factors between a and b.
$2 = \frac{a^3}{b^3}$ or $a^3 = 2b^3$
This means that a^3 is even, so a must also be even.
If a is even, $a = 2n$.
So $a^3 = 2b^3$ becomes $(2n)^3 = 2b^3$ which means $8n^3 = 2b^3$ or $4n^3 = b^3$ or $2(2n^3) = b^3$.
This means that b^3 must be even, so b is also even.
If a and b are both even, they will have a common factor of 2.
This contradicts the statement that a and b have no common factors.
We can conclude the original statement is true: $\sqrt[3]{2}$ is an irrational number.

10 a $n - 1$ could be non-positive, e.g. if $n = \frac{1}{2}$

b Assumption: There is a least positive rational number, n.
$n = \frac{a}{b}$ where a and b are integers.
Let $m = \frac{a}{2b}$. Since a and b are integers, m is rational and $m < n$.
This contradicts the statement that n is the least positive rational number.
Therefore, there is no least positive rational number.

Exercise 1B

1 a $\frac{a^2}{cd}$ **b** a **c** $\frac{1}{2}$ **d** $\frac{1}{2}$ **e** $\frac{4}{x^2}$ **f** $\frac{r^5}{10}$

2 a $\frac{1}{x-2}$ **b** $\frac{a-3}{2(a+3)}$ **c** $\frac{x-3}{y}$ **d** $\frac{y+1}{y}$

e $\frac{x}{6}$ **f** 4 **g** $\frac{1}{x+5}$ **h** $\frac{3y-2}{2}$ **i** $\frac{2(x+y)^2}{(x-y)^2}$

3 All factors cancel exactly except $\frac{x-8}{8-x} = \frac{x-8}{-(x-8)} = -1$

4 $a = 5, b = 12$

5 a $\frac{x-4}{2x+10}$ **b** $x = \frac{10e^2 + 4}{1 - 2e^2}$

6 a $\frac{2x^2 - 3x - 2}{6x - 8} \div \frac{x-2}{3x^2 + 14x - 24} = \frac{2x^2 - 3x - 2}{6x - 8}$
$\times \frac{3x^2 + 14x - 24}{x - 2} = \frac{(2x+1)(x-2)}{2(3x-4)} \times \frac{(3x-4)(x+6)}{x-2}$
$= \frac{(2x+1)(x+6)}{2} = \frac{2x^2 + 13x + 6}{2}$

b $f'(x) = 2x + \frac{13}{2}$; $f'(4) = \frac{29}{2}$

Exercise 1C

1 a $\frac{7}{12}$ **b** $\frac{7}{20}$ **c** $\frac{p+q}{pq}$ **d** $\frac{7}{8x}$ **e** $\frac{3-x}{x^2}$ **f** $\frac{2a-15}{10b}$

2 a $\frac{x+3}{x(x+1)}$ **b** $\frac{-x+7}{(x-1)(x+2)}$ **c** $\frac{8x-2}{(2x+1)(x-1)}$

d $\frac{-x-5}{6}$ **e** $\frac{2x-4}{(x+4)^2}$ **f** $\frac{23x+9}{6(x+3)(x-1)}$

3 a $\frac{x+3}{(x+1)^2}$ **b** $\frac{3x+1}{(x-2)(x+2)}$ **c** $\frac{-x-7}{(x+1)(x+3)^2}$

d $\frac{3x + 3y + 2}{(y-x)(y+x)}$ **e** $\frac{2x+5}{(x+2)^2(x+1)}$ **f** $\frac{7x+8}{(x+2)(x+3)(x-4)}$

4 $\frac{2x-19}{(x+5)(x-3)}$

5 a $\frac{6x^2 + 14x + 6}{x(x+1)(x+2)}$ **b** $\frac{-x^2 - 24x - 8}{3x(x-2)(2x+1)}$

c $\frac{9x^2 - 14x - 7}{(x-1)(x+1)(x-3)}$

6 $\frac{50x + 3}{(6x+1)(6x-1)}$

7 a $x + \frac{6}{x+2} + \frac{36}{x^2 - 2x - 8}$
$= \frac{x(x+2)(x-4)}{(x+2)(x-4)} + \frac{6(x-4)}{(x+2)(x-4)} + \frac{36}{(x+2)(x-4)}$
$= \frac{x^3 - 2x^2 - 2x + 12}{(x+2)(x-4)}$

b Divide $x^3 - 2x^2 - 2x + 12$ by $(x+2)$ to give $x^2 - 4x + 6$

Exercise 1D

1 a $\frac{4}{x+3} + \frac{2}{x-2}$ **b** $\frac{3}{x+1} - \frac{1}{x+4}$

c $\frac{3}{2x} - \frac{5}{x-4}$ **d** $\frac{4}{2x+1} - \frac{1}{x-3}$

e $\frac{2}{x+3} + \frac{4}{x-3}$ **f** $-\frac{2}{x+1} - \frac{1}{x-4}$

g $\frac{2}{x} - \frac{3}{x+4}$ **h** $\frac{3}{x+5} - \frac{1}{x-3}$

2 $A = \frac{1}{2}, B = -\frac{3}{2}$

3 $A = 24, B = -2$

4 $A = 1, B = -2, C = 3$

5 $D = -1, E = 2, F = -5$

6 $\frac{3}{x+1} - \frac{2}{x+2} - \frac{6}{x-5}$

7 a $\frac{3}{x} - \frac{2}{x+1} + \frac{5}{x-1}$

b $\frac{-1}{5x+4} + \frac{2}{2x-1}$

Challenge

$\frac{6}{x-2} + \frac{1}{x+1} - \frac{2}{x-3}$

Exercise 1E
1. $A = 0, B = 1, C = 3$
2. $D = 3, E = -2, F = -4$
3. $P = -2, Q = 4, R = 2$
4. $C = 3, D = 1, E = 2$
5. $A = 2, B = -4$
6. $A = 2, B = 4, C = 11$
7. $A = 4, B = 1$ and $C = 12$.
8. a $\dfrac{4}{x+5} - \dfrac{19}{(x+5)^2}$ b $\dfrac{2}{x} - \dfrac{1}{2x-1} + \dfrac{6}{(2x-1)^2}$

Exercise 1F
1. $A = 1, B = 1, C = 2, D = -6$
2. $a = 2, b = -3, c = 5, d = -10$
3. $p = 1, q = 2, r = 4$
4. $m = 2, n = 4, p = 7$
5. $A = 4, B = 1, C = -8$ and $D = 3$.
6. $A = 4, B = -13, C = 33$ and $D = -27$
7. $p = 1, q = 0, r = 2, s = 0$ and $t = -6$
8. $a = 2, b = 1, c = 1, d = 5$ and $e = -4$
9. $A = 3, B = -4, C = 1, D = 4, E = 1$
10. a $(x^2 - 1)(x^2 + 1) = (x - 1)(x + 1)(x^2 + 1)$
    b $(x - 1)(x^2 + 1), a = 1, b = -1, c = 1, d = 0$ and $e = 1$.

Exercise 1G
1. $A = 1, B = -2, C = 8$
2. $A = 1, B = -2, C = 3$
3. $A = 1, B = 0, C = 3, D = -4$
4. $A = 2, B = -3, C = 5, D = 1$
5. $A = 1, B = 5, C = -5$
6. $A = 2, B = -4, C = 1$
7. a $4 + \dfrac{2}{(x-1)} + \dfrac{3}{(x+4)}$ b $x + \dfrac{3}{x} + \dfrac{2}{(x-2)} - \dfrac{1}{(x-2)^2}$
8. $A = 2, B = -3, C = \dfrac{34}{11}, D = \dfrac{73}{11}$
9. $A = 2, B = 2, C = 3, D = 2$.
10. $A = 1, B = -1, C = 5, D = -\dfrac{38}{3}, E = \dfrac{8}{3}$.

Mixed exercise 1
1. Assume $\sqrt{\dfrac{1}{2}}$ is a rational number.
   Then $\sqrt{\dfrac{1}{2}} = \dfrac{a}{b}$ for some integers a and b.
   Further assume that this fraction is in its simplest terms: there are no common factors between a and b.
   So $0.5 = \dfrac{a^2}{b^2}$ or $2a^2 = b^2$.
   Therefore b^2 must be a multiple of 2.
   We know that this means b must also be a multiple of 2.
   Write $b = 2c$, which means $b^2 = (2c)^2 = 4c^2$.
   Now $4c^2 = 2a^2$, or $2c^2 = a^2$.
   Therefore a^2 must be a multiple of 2, which implies a is also a multiple of 2.
   If a and b are both multiples of 2, this contradicts the statement that there are no common factors between a and b.
   Therefore, $\sqrt{\dfrac{1}{2}}$ is an irrational number.

2. Assume there exists a rational number q such that q^2 is irrational.
   So write $q = \dfrac{a}{b}$ where a and b are integers.
   $q^2 = \dfrac{a^2}{b^2}$
   As a and b are integers a^2 and b^2 are integers.
   So q^2 is rational.
   This contradicts assumption that q^2 is irrational.
   Therefore if q^2 is irrational then q is irrational.

3. a $\dfrac{1}{3}$ b $\dfrac{2(x^2 + 4)(x - 5)}{(x^2 - 7)(x + 4)}$ c $\dfrac{2x + 3}{x}$

4. a $\dfrac{2x - 4}{x - 4}$ b $\dfrac{4(e^6 - 1)}{e^6 - 2}$

5. a $a = \dfrac{3}{4}, b = -\dfrac{13}{8}, c = -\dfrac{5}{8}$
   b $g'(x) = \dfrac{3}{2}x - \dfrac{13}{8}, g'(-2) = -\dfrac{37}{8}$

6. $\dfrac{6x^2 + 18x + 5}{x^2 - 3x - 10}$

7. $x + \dfrac{3}{x-1} - \dfrac{12}{x^2 + 2x - 3}$
   $= \dfrac{x(x+3)(x-1)}{(x+3)(x-1)} + \dfrac{3(x+3)}{(x+3)(x-1)} - \dfrac{12}{(x+3)(x-1)}$
   $= \dfrac{(x^2 + 3x + 3)(x-1)}{(x+3)(x-1)} = \dfrac{x^2 + 3x + 3}{x+3}$

8. $A = 3, B = -2$
9. $P = 1, Q = 2, R = -3$
10. $D = 5, E = 2$
11. $A = 4, B = -2, C = 3$
12. $D = 2, E = 1, F = -2$
13. $A = 1, B = -4, C = 3, D = 8$
14. $A = 2, B = -4, C = 6, D = -11$
15. $A = 1, B = 0, C = 1, D = 3$
16. $A = 1, B = 2, C = 3, D = 4, E = 1$.
17. $A = 2, B = -\dfrac{9}{4}, C = \dfrac{1}{4}$
18. $P = 1, Q = -\dfrac{1}{2}, R = \dfrac{5}{2}$
19. a $f(-3) = 0$ or $f(x) = (x + 3)(2x^2 + 3x + 1)$
    b $\dfrac{1}{(x+3)} + \dfrac{8}{(2x+1)} - \dfrac{5}{(x+1)}$

Challenge
Assume L is not perpendicular to OA. Draw the line through O which is perpendicular to L. This line meets L at a point B, outside the circle. Triangle OBA is right-angled at B, so OA is the hypotenuse of this triangle, so $OA > OB$. This gives a contradiction, as B is outside the circle, so $OA < OB$. Therefore L is perpendicular to OA.

CHAPTER 2
Prior knowledge 2
1. a $y = \dfrac{9 - 5x}{7}$ b $y = \dfrac{5p - 8x}{2}$ c $y = \dfrac{5x - 4}{8 + 9x}$
2. a $25x^2 - 30x + 5$
   b $\dfrac{1}{6x - 14}$ c $\dfrac{3x + 7}{-x - 1}$
3. a 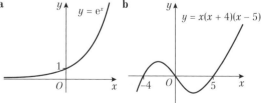   b

Answers

c [graph: $y = \sin x$, with markings at 90°, 180°, 270°, 360°]

4 a 28   b 0   c 18

Exercise 2A
1 a $\frac{3}{4}$ b 0.28   c 8   d $\frac{19}{56}$ e 4   f 11
2 a 5   b 46   c 40
3 a 16   b 65   c 0
4 a Positive $|x|$ graph with vertex at (1, 0), y-intercept at (0, 1)
   b Positive $|x|$ graph with vertex at $(-1\frac{1}{2}, 0)$, y-intercept at (0, 3)
   c Positive $|x|$ graph with vertex at $(\frac{7}{4}, 0)$, y-intercept at (0, 7)
   d Positive $|x|$ graph with vertex at (10, 0), y-intercept at (0, 5)
   e Positive $|x|$ graph with vertex at (7, 0), y-intercept at (0, 7)
   f Positive $|x|$ graph with vertex at $(\frac{3}{2}, 0)$, y-intercept at (0, 6)
   g Negative $|x|$ graph with vertex and y-intercept at (0, 0)
   h Negative $|x|$ graph with vertex at $(\frac{1}{3}, 0)$, y-intercept at (0, –1)

5 a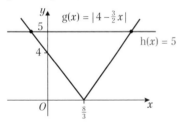

   b $x = -\frac{2}{3}$ and $x = 6$
6 a $x = 2$ and $x = -\frac{4}{3}$ b $x = 7$ or $x = 3$
   c No solution   d $x = 1$ and $x = -\frac{1}{7}$
   e $x = -\frac{2}{5}$ or $x = 2$ f $x = 24$ or $x = -12$
7 a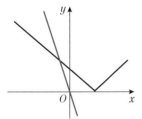

   b $x = -\frac{4}{3}$
8 $x = -3, x = 4$

9 a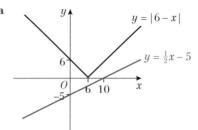

   b The two graphs do not intersect, therefore there are no solutions to the equation $|6 - x| = \frac{1}{2}x - 5$.
10 Value for x cannot be negative as it equals a modulus.
11 a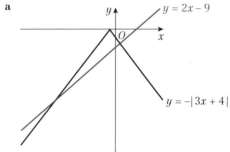

   b $x < -13$ and $x > 1$
12 $-23 < x < \frac{5}{3}$
13 a $k = -3$ b Solution is $x = 6$.

Challenge
a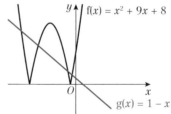

b There are 4 solutions: $x = -5 \pm 3\sqrt{2}$ and $x = -4 \pm \sqrt{7}$

Exercise 2B
1 a i   b i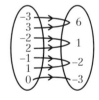

      ii one-to-one   ii many-to-one
      iii {f(x) = 12, 17, 22, 27}   iii {g(x) = –3, –2, 1, 6}

   c i [mapping: –1 → 1, 0 → $\frac{7}{4}$, 1 → 7]

      ii one-to-one
      iii {h(x) = 1, $\frac{7}{4}$, 7}

Answers

2 **a i** one-to-one **ii** function
 b i one-to-one **ii** function
 c i one-to-many **ii** not a function
 d i one to many **ii** not a function
 e i many-to-one
   ii not valid at the asymptote, so not a function.
 f i many to one **ii** function

3 **a** 6 **b** $\pm 2\sqrt{5}$ **c** 4 **d** 2, −3

4 **a i** 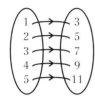 **b i**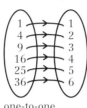
  ii one-to-one **ii** one-to-one
 c i 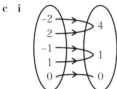 **d i**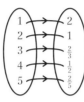
  ii many-to-one **ii** one-to-one
 e i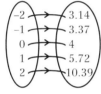
  ii one-to-one

5 **a i** 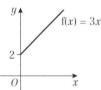 **ii** $f(x) \geq 2$
   iii one-to-one
 b i 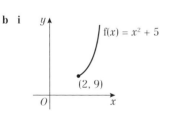 **ii** $f(x) \geq 9$
   iii one-to-one
 c i 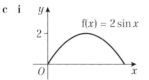 **ii** $0 \leq f(x) \leq 2$
   iii many-to-one
 d i 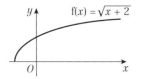 **ii** $f(x) \geq 0$
   iii one-to-one
 e i 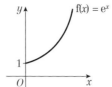 **ii** $f(x) \geq 1$
   iii one-to-one
 f i 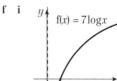 **ii** $f(x) \in \mathbb{R}$
   iii one-to-one

6 **a** $g(x)$ is not a function because it is not defined for $x = 4$
 b **c i** 1 **ii** 109
  d $a = -86$ or $a = 9$

7 **a** 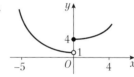 **b** −7
  c −2 and 5

8 **a**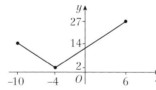
 b $a = -3.91$ or $a = 3.58$

9 **a**

$$\begin{array}{c}\text{(graph with points at } (-10, 27), (-4, 2), (6, 27)\text{, min at } (-4, 2)\text{, value 14 marked)}\end{array}$$

 b Range $\{2 \leq h(x) \leq 27\}$ **c** $a = -9, a = 0$

10 $c = \frac{2}{5}, d = \frac{44}{5}$ **11** $a = 2, b = -1$ **12** $a = 3$

Exercise 2C

1 **a** 7 **b** $\frac{9}{4}$ or 2.25 **c** 0.25 **d** −47 **e** −26

2 **a** $fg(x) = 4x^2 - 15, x \in \mathbb{R}$ **b** $gf(x) = 16x^2 + 8x - 3, x \in \mathbb{R}$
 c $gh(x) = \frac{1}{x^2} - 4, x \in \mathbb{R}, x \neq 0$ **d** $fh(x) = \frac{4}{x} + 1, x \in \mathbb{R}, x \neq 0$
 e $f^2(x) = 16x + 5, x \in \mathbb{R}$

3 **a** $fg(x) = 3x^2 - 2, x \in \mathbb{R}$ **b** $x = 1$

4 **a** $3\left(\dfrac{1}{x-2}\right) + 4 = \dfrac{3}{x-2} + \dfrac{4(x-2)}{x-2} = \dfrac{3 + 4x - 8}{x-2}$

 So $qp(x) = \dfrac{4x - 5}{x - 2}, x \in \mathbb{R}, x \neq 2$

 b $x = \frac{9}{4}$

5 **a** 23 **b** $x = \frac{13}{7}$ and $x = \frac{13}{5}$

Answers

6 a $f^2(x) = f\left(\dfrac{1}{x+1}\right) = \dfrac{1}{\left(\dfrac{1}{x+1}\right)+1} = \dfrac{x+1}{x+2}$, $x \neq -1$, $x \neq -2$

  b $f^3(x) = \dfrac{x+2}{2x+3}$, $x \neq -1$, $x \neq -2$, $x \neq -\dfrac{3}{2}$

7 a $st(x) = 2^{x+3}$, $x \in \mathbb{R}$ **b** $ts(x) = 2^x + 3$, $x \in \mathbb{R}$

  c $\dfrac{\ln\left(\dfrac{3}{7}\right)}{\ln(2)}$

8 a $gf(x) = 20x$, $x \in \mathbb{R}$ **b** $fg(x) = x^{20}$, $x \in \mathbb{R}$, $x > 0$

9 a $qp(x) = (x+3)^3 - 1$, $x \in \mathbb{R}$, $x > -3$, $qp(x) > -1$
  b 999 **c** $x = 2$

10 $3 \pm \dfrac{\sqrt{6}}{2}$

11 a $-8 \leq g(x) \leq 12$ **b** 6 **c** 10.5

Exercise 2D

1 a i $\{y \in \mathbb{R}\}$ **ii** $f^{-1}(x) = \dfrac{x-3}{2}$
  iii Domain: $\{x \in \mathbb{R}\}$, Range: $\{y \in \mathbb{R}\}$
  iv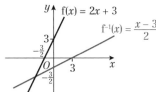

  b i $\{y \in \mathbb{R}\}$ **ii** $f^{-1}(x) = 2x - 5$
  iii Domain: $\{x \in \mathbb{R}\}$, Range: $\{y \in \mathbb{R}\}$
  iv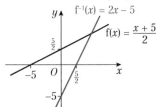

  c i $\{y \in \mathbb{R}\}$ **ii** $f^{-1}(x) = \dfrac{4-x}{3}$
  iii Domain: $\{x \in \mathbb{R}\}$, Range: $\{y \in \mathbb{R}\}$
  iv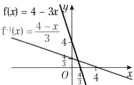

  d i $\{y \in \mathbb{R}\}$ **ii** $f^{-1}(x) = \sqrt[3]{x+7}$
  iii Domain: $\{x \in \mathbb{R}\}$, Range: $\{y \in \mathbb{R}\}$
  iv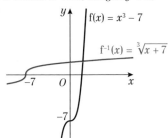

2 a $f^{-1}(x) = 10 - x$, $\{x \in \mathbb{R}\}$ **b** $g^{-1}(x) = 5x$, $\{x \in \mathbb{R}\}$
  c $h^{-1}(x) = \dfrac{3}{x}$, $\{x \neq 0\}$ **d** $k^{-1}(x) = x + 8$, $\{x \in \mathbb{R}\}$

3 Domain becomes $x < 4$

4 a i $0 < g(x) \leq \dfrac{1}{3}$ **ii** $g^{-1}(x) = \dfrac{1}{x}$
  iii $\{x \in \mathbb{R}, 0 < x \leq \dfrac{1}{3}\}$, $g^{-1}(x) \geq 3$
  iv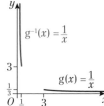

  b i $g(x) \geq -1$ **ii** $g^{-1}(x) = \dfrac{x+1}{2}$
  iii $\{x \in \mathbb{R}, x \geq -1\}$, $g^{-1}(x) \geq 0$
  iv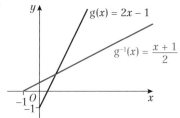

  c i $g(x) > 0$ **ii** $g^{-1}(x) = \dfrac{2x+3}{x}$
  iii $\{x \in \mathbb{R}, x > 0\}$, $g^{-1}(x) > 2$
  iv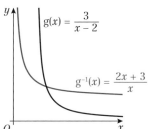

  d i $g(x) \geq 2$ **ii** $g^{-1}(x) = x^2 + 3$
  iii $\{x \in \mathbb{R}, x \geq 2\}$, $g^{-1}(x) \geq 7$
  iv

  e i $g(x) > 6$ **ii** $g^{-1}(x) = \sqrt{x-2}$
  iii $\{x \in \mathbb{R}, x > 6\}$, $g^{-1}(x) > 2$

iv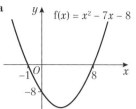

f   **i** $g(x) \geq 0$ **ii** $g^{-1}(x) = \sqrt[3]{x+8}$
   iii $\{x \in \mathbb{R}, x \geq 0\}$, $g^{-1}(x) \geq 2$
   iv

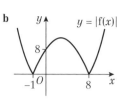

5 $t^{-1}(x) = \sqrt{x+4} + 3$, $\{x \in \mathbb{R}, x \geq 0\}$
6   **a** -2 **b** $m^{-1}(x) = \sqrt{x-5} - 2$ **c** $x > 5$
7   **a**   tends to $\pm\infty$
   b   7
   c $h^{-1}(x) = \dfrac{2x+1}{x-2}$ $\{x \in \mathbb{R}, x \neq 2\}$
   d $2 + \sqrt{5}$, $2 - \sqrt{5}$
8   **a** $nm(x) = x$, $x \in \mathbb{R}$
   b   The functions m and n are inverse of one another as $mn(x) = nm(x) = x$.
9 $st(x) = \dfrac{3}{\dfrac{3-x}{x}+1} = x$, $ts(x) = \dfrac{3 - \dfrac{3}{x-1}}{\dfrac{3}{x+1}} = x$

10   **a** $f^{-1}(x) = -\sqrt{\dfrac{x+3}{2}}$ $\{x \in \mathbb{R}, x > -3\}$
    b $a = -1$
11   **a** $f(x) > -5$ **b** $f^{-1}(x) = \ln(x+5)$ $\{x \in \mathbb{R}, x > -5\}$
    c

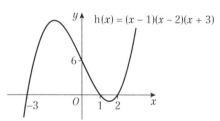

    d $g^{-1}(x) = e^x + 4$, $x \in \mathbb{R}$ **e** $x = 1.95$
12   **a** $f(x) = \dfrac{3(x+2)}{x^2 + x - 20} - \dfrac{2}{x-4}$

$= \dfrac{3(x+2)}{(x+5)(x-4)} - \dfrac{2(x+5)}{(x+5)(x-4)} = \dfrac{x-4}{(x+5)(x-4)}$

$= \dfrac{1}{x+5}$

    b $\{y \in \mathbb{R}, 0 < y < \tfrac{1}{9}\}$
    c $f^{-1}: x \to \dfrac{1}{x} - 5$. Domain is $\{x \in \mathbb{R}, 0 < x < \tfrac{1}{9} \text{ and } x \neq 0\}$

Exercise 2E

1   **a** 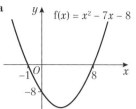   **b**

c

2   **a**

   b

   c

3   **a**

   b

   c

4 a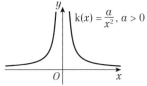

b Both these graphs would match the original graph.

c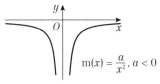

d i True, $|k(x)| = \left|\dfrac{a}{x^2}\right| = \left|\dfrac{-a}{x^2}\right| = |m(x)|$

ii False, $k(|x|) = \dfrac{a}{|x|^2} \ne \dfrac{-a}{|x|^2} = m(|x|)$

iii True, $m(|x|) = \dfrac{-a}{|x|^2} = \dfrac{-a}{x^2} = m(x)$

5 a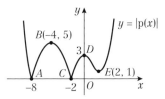

b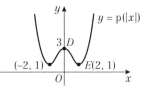

6 a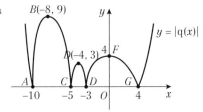

b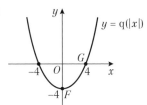

7 a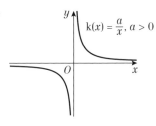

b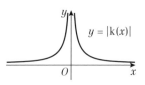

c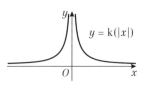

8 a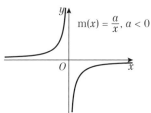

b They are reflections of each other in the x-axis. $|m(x)| = -m(|x|)$

9 a

b They would be the same as the original graph.

c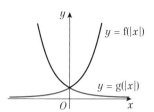

10 a $-4 < f(x) \le 9$

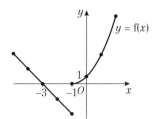

b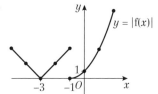

c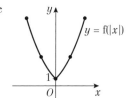

Answers

Exercise 2F

1 a 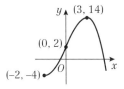 b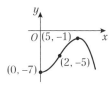

  c 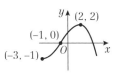 d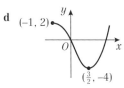

  e 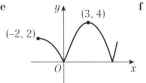 f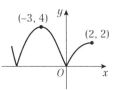

2 a 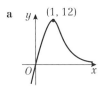 b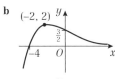

  c 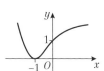 d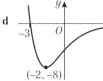

  e 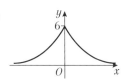 f

3 a 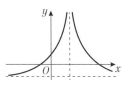 b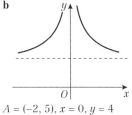

  $A = (0, 2)$, $x = 2$, $y = -1$ $A = (-2, 5)$, $x = 0$, $y = 4$

  c 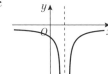 d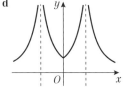

  $A = (0, -1)$, $x = 1$, $y = 0$ $A = (0, 1)$, $x = 2$, $x = -2$, $y = 0$

4 a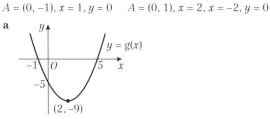

  b i $(6, -18)$ ii $(1, -9)$ iii $(2, 9)$

  c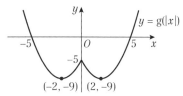

5 a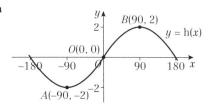

  b $A(-90, -2)$ and $B(90, 2)$

  c i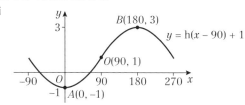

   ii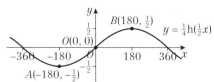

   iii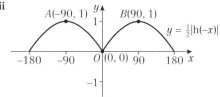

6 $O(-6, 0)$, $A(6, -2)$ and $B(15, 0)$

Exercise 2G

1 a Range $f(x) \geq -3$

   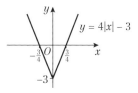

  b Range $f(x) \geq -1$

Answers

c Range f(x) ⩽ 6

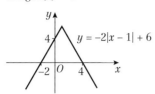

d Range f(x) ⩽ 4

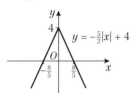

2 a, b

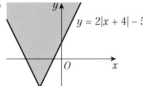

3 a, b

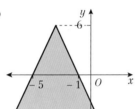

4 a

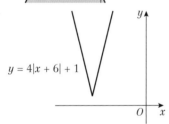

b f(x) ⩾ 1   **c** $x = -\frac{16}{3}$ and $x = -\frac{48}{7}$

5 a

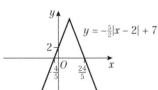

b g(x) ⩽ 7   **c** $x = -\frac{2}{3}$ and $x = \frac{22}{7}$

6 k < 14
7 b = 2
8 a h(x) ⩾ −7
 b Original function is many-to-one, therefore the inverse is one-to-many, which is not a function.
 c $-\frac{1}{2} < x < \frac{5}{2}$ **d** $k < -\frac{23}{3}$
9 a a = 10   **b** P(−3, 10) and Q(2, 0)
 c $x = -\frac{6}{7}$ and $x = -6$
10 a m(x) ⩽ 7   **b** $x = -\frac{35}{23}$ and $x = -5$
 c k < 7

Challenge
1 a A(3, −6) and B(7, −2)
 b 6 units².
2 Graphs intersect at $x = \frac{1}{3}$ and $x = \frac{17}{3}$,
 Maximum point of f(x) is (3, 10). Minimum point of g(x) is (3, 2). Using area of a kite, area $= \frac{64}{3}$

Mixed exercise 2

1 a

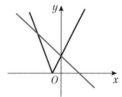

b x = 0, x = −4

2 $k > -\frac{11}{4}$

3 $x = -\frac{24}{19}$ and $x = \frac{40}{21}$

4 a

b The graphs do not intersect, so there are no solutions.

5 a i one-to-many   **ii** not a function
 b i one-to-one   **ii** function
 c i many-to-one   **ii** function
 d i many-to-one   **ii** function
 e i one-to-one
  ii not a function
 f i one-to-one
  ii not a function

6 a

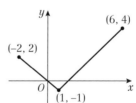

b $\frac{1}{2}$ and $1\frac{1}{2}$

7 a pq(x) = 4x² + 10x, x ∈ ℝ   **b** $x = \frac{-3 \pm \sqrt{21}}{4}$

8 a Range g(x) ⩾ 7

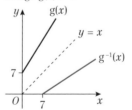

b $g^{-1}(x) = \frac{x - 7}{2}$, {x ∈ ℝ, x ⩾ 7}

c $g^{-1}(x)$ is a reflection of g(x) in the line y = x

9 a $f^{-1}(x) = \frac{x + 3}{x - 2}$, {x ∈ ℝ, x > 2}

b Range $f^{-1}(x) > 1$

10 a $f(x) = \dfrac{x}{x^2-1} - \dfrac{1}{x+1} = \dfrac{x}{(x-1)(x+1)} - \dfrac{1}{x+1}$
$= \dfrac{x}{(x-1)(x+1)} - \dfrac{x-1}{(x-1)(x+1)} = \dfrac{1}{(x-1)(x+1)}$

   b $f(x) > 0$ c $x = 6$

11 a $20, 28, \dfrac{1}{9}$ b $f(x) \geq -8, g(x) \in \mathbb{R}$

   c $g^{-1}(x) = \sqrt[3]{x-1}, \{x \in \mathbb{R}\}$

   d $fg(x) = 4(x^3 - 1), x \in \mathbb{R}, x \geq -1$

   e $a = \dfrac{5}{3}$ f $x = -10$

12 a $a = -3$ b $f^{-1}: x \mapsto \sqrt{x+13} - 3, x > -4$

13 a $f^{-1}(x) = \dfrac{x+1}{4}, \{x \in \mathbb{R}\}$

   b $gf(x) = \dfrac{3}{8x-3}, \{x \in \mathbb{R}, x \neq \tfrac{3}{8}\}$

   c -0.076 and 0.826 (3 d.p.)

14 a $f^{-1}(x) = \dfrac{2x}{x-1}, \{x \in \mathbb{R}, x \neq 1\}$

   b Range $f^{-1}(x) \in \mathbb{R}, f^{-1}(x) \neq 2$

   c -1 d $1, \tfrac{6}{5}$

15 a $8, 9$ b $-4\sqrt{5}$ and $5\sqrt{2}$

16 a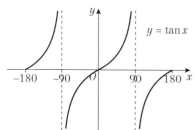

   b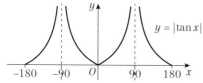

   c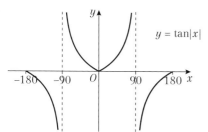

17 a 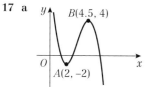   b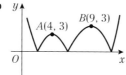

   c 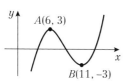   d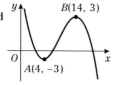

18 a $g(x) \geq 0$ b $x = 0, x = 8$

   c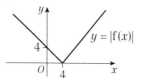

19 a Positive $|x|$ graph with vertex at $\left(\dfrac{a}{2}, 0\right)$ and y-intercept at $(0, a)$.

   b Positive $|x|$ graph with vertex at $\left(\dfrac{a}{4}, 0\right)$ and y-intercept at $(0, a)$.

   c $a = 6, a = 10$

20 a Positive $|x|$ graph with vertex at $(2a, 0)$ and y-intercept at $(0, 2a)$.

   b $x = \dfrac{3a}{2}, x = 3a$

   c Negative $|x|$ graph with x-intercepts at $(a, 0)$ and $(3a, 0)$ and y-intercept at $(0, -a)$.

21 a, b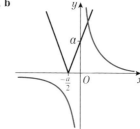

   c One intersection point   d $x = \dfrac{-a + \sqrt{(a^2+8)}}{4}$

22 a $(1, 2), (\tfrac{5}{2}, 5\ln\tfrac{5}{2} - \tfrac{13}{4})$

   b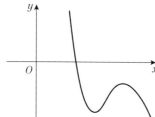

   c $(3, -6)$, Minimum

     $(\tfrac{9}{2}, \tfrac{39}{4} - 15\ln\tfrac{5}{2})$, Maximum

23 a $-2 \leq f(x) \leq 18$ b 0

   c 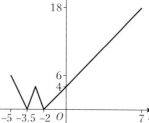   d $x = \dfrac{7 \pm \sqrt{5}}{2}$

24 a $p(x) \leq 10$

   b Original function is many-to-one, therefore the inverse is one-to-many, which is not a function.

   c $-11 < x < 3$ d $k > 8$

25 a $f(x) = 3(x-2)^2 + 8$ b $0 < g(x) \leq \tfrac{1}{8}$

Answers

Challenge
a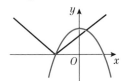

b $(-a, 0), (a, 0), (0, a^2)$ **c** $a = 5$

CHAPTER 3

Prior knowledge 3
1 **a** 22, 27, 32 **b** −1, −4, −7 **c** 9, 15, 21
  d 48, 96, 192 **e** $\frac{1}{32}, \frac{1}{64}, \frac{1}{128}$ **f** −16, 64, −256
2 **a** $x = 5.64$ **b** $x = 3.51$ **c** $x = 9.00$

Exercise 3A
1 **a i** 7, 12, 17, 22 **ii** $a = 7, d = 5$
  b i 7, 5, 3, 1 **ii** $a = 7, d = -2$
  c i 7.5, 8, 8.5, 9 **ii** $a = 7.5, d = 0.5$
  d i −9, −8, −7, −6 **ii** $a = -9, d = 1$
2 **a** $2n + 3, 23$ **b** $3n + 2, 32$
  c $27 - 3n, -3$ **d** $4n - 5, 35$
  e $nx, 10x$ **f** $a + (n - 1)d, a + 9d$
3 **a** 22 **b** 40 **c** 39 **d** 46 **e** 18 **f** n
4 $d = 6$ 5 $p = \frac{2}{3}, q = 5$ 6 −1.5
7 24 8 −70 9 $k = \frac{1}{2}, k = 8$
10 $-2 + 3\sqrt{5}$

Challenge
$a = 4, b = 2$

Exercise 3B
1 **a** 820 **b** 450 **c** −1140
  d −294 **e** 1440 **f** 1425
  g −1155 **h** $231x + 21$
2 **a** 20 **b** 25 **c** 65 **d** 4 or 14
3 2550 4 20
5 $d = -\frac{1}{2}$, 20th term = −5.5 6 $a = 6, d = -2$
7 $S_{50} = 1 + 2 + 3 + \ldots + 50$
  $S_{50} = 50 + 49 + 48 + \ldots + 1$
  $2 \times S_{50} = 50(51) \Rightarrow S_{50} = 1275$
8 $S_{2n} = 1 + 2 + 3 + \ldots + 2n$
  $S_{2n} = 2n + (2n - 1) + (2n - 2) + \ldots + 1$
  $2 \times S_n = 2n(2n + 1) \Rightarrow S_n = n(2n + 1)$
9 $S_n = 1 + 3 + 5 + \ldots + (2n - 3) + (2n - 1)$
  $S_n = (2n - 1) + (2n - 3) + \ldots + 5 + 3 + 1$
  $2 \times S_n = n(2n) \Rightarrow S_n = n^2$
10 **a** $a + 4d = 33, a + 9d = 68$
   $d = 7, a = 5$ so $S_n = \frac{n}{2}[2(5) + (n - 1)7]$
   $\Rightarrow 2225 = \frac{n}{2}(7n + 3) \Rightarrow 7n^2 + 3n - 4450 = 0$
   b 25
11 **a** $\frac{304}{k+2}$
   b $S_n = \frac{152}{k+2}(k + 1 + 303) = \frac{152k + 46208}{k+2}$
   c 17

12 **a** 1683
   b i $\frac{100}{p}$
     ii $S_{\frac{100}{p}} = \frac{50}{p}\left[8p + \left(\frac{100-p}{p}\right)4p\right]$
     $S_{\frac{100}{p}} = \frac{50}{p}[4p + 400] = 200\left[1 + \frac{100}{p}\right]$
   c $161p + 81$
13 **a** $5n + 1$ **b** 285
   c $S_k = \frac{k}{2}[2(6) + (k - 1)5] = \frac{k}{2}(5k + 7)$
   $\frac{k}{2}(5k + 7) \leq 1029$
   $5k^2 + 7k - 2058 \leq 0$
   $(5k - 98)(k + 21) \leq 0$
   d $k = 19$

Challenge
$S_n = \frac{n}{2}(2\ln 9 + (n - 1)\ln 3) = \frac{n}{2}(\ln 81 - \ln 3 + n\ln 3)$
$= \frac{n}{2}(\ln 27 + n\ln 3) = \frac{n}{2}(\ln 3^3 + \ln 3^n)$
$= \frac{n}{2}(\ln 3^{n+3}) = \frac{1}{2}(\ln 3^{n^2+3n}) \Rightarrow a = \frac{1}{2}$

Exercise 3C
1 **a** Geometric, $r = 2$ **b** Not geometric
  c Not geometric **d** Geometric, $r = 3$
  e Geometric, $r = \frac{1}{2}$ **f** Geometric, $r = -1$
  g Geometric, $r = 1$ **h** Geometric, $r = -\frac{1}{4}$
2 **a** 135, 405, 1215 **b** −32, 64, −128
  c 7.5, 3.75, 1.875 **d** $\frac{1}{64}, \frac{1}{256}, \frac{1}{1024}$
  e p^3, p^4, p^5 **f** $-8x^4, 16x^5, -32x^6$
3 **a** $x = 3\sqrt{3}$ **b** $9\sqrt{3}$
4 **a** $486, 2 \times 3^{n-1}$ **b** $\frac{25}{8}, 100 \times \left(\frac{1}{2}\right)^{n-1}$
  c $-32, (-2)^{n-1}$ **d** $1.61051, (1.1)^{n-1}$
5 10, 6250 6 $a = 1, r = 2$ 7 $\frac{1}{8}, -\frac{1}{8}$
8 **a** $\frac{x^2}{2x} = \frac{2x}{8-x} \Rightarrow x^2(8 - x) = 4x^2 \Rightarrow x^3 - 4x^2 = 0$
  b 2 097 152
  c Yes, 4096 is in sequence as n is integer, $n = 11$
9 **a** $ar^5 = 40 \Rightarrow 200p^5 = 40$
  $\Rightarrow p^5 = \frac{1}{5} \Rightarrow \log p^5 = \log\left(\frac{1}{5}\right)$
  $\Rightarrow 5 \log p = \log 1 - \log 5 \Rightarrow 5 \log p + \log 5 = 0$
  b $p = 0.725$
10 $k = 12$
11 $n = 8.69$, so not a sequence as n not an integer
12 No, −49152 is in sequence
13 $n = 11, 3\,145\,728$

Exercise 3D
1 **a** 255 **b** 63.938 **c** 1.110
  d −728 **e** $546\frac{2}{3}$ **f** −1.667
2 4.9995 3 14.4147 4 $\frac{5}{4}, -\frac{9}{4}$
5 19 terms 6 22 terms

Answers

7 a $\dfrac{25\left(1-\left(\frac{3}{5}\right)^k\right)}{\left(1-\frac{3}{5}\right)} > 61 \Rightarrow 1-\left(\frac{3}{5}\right)^k > \dfrac{122}{125} \Rightarrow \left(\frac{3}{5}\right)^k < \dfrac{3}{125}$

$\Rightarrow k\log\left(\frac{3}{5}\right) < \log\left(\frac{3}{125}\right) \Rightarrow k > \dfrac{\log(0.024)}{\log(0.6)}$

b $k = 8$

8 $r = \pm 0.4$

9 $S_{10} = \dfrac{a\left[(\sqrt{3})^{10} - 1\right]}{\sqrt{3} - 1} = \dfrac{a(243-1)}{\sqrt{3}-1}$

$= \dfrac{242a(\sqrt{3}+1)}{(\sqrt{3}-1)(\sqrt{3}+1)} = 121a(\sqrt{3}+1)$

10 $\dfrac{a(2^4 - 1)}{1} = \dfrac{b(3^4 - 1)}{2}$

$15a = 40b \Rightarrow a = \dfrac{8}{3}b$

11 a $\dfrac{2k+5}{k} = \dfrac{k}{k-6} \Rightarrow (k-6)(2k+5) = k^2$

$k^2 - 7k - 30 = 0$

b $k = 10$ **c** 2.5 **d** 25429

Exercise 3E

1 a Yes as $|r| < 1$, $\frac{10}{9}$ **b** No as $|r| \geqslant 1$

c Yes as $|r| < 1$, $6\frac{2}{3}$

d No; arithmetic series does not converge.

e No as $|r| \geqslant 1$ **f** Yes as $|r| < 1$, $4\frac{1}{2}$

g No; arithmetic series does not converge.

h Yes as $|r| < 1$, 90

2 $\frac{2}{3}$ **3** $-\frac{2}{3}$ **4** 20 **5** $13\frac{1}{3}$

6 $\frac{23}{99}$ **7** $r = -\frac{1}{2}$, $a = 12$

8 a $-\frac{1}{2} < x < \frac{1}{2}$ **b** $S_\infty = \dfrac{1}{1+2x}$

9 a 0.9787 **b** 1.875

10 a $\dfrac{30}{1-r} = 240 \Rightarrow 1 - r = \dfrac{1}{8} \Rightarrow r = \dfrac{7}{8}$

b 2.51 **c** 99.3 **d** 11

11 a $ar = \dfrac{15}{8} \Rightarrow a = \dfrac{15}{8r}$

$\dfrac{a}{1-r} = 8 \Rightarrow a = 8(1-r)$

$\dfrac{15}{8r} = 8(1-r) \Rightarrow 15 = 64r - 64r^2$

$\Rightarrow 64r^2 - 64r + 15 = 0$

b $\frac{3}{8}, \frac{5}{8}$ **c** $5, 3$ **d** 7

Challenge

a First series: $a + ar + ar^2 + ar^3 + \ldots$
Second series: $a^2 + a^2r^2 + a^2r^4 + a^2r^6 + \ldots$
Second series is geometric with common ratio is r^2 and first term a^2.

b $\dfrac{a}{1-r} = 7 \Rightarrow a = 7(1-r) \Rightarrow a^2 = 49(1-r)(1-r)$

$\dfrac{a^2}{1-r^2} = 35 \Rightarrow \dfrac{49(1-r)(1-r)}{(1-r)(1+r)} = 35$

$49(1-r) = 35(1+r) \Rightarrow 49 - 49r = 35 + 35r \Rightarrow r = \dfrac{1}{6}$

Exercise 3F

1 a i $4 + 7 + 10 + 13 + 16$ **ii** 50
  b i $3 + 12 + 27 + 48 + 75 + 108$ **ii** 273
  c i $1 + 0 + (-1) + 0 + 1$ **ii** 1
  d i $-\dfrac{2}{243} + \dfrac{2}{729} - \dfrac{2}{2187} + \dfrac{2}{6561}$ **ii** $-\dfrac{40}{6561}$

2 a i $\sum\limits_{r=1}^{4} 2r$ **ii** 20

  b i $\sum\limits_{r=1}^{5} (2 \times 3^{r-1})$ **ii** 242

  c i $\sum\limits_{r=1}^{6} \left(-\dfrac{3}{2}r + \dfrac{15}{2}\right)$ **ii** 13.5

3 a i 26 **ii** $\sum\limits_{r=1}^{26}(6r+1)$

  b i 7 **ii** $\sum\limits_{r=1}^{7}\left(\dfrac{1}{3} \times \left(\dfrac{2}{5}\right)^{r-1}\right)$

  c i 16 **ii** $\sum\limits_{r=1}^{16}(17-9r)$

4 a -280 **b** $4\,194\,300$
  c 9300 **d** $-\dfrac{7}{4}$

5 a 2134 **b** 45854 **c** $\dfrac{3}{16}$ **d** 96

6 $2\,147\,484\,111$

7 $797\,256$

8 a $\dfrac{8}{3}((-2)^k - 1)$ **b** $99k - k^2$
  c $6k - k^2 + 27$

9 $\dfrac{25}{98304}$

10 a $a = 11, d = 3$

$377 = \dfrac{k}{2}(2(11) + (k-1)(3)) = \dfrac{k}{2}(19 + 3k)$

$3k^2 + 19k - 754 = 0 \Rightarrow (3k + 58)(k - 13) = 0$

b $k = 13$

11 a $a = 6, r = 3; S_k = \dfrac{6(3^k - 1)}{3 - 1} = 3(3^k - 1)$

$\Rightarrow 3(3^k - 1) = 59046 \Rightarrow 3^k = 19683$

$\Rightarrow k\log 3 = \log 19683 \Rightarrow k = \dfrac{\log 19683}{\log 3}$

b 4723920

12 a $|x| < \dfrac{1}{3}$ **b** $\dfrac{1}{6}$

Challenge

$\sum\limits_{r=1}^{10}[a + (r-1)d]$
$S_{10} = 5(2a + 9d)$
$\sum\limits_{r=11}^{14}[a + (r-1)d] = \sum\limits_{r=1}^{14}[a + (r-1)d] - \sum\limits_{r=1}^{10}[a + (r-1)d]$
$= [7(2a + 13d) - 5(2a + 9d)] = 4a + 46d$
$4a + 46d = 10a + 45d \Rightarrow 6a = d$

Answers

Exercise 3G
1. **a** 1, 4, 7, 10 **b** 9, 4, −1, −6
   c 3, 6, 12, 24 **d** 2, 5, 11, 23
   e 10, 5, 2.5, 1.25 **f** 2, 3, 8, 63
2. **a** $u_{n+1} = u_n + 2, u_1 = 3$ **b** $u_{n+1} = u_n - 3, u_1 = 20$
   c $u_{n+1} = 2u_n, u_1 = 1$ **d** $u_{n+1} = \frac{u_n}{4}, u_1 = 100$
   e $u_{n+1} = -1 \times u_n, u_1 = 1$ **f** $u_{n+1} = 2u_n + 1, u_1 = 3$
   g $u_{n+1} = (u_n)^2 + 1, u_1 = 0$ **h** $u_{n+1} = \frac{u_n + 2}{2}, u_1 = 26$
3. **a** $u_{n+1} = u_n + 2, u_1 = 1$ **b** $u_{n+1} = u_n + 3, u_1 = 5$
   c $u_{n+1} = u_n + 1, u_1 = 3$ **d** $u_{n+1} = u_n + \frac{1}{2}, u_1 = 1$
   e $u_{n+1} = u_n + 2n + 1, u_1 = 1$ **f** $u_{n+1} = 3u_n + 2, u_1 = 2$
4. **a** $3k + 2$ **b** $3k^2 + 2k + 2$ **c** $\frac{10}{3}, -4$
5. $p = -4, q = 7$
6. **a** $x_2 = x_1(p - 3x_1) = 2(p - 3(2)) = 2p - 12$
   $x_3 = (2p - 12)(p - 3(2p - 12)) = (2p - 12)(-5p + 36)$
   $= -10p^2 + 132p - 432$
   b 12 **c** −252 288
7. **a** $16k + 25$
   b $a_4 = 4(16k + 25) + 5 = 64k + 105$
   $\sum_{r=1}^{4} a_r = k + 4k + 5 + 16k + 25 + 64k + 105$
   $= 85k + 135 = 5(17k + 27)$

Exercise 3H
1. **a i** increasing
   b i decreasing
   c i increasing
   d i periodic **ii** 2
2. **a i** 17, 14, 11, 8, 5 **ii** decreasing
   b i 1, 2, 4, 8, 16 **ii** increasing
   c i −1, 1, −1, 1, −1 **ii** periodic
   iii 2
   d i −1, 1, −1, 1, −1 **ii** periodic
   iii 2
   e i 20, 15, 10, 5, 0 **ii** decreasing
   f i 20, −15, 20, −15, 20 **ii** periodic
   iii 2
   g i $k, \frac{2k}{3}, \frac{4k}{9}, \frac{8k}{27}, \frac{16k}{81}$
   ii dependent on value of k
3. $0 < k < 1$ 4. $p = -1$
5. **a** 4 **b** 0

Challenge
a $u_3 = \frac{1 + b}{a}, u_4 = \frac{a + b + 1}{ab}, u_5 = \frac{a + 1}{b}, u_6 = a, u_7 = b$
b Order is 5 as $u_6 = u_1$ and $u_7 = u_2$
c 340

Exercise 3I
1. **a** £5800 **b** £(3800 + 200m)
2. **a** £222 500 **b** £347 500
   c It is unlikely her salary will rise by the same amount each year.
3. **a** £9.03 **b** 141 days
4. **a** 220 **b** 242 **c** 266 **d** 519
5. 57.7, 83.2
6. **a** £18 000 **b** after 7.88 years
7. **a** £13 780
   b Let a denote term of first year and u denote term of second year

$a_{52} = 10 + 51(10) = 520$
$u_1 = 520 + 11$
$u_2 = 531 + 11 = 542$
   c £42 198
8. **a** 500 m is 10 terms,
   $S_{10} = \frac{10}{2}(1000 + 9(140)) = 11\,300$
   b 1500 m
9. **a** £2450 **b** £59 000 **c** $d = 30$
10. 59 days 11. 20.15 years
12. 11.2 years 13. $2^{64} - 1 = 1.84 \times 10^{19}$
14. **a** 2.401 m **b** 48.8234 m
15. **a** 26 days **b** 98.5 miles on 25th day
16. 25 years

Mixed exercise 3
1. **a** $ar^2 = 27, ar^5 = 8 \Rightarrow r^3 = \frac{8}{27} \Rightarrow r = \frac{2}{3}$
   b 60.75 **c** 182.25 **d** 3.16
2. **a** $ar = 80, ar^4 = 5.12$
   $\Rightarrow r^3 = \frac{8}{125} \Rightarrow r = \frac{2}{5} = 0.4$
   b 200 **c** $333\frac{1}{3}$ **d** 8.95×10^{-4}
3. **a** 76, 60.8 **b** 0.876 **c** 367 **d** 380
4. **a** $1, \frac{1}{3}, -\frac{1}{9}$
   b $\sum_{n=1}^{15}\left(3\left(\frac{2}{3}\right)^n - 1\right) = \sum_{n=1}^{15} 3\left(\frac{2}{3}\right)^n - \sum_{r=1}^{15} 1$
   $\sum_{n=1}^{15} 3\left(\frac{2}{3}\right)^n = \frac{2\left(1 - \left(\frac{2}{3}\right)^{15}\right)}{\frac{1}{3}} = 5.9863$
   $\sum_{r=1}^{15} 1 = 15$
   $5.9863 - 15 = -9.014$
   c $u_{n+1} = 3\left(\frac{2}{3}\right)^{n+1} - 1 = 3 \times \frac{2}{3}\left(\frac{2}{3}\right)^n - 1 = \frac{1}{3}\left(2 \times 3\left(\frac{2}{3}\right)^n - 3\right)$
   $= \frac{2u_n - 1}{3}$
5. **a** 0.8 **b** 10 **c** 50 **d** 0.189
6. **a** £8874.11 **b** after 9.9 years
7. **a** $\frac{p(2q + 2)}{p(3q + 1)} = \frac{p(2q - 1)}{p(2q + 2)}$
   $(2q + 2)^2 = (2q - 1)(3q + 1)$
   $4q^2 + 8q + 4 = 6q^2 - q - 1$
   $0 = 2q^2 - 9q - 5 = (q - 5)(2q + 1) \Rightarrow q = 5$ or $-\frac{1}{2}$
   b 867.62
8. **a** $S_n = a + (a + d) + (a + 2d) + \ldots + (a + (n - 2)d)$
   $+ (a + (n - 1)d)$ (1)
   $S_n = (a + (n - 1)d) + (a + (n - 2)d) + \ldots + (a + 2d)$
   $+ (a + d) + a$ (2)
   Adding (1) and (2):
   $2 \times S_n = n(2a + (n - 1)d) \Rightarrow S_n = \frac{n}{2}(2a + (n - 1)d)$
   b 5050
9. 32
10. **a** $a = 25, d = -3$ **b** −3810
11. **a** 26733 **b** 53467
12. 45 cm
13. $S_{2n} = \frac{2n}{2}(2(4) + (2n - 1)4) = n(4 + 8n) = 4n(2n + 1)$
14. **a** $u_2 = 2k - 4, u_3 = 2k^2 - 4k - 4$ **b** 5, −3

15 a $a + 4d = 14$, $\frac{3}{2}(2a + 2d) = -3$
$3a + 3d = -3$, $3a + 12d = 42$
$9d = 45 \Rightarrow d = 5 \Rightarrow a = -6$
b 59

16 a $a + 3d = 3k$, $3(2a + 5d) = 7k + 9 \Rightarrow$
$6a + 15d = 7k + 9$
$6a + 15\left(\frac{3k-a}{3}\right) = 7k + 9$
$6a + 15k - 5a = 7k + 9 \Rightarrow a = 9 - 8k$
b $\frac{11k - 9}{3}$ **c** 1.5 **d** 415

17 a $a_1 = p$, $a_2 = \frac{1}{p}$, $a_3 = \frac{1}{\frac{1}{p}} = 1 \times \frac{p}{1} = p$
$a_1 = a_3 \Rightarrow$ Sequence is periodic, order 2
b $500\left(p + \frac{1}{p}\right)$

18 a $a_1 = k$, $a_2 = 2k + 6$, $a_3 = 2(2k + 6) + 6 = 4k + 18$
$a_1 < a_2 < a_3 \Rightarrow k < 2k + 6 < 4k + 18 \Rightarrow k > -6$
b $a_4 = 8k + 42$
c $a_4 = 8k + 42$
$\sum_{r=1}^{4} a_r = k + 2k + 6 + 4k + 18 + 8k + 42$
$= 15k + 66 = 3(5k + 22)$
therefore divisible by 3

19 a $a = 130$
$S_\infty = \frac{130}{1-r} = 650 \Rightarrow 130 = 650 - 650r$
$-520 = -650r \Rightarrow r = \frac{-520}{-650} = \frac{4}{5}$
b 6.82
c 513.69 (2 d.p.)
d $\frac{130(1 - (0.8)^n)}{0.2} > 600 \Rightarrow 1 - (0.8)^n > \frac{12}{13}$
$(0.8)^n < \frac{1}{13} \Rightarrow n\log(0.8) < -\log 13 \Rightarrow n > \frac{-\log 13}{\log 0.8}$

20 a $25000 \times 1.02^2 = 26010$
b $25000 \times 1.02^n > 50000$
$1.02^n > 2 \Rightarrow n\log 1.02 > \log 2 \Rightarrow n > \frac{\log 2}{\log 1.02}$
c 2047
d 214574
e People may visit the doctor more frequently than once a year, some may not visit at all, depends on state of health

21 a $2n + 1$ **b** 150
c i $S_q = \frac{q}{2}(2(3) + (q-1)2) = 2q + q^2$
$S_q = p \Rightarrow q^2 + 2q - p = 0$
ii 39

22 a $ar = -3$, $\frac{a}{1-r} = 6.75$
$\Rightarrow -\frac{3}{r} \times \frac{1}{1-r} = 6.75 \Rightarrow \frac{-3}{r - r^2} = 6.75$
$6.75r - 6.75r^2 + 3 = 0$
$27r^2 - 27r - 12 = 0$
b $-\frac{1}{3}$ series is convergent so $|r| < 1$
c 6.78

Challenge
a $u_{n+2} = 5u_{n+1} - 6u_n$
$= 5[p(3^{n+1}) + q(2^{n+1})] - 6[p(3^n) + q(2^n)]$
$= 5\left[p\left(\frac{1}{3}\right)(3^{n+2}) + q\left(\frac{1}{2}\right)(2^{n+2})\right]$
$\quad -6\left[p\left(\frac{1}{3}\right)^2(3^{n+2}) + q\left(\frac{1}{2}\right)^2(2^{n+2})\right]$
$= \left(\frac{5}{3}p - \frac{6}{9}p\right)(3^{n+2}) + \left(\frac{5}{2}q - \frac{6}{4}q\right)(2^{n+2})$
$= p(3^{n+2}) + q(2^{n+2})$

b $u_n = \left(\frac{2}{3}\right)(3^n) + \left(\frac{3}{2}\right)(2^n)$ or e.g. $u_n = 2(3^{n-1}) + 3(2^{n-1})$

c $u_{100} = 3.436 \times 10^{47}$ (4 s.f.) so contains 48 digits.

CHAPTER 4

Prior knowledge 4

1 a $1 + 35x + 525x^2 + 4375x^3$
b $9765625 - 39062500x + 70312500x^2 - 75000000x^3$
c $64 + 128x + 48x^2 - 80x^3$

2 a $\frac{4}{1 + 2x} + \frac{3}{1 - 5x}$ **b** $\frac{12}{1 + 2x} - \frac{13}{(1 + 2x)^2}$
c $\frac{8}{3x - 4} + \frac{56}{(3x - 4)^2}$

Exercise 4A

1 a i $1 - 4x + 10x^2 - 20x^3...$ **ii** $|x| < 1$
b i $1 - 6x + 21x^2 - 56x^3...$ **ii** $|x| < 1$
c i $1 + \frac{x}{2} - \frac{x^2}{8} + \frac{x^3}{16}...$ **ii** $|x| < 1$
d i $1 + \frac{5x}{3} + \frac{5x^2}{9} - \frac{5x^3}{81}...$ **ii** $|x| < 1$
e i $1 - \frac{x}{4} + \frac{5x^2}{32} - \frac{15x^3}{128}...$ **ii** $|x| < 1$
f i $1 - \frac{3x}{2} + \frac{15x^2}{8} - \frac{35x^3}{16}...$ **ii** $|x| < 1$

2 a i $1 - 9x + 54x^2 - 270x^3...$ **ii** $|x| < \frac{1}{3}$
b i $1 - \frac{5x}{2} + \frac{15x^2}{4} - \frac{35x^3}{8}...$ **ii** $|x| < 2$
c i $1 + \frac{3x}{2} - \frac{3x^2}{8} + \frac{5x^3}{16}...$ **ii** $|x| < \frac{1}{2}$
d i $1 - \frac{35x}{3} + \frac{350x^2}{9} - \frac{1750x^3}{81}...$ **ii** $|x| < \frac{1}{5}$
e i $1 - 4x + 20x^2 - \frac{320x^3}{3}...$ **ii** $|x| < \frac{1}{6}$
f i $1 + \frac{5x}{4} + \frac{5x^2}{4} + \frac{55x^3}{48}...$ **ii** $|x| < \frac{4}{3}$

3 a i $1 - 2x + 3x^2 - 4x^3...$ **ii** $|x| < 1$
b i $1 - 12x + 90x^2 - 540x^3...$ **ii** $|x| < \frac{1}{3}$
c i $1 - \frac{x}{2} - \frac{x^2}{8} - \frac{x^3}{16}...$ **ii** $|x| < 1$
d i $1 - x - x^2 - \frac{5x^3}{3}...$ **ii** $|x| < \frac{1}{3}$
e i $1 - \frac{x}{4} + \frac{3x^2}{32} - \frac{5x^3}{128}...$ **ii** $|x| < 2$
f i $1 + \frac{4x}{3} + \frac{20x^2}{9} + \frac{320x^3}{81}...$ **ii** $|x| < \frac{1}{2}$

Answers

4 a Expansion of $(1 - 2x)^{-1} = 1 + 2x + 4x^2 + 8x^3 + \ldots$
Multiply by $(1 + x) = 1 + 3x + 6x^2 + 12x^3 + \ldots$
  b $|x| < \frac{1}{2}$

5 a $1 + \frac{3}{2}x - \frac{9}{8}x^2 + \frac{27}{16}x^3$
  b $f(x) = \sqrt{\frac{103}{100}} = \frac{\sqrt{103}}{\sqrt{100}} = \frac{\sqrt{103}}{10}$
  c $3.1 \times 10^{-6}\%$

6 a $a = \pm 8$ **b** ± 160

7 For small values of x ignore powers of x^3 and higher.
$(1 + x)^{\frac{1}{2}} = 1 + \frac{x}{2} - \frac{x^2}{8} + \ldots$, $(1 - x)^{-\frac{1}{2}} = 1 + \frac{x}{2} + \frac{3x^2}{8} + \ldots$
$\sqrt{\frac{1+x}{1-x}} = 1 + \frac{x}{2} - \frac{x^2}{8} + \frac{x}{2} + \frac{x^2}{4} + \frac{3x^2}{8} + \ldots = 1 + x + \frac{x^2}{2}$

8 a $2 - 42x + 114x^2$
  b 0.052%
  c The expansion is only valid for $|x| < \frac{1}{5}$. $|0.5|$ is not less than $\frac{1}{5}$.

9 a $1 - \frac{9}{2}x + \frac{27}{8}x^2 + \frac{27}{16}x^3$
  b $0.97^{\frac{3}{2}} = \left(\frac{\sqrt{97}}{10}\right)^3 = \frac{97\sqrt{97}}{1000}$
  c 9.84886
  d Use more terms from the binomial expansion of $(1 - 3x)^{\frac{3}{2}}$

Challenge

a $1 - \frac{1}{2x} + \frac{3}{8x^2}$

b $h(x) = \left(\frac{10}{9}\right)^{-\frac{1}{2}} = \left(\frac{9}{10}\right)^{\frac{1}{2}} = \frac{3}{\sqrt{10}} = \frac{3\sqrt{10}}{10}$

c 3.16

Exercise 4B

1 a i $2 + \frac{x}{2} - \frac{x^2}{16} + \frac{x^3}{64}$ **ii** $|x| < 2$
  b i $\frac{1}{2} - \frac{x}{4} + \frac{x^2}{8} - \frac{x^3}{16}$ **ii** $|x| < 2$
  c i $\frac{1}{16} + \frac{x}{32} + \frac{3x^2}{256} + \frac{x^3}{256}$ **ii** $|x| < 4$
  d i $3 + \frac{x}{6} - \frac{x^2}{216} + \frac{x^3}{3888}$ **ii** $|x| < 9$
  e i $\frac{\sqrt{2}}{2} - \frac{\sqrt{2}}{8}x + \frac{3\sqrt{2}}{64}x^2 - \frac{5\sqrt{2}}{256}x^3$ **ii** $|x| < 2$
  f $\frac{5}{3} - \frac{10}{9}x + \frac{20}{27}x^2 - \frac{40}{81}x^3$ **ii** $|x| < \frac{3}{2}$
  g i $\frac{1}{2} + \frac{1}{4}x - \frac{1}{8}x^2 + \frac{1}{16}x^3$ **ii** $|x| < 2$
  h i $\sqrt{2} + \frac{3\sqrt{2}}{4}x + \frac{15\sqrt{2}}{32}x^2 + \frac{51\sqrt{2}}{128}x^3$ **ii** $|x| < 1$

2 $\frac{1}{25} - \frac{8}{125}x + \frac{48}{625}x^2 - \frac{256}{3125}x^3$

3 a $2 - \frac{x}{4} - \frac{x^2}{64}$
  b $m(x) = \sqrt{\frac{35}{9}} = \frac{\sqrt{35}}{\sqrt{9}} = \frac{\sqrt{35}}{3}$
  c 5.91609 (correct to 5 decimal places), % error $= 1.38 \times 10^{-4}\%$

4 a $a = \frac{1}{9}, b = -\frac{2}{81}$ **b** $\frac{5}{486}$

5 For small values of x ignore powers of x^3 and higher.
$(4 - x)^{-1} = \frac{1}{4} + \frac{x}{16} + \frac{x^2}{64} + \ldots$
Multiply by $(3 + 2x - x^2) = \frac{3}{4} + \frac{x}{2} - \frac{x^2}{4} + \frac{3x}{16} + \frac{x^2}{8} + \frac{3x^2}{64}$
$= \frac{3}{4} + \frac{11}{16}x - \frac{5}{64}x^2$

6 a $\frac{1}{\sqrt{5}} - \frac{x}{5\sqrt{5}} + \frac{3x^2}{50\sqrt{5}}$ **b** $-\frac{1}{\sqrt{5}} + \frac{11x}{5\sqrt{5}} - \frac{23x^2}{50\sqrt{5}}$

7 a $2 - \frac{3}{32}x - \frac{27}{4096}x^2$ **b** 1.991

8 a $\frac{3}{4 - 2x} = \frac{3}{4} + \frac{3x}{8} + \frac{3x^2}{16}$, $\frac{2}{3 + 5x} = \frac{2}{3} - \frac{10x}{9} + \frac{50x^2}{27}$
$\frac{3}{4 - 2x} - \frac{2}{3 + 5x} = \frac{1}{12} + \frac{107}{72}x - \frac{719}{432}x^2$
  b 0.0980311
  c 0.0032%

Exercise 4C

1 a $\frac{4}{1-x} - \frac{4}{2+x}$ **b** $2 + 5x + \frac{7}{2}x^2$
  c valid $|x| < 1$

2 a $-\frac{2}{2+x} + \frac{4}{(2+x)^2}$ **b** $B = \frac{1}{2}, C = -\frac{3}{8}$
  c $|x| < 2$

3 a $\frac{2}{1+x} + \frac{3}{1-x} - \frac{4}{2+x}$ **b** $3 + 2x + \frac{9}{2}x^2 + \frac{5}{4}x^3$
  c $|x| < 1$

4 a $A = -\frac{14}{5}$ and $B = \frac{9}{5}$ **b** $-1 + 11x + 5x^2$

5 a $2 - \frac{1}{x+5} + \frac{6}{x-4}$ **b** $\frac{3}{10} - \frac{67}{200}x - \frac{407}{4000}x^2$
  c $|x| < 4$

6 a $A = 3, B = -2$ and $C = 3$ **b** $\frac{5}{6} - \frac{19}{36}x - \frac{97}{216}x^2$

7 a $A = -\frac{7}{9}, B = \frac{28}{3}$ and $C = \frac{8}{9}$ **b** $11 + 38x + 116x^2$
  c 0.33%

Mixed exercise 4

1 a i $1 - 12x + 48x^2 - 64x^3$ **ii** all x
  b i $4 + \frac{x}{8} - \frac{x^2}{512} + \frac{x^3}{16384}$ **ii** $|x| < 16$
  c i $1 + 2x + 4x^2 + 8x^3$ **ii** $|x| < \frac{1}{2}$
  d i $2 - 3x + \frac{9x^2}{2} - \frac{27x^3}{4}$ **ii** $|x| < \frac{2}{3}$
  e i $2 + \frac{x}{4} + \frac{3x^2}{64} + \frac{5x^3}{512}$ **ii** $|x| < 4$
  f i $1 - 2x + 6x^2 - 18x^3$ **ii** $|x| < \frac{1}{3}$
  g i $1 + 4x + 8x^2 + 12x^3$ **ii** $|x| < 1$
  h i $-3 - 8x - 18x^2 - 38x^3$ **ii** $|x| < \frac{1}{2}$

2 $1 - \frac{x}{4} - \frac{x^2}{32} - \frac{x^3}{128}$

3 a $1 + \frac{x}{2} - \frac{x^2}{8} + \frac{x^3}{16}$ **b** $\frac{1145}{512}$

4 a $c = -9, d = 36$ **b** 1.282
  c calculator $= 1.28108713$, approximation is correct to 2 decimal places.

5 a $a = 4$ or $a = -4$
  b coefficient of $x^3 = 4$, coefficient of $x^3 = -4$.

Answers

6   **a** $1 - 3x + 9x^2 - 27x^3$
    b $(1 + x)(1 - 3x + 9x^2 - 27x^3)$
      $= 1 - 3x + 9x^2 - 27x^3 + x - 3x^2 + 9x^3$
      $= 1 - 2x + 6x^2 - 18x^3$
    c $x = 0.01, 0.98058$

7   **a** $n = -2, a = 3$ **b** -108
    c $|x| < \frac{1}{3}$

8   For small values of x ignore powers of x^3 and higher.
    $\frac{1}{\sqrt{4+x}} = \frac{1}{2} - \frac{x}{16} + \frac{3x^2}{256}$, $\frac{3}{\sqrt{4+x}} = \frac{3}{2} - \frac{3}{16}x + \frac{9}{256}x^2$

9   **a** $\frac{1}{2} + \frac{x}{16} + \frac{3}{256}x^2$ **b** $\frac{1}{2} + \frac{17}{16}x + \frac{35}{256}x^2$

10 **a** $\frac{1}{2} - \frac{3}{4}x + \frac{9}{8}x^2 - \frac{27}{16}x^3$ **b** $\frac{1}{2} - \frac{x}{4} + \frac{3}{8}x^2 - \frac{9}{16}x^3$

11 **a** $\frac{1}{2} - \frac{x}{16} + \frac{3x^2}{256} - \frac{5x^3}{2048}$ **b** **i** 1.3828   **ii** 1.4141
    c $x = \frac{1}{2}$, because it is closer to 0.

12 $\frac{1}{27} - \frac{4}{27}x + \frac{32}{81}x^2$

13 **a** $A = 1, B = -4, C = 3$ **b** $-\frac{3}{8} - \frac{51}{64}x + \frac{477}{512}x^2$

14 **a** $A = 3$ and $B = 2$ **b** $5 - 28x + 144x^2$

15 **a** $10 - 2x + \frac{5}{2}x^2 - \frac{11}{4}x^3$, so $B = \frac{5}{2}$ and $C = -\frac{11}{4}$
    b Percent error = 0.0027%

Challenge

$1 - \frac{3x^2}{2} + \frac{27x^4}{8} - \frac{135x^6}{16}$

Review exercise 1

1 Assumption: there are finitely many prime numbers, p_1, p_2, p_3 up to p_n. Let $X = (p_1 \times p_2 \times p_3 \times ... \times p_n) + 1$
None of the prime numbers $p_1, p_2, ... p_n$ can be a factor of X as they all leave a remainder of 1 when X is divided by them. But X must have at least one prime factor. This is a contradiction.
So there are infinitely many prime numbers.

2 Assumption: $x = \frac{a}{b}$ is a solution to the equation, where a and b are integers with no common factors.
$\left(\frac{a}{b}\right)^2 - 2 = 0 \Rightarrow \frac{a^2}{b^2} = 2 \Rightarrow a^2 = 2b^2$
So a^2 is even, which implies that a is even.
Write $a = 2n$ for some integer n.
$(2n)^2 = 2b^2 \Rightarrow 4n^2 = 2b^2 \Rightarrow 2n^2 = b^2$
So b^2 is even, which implies that b is even.
This contradicts the assumption that a and b have no common factor.
Hence there are no rational solutions to the equation.

3 $\frac{4x - 3}{x(x - 3)}$

4 **a** $f(x) = \frac{(x+2)^2 - 3(x+2) + 3}{(x+2)^2} = \frac{x^2 + x + 1}{(x+2)^2}$
    b $(x + \frac{1}{2})^2 + \frac{3}{4} > 0$
    c $x^2 + x + 1 > 0$ from **b** and $(x+2)^2 > 0$ as $x \neq -2$

5 $\frac{-1}{x - 1} + \frac{4}{2x - 3}$

6 $P = 2, Q = -1, R = -1$

7 $A = \frac{2}{9}, B = \frac{2}{9}, C = \frac{2}{3}$

8 $A = 3, B = 1, C = -2$

9 $d = 3, e = 6, f = -14$

10 $p(x) = 6 - \frac{2}{1 - x} + \frac{5}{1 + 2x}$

11 $x > \frac{2}{3}$ or $x < -5$

12 **a** Range: $p(x) \leq 4$

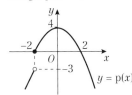

    b $a = -\frac{25}{4}$ or $a = 2\sqrt{6}$

13 **a** $2\left(\frac{1}{x+4}\right) - 5 = \frac{2}{x+4} - \frac{5(x+4)}{x+4} = \frac{2 - 5x - 20}{x+4}$
So $qp(x) = \frac{-5x - 18}{x + 4}, x \in \mathbb{R}, x \neq -4$
    b $x = -\frac{39}{10}$
    c $r^{-1}(x) = \frac{-4x - 18}{x + 5}, x \in \mathbb{R}, x \neq -5$

14 **a**

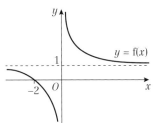

    b $\frac{\left(\frac{x+2}{x}\right) + 2}{\left(\frac{x+2}{x}\right)} = \frac{x + 2 + 2x}{x + 2} = \frac{3x + 2}{x + 2}$
So $f^2(x) = \frac{3x + 2}{x + 2}, x \in \mathbb{R}, x \neq 0, x \neq -2$
    c $\ln 13$ **d** $g^{-1}(x) = \frac{e^x + 5}{2}, x \in \mathbb{R}$

15 **a** $3(1 - 2x) + b = 1 - 2(3x + b), b = -\frac{2}{3}$
    b $p^{-1}(x) = \frac{3x + 2}{9}, x \in \mathbb{R}, q^{-1}(x) = \frac{1 - x}{2}, x \in \mathbb{R}$
    c $p^{-1}(x)q^{-1}(x) = q^{-1}(x)p^{-1}(x) = \frac{-3x + 7}{18}, x \in \mathbb{R}$
    $a = -3, b = 7, c = 18$

16 **a**   **b**

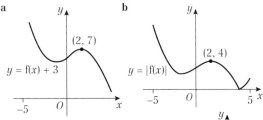

    c   **d**

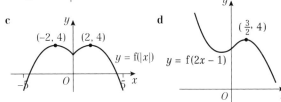

381

Answers

17 a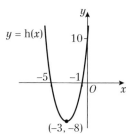

  b i $(-5, -24)$ **ii** $(3, -8)$ **iii** $(-3, 8)$

  c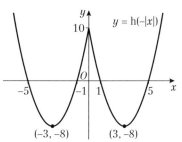

18 a i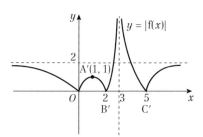

  ii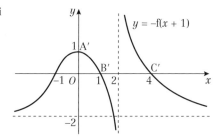

  iii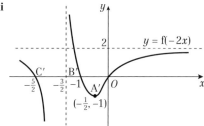

  b i 6   **ii** 4

19 a $b = -9$ **b** $A(9, -3), B(15, 0)$

  c $x = 15, x = -21$

20 a $f(x) \leq 8$

  b The function is not one-to-one.

  c $-\frac{32}{3} < x < -\frac{8}{7}$

  d $k > \frac{44}{3}$

21 a $k = 0.6, k = -4$ **b** $a = 16, d = 8$

22 a Solve $a + 3d = 72, a + 10d = 51$ simultaneously to obtain $a = 81, d = -3$
$$1125 = \frac{n}{2}(162 + (n-1)(-3))$$
$$2250 = 165n - 3n^2$$
Therefore $3n^2 - 165n + 2250 = 0$

  b $n = 25, n = 30$

23 a $a = 19p - 18, d = 10 - 2p$, 30th term $= 272 - 39p$

  b $p = 12$

24 a $r^6 = \frac{225}{64} \Rightarrow \ln r^6 = \ln\left(\frac{225}{64}\right) \Rightarrow 6 \ln r - \ln\left(\frac{225}{64}\right) = 0$
$$\Rightarrow 6 \ln r + \ln\left(\frac{64}{225}\right) = 0$$

  b $r = 1.23$

25 a 60

  b $a = 10, r = \frac{5}{6}$

$$\frac{10\left(1 - \left(\frac{5}{6}\right)^k\right)}{1 - \frac{5}{6}} > 55 \Rightarrow 1 - \left(\frac{5}{6}\right)^k > \frac{11}{12}$$

$$\Rightarrow \frac{1}{12} > \left(\frac{5}{6}\right)^k \Rightarrow \log\left(\frac{1}{12}\right) > \log\left(\left(\frac{5}{6}\right)^k\right)$$

$$\Rightarrow \log\left(\frac{1}{12}\right) > k \log\left(\frac{5}{6}\right) \Rightarrow \frac{\log\left(\frac{1}{12}\right)}{\log\left(\frac{5}{6}\right)} < k$$

  c $k = 14$

26 a $4 + 4r + 4r^2 = 7 \Rightarrow 4r^2 + 4r - 3 = 0$

  b $r = \frac{1}{2}$ or $r = -\frac{3}{2}$ **c** 8

27 a $x = 1, r = 3$ and $x = -9, r = -\frac{1}{3}$

  b 243   **c** 182.25

28 a $a_1 = k, a_2 = 3k + 5$

  b $a_3 = 3a_2 + 5 = 9k + 20$

  c i $40k + 90$ **ii** $10(4k + 9)$

29 a 2860

  b $2400 \times 1.06^{N-1} > 6000 \Rightarrow 1.06^{N-1} > 2.5$
$$\Rightarrow \log 1.06^{N-1} > \log 2.5 \Rightarrow (N - 1) \log 1.06 > \log 2.5$$

  c $N = 16.7...$, therefore $N = 17$

  d $S_n = \frac{2400(1.06^{10} - 1)}{1.06 - 1} = 31633.90...$ donations.
Total value of donations is 5 times this, so £158,000 over the 10-year period.

30 a $|x| < \frac{1}{4}$

  b $\frac{6}{1 + 4x} = \frac{24}{5} \Rightarrow x = \frac{1}{16}$

31 a $(1 - x)^{-\frac{1}{2}} = 1 + \left(-\frac{1}{2}\right)(-x) + \frac{\left(-\frac{1}{2}\right)\left(-\frac{3}{2}\right)}{2!}(-x)^2$

$$+ \frac{\left(-\frac{1}{2}\right)\left(-\frac{3}{2}\right)\left(-\frac{5}{2}\right)}{3!}(-x)^3 + ...$$

$$= 1 + \frac{1}{2}x + \frac{3}{8}x^2 + \frac{5}{16}x^3 + ...$$

  b $|x| < 1$. Accept $-1 < x < 1$.

32 a $a = 9$ $n = -\frac{36}{54} = -\frac{2}{3}$

  b -360

  c $-\frac{1}{9} < x < \frac{1}{9}$

Answers

33 a $1 + 6x + 6x^2 - 4x^3$

  b $\left(1 + 4\left(\dfrac{3}{100}\right)\right)^{\frac{3}{2}} = \left(\dfrac{112}{100}\right)^{\frac{3}{2}} = \left(\sqrt{\dfrac{112}{100}}\right)^3 = \left(\dfrac{\sqrt{112}}{10}\right)^3$
  $= \dfrac{112\sqrt{112}}{1000}$

  c 10.58296   **d** 0.00039%

34 $\dfrac{1}{27} - \dfrac{x}{27} + \dfrac{2x^2}{81} - \dfrac{8x^3}{729}$

35 a $(4 - 9x)^{\frac{1}{2}} = 2\left(1 - \dfrac{9}{4}x\right)^{\frac{1}{2}} = 2 - \dfrac{9}{4}x - \dfrac{81}{64}x^2$

  b $\sqrt{4 - 9\left(\dfrac{1}{100}\right)} = \sqrt{\dfrac{391}{100}} = \dfrac{\sqrt{391}}{10}$

  c Approximate: 1.97737 correct to 5 decimal places.

36 a $a = 2, b = -1, c = \dfrac{3}{16}$ **b** $\dfrac{1}{8}$

  $a = -2, b = 1, c = \dfrac{3}{16}$

37 a $A = 1, B = 2$ **b** $3 - x + 11x^2 - \ldots$

38 a $A = -\dfrac{3}{2}, B = \dfrac{1}{2}$ **b** $-1 - x + 4x^3 + \ldots$

39 a $A = 2, B = 10, C = 1$ **b** $\dfrac{25}{9} - \dfrac{25}{27}x + \dfrac{25}{9}x^2 + \ldots$

40 a $A = 2, B = 5, C = -2$ **b** $\dfrac{31}{12} + \dfrac{19}{144}x - \dfrac{377}{1728}x^2$

Challenge

1 a $(x + 2)^2 + (y - 3)^2 = 25$ **b** 15

2 $a_1 = m, a_2 = m + k, a_3 = m + 2k, \ldots$
  $6m + 45k = 4m + 50k \Rightarrow 2m = 5k \Rightarrow m = \dfrac{5}{2}k$

3 $A: x = \dfrac{19 - \sqrt{41}}{4}$, $B: x = \dfrac{16}{3}$, $C: x = \dfrac{19 + \sqrt{41}}{4}$

4 $\sum_{r=1}^{40} \log_3\left(\dfrac{2n+1}{2n-1}\right) = \log_3\left(\dfrac{3}{1}\right) + \log_3\left(\dfrac{5}{3}\right) + \ldots + \log_3\left(\dfrac{79}{77}\right) + \log_3\left(\dfrac{81}{79}\right)$
  $= \log_3\left(\dfrac{3}{1} \times \dfrac{5}{3} \times \ldots \times \dfrac{79}{77} \times \dfrac{81}{79}\right) = \log_3 81 = 4$

5 $y = f(ax + b)$ is a stretch by horizontal scale factor $\dfrac{1}{a}$ followed by a translation $\begin{pmatrix} -\dfrac{b}{a} \\ 0 \end{pmatrix}$. Point (x, y) maps to point $\left(\dfrac{x}{a} - \dfrac{b}{a}, y\right)$

So (x, y) invariant implies that: $\dfrac{x}{a} - \dfrac{b}{a} = x \Rightarrow x = \dfrac{b}{1 - a}$

CHAPTER 5

Prior knowledge 5

1 a $-\dfrac{1}{2}$ **b** $-\dfrac{\sqrt{2}}{2}$ **c** $\sqrt{3}$ **d** $-\dfrac{\sqrt{3}}{2}$

2 a 1   **b** $-\tan^2 \theta$ **c** $|\sin \theta|$

3 a $(\sin 2\theta + \cos 2\theta)^2 = \sin^2 2\theta + 2 \sin 2\theta \cos 2\theta + \cos^2 2\theta$
  $= 1 + 2 \sin 2\theta \cos 2\theta$

  b $\dfrac{2}{\sin \theta} - 2 \sin \theta = \dfrac{2 - 2 \sin^2 \theta}{\sin \theta} = \dfrac{2(1 - \sin^2 \theta)}{\sin \theta} = \dfrac{2 \cos^2 \theta}{\sin \theta}$

4 a 75.5°, 284°   **b** 15°, 75°, 195°, 255°
  c 19.7°, 118°, 200°, 298°   **d** 75.5°, 132°, 228°, 284°

Exercise 5A

1 a 9°   **b** 12°   **c** 75°   **d** 225°
  e 270°   **f** 540°

2 a 26.4°   **b** 57.3°   **c** 65.0°   **d** 99.2°

3 a 0.479   **b** 0.156   **c** 1.74   **d** 0.909   **e** −0.443

4 a $\dfrac{2\pi}{45}$ **b** $\dfrac{\pi}{18}$ **c** $\dfrac{\pi}{8}$ **d** $\dfrac{\pi}{6}$
  e $\dfrac{5\pi}{8}$ **f** $\dfrac{4\pi}{3}$ **g** $\dfrac{3\pi}{2}$ **h** $\dfrac{7\pi}{4}$
  i $\dfrac{11\pi}{6}$

5 a 0.873 rad   **b** 1.31 rad   **c** 1.75 rad   **d** 2.79 rad
  e 4.01 rad   **f** 5.59 rad

6 a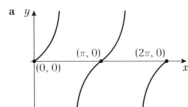

  b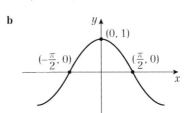

7 a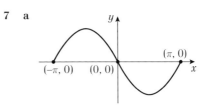

  b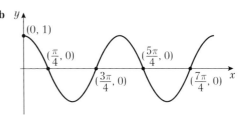

  c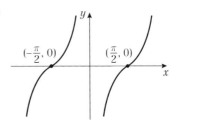

  d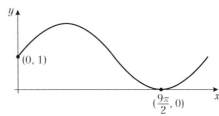

8 $(0, -0.5)$
  $\left(-\dfrac{11\pi}{6}, 0\right), \left(-\dfrac{5\pi}{6}, 0\right), \left(\dfrac{\pi}{6}, 0\right), \left(\dfrac{7\pi}{6}, 0\right)$

Answers

Challenge
a $2\pi n, n \in \mathbb{Z}$ b $\frac{3\pi}{2} + 2\pi n, n \in \mathbb{Z}$ c $\frac{\pi}{2} + \pi n, n \in \mathbb{Z}$

Exercise 5B

1 a $\sin\left(\pi - \frac{\pi}{4}\right) = \frac{\sqrt{2}}{2}$ b $-\sin\left(\frac{\pi}{3}\right) = -\frac{\sqrt{3}}{2}$
   c $\sin\left(2\pi - \frac{\pi}{6}\right) = -\frac{1}{2}$ d $\cos\left(\pi - \frac{\pi}{3}\right) = -\frac{1}{2}$
   e $\cos\left(2\pi - \frac{\pi}{3}\right) = \frac{1}{2}$ f $\cos\left(\pi + \frac{\pi}{4}\right) = -\frac{\sqrt{2}}{2}$
   g $\tan\left(\pi - \frac{\pi}{4}\right) = -1$ h $-\tan\left(\pi + \frac{\pi}{4}\right) = -1$
   i $\tan\left(\pi + \frac{\pi}{6}\right) = \frac{\sqrt{3}}{3}$

2 a $\frac{\sqrt{3}}{2}$ b $\frac{\sqrt{3}}{2}$ c $-\frac{\sqrt{3}}{2}$
   d $-\frac{\sqrt{2}}{2}$ e $-\sqrt{3}$ f $\sqrt{3}$

3 $AC = \frac{2}{\sin\left(\frac{\pi}{3}\right)} = \frac{4\sqrt{3}}{3}$
   $DC^2 = AD^2 + AC^2 = \left(\frac{2\sqrt{6}}{3}\right)^2 + \left(\frac{4\sqrt{3}}{3}\right)^2 = 8$
   $DC = 2\sqrt{2}$

Exercise 5C

1 a i 2.7   ii 2.025   iii 7.5π
   b i $\frac{50}{3}$ ii 1.8   iii 3.6
   c i $\frac{4}{3}$ ii 0.8   iii 2

2 $\frac{10}{3}\pi$ cm   3 2π 4 $5\sqrt{2}$ cm

5 a 10.4 cm   b 1.25 rad
6 7.5   7 0.8
8 a $\frac{1}{3}\pi$ b $6 + \frac{4}{3}\pi$ cm
9 6.8 cm
10 a $R - r$
   b $\sin\theta = \frac{r}{R-r} \Rightarrow (R-r)\sin\theta = r \Rightarrow (R\sin\theta - r\sin\theta) = r$
      $\Rightarrow R\sin\theta = r + r\sin\theta \Rightarrow R\sin\theta = r(1 + \sin\theta)$.
   c 2.43 cm
11 2 rad
12 a 36 m   b 13.6 km/h
13 a 3.5 m   b 15.3 m
14 a 2.59 rad b 44 mm

Exercise 5D

1 a 19.2 cm^2 b $\frac{27}{4}\pi$ cm^2 c $\frac{162}{125}\pi$ cm^2
   d 25.1 cm^2 e $6\pi - 9\sqrt{3}$ cm^2 f $\frac{63}{2}\pi + 9\sqrt{2}$ cm^2

2 a $\frac{16}{3}\pi$ cm^2 b 5 cm^2

3 a 4.47   b 3.96   c 1.98
4 12 cm^2
5 a $\cos\theta = \frac{10^2 + 10^2 - 18.65^2}{2 \times 10 \times 10} = -0.739\ldots$
   b 120 cm^2
6 $40\frac{2}{3}$ cm
7 a 12
   b $A = \frac{1}{2}r^2\theta = \frac{1}{2} \times 12^2 \times 0.5 = 36$ cm^2
   c 1.48 cm^2

8 a $l = r\theta = \frac{x\pi}{12}, x = \frac{12l}{\pi}$
   $A = \frac{1}{2}r^2\theta = \frac{1}{2}\left(\frac{12l}{\pi}\right)^2 \frac{\pi}{12} = \frac{\pi}{24}\left(\frac{144l^2}{\pi^2}\right) = \frac{6l^2}{\pi}$
   b 5π cm   c 60

9 $\triangle COB = \frac{1}{2}r^2\sin\theta$
   Shaded area $= \frac{1}{2}r^2(\pi - \theta) - \frac{1}{2}r^2\sin(\pi - \theta)$
   $= \frac{1}{2}r^2\pi - \frac{1}{2}r^2\theta - \frac{1}{2}r^2\sin\theta$
   Since $\triangle COB =$ shaded area,
   $\frac{1}{2}r^2\sin\theta = \frac{1}{2}r^2\pi - \frac{1}{2}r^2\theta - \frac{1}{2}r^2\sin\theta$
   $\sin\theta = \pi - \theta - \sin\theta$
   $\theta + 2\sin\theta = \pi$

10 38.7 cm^2
11 8.88 cm^2
12 a $OAD = \frac{1}{2}r^2\theta$, $OBC = \frac{1}{2}(r+8)^2\theta$
   $ABCD = \frac{1}{2}(r+8)^2\theta - \frac{1}{2}r^2\theta = 48$
   $\frac{1}{2}(r^2 + 16r + 64)\theta - \frac{1}{2}r^2\theta = 48$
   $(r^2 + 16r + 64)\theta - r^2\theta = 96$
   $16r + 64 = \frac{96}{\theta} \Rightarrow r = \frac{6}{\theta} - 4$
   b 28 cm
13 78.4 ($\theta = 0.8$)
14 a $14^2 = 12^2 + 10^2 - 2 \times 12 \times 10 \cos A$
   $196 = 144 + 100 - 240 \cos A$
   $-48 = -240 \cos A$
   $0.2 = \cos A$
   $A = \cos^{-1}(0.2) = 1.369438406\ldots = 1.37$ (3 s.f.)
   b 34.1 m^2
15 a 18.1 cm   b 11.3 cm^2
16 a 98.79 cm^2 b 33.24 cm
17 4.62 cm^2

Challenge
Area $= \frac{1}{2}r^2\theta$, arc length, $l = r\theta$
Area $= \frac{1}{2}rl$

Exercise 5E

1 a 0.795, 5.49   b 3.34, 6.08
   c 1.37, 4.51   d π
2 a 0.848, 2.29   b 0.142, 3.28
   c 1.08, 4.22   d 0.886, 5.40
3 a 1.16, 5.12   b 3.61, 5.82
   c 0.896, 4.04   d 0.421, 5.86
4 a $-\frac{5\pi}{6}, \frac{\pi}{6}$ b 0.201, 2.94
   c $-5.39, -0.896, 0.896, 5.39$
   d $-1.22, 1.22, 5.06, 7.51$
   e $1.77, 4.91, 8.05, 11.2$
   f 4.89
5 a $0.322, 2.82, 3.46, 5.96$
   b $1.18, 1.96, 3.28, 4.05, 5.37, 6.15$
   c $\frac{\pi}{24}, \frac{7\pi}{24}, \frac{13\pi}{24}, \frac{19\pi}{24}, \frac{25\pi}{24}, \frac{31\pi}{24}, \frac{37\pi}{24}, \frac{43\pi}{24}$
   d $0.232, 2.91, 3.37, 6.05$
6 a $-\frac{7\pi}{12}, -\frac{5\pi}{12}, \frac{\pi}{12}, \frac{\pi}{4}, \frac{3\pi}{4}, \frac{11\pi}{12}$
   b $-\frac{5\pi}{6}, -\frac{2\pi}{3}, -\frac{\pi}{3}, -\frac{\pi}{6}, \frac{\pi}{6}, \frac{\pi}{3}, \frac{2\pi}{3}, \frac{5\pi}{6}, \frac{7\pi}{6}, \frac{4\pi}{3}, \frac{5\pi}{3}, \frac{11\pi}{6}$

c −5.92, −4.35, −2.78, −1.21, 0.359, 1.93, 3.50, 5.07
   d −2.46, −0.685, 0.685, 2.46, 3.83, 5.60, 6.97, 8.74
7 a $\frac{\pi}{4}, \frac{3\pi}{4}, \frac{5\pi}{4}, \frac{7\pi}{4}$ b 0, 2.82, π, 5.96, 2π
   c π   d 0.440, 2.70, 3.58, 5.84
8 a π   b 0.501, 2.64, 3.64, 5.78
   c No solutions   d 1.10, 5.18
9 a $\frac{\pi}{3}, \frac{11\pi}{6}$ b $\frac{7\pi}{18}, \frac{11\pi}{18}, \frac{19\pi}{18}, \frac{23\pi}{18}, \frac{31\pi}{18}, \frac{35\pi}{18}$
   c −0.986, 0.786   d $0, \frac{\pi}{2}, \pi, \frac{3\pi}{2}, 2\pi$
10 a $-\frac{\pi}{4}, \frac{3\pi}{4}$, 0.412, 2.73   b 0, 0.644, π, 5.64
11 0.3, 0.5, 2.6, 2.9
12 0.7, 2.4, 3.9, 5.6
13 $8\sin^2 x + 4\sin x - 20 = 4$
   $8\sin^2 x + 4\sin x - 24 = 0$
   $2\sin^2 x + \sin x - 6 = 0$
   Let $Y = \sin x \Rightarrow 2Y^2 + Y - 6 = 0$
   $\Rightarrow (2Y - 3)(Y + 2) = 0$ So $Y = 1.5$ or $Y = -2$
   Since $Y = \sin x$, $\sin x = 1.5 \to$ No Solutions,
   $\sin x = -2 \to$ No Solutions
14 a Using the quadratic formula with $a = 1, b = -2$ and
       $c = -6$ (can complete the square as well)
       $\tan x = \frac{2 \pm \sqrt{(-2)^2 - 4 \times 1 \times (-6)}}{2 \times 1}$
       $\tan x = \frac{2 \pm \sqrt{4 + 24}}{2} = \frac{2 \pm \sqrt{28}}{2} = \frac{2 \pm 2\sqrt{7}}{2} = 1 \pm \sqrt{7}$
   b 1.3, 2.1, 4.4, 5.3, 7.6, 8.4
15 a $\sin x = 0.599$ (3 d.p.)
   b 0.64, 2.50

Exercise 5F

1 a $\frac{2}{3}$ b 1   c 1
2 a $\frac{\sin 3\theta}{\theta \sin 4\theta} \approx \frac{3\theta}{\theta \times 4\theta} = \frac{3\theta}{4\theta^2} = \frac{3}{4\theta}$
   b $\frac{\cos\theta - 1}{\tan 2\theta} \approx \frac{1 - \frac{\theta^2}{2} - 1}{2\theta} = \frac{-\frac{\theta^2}{2}}{2\theta} = -\frac{\theta}{4}$
   c $\frac{\tan 4\theta + \theta^2}{3\theta - \sin 2\theta} \approx \frac{4\theta + \theta^2}{3\theta - 2\theta} = \frac{4\theta + \theta^2}{\theta} = 4 + \theta$
3 a 0.970379   b 0.970232
   c −0.015%   d −1.77%
   e The larger the value of θ the less accurate the approximation is.
4 $\frac{\theta - \sin\theta}{\sin\theta} \times 100 = 1 \Rightarrow (\theta - \sin\theta) \times 100 = \sin\theta$
   $\Rightarrow 100\theta - 100\sin\theta = \sin\theta \Rightarrow 100\theta = 101\sin\theta$.
5 a $\frac{4\cos 3\theta - 2 + 5\sin\theta}{1 - \sin 2\theta} \approx \frac{4\left(1 - \frac{(3\theta)^2}{2}\right) - 2 + 5\theta}{1 - 2\theta}$
      $= \frac{4\left(1 - \frac{9\theta^2}{2}\right) - 2 + 5\theta}{1 - 2\theta} = \frac{4 - 18\theta^2 - 2 + 5\theta}{1 - 2\theta}$
      $= \frac{(1 - 2\theta)(9\theta + 2)}{1 - 2\theta} = 9\theta + 2$
   b 2

Challenge

1 a $CD = AC\theta$
   b $\sin\theta \approx \frac{CD}{AD} = \frac{r\theta}{r} = \theta$
      $\tan\theta \approx \frac{CD}{AC} = \frac{r\theta}{r} = \theta$
2 a $1 - \frac{x^2}{2}$
   b $\cos\theta = \sqrt{1 - \sin^2\theta} \approx 1 - \frac{\sin^2\theta}{2}$, if $\sin\theta \approx \theta$ then this becomes $\cos\theta \approx 1 - \frac{\theta^2}{2}$

Mixed exercise 5

1 a $\frac{\pi}{3}$ b $8.56\,\text{cm}^2$
2 a $120\,\text{cm}^2$ b $161.07\,\text{cm}^2$
3 a 1.839   b 11.03 cm
4 a $\frac{p}{r}$
   b Area $= \frac{1}{2}r^2\theta = \frac{1}{2}r^2\frac{p}{r} = \frac{1}{2}pr\,\text{cm}^2$
   c $12.206\,\text{cm}^2$
   d $1.105 \leq \theta \leq 1.151$ (3 d.p.)
5 a 1.28   b 16   c $1:3.91$
6 a Area of shape $X = 2d^2 + \frac{1}{2}d^2\pi$
      Area of shape $Y = \frac{1}{2}(2d)^2\theta$
      $2d^2 + \frac{1}{2}d^2\pi = \frac{1}{2}(2d)^2\theta$
      $2d^2 + \frac{1}{2}d^2\pi = 2d^2\theta \Rightarrow 1 + \frac{1}{4}\pi = \theta$
   b $(3\pi + 12)\,\text{cm}$ c $\left(18 + \frac{3\pi}{2}\right)\,\text{cm}$ d 12.9 mm
7 a $A_1 = \frac{1}{2} \times 6^2 \times \theta - \frac{1}{2} \times 6^2 \times \sin\theta = 18(\theta - \sin\theta)$
   b $A_2 = \pi \times 6^2 - 18(\theta - \sin\theta) = 36\pi - 18(\theta - \sin\theta)$
      Since $A_2 = 3A_1$
      $36\pi - 18(\theta - \sin\theta) = 3 \times 18(\theta - \sin\theta)$
      $36\pi - 18(\theta - \sin\theta) = 54(\theta - \sin\theta)$
      $36\pi = 72(\theta - \sin\theta)$
      $\frac{1}{2}\pi = \theta - \sin\theta$
      $\sin\theta = \theta - \frac{\pi}{2}$
8 a $10^2 = 5^2 + 9^2 - 2 \times 5 \times 9\cos A$
      $100 = 25 + 81 - 90\cos A$
      $-6 = -90\cos A$
      $\frac{1}{15} = \cos A$
      $A = \cos^{-1}\left(\frac{1}{15}\right) = 1.504$
   b i $6.77\,\text{cm}^2$ ii $15.7\,\text{cm}^2$ iii 22.5 cm
9 a $\frac{1}{2}r^2 \times 1.5 = 15 \Rightarrow r^2 = 20$
      $r = \sqrt{20} = 2\sqrt{5}$
   b 15.7 cm   c $5.025\,\text{cm}^2$
10 a $2\sqrt{3}$ cm   b $2\pi\,\text{cm}^2$
   c Perimeter $= 2\sqrt{3} + 2\sqrt{3} + 2\sqrt{3} \times \frac{\pi}{3} = \frac{2\sqrt{3}}{3}(\pi + 6)$
11 a $70^2 = 44^2 + 44^2 - 2 \times 44 \times 44\cos C$
      $\cos C = -\frac{257}{968}$
      $C = \cos^{-1}\left(-\frac{257}{968}\right) = 1.84$
   b i 80.9 m   ii 26.7 m   iii $847\,\text{m}^2$
12 a Arc $AB = 6 \times 2\theta = 12\theta$
      Length DC = Chord AB
      Chord $AB = 2 \times 6\sin\theta = 12\sin\theta$

Answers

Perimeter $ABCD = 12\theta + 4 + 12 \sin \theta + 4 = 2(7 + \pi)$
$12\theta + 12 \sin \theta + 8 = 2(7 + \pi)$
$6\theta + 6 \sin \theta - 3 = \pi$
$2\theta + 2 \sin \theta - 1 = \dfrac{\pi}{3}$

b $2 \times \dfrac{\pi}{6} + 2 \sin\left(\dfrac{\pi}{6}\right) - 1 = \dfrac{\pi}{3} + 2 \times \dfrac{1}{2} - 1 = \dfrac{\pi}{3}$

c 20.7 cm²

13 a $O_1A = O_2A = 12$, as they are radii of their respective circles.
$O_1O_2 = 12$, as O_2 is on the circumference of C_1 and hence is a radius (and vice versa).
Therefore,
O_1AO_2 is an equilateral triangle $\Rightarrow \angle AO_1O_2 = \dfrac{\pi}{3}$.

By symmetry, $\angle BO_1O_2$ is $\dfrac{\pi}{3} \Rightarrow \angle AO_1B = \dfrac{\pi}{3} + \dfrac{\pi}{3} = \dfrac{2\pi}{3}$

b 16π cm **c** 177 cm²

14 a Student has used an angle measured in degrees – it needs to be measured in radians to use that formula.

b $\dfrac{5\pi}{4}$ cm²

15 a $-\dfrac{1}{4}$ **b** $\theta + 1$

16 a $\dfrac{7 + 2\cos 2\theta}{\tan 2\theta + 3} \approx \dfrac{7 + 2\left(1 - \dfrac{(2\theta)^2}{2}\right)}{2\theta + 3}$

$= \dfrac{7 + 2\left(1 - \dfrac{4\theta^2}{2}\right)}{2\theta + 3} = \dfrac{9 - 4\theta^2}{2\theta + 3}$

$= \dfrac{(3 + 2\theta)(3 - 2\theta)}{2\theta + 3} = 3 - 2\theta$

b 3

17 a $32 \cos 5\theta + 203 \tan 10\theta = 182$

$32\left(1 - \dfrac{(5\theta)^2}{2}\right) + 203(10\theta) = 182$

$32 - 16(25\theta^2) + 2030\theta = 182$
$0 = 400\theta^2 - 2030\theta + 150$
$0 = 40\theta^2 - 203\theta + 15$

b $5, \dfrac{3}{40}$

c 5 is not valid as it is not "small". $\dfrac{3}{40}$ is "small" so is valid.

18 $1 - 2\theta^2$

19 a 0.730, 2.41 **b** $-\dfrac{\pi}{4}, \dfrac{3\pi}{4}$

c $\dfrac{\pi}{4}, \dfrac{5\pi}{4}$ **d** $-2.48, -0.666$

20 a

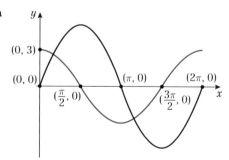

b 2 solutions **c** 0.540, 3.68
21 a $3 \sin \theta$ **b** 0.340, 2.80
22 $\dfrac{\pi}{12}, \dfrac{5\pi}{12}, \dfrac{13\pi}{12}, \dfrac{17\pi}{12}$

23 a Cosine can be negative so do not reject $-\dfrac{1}{\sqrt{2}}$. Cosine squared cannot be negative but the student has already square rooted it so no need to reject $-\dfrac{1}{\sqrt{2}}$.

b Rearranged incorrectly – square rooted incorrectly

c $-\dfrac{3\pi}{4}, -\dfrac{\pi}{4}, \dfrac{\pi}{4}, \dfrac{3\pi}{4}$

24 a Not found all the solutions
b 0.595, 2.17, 3.74, 5.31
25 a $5 \sin x = 1 + 2 \cos^2 x \Rightarrow 5 \sin x = 1 + 2(1 - \sin^2 x) \Rightarrow$
$2 \sin^2 x + 5 \sin x - 3 = 0$

b $\dfrac{\pi}{6}, \dfrac{5\pi}{6}$

26 a $4 \sin^2 x + 9 \cos x - 6 = 0 \Rightarrow$
$4(1 - \cos^2 x) + 9 \cos x - 6 = 0 \Rightarrow$
$4 \cos^2 x - 9 \cos x + 2 = 0$

b 1.3, 5.0, 7.6, 11.2

27 a $\tan 2x = 5 \sin 2x \Rightarrow \dfrac{\sin 2x}{\cos 2x} = 5 \sin 2x \Rightarrow$
$(1 - 5 \cos 2x) \sin 2x = 0$

b $0, 0.7, \dfrac{\pi}{2}, 2.5, \pi$

28 a

b $\left(0, \dfrac{\sqrt{3}}{2}\right)\left(\dfrac{\pi}{3}, 0\right), \left(\dfrac{4\pi}{3}, 0\right)$ **c** 0.34, 4.90

29 $x = 0.54, 1.90$ or 2.64 (2 d.p)

Challenge

a $\theta = \dfrac{2}{9}$ or $\theta = -3$

$\theta = \dfrac{2}{9}$ is small, so this value is valid. $\theta = -3$ is not small so this value is not valid. Small in this context is "close to 0".

b $\theta = -\dfrac{1}{4}$ or $\theta = \dfrac{1}{5}$
Both θ could be considered "small" in this case so both are valid.

c No solutions

CHAPTER 6
Prior knowledge 6
1

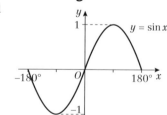

a 53.1°, 126.9° (1 d.p.) **b** −23.6°, −156.4° (1 d.p.)

2 $\dfrac{1}{\sin x \cos x} - \dfrac{1}{\tan x} = \dfrac{1}{\sin x \cos x} - \dfrac{\cos x}{\sin x} = \dfrac{1 - \cos^2 x}{\sin x \cos x}$

$= \dfrac{\sin^2 x}{\sin x \cos x} = \dfrac{\sin x}{\cos x} = \tan x$

3 0.308, 1.26, 1.88, 2.83, 3.45, 4.40, 5.02, 5.98 (3 s.f.)

Answers

Exercise 6A

1. **a** +ve **b** −ve **c** −ve **d** +ve
   e −ve
2. **a** −5.76 **b** −1.02 **c** −1.02 **d** 5.67
   e 0.577 **f** −1.36 **g** −3.24 **h** 1.04
3. **a** 1 **b** −1 **c** −1 **d** −2
   e $-\dfrac{2\sqrt{3}}{3}$ **f** −1 **g** 2 **h** 2
   i $-\sqrt{2}$ **j** $\dfrac{\sqrt{3}}{3}$ **k** $\dfrac{2\sqrt{3}}{3}$ **l** $-\sqrt{2}$

4. $\operatorname{cosec}(\pi - x) = \dfrac{1}{\sin(\pi - x)} = \dfrac{1}{\sin x} = \operatorname{cosec} x$

5. $\cot 30° \sec 30° = \dfrac{1}{\tan 30°} \times \dfrac{1}{\cos 30°} = \dfrac{\sqrt{3}}{1} \times \dfrac{2}{\sqrt{3}} = 2$

6. $\operatorname{cosec}\left(\dfrac{2\pi}{3}\right) + \sec\left(\dfrac{2\pi}{3}\right) = \dfrac{1}{\sin\left(\dfrac{2\pi}{3}\right)} + \dfrac{1}{\cos\left(\dfrac{2\pi}{3}\right)}$

   $= \dfrac{1}{\frac{\sqrt{3}}{2}} + \dfrac{1}{-\frac{1}{2}}$

   $= -2 + \dfrac{2}{\sqrt{3}} = -2 + \dfrac{2}{3}\sqrt{3}$

Challenge

a Using triangle OBP, $OB \cos\theta = 1$
  $\Rightarrow OB = \dfrac{1}{\cos\theta} = \sec\theta$

b Using triangle OAP, $OA \sin\theta = 1$
  $\Rightarrow OA = \dfrac{1}{\sin\theta} = \operatorname{cosec}\theta$

c Using Pythagoras' theorem, $AP^2 = OA^2 - OP^2$
  So $AP^2 = \operatorname{cosec}^2\theta - 1 = \dfrac{1}{\sin^2\theta} - 1$
  $= \dfrac{1 - \sin^2\theta}{\sin^2\theta} = \dfrac{\cos^2\theta}{\sin^2\theta} = \cot^2\theta$

  Therefore $AP = \cot\theta$.

Exercise 6B

1. **a**
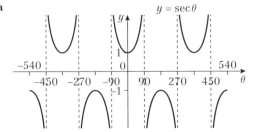

   b
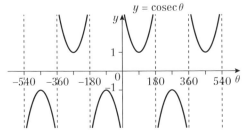

   c
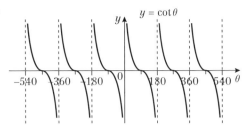

2. **a**
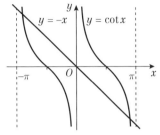

   b 2 solutions

3. **a**
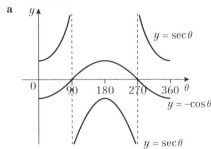

   b The solutions of $\sec\theta = -\cos\theta$ are the θ values of the points of intersection of $y = \sec\theta$ and $y = -\cos\theta$. As they do not meet, there are no solutions.

4. **a**
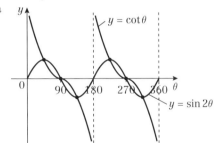

   b 6

5. **a**
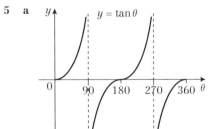

387

Answers

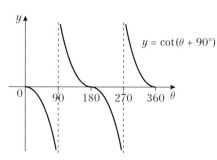

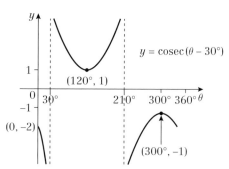

b $\cot(90° + \theta) = -\tan \theta$

6 a i The graph of $y = \tan\left(\theta + \dfrac{\pi}{2}\right)$ is the same as that of $y = \tan \theta$ translated by $\dfrac{\pi}{2}$ to the left.

  ii The graph of $y = \cot(-\theta)$ is the same as that of $y = \cot \theta$ reflected in the y-axis.

  iii The graph of $y = \csc\left(\theta + \dfrac{\pi}{4}\right)$ is the same as that of $y = \csc \theta$ translated by $\dfrac{\pi}{4}$ to the left.

  iv The graph of $y = \sec\left(\theta - \dfrac{\pi}{4}\right)$ is the same as that of $y = \sec \theta$ translated by $\dfrac{\pi}{4}$ to the right.

b $\tan\left(\theta + \dfrac{\pi}{2}\right) = \cot(-\theta)$; $\csc\left(\theta + \dfrac{\pi}{4}\right) = \sec\left(\theta - \dfrac{\pi}{4}\right)$

7 a

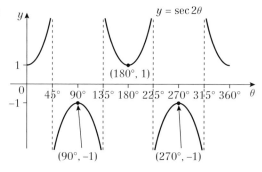

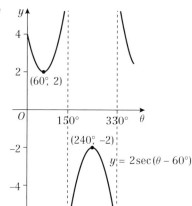

b

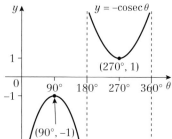

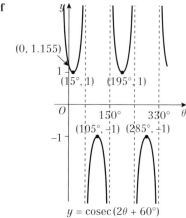

c

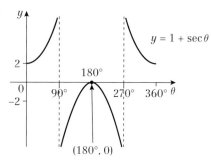

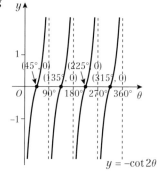

h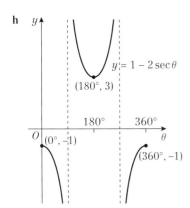

8 a $\dfrac{2\pi}{3}$ b 4π c π d 2π

9 a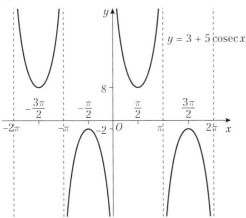

   b $-2 < k < 8$

10 a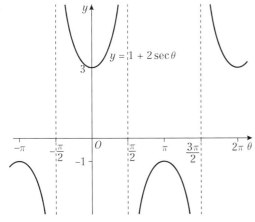

   b $\theta = -\pi, 0, \pi, 2\pi$
   c Max $= \tfrac{1}{3}$, first occurs at $\theta = 2\pi$
      Min $= -1$, first occurs at $\theta = \pi$

Exercise 6C

1 a $\operatorname{cosec}^3 \theta$ b $4 \cot^6 \theta$ c $\tfrac{1}{2} \sec^2 \theta$
   d $\cot^2 \theta$ e $\sec^5 \theta$ f $\operatorname{cosec}^2 \theta$
   g $2 \cot^{\tfrac{1}{2}} \theta$ h $\sec^3 \theta$
2 a $\tfrac{5}{4}$ b $-\tfrac{1}{2}$ c $\pm\sqrt{3}$
3 a $\cos \theta$ b 1 c $\sec 2\theta$
   d 1 e 1 f $\cos A$
   g $\cos x$

4 a L.H.S. $= \cos\theta + \sin\theta \dfrac{\sin\theta}{\cos\theta} = \dfrac{\cos^2\theta + \sin^2\theta}{\cos\theta}$
      $= \dfrac{1}{\cos\theta} = \sec\theta = $ R.H.S.

   b L.H.S. $= \dfrac{\cos\theta}{\sin\theta} + \dfrac{\sin\theta}{\cos\theta} \equiv \dfrac{\cos^2\theta + \sin^2\theta}{\sin\theta\cos\theta}$
      $\equiv \dfrac{1}{\sin\theta\cos\theta} = \dfrac{1}{\sin\theta} \times \dfrac{1}{\cos\theta}$
      $\equiv \operatorname{cosec}\theta \sec\theta = $ R.H.S.

   c L.H.S. $= \dfrac{1}{\sin\theta} - \sin\theta \equiv \dfrac{1 - \sin^2\theta}{\sin\theta} \equiv \dfrac{\cos^2\theta}{\sin\theta}$
      $\equiv \cos\theta \times \dfrac{\cos\theta}{\sin\theta} \equiv \cos\theta \cot\theta = $ R.H.S.

   d L.H.S. $= (1 - \cos x)\left(1 + \dfrac{1}{\cos x}\right) \equiv 1 - \cos x + \dfrac{1}{\cos x} - 1$
      $\equiv \dfrac{1}{\cos x} - \cos x \equiv \dfrac{1 - \cos^2 x}{\cos x} \equiv \dfrac{\sin^2 x}{\cos x}$
      $\equiv \sin x \times \dfrac{\sin x}{\cos x} \equiv \sin x \tan x = $ R.H.S.

   e L.H.S. $= \dfrac{\cos^2 x + (1 - \sin x)^2}{(1 - \sin x)\cos x}$
      $\equiv \dfrac{\cos^2 x + 1 - 2\sin x + \sin^2 x}{(1 - \sin x)\cos x}$
      $\equiv \dfrac{2 - 2\sin x}{(1 - \sin x)\cos x} \equiv \dfrac{2(1 - \sin x)}{(1 - \sin x)\cos x}$
      $\equiv 2 \sec x = $ R.H.S.

   f L.H.S. $= \dfrac{\cos\theta}{1 + \dfrac{1}{\tan\theta}} \equiv \dfrac{\cos\theta}{\left(\dfrac{\tan\theta + 1}{\tan\theta}\right)}$
      $\equiv \dfrac{\cos\theta \tan\theta}{\tan\theta + 1} \equiv \dfrac{\sin\theta}{1 + \tan\theta} = $ R.H.S.

5 a $45°, 315°$ b $199°, 341°$
   c $112°, 292°$ d $30°, 150°$
   e $30°, 150°, 210°, 330°$ f $36.9°, 90°, 143°, 270°$
   g $26.6°, 207°$ h $45°, 135°, 225°, 315°$
6 a $90°$ b $\pm 109°$
   c $-164°, 16.2°$ d $41.8°, 138°$
   e $\pm 45°, \pm 135°$ f $\pm 60°$
   g $-173°, -97.2°, 7.24°, 82.8°$
   h $-152°, -36.5°, 28.4°, 143°$
7 a π b $\dfrac{5\pi}{6}, \dfrac{11\pi}{6}$ c $\dfrac{2\pi}{3}, \dfrac{4\pi}{3}$ d $\dfrac{\pi}{4}, \dfrac{3\pi}{4}$

8 a $\dfrac{AB}{AD} = \cos\theta \Rightarrow AD = 6\sec\theta$
      $\dfrac{AC}{AB} = \cos\theta \Rightarrow AC = 6\cos\theta$
      $CD = AD - AC \Rightarrow CD = 6\sec\theta - 6\cos\theta$
      $= 6(\sec\theta - \cos\theta)$
   b 2 cm

9 a $\dfrac{\operatorname{cosec} x - \cot x}{1 - \cos x} \equiv \dfrac{\dfrac{1}{\sin x} - \dfrac{\cos x}{\sin x}}{1 - \cos x} \equiv \dfrac{1}{\sin x} \times \dfrac{1 - \cos x}{1 - \cos x}$
      $\equiv \operatorname{cosec} x$
   b $x = \dfrac{\pi}{6}, \dfrac{5\pi}{6}$

10 a $\dfrac{\sin x \tan x}{1 - \cos x} - 1 \equiv \dfrac{\sin^2 x}{\cos x(1 - \cos x)} - 1$
      $\equiv \dfrac{\sin^2 x - \cos x + \cos^2 x}{\cos x(1 - \cos x)} \equiv \dfrac{1 - \cos x}{\cos x(1 - \cos x)}$
      $\equiv \dfrac{1}{\cos x} \equiv \sec x$

Answers

b Would need to solve $\sec x = -\frac{1}{2}$, which is equivalent to $\cos x = -2$, which has no solutions.

11 $x = 11.3°, 191.3°$ (1 d.p.)

Exercise 6D

1 **a** $\sec^2\left(\frac{1}{2}\theta\right)$ **b** $\tan^2\theta$ **c** 1
   d $\tan\theta$ **e** 1 **f** 3
   g $\sin\theta$ **h** 1 **i** $\cos\theta$
   j 1 **k** $4\operatorname{cosec}^4 2\theta$

2 $\pm\sqrt{k-1}$

3 **a** $\frac{1}{2}$ **b** $-\frac{\sqrt{3}}{2}$

4 **a** $-\frac{5}{4}$ **b** $-\frac{4}{5}$ **c** $-\frac{3}{5}$

5 **a** $-\frac{7}{24}$ **b** $-\frac{25}{7}$

6 **a** L.H.S. $\equiv (\sec^2\theta - \tan^2\theta)(\sec^2\theta + \tan^2\theta)$
      $\equiv 1(\sec^2\theta + \tan^2\theta) = $ R.H.S.
   b L.H.S. $\equiv (1 + \cot^2 x) - (1 - \cos^2 x)$
      $\equiv \cot^2 x + \cos^2 x = $ R.H.S.
   c L.H.S. $\equiv \dfrac{1}{\cos^2 A}\left(\dfrac{\cos^2 A}{\sin^2 A} - \cos^2 A\right) \equiv \dfrac{1}{\sin^2 A} - 1$
      $\equiv \operatorname{cosec}^2 A - 1 = \cot^2 A = $ R.H.S.
   d R.H.S. $\equiv \tan^2\theta \times \cos^2\theta \equiv \dfrac{\sin^2\theta}{\cos^2\theta} \times \cos^2\theta \equiv \sin^2\theta$
      $\equiv 1 - \cos^2\theta = $ L.H.S.
   e L.H.S. $= \dfrac{1 - \tan^2 A}{\sec^2 A} \equiv \cos^2 A\left(1 - \dfrac{\sin^2 A}{\cos^2 A}\right)$
      $\equiv \cos^2 A - \sin^2 A \equiv (1 - \sin^2 A) - \sin^2 A$
      $\equiv 1 - 2\sin^2 A = $ R.H.S.
   f L.H.S. $= \dfrac{1}{\cos^2\theta} + \dfrac{1}{\sin^2\theta} \equiv \dfrac{\sin^2\theta + \cos^2\theta}{\cos^2\theta \sin^2\theta}$
      $\equiv \dfrac{1}{\cos^2\theta \sin^2\theta} \equiv \sec^2\theta \operatorname{cosec}^2\theta = $ R.H.S.
   g L.H.S. $= \operatorname{cosec} A(1 + \tan^2 A) \equiv \operatorname{cosec} A\left(1 + \dfrac{\sin^2 A}{\cos^2 A}\right)$
      $\equiv \operatorname{cosec} A + \dfrac{1}{\sin A} \cdot \dfrac{\sin^2 A}{\cos^2 A} \equiv \operatorname{cosec} A + \dfrac{\sin A}{\cos A} \cdot \dfrac{1}{\cos A}$
      $\equiv \operatorname{cosec} A + \tan A \sec A = $ R.H.S.
   h L.H.S. $= \sec^2\theta - \sin^2\theta \equiv (1 + \tan^2\theta) - (1 - \cos^2\theta)$
      $\equiv \tan^2\theta + \cos^2\theta \equiv $ R.H.S.

7 $\dfrac{\sqrt{2}}{4}$

8 **a** $20.9°, 69.1°, 201°, 249°$ **b** $\pm\dfrac{\pi}{3}$
   c $-153°, -135°, 26.6°, 45°$ **d** $\dfrac{\pi}{2}, \dfrac{3\pi}{4}, \dfrac{3\pi}{2}, \dfrac{7\pi}{4}$
   e $120°$ **f** $0, \dfrac{\pi}{4}, \pi$
   g $0°, 180°$ **h** $\dfrac{\pi}{4}, \dfrac{\pi}{3}, \dfrac{5\pi}{4}, \dfrac{4\pi}{3}$

9 **a** $1 + \sqrt{2}$
   b $\cos k = \dfrac{1}{1+\sqrt{2}} = \dfrac{\sqrt{2}-1}{(\sqrt{2}-1)(\sqrt{2}+1)} = \sqrt{2}-1$
   c $65.5°, 294.5°$

10 **a** $b = \dfrac{4}{a}$
   b $c^2 = \cot^2 x = \dfrac{\cos^2 x}{\sin^2 x} = \dfrac{b^2}{1-b^2} = \dfrac{\left(\frac{4}{a}\right)^2}{1-\left(\frac{4}{a}\right)^2}$
      $= \dfrac{16}{a^2} \times \dfrac{a^2}{(a^2-16)} = \dfrac{16}{a^2-16}$

11 **a** $\dfrac{1}{x} = \dfrac{1}{\sec\theta + \tan\theta} = \dfrac{\sec\theta - \tan\theta}{(\sec\theta - \tan\theta)(\sec\theta + \tan\theta)}$
      $= \dfrac{\sec\theta - \tan\theta}{(\sec^2\theta - \tan^2\theta)} = \dfrac{\sec\theta - \tan\theta}{1}$
   b $x^2 + \dfrac{1}{x^2} + 2 = \left(x + \dfrac{1}{x}\right)^2 = (2\sec\theta)^2 = 4\sec^2\theta$

12 $p = 2(1 + \tan^2\theta) - \tan^2\theta = 2 + \tan^2\theta$
   $\Rightarrow \tan^2\theta = p - 2 \Rightarrow \cot^2\theta = \dfrac{1}{p-2}$
   $\operatorname{cosec}^2\theta = 1 + \cot^2\theta = 1 + \dfrac{1}{p-2} = \dfrac{(p-2)+1}{p-2} = \dfrac{p-1}{p-2}$

Exercise 6E

1 **a** $\dfrac{\pi}{2}$ **b** $\dfrac{\pi}{2}$ **c** $-\dfrac{\pi}{4}$ **d** $-\dfrac{\pi}{6}$
   e $\dfrac{3\pi}{4}$ **f** $-\dfrac{\pi}{6}$ **g** $\dfrac{\pi}{3}$ **h** $\dfrac{\pi}{3}$

2 **a** 0 **b** $-\dfrac{\pi}{3}$ **c** $\dfrac{\pi}{2}$

3 **a** $\dfrac{1}{2}$ **b** $-\dfrac{1}{2}$ **c** -1 **d** 0

4 **a** $\dfrac{\sqrt{3}}{2}$ **b** $\dfrac{\sqrt{3}}{2}$ **c** -1 **d** 2
   e -1 **f** 1

5 $\alpha, \pi - \alpha$

6 **a** $0 < x < 1$
   b **i** $\sqrt{1-x^2}$ **ii** $\dfrac{x}{\sqrt{1-x^2}}$
   c **i** no change **ii** no change

7 **a**
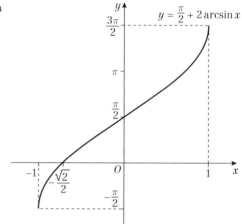

   b
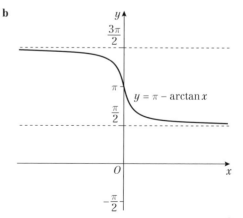

Answers

c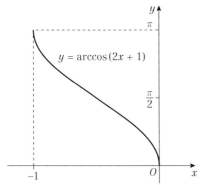

d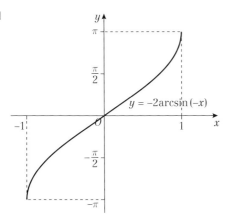

8 a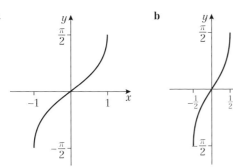

Range: $-\frac{\pi}{2} \leq f(x) \leq \frac{\pi}{2}$

c $g: x \to \arcsin 2x, -\frac{1}{2} \leq x \leq \frac{1}{2}$

d $g^{-1}: x \to \frac{1}{2}\sin x, -\frac{\pi}{2} \leq x \leq \frac{\pi}{2}$

9 a Let $y = \arccos x$. $x \in [0,1] \Rightarrow y \in \left[0, \frac{\pi}{2}\right]$
$\cos y = x$, so $\sin y = \sqrt{1 - \cos^2 y} = \sqrt{1 - x^2}$
(Note, $\sin y \neq -\sqrt{1 - x^2}$ since $y \in \left[0, \frac{\pi}{2}\right]$, so $\sin y \geq 0$)
$y = \arcsin\sqrt{1 - x^2}$
Therefore, $\arccos x = \arcsin\sqrt{1 - x^2}$ for $x \in [0,1]$.

b For $x \in [-1,0]$, $\arccos x \in \left(\frac{\pi}{2}, \pi\right)$, but arcsin only has range $\left[-\frac{\pi}{2}, \frac{\pi}{2}\right]$.

Challenge

a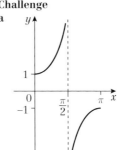

b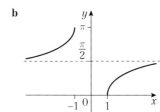

Range: $0 \leq \text{arcsec}\, x \leq \pi$, $\text{arcsec}\, x \neq \frac{\pi}{2}$

Mixed exercise 6

1 $-125.3°, \pm 54.7°$

2 $p = \frac{8}{q}$

3 $p^2 q^2 = \sin^2\theta \times 4^2 \cot^2\theta = 16\sin^2\theta \times \frac{\cos^2\theta}{\sin^2\theta}$
$= 16\cos^2\theta = 16(1 - \sin^2\theta) = 16(1 - p^2)$

4 a i $60°$
     ii $30°, 41.8°, 138.2°, 150°$
   b i $30°, 165°, 210°, 345°$
     ii $45°, 116.6°, 225°, 296.6°$
   c i $\frac{71\pi}{60}, \frac{101\pi}{60}$
     ii $\frac{\pi}{6}, \frac{5\pi}{6}, \frac{7\pi}{6}, \frac{11\pi}{6}$

5 $-\frac{8}{5}$

6 a L.H.S. $\equiv \left(\frac{\sin\theta}{\cos\theta} + \frac{\cos\theta}{\sin\theta}\right)(\sin\theta + \cos\theta)$
$\equiv \frac{(\sin^2\theta + \cos^2\theta)}{\cos\theta\sin\theta}(\sin\theta + \cos\theta)$
$\equiv \frac{\sin\theta}{\sin\theta\cos\theta} + \frac{\cos\theta}{\cos\theta\sin\theta}$
$\equiv \sec\theta + \text{cosec}\,\theta \equiv$ R.H.S.

b L.H.S. $\equiv \dfrac{\dfrac{1}{\sin x}}{\dfrac{1}{\sin x} - \sin x}$
$\equiv \dfrac{\dfrac{1}{\sin x}}{\dfrac{1 - \sin^2 x}{\sin x}} \equiv \dfrac{1}{\sin x} \times \dfrac{\sin x}{\cos^2 x} \equiv \dfrac{1}{\cos^2 x} \equiv \sec^2 x \equiv$ R.H.S.

c L.H.S. $\equiv 1 - \sin x + \text{cosec}\,x - 1 \equiv \dfrac{1}{\sin x} - \sin x$
$\equiv \dfrac{1 - \sin^2 x}{\sin x} \equiv \dfrac{\cos^2 x}{\sin x} \equiv \cos x \dfrac{\cos x}{\sin x} \equiv \cos x \cot x$
$\equiv$ R.H.S.

d L.H.S. $\equiv \dfrac{\cot x(1 + \sin x) - \cos x(\text{cosec}\,x - 1)}{(\text{cosec}\,x - 1)(1 + \sin x)}$
$\equiv \dfrac{\cot x + \cos x - \cot x + \cos x}{\text{cosec}\,x - 1 + 1 - \sin x} \equiv \dfrac{2\cos x}{\text{cosec}\,x - \sin x}$
$\equiv \dfrac{2\cos x}{\dfrac{1}{\sin x} - \sin x} \equiv \dfrac{2\cos x}{\left(\dfrac{1 - \sin^2 x}{\sin x}\right)} \equiv \dfrac{2\cos x \sin x}{\cos^2 x}$
$\equiv 2\tan x \equiv$ R.H.S.

391

Answers

e L.H.S. $\equiv \dfrac{\csc\theta + 1 + \csc\theta - 1}{(\csc^2\theta - 1)} \equiv \dfrac{2\csc\theta}{\cot^2\theta}$

$\equiv \dfrac{2}{\sin\theta} \cdot \dfrac{\sin^2\theta}{\cos^2\theta} \equiv \dfrac{2\sin\theta}{\cos^2\theta} \equiv \dfrac{2}{\cos\theta} \cdot \dfrac{\sin\theta}{\cos\theta}$

$\equiv 2\sec\theta\tan\theta \equiv$ R.H.S.

f L.H.S. $\equiv \dfrac{\sec^2\theta - \tan^2\theta}{\sec^2\theta} \equiv \dfrac{1}{\sec^2\theta} \equiv \cos^2\theta \equiv$ R.H.S.

7 a L.H.S. $\equiv \dfrac{\sin^2 x + (1 + \cos x)^2}{(1 + \cos x)\sin x}$

$\equiv \dfrac{\sin^2 x + 1 + 2\cos x + \cos^2 x}{(1 + \cos x)\sin x} \equiv \dfrac{2 + 2\cos x}{(1 + \cos x)\sin x}$

$\equiv \dfrac{2(1 + \cos x)}{(1 + \cos x)\sin x} \equiv \dfrac{2}{\sin x} \equiv 2\csc x$

b $-\dfrac{\pi}{3}, -\dfrac{2\pi}{3}, \dfrac{4\pi}{3}, \dfrac{5\pi}{3}$

8 R.H.S. $\equiv \left(\dfrac{1}{\sin\theta} + \dfrac{\cos\theta}{\sin\theta}\right)^2 \equiv \dfrac{(1+\cos\theta)^2}{\sin^2\theta} \equiv \dfrac{(1+\cos\theta)^2}{1-\cos^2\theta}$

$\equiv \dfrac{(1+\cos\theta)^2}{(1-\cos\theta)(1+\cos\theta)} \equiv \dfrac{1+\cos\theta}{1-\cos\theta} \equiv$ L.H.S.

9 a $-2\sqrt{2}$

b $\csc^2 A = 1 + \cot^2 A = 1 + \dfrac{1}{8} = \dfrac{9}{8}$

$\Rightarrow \csc A = \pm\dfrac{3}{2\sqrt{2}} = \pm\dfrac{3\sqrt{2}}{4}$

As A is obtuse, $\csc A$ is +ve, $\Rightarrow \csc A = \dfrac{3\sqrt{2}}{4}$

10 a $\dfrac{1}{k}$ **b** $k^2 - 1$ **c** $-\dfrac{1}{\sqrt{k^2-1}}$ **d** $-\dfrac{k}{\sqrt{k^2-1}}$

11 $\dfrac{\pi}{12}, \dfrac{17\pi}{12}$ **12** $\dfrac{\pi}{3}$ **13** $\dfrac{\pi}{3}, \dfrac{5\pi}{6}, \dfrac{4\pi}{3}, \dfrac{11\pi}{6}$

14 a $(\sec x - 1)(\csc x - 2)$ **b** $30°, 150°$

15 $2 - \sqrt{3}$

16

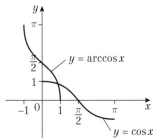

17 a $-\dfrac{1}{3}$ **b i** $-\dfrac{5}{3}$, **ii** $-\dfrac{4}{3}$ **c** $126.9°$

18 $pq = (\sec\theta - \tan\theta)(\sec\theta + \tan\theta) = \sec^2\theta - \tan^2\theta$

$= 1 \Rightarrow p = \dfrac{1}{q}$

19 a L.H.S. $= (\sec^2\theta - \tan^2\theta)(\sec^2\theta + \tan^2\theta)$
$= 1 \times (\sec^2\theta + \tan^2\theta) = \sec^2\theta + \tan^2\theta =$ R.H.S.

b $-153.4°, -135°, 26.6°, 45°$

20 a

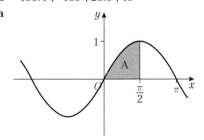

b

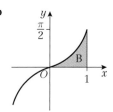

c The regions A and B fit together to make a rectangle.

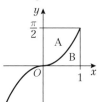

Area $= 1 \times \dfrac{\pi}{2} = \dfrac{\pi}{2}$

21 $\cot 60° \sec 60° = \dfrac{1}{\tan 60°} \times \dfrac{1}{\cos 60°} = \dfrac{2}{\sqrt{3}} = \dfrac{2\sqrt{3}}{3}$

22 a

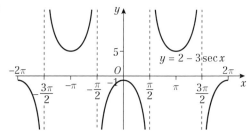

b $-1 < k < 5$

23 a

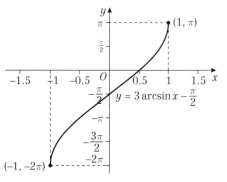

b $\left(\dfrac{1}{2}, 0\right)$

24 a Let $y = \arccos x$. So $\cos y = x$, $\sin y = \sqrt{1-x^2}$.

Thus $\tan y = \dfrac{\sqrt{1-x^2}}{x}$, which is valid for $x \in (0, 1]$.

Therefore $\arccos x = \arctan\dfrac{\sqrt{1-x^2}}{x}$ for $0 < x \le 1$.

b Letting $y = \arccos x$, $x \in [-1, 0) \Rightarrow y \in \left(\dfrac{\pi}{2}, \pi\right]$

$\tan y = \dfrac{\sin y}{\cos y} = \dfrac{\sqrt{1-x^2}}{x}$

$\arctan\dfrac{\sqrt{1-x^2}}{x}$ gives values in the range $\left(-\dfrac{\pi}{2}, 0\right)$,

so for $y \in \left(\dfrac{\pi}{2}, \pi\right]$ you need to add π:

$y = \pi + \arctan\dfrac{\sqrt{1-x^2}}{x}$

Therefore $\arccos x = \pi + \arctan\dfrac{\sqrt{1-x^2}}{x}$

CHAPTER 7

Prior knowledge 7

1 **a** $\dfrac{1}{\sqrt{2}}$ **b** $\dfrac{\sqrt{3}}{2}$ **c** $\sqrt{3}$

2 **a** 194.2°, 245.8°   **b** 45°, 165°, 225°, 345°   **c** 270°

3 **a** LHS $\equiv \cos x + \sin x \tan x \equiv \cos x + \sin x \left(\dfrac{\sin x}{\cos x}\right)$
   $\equiv \dfrac{\cos^2 x + \sin^2 x}{\cos x} \equiv \dfrac{1}{\cos x} \equiv \sec x \equiv$ RHS

  b LHS $\equiv \cot x \sec x \sin x \equiv \left(\dfrac{\cos x}{\sin x}\right)\left(\dfrac{1}{\cos x}\right)\left(\dfrac{\sin x}{1}\right)$
   $\equiv 1 \equiv$ RHS

  c LHS $\equiv \dfrac{\cos^2 x + \sin^2 x}{1 + \cot^2 x} \equiv \dfrac{1}{\operatorname{cosec}^2 x} \equiv \sin^2 x \equiv$ RHS

Exercise 7A

1 **a** **i** $(\alpha - \beta) + \beta = \alpha$. So $\angle FAB = \alpha$.
   ii $\angle FAB = \angle ABD$ (alternate angles)
   $\angle CBE = 90 - \alpha$, so $\angle BCE = 90 - (90 - \alpha) = \alpha$.
   iii $\cos \beta = \dfrac{AB}{1} \Rightarrow AB = \cos \beta$
   iv $\sin \beta = \dfrac{BC}{1} \Rightarrow BC = \sin \beta$

  b **i** $\sin \alpha = \dfrac{AD}{\cos \beta} \Rightarrow AD = \sin \alpha \cos \beta$
   ii $\cos \alpha = \dfrac{BD}{\cos \beta} \Rightarrow BD = \cos \alpha \cos \beta$

  c **i** $\cos \alpha = \dfrac{CE}{\sin \beta} \Rightarrow CE = \cos \alpha \sin \beta$
   ii $\sin \alpha = \dfrac{BE}{\sin \beta} \Rightarrow BE = \sin \alpha \sin \beta$

  d **i** $\sin(\alpha - \beta) = \dfrac{FC}{1} \Rightarrow FC = \sin(\alpha - \beta)$
   ii $\cos(\alpha - \beta) = \dfrac{FA}{1} \Rightarrow FA = \cos(\alpha - \beta)$

  e **i** $FC + CE = AD$, so $FC = AD - CE$
   $\sin(\alpha - \beta) = \sin \alpha \cos \beta - \cos \alpha \sin \beta$
   ii $AF = DB + BE$
   $\cos(\alpha - \beta) = \cos \alpha \cos \beta + \sin \alpha \sin \beta$

2 $\tan(A - B) = \dfrac{\sin(A - B)}{\cos(A - B)} = \dfrac{\sin A \cos B - \cos A \sin B}{\cos A \cos B + \sin A \sin B}$
   $= \dfrac{\dfrac{\sin A \cos B}{\cos A \cos B} - \dfrac{\cos A \sin B}{\cos A \cos B}}{\dfrac{\cos A \cos B}{\cos A \cos B} + \dfrac{\sin A \sin B}{\cos A \cos B}} = \dfrac{\tan A - \tan B}{1 + \tan A \tan B}$

3 $\sin(A + B) = \sin A \cos B + \cos A \sin B$
   $\sin(P + (-Q)) = \sin P \cos(-Q) + \cos P \sin(-Q)$
   $\sin(P - Q) = \sin P \cos Q - \cos P \sin Q$

4 Example: with $A = 60°, B = 30°$,
   $\sin(A + B) = \sin 90° = 1$; $\sin A + \sin B = \dfrac{\sqrt{3}}{2} + \dfrac{1}{2} \neq 1$
   [You can find examples of A and B for which the statement is true, e.g. $A = 30°, B = -30°$, but one counter-example shows that it is not an identity.]

5 $\cos(\theta - \theta) \equiv \cos \theta \cos \theta + \sin \theta \sin \theta$
   $\Rightarrow \sin^2 \theta + \cos^2 \theta \equiv 1$ as $\cos 0 = 1$

6 **a** $\sin\left(\dfrac{\pi}{2} - \theta\right) \equiv \sin \dfrac{\pi}{2} \cos \theta - \cos \dfrac{\pi}{2} \sin \theta$
   $\equiv (1) \cos \theta - (0) \sin \theta = \cos \theta$

  b $\cos\left(\dfrac{\pi}{2} - \theta\right) \equiv \cos \dfrac{\pi}{2} \cos \theta - \sin \dfrac{\pi}{2} \sin \theta$
   $\equiv (0) \cos \theta - (1) \sin \theta = \sin \theta$

7 $\sin\left(x + \dfrac{\pi}{6}\right) = \sin x \cos \dfrac{\pi}{6} + \cos x \sin \dfrac{\pi}{6} = \dfrac{\sqrt{3}}{2} \sin x + \dfrac{1}{2} \cos x$

8 $\cos\left(x + \dfrac{\pi}{3}\right) = \cos x \cos \dfrac{\pi}{3} - \sin x \sin \dfrac{\pi}{3} = \dfrac{1}{2} \cos x - \dfrac{\sqrt{3}}{2} \sin x$

9 **a** $\sin 35°$ **b** $\sin 35°$ **c** $\cos 210°$ **d** $\tan 31°$
   e $\cos \theta$ **f** $\cos 7\theta$ **g** $\sin 3\theta$ **h** $\tan 5\theta$
   i $\sin A$ **j** $\cos 3x$

10 **a** $\sin\left(x + \dfrac{\pi}{4}\right)$ or $\cos\left(x - \dfrac{\pi}{4}\right)$ **b** $\cos\left(x + \dfrac{\pi}{4}\right)$
   c $\sin\left(x + \dfrac{\pi}{3}\right)$ or $\cos\left(x - \dfrac{\pi}{6}\right)$ **d** $\sin\left(x - \dfrac{\pi}{4}\right)$

11 $\cos y = \sin x \cos y + \sin y \cos x$
   Divide by $\cos x \cos y \Rightarrow \sec x = \tan x + \tan y$,
   so $\tan y = \sec x - \tan x$

12 $\dfrac{\tan x - 3}{3 \tan x + 1}$ 13 2

14 **a** $\dfrac{5}{3}$ **b** $\sqrt{3}$ **c** $-\left(\dfrac{8 + 5\sqrt{3}}{11}\right)$

15 $\dfrac{\tan x + \sqrt{3}}{1 - \sqrt{3} \tan x} = \dfrac{1}{2} \Rightarrow (2 + \sqrt{3}) \tan x = 1 - 2\sqrt{3}$, so
   $\tan x = \dfrac{1 - 2\sqrt{3}}{2 + \sqrt{3}} = \dfrac{(1 - 2\sqrt{3})(2 - \sqrt{3})}{1} = 8 - 5\sqrt{3}$

16 Write θ as $\left(\theta + \dfrac{2\pi}{3}\right) - \dfrac{2\pi}{3}$ and $\theta + \dfrac{4\pi}{3}$ as $\left(\theta + \dfrac{2\pi}{3}\right) + \dfrac{2\pi}{3}$.
   Use the addition formulae for cos and simplify.

Challenge

a **i** Area $= \dfrac{1}{2} ab \sin \theta = \dfrac{1}{2} x(y \cos B)(\sin A) = \dfrac{1}{2} xy \sin A \cos B$
   ii Area $= \dfrac{1}{2} ab \sin \theta = \dfrac{1}{2} y(x \cos A)(\sin B) = \dfrac{1}{2} xy \cos A \sin B$
   iii Area $= \dfrac{1}{2} ab \sin \theta = \dfrac{1}{2} xy \sin(A + B)$

b Area of large triangle $=$ area $T_1 +$ area T_2
   $\dfrac{1}{2} xy \sin(A + B) = \dfrac{1}{2} xy \sin A \cos B + \dfrac{1}{2} xy \cos A \sin B$
   $\sin(A + B) = \sin A \cos B + \cos A \sin B$

Exercise 7B

1 **a** $\dfrac{\sqrt{2}(\sqrt{3} + 1)}{4}$ **b** $\dfrac{\sqrt{2}(\sqrt{3} + 1)}{4}$ **c** $\dfrac{\sqrt{2}(\sqrt{3} - 1)}{4}$ **d** $\sqrt{3} - 2$

2 **a** 1   **b** 0   **c** $\dfrac{\sqrt{3}}{2}$ **d** $\dfrac{\sqrt{2}}{2}$ **e** $\dfrac{\sqrt{2}}{2}$
   f $-\dfrac{1}{2}$ **g** $\sqrt{3}$ **h** $\dfrac{\sqrt{3}}{3}$ **i** 1   **j** $\sqrt{2}$

3 **a** $\tan(45° + 30°) = \dfrac{\tan 45° + \tan 30°}{1 - \tan 45° \tan 30°}$

  b $\tan 75° = \dfrac{1 + \dfrac{\sqrt{3}}{3}}{1 - \dfrac{\sqrt{3}}{3}} = \dfrac{3 + \sqrt{3}}{3 - \sqrt{3}} = \dfrac{(3 + \sqrt{3})(3 + \sqrt{3})}{(3 - \sqrt{3})(3 + \sqrt{3})}$
   $= \dfrac{12 + 6\sqrt{3}}{9 - 3} = 2 + \sqrt{3}$

4 $-\dfrac{6}{7}$

Answers

5 **a** $\cos 105° = \cos(45° + 60°)$
$= \cos 45° \cos 60° - \sin 45° \sin 60°$
$= \dfrac{1}{\sqrt{2}} \times \dfrac{1}{2} - \dfrac{1}{\sqrt{2}} \times \dfrac{\sqrt{3}}{2} = \dfrac{1-\sqrt{3}}{2\sqrt{2}} = \dfrac{\sqrt{2}-\sqrt{6}}{4}$

b $a = 2, b = 3$

6 **a** $\dfrac{3+4\sqrt{3}}{10}$ **b** $\dfrac{4+3\sqrt{3}}{10}$ **c** $\dfrac{10(3\sqrt{3}-4)}{11}$

7 **a** $\dfrac{3}{5}$ **b** $\dfrac{4}{5}$ **c** $\dfrac{3-4\sqrt{3}}{10}$ **d** $\dfrac{1}{7}$

8 **a** $-\dfrac{77}{85}$ **b** $-\dfrac{36}{85}$ **c** $\dfrac{36}{77}$

9 **a** $-\dfrac{36}{325}$ **b** $\dfrac{204}{253}$ **c** $-\dfrac{325}{36}$

10 **a** $45°$ **b** $225°$

Exercise 7C

1 $\sin 2A = \sin A \cos A + \cos A \sin A = 2\sin A \cos A$

2 **a** $\cos 2A = \cos A \cos A - \sin A \sin A = \cos^2 A - \sin^2 A$
b **i** $\cos 2A = \cos^2 A - \sin^2 A = \cos^2 A - (1-\cos^2 A)$
$= 2\cos^2 A - 1$
ii $\cos 2A = (1-\sin^2 A) - \sin^2 A = 1 - 2\sin^2 A$

3 $\tan 2A = \dfrac{\tan A + \tan A}{1 - \tan A \tan A} = \dfrac{2\tan A}{1-\tan^2 A}$

4 **a** $\sin 20°$ **b** $\cos 50°$ **c** $\cos 80°$
d $\tan 10°$ **e** $\operatorname{cosec} 49°$ **f** $3\cos 60°$
g $\tfrac{1}{2}\sin 16°$ **h** $\cos\left(\dfrac{\pi}{8}\right)$

5 **a** $\dfrac{\sqrt{2}}{2}$ **b** $\dfrac{\sqrt{3}}{2}$ **c** $\dfrac{1}{2}$ **d** 1

6 **a** $(\sin A + \cos A)^2 = \sin^2 A + 2\sin A \cos A + \cos^2 A$
$= 1 + \sin 2A$
b $\left(\sin\dfrac{\pi}{8} + \cos\dfrac{\pi}{8}\right)^2 = 1 + \sin\dfrac{\pi}{4} = 1 + \dfrac{\sqrt{2}}{2} = \dfrac{2+\sqrt{2}}{2}$

7 **a** $\cos 6\theta$ **b** $3\sin 4\theta$ **c** $\tan\theta$
d $2\cos\theta$ **e** $\sqrt{2}\cos\theta$ **f** $\tfrac{1}{4}\sin^2 2\theta$
g $\sin 4\theta$ **h** $-\tfrac{1}{2}\tan 2\theta$ **i** $\cos^2 2\theta$

8 $q = \dfrac{p^2}{2} - 1$

9 **a** $y = 2(1-x)$ **b** $2xy = 1 - x^2$
c $y^2 = 4x^2(1-x^2)$ **d** $y^2 = \dfrac{2(4-x)}{3}$

10 $-\dfrac{7}{8}$ **11** $\pm\dfrac{1}{5}$

12 **a** **i** $\dfrac{24}{7}$ **ii** $\dfrac{7}{25}$ **iii** $\dfrac{7}{25}$ **b** $\dfrac{336}{625}$

13 **a** **i** $-\dfrac{7}{9}$ **ii** $\dfrac{2\sqrt{2}}{3}$ **iii** $-\dfrac{9\sqrt{2}}{8}$
b $\tan 2A = \dfrac{\sin 2A}{\cos 2A} = -\dfrac{4\sqrt{2}}{9} \times -\dfrac{9}{7} = \dfrac{4\sqrt{2}}{7}$

14 -3 **15** mn

16 **a** $\cos 2\theta = \dfrac{3^2+6^2-5^2}{2\times 3\times 6} = \dfrac{20}{36} = \dfrac{5}{9}$ **b** $\dfrac{\sqrt{2}}{3}$

17 **a** $\dfrac{3}{4}$ **b** $m = \tan 2\theta = \dfrac{2\left(\tfrac{3}{4}\right)}{1-\left(\tfrac{3}{4}\right)^2} = \dfrac{3}{2} \times \dfrac{16}{7} = \dfrac{24}{7}$

18 **a** $\cos 2A = \cos A \cos A - \sin A \sin A = \cos^2 A - \sin^2 A$
$= \cos^2 A - (1-\cos^2 A) = 2\cos^2 A - 1$
b $4\cos 2x = 6\cos^2 x - 3\sin 2x$
$\cos 2x + 3\cos 2x - 6\cos^2 x + 3\sin 2x = 0$
$\cos 2x + 3(2\cos^2 x - 1) - 6\cos^2 x + 3\sin 2x = 0$
$\cos 2x - 3 + 3\sin 2x = 0$
$\cos 2x + 3\sin 2x - 3 = 0$

19 $\tan 2A \equiv \dfrac{\sin 2A}{\cos 2A} \equiv \dfrac{2\sin A \cos A}{\cos^2 A - \sin^2 A}$
$\equiv \dfrac{\dfrac{2\sin A \cos A}{\cos^2 A}}{\dfrac{\cos^2 A - \sin^2 A}{\cos^2 A}} \equiv \dfrac{2\tan A}{1-\tan^2 A}$

Exercise 7D

1 **a** $51.7°, 231.7°$ **b** $170.1°, 350.1°$
c $56.5°, 303.5°$ **d** $150°, 330°$

2 **a** $\sin\left(\theta+\dfrac{\pi}{4}\right) \equiv \sin\theta\cos\dfrac{\pi}{4} + \cos\theta\sin\dfrac{\pi}{4}$
$\equiv \dfrac{1}{\sqrt{2}}\sin\theta + \dfrac{1}{\sqrt{2}}\cos\theta \equiv \dfrac{1}{\sqrt{2}}(\sin\theta+\cos\theta)$
b $0, \dfrac{\pi}{2}, 2\pi$ **c** $0, \dfrac{\pi}{2}, 2\pi$

3 **a** $30°, 270°$ **b** $30°, 270°$

4 **a** $3(\sin x \cos y - \cos x \sin y)$
$- (\sin x \cos y + \cos x \sin y) = 0$
$\Rightarrow 2\sin x \cos y - 4\cos x \sin y = 0$
Divide throughout by $2\cos x \cos y$
$\Rightarrow \tan x - 2\tan y = 0$, so $\tan x = 2\tan y$
b Using **a** $\tan x = 2\tan y = 2\tan 45° = 2$
so $x = 63.4°, 243.4°$

5 **a** $0, \dfrac{\pi}{3}, \pi, \dfrac{5\pi}{3}, 2\pi$ **b** $\pm 38.7°$
c $30°, 150°, 210°, 330°$ **d** $\dfrac{\pi}{12}, \dfrac{\pi}{4}, \dfrac{5\pi}{12}, \dfrac{3\pi}{4}$
e $60°, 300°, 443.6°, 636.4°$ **f** $\dfrac{\pi}{8}, \dfrac{5\pi}{8}$
g $\dfrac{\pi}{4}, \dfrac{5\pi}{4}$
h $0°, 30°, 150°, 180°, 210°, 330°$ **i** $\dfrac{\pi}{6}, \dfrac{2\pi}{3}, \dfrac{7\pi}{6}, \dfrac{5\pi}{3}$
j $-104.0°, 0°, 76.0°$
k $0°, 35.3°, 144.7°, 180°, 215.3°, 324.7°, 360°$

6 $51.3°$

7 **a** $5\sin 2\theta = 10\sin\theta\cos\theta$, so equation becomes
$10\sin\theta\cos\theta + 4\sin\theta = 0$, or $2\sin\theta(5\cos\theta+2) = 0$
b $0°, 180°, 113.6°, 246.4°$

8 **a** $2\sin\theta\cos\theta + \cos^2\theta - \sin^2\theta = 1$
$\Rightarrow 2\sin\theta\cos\theta - 2\sin^2\theta = 0$
$\Rightarrow 2\sin\theta(\cos\theta - \sin\theta) = 0$
b $0°, 180°, 45°, 225°$

9 **a** L.H.S. $= \cos^2 2\theta + \sin^2 2\theta - 2\sin 2\theta\cos 2\theta$
$= 1 - \sin 4\theta = $ R.H.S.
b $\dfrac{\pi}{24}, \dfrac{17\pi}{24}$

10 **a** **i** R.H.S. $= \dfrac{2\tan\left(\tfrac{\theta}{2}\right)}{\sec^2\left(\tfrac{\theta}{2}\right)} = 2\dfrac{\sin\left(\tfrac{\theta}{2}\right)}{\cos\left(\tfrac{\theta}{2}\right)} \times \dfrac{\cos^2\left(\tfrac{\theta}{2}\right)}{1}$
$= 2\sin\left(\dfrac{\theta}{2}\right)\cos\left(\dfrac{\theta}{2}\right) = \sin\theta$
ii R.H.S. $= \dfrac{1-\tan^2\left(\tfrac{\theta}{2}\right)}{1+\tan^2\left(\tfrac{\theta}{2}\right)} = \dfrac{1-\tan^2\left(\tfrac{\theta}{2}\right)}{\sec^2\left(\tfrac{\theta}{2}\right)}$
$= \cos^2\left(\dfrac{\theta}{2}\right)\left\{1 - \tan^2\left(\dfrac{\theta}{2}\right)\right\} = \cos^2\left(\dfrac{\theta}{2}\right) - \sin^2\left(\dfrac{\theta}{2}\right)$
$= \cos\theta = $ L.H.S.
b **i** $90°, 323.1°$ **ii** $13.3°, 240.4°$

11 a L.H.S. $\equiv \dfrac{3(1 + \cos 2x)}{2} - \dfrac{(1 - \cos 2x)}{2}$
$\equiv 1 + 2\cos 2x$

b

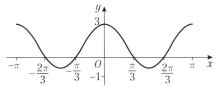

Crosses y-axis at $(0, 3)$

Crosses x-axis at $\left(-\dfrac{2\pi}{3}, 0\right), \left(-\dfrac{\pi}{3}, 0\right), \left(\dfrac{\pi}{3}, 0\right), \left(\dfrac{2\pi}{3}, 0\right)$

12 a $2\cos^2\left(\dfrac{\theta}{2}\right) - 4\sin^2\left(\dfrac{\theta}{2}\right) = 2\left(\dfrac{1+\cos\theta}{2}\right) - 4\left(\dfrac{1-\cos\theta}{2}\right)$
$= 1 + \cos\theta - 2 + 2\cos\theta = 3\cos\theta - 1$
b $131.8°, 228.2°$

13 a $(\sin^2 A + \cos^2 A)^2 \equiv \sin^4 A + \cos^4 A + 2\sin^2 A \cos^2 A$

So $\qquad 1 \equiv \sin^4 A + \cos^4 A + \dfrac{(2\sin A \cos A)^2}{2}$

$\Rightarrow \qquad 2 = 2(\sin^4 A + \cos^4 A) + \sin^2 2A$

$\sin^4 A + \cos^4 A \equiv \tfrac{1}{2}(2 - \sin^2 2A)$

b Using **a**: $\sin^4 A + \cos^4 A \equiv \tfrac{1}{2}(2 - \sin^2 2A)$

$\equiv \tfrac{1}{2}\left\{2 - \dfrac{(1-\cos 4A)}{2}\right\} \equiv \dfrac{(4-1+\cos 4A)}{4} \equiv \dfrac{3+\cos 4A}{4}$

c $\dfrac{\pi}{12}, \dfrac{5\pi}{12}, \dfrac{7\pi}{12}, \dfrac{11\pi}{12}$

14 a $\cos 3\theta \equiv \cos(2\theta + \theta) \equiv \cos 2\theta \cos\theta - \sin 2\theta \sin\theta$
$\equiv (\cos^2\theta - \sin^2\theta)\cos\theta - 2\sin\theta\cos\theta\sin\theta$
$\equiv \cos^3\theta - 3\sin^2\theta\cos\theta$
$\equiv \cos^3\theta - 3(1-\cos^2\theta)\cos\theta$
$\equiv 4\cos^3\theta - 3\cos\theta$

b $\dfrac{\pi}{9}, \dfrac{5\pi}{9}$ and $\dfrac{7\pi}{9}$

Exercise 7E

1 $R = 13$; $\tan\alpha = \dfrac{12}{5}$ **2** $35.3°$ **3** $41.8°$

4 a $\cos\theta - \sqrt{3}\sin\theta \equiv R\cos(\theta + \alpha)$ gives $R = 2, \alpha = \dfrac{\pi}{3}$

b $y = 2\cos\left(\theta + \dfrac{\pi}{3}\right)$

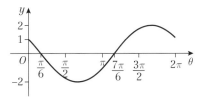

5 a $25\cos(\theta + 73.7°)$ **b** $(0, 7)$
c $25, -25$ **d i** 2 **ii** 0 **iii** 1
6 a $R = \sqrt{10}, \alpha = 71.6°$ **b** $\theta = 69.2°, 327.7°$
7 a $\sqrt{5}\cos(2\theta + 1.107)$ **b** $\theta = 0.60, 1.44$
8 a $6.9°, 66.9°$ **b** $16.6°, 65.9°$
c $8.0°, 115.9°$ **d** $-165.2°, 74.8°$
9 a $5\sin(3\theta - 53.1°)$
b Minimum value is -5,
when $3\theta - 53.1° = 270° \Rightarrow \theta = 107.7°$
c $21.6°, 73.9°, 141.6°$

10 a $5\left(\dfrac{1-\cos 2\theta}{2}\right) - 3\left(\dfrac{1+\cos 2\theta}{2}\right) + 3\sin 2\theta$
$\equiv 1 + 3\sin 2\theta - 4\cos 2\theta$, so $a = 3, b = -4, c = 1$
b Maximum $= 6$, minimum $= -4$ **c** $14.8°, 128.4°$

11 a $R = \sqrt{10}, \alpha = 18.4°, \theta = 69.2°, 327.7°$
b $9\cos^2\theta = 4 - 4\sin\theta + \sin^2\theta$
$\Rightarrow 9(1 - \sin^2\theta) = 4 - 4\sin\theta + \sin^2\theta$
So $10\sin^2\theta - 4\sin\theta - 5 = 0$
c $69.2°, 110.8°, 212.3°, 327.7°$
d When you square you are also solving
$3\cos\theta = -(2 - \sin\theta)$. The other two solutions are for this equation.

12 a $\dfrac{\cos\theta}{\sin\theta} \times \sin\theta + 2\sin\theta = \dfrac{1}{\sin\theta} \times \sin\theta \Rightarrow$
$\cos\theta + 2\sin\theta = 1$
b $\theta = 126.9°$ (1 d.p.)

13 a $\sqrt{2}\cos\theta\cos\dfrac{\pi}{4} + \sqrt{2}\sin\theta\sin\dfrac{\pi}{4} + \sqrt{3}\sin\theta - \sin\theta = 2$
$\Rightarrow \cos\theta + \sin\theta - \sin\theta + \sqrt{3}\sin\theta = 2$
$\Rightarrow \cos\theta + \sqrt{3}\sin\theta = 2$
b $\dfrac{\pi}{3}$

14 a $R = 41, \alpha = 77.320°$ **b i** $\dfrac{18}{91}$ **ii** $77.320°$
15 a $R = 13, \alpha = 22.6°$ **b** $\theta = 48.7°, 108.7°$
c $a = 12, b = -5, c = 12$ **d** minimum value $= -1$

Exercise 7F

1 a L.H.S. $= \dfrac{\cos^2 A - \sin^2 A}{\cos A + \sin A} = \dfrac{(\cos A + \sin A)(\cos A - \sin A)}{\cos A + \sin A}$
$= \cos A - \sin A = $ R.H.S.

b R.H.S. $= \dfrac{2}{2\sin A \cos A}\{\sin B \cos A - \cos B \sin A\}$
$= \dfrac{\sin B}{\sin A} - \dfrac{\cos B}{\cos A} = $ L.H.S.

c L.H.S. $= \dfrac{1 - (1 - 2\sin^2\theta)}{2\sin\theta\cos\theta} = \dfrac{2\sin^2\theta}{2\sin\theta\cos\theta} = \tan\theta = $ R.H.S.

d L.H.S. $= \dfrac{1 + \tan^2\theta}{1 - \tan^2\theta} = \dfrac{1 + \dfrac{\sin^2\theta}{\cos^2\theta}}{1 - \dfrac{\sin^2\theta}{\cos^2\theta}}$
$= \dfrac{\cos^2\theta + \sin^2\theta}{\cos^2\theta - \sin^2\theta} = \dfrac{1}{\cos 2\theta} = \sec 2\theta = $ R.H.S.

e L.H.S. $= 2\sin\theta\cos\theta(\sin^2\theta + \cos^2\theta)$
$= 2\sin\theta\cos\theta = \sin 2\theta = $ R.H.S.

f L.H.S. $= \dfrac{\sin 3\theta\cos\theta - \cos 3\theta\sin\theta}{\sin\theta\cos\theta} = \dfrac{\sin(3\theta - \theta)}{\sin\theta\cos\theta}$
$= \dfrac{\sin 2\theta}{\sin\theta\cos\theta} = \dfrac{2\sin\theta\cos\theta}{\sin\theta\cos\theta} = 2 = $ R.H.S.

g L.H.S. $= \dfrac{1}{\sin\theta} - \dfrac{2\cos 2\theta\cos\theta}{\sin 2\theta} = \dfrac{1}{\sin\theta} - \dfrac{2\cos 2\theta\cos\theta}{2\sin\theta\cos\theta}$
$= \dfrac{1 - \cos 2\theta}{\sin\theta} = \dfrac{1 - (1 - 2\sin^2\theta)}{\sin\theta} = 2\sin\theta = $ R.H.S.

h L.H.S. $= \dfrac{\dfrac{1}{\cos\theta} - 1}{\dfrac{1}{\cos\theta} + 1} = \dfrac{1 - \cos\theta}{1 + \cos\theta} = \dfrac{1 - \left(1 - 2\sin^2\dfrac{\theta}{2}\right)}{1 + \left(2\cos^2\dfrac{\theta}{2} - 1\right)}$
$= \dfrac{2\sin^2\dfrac{\theta}{2}}{2\cos^2\dfrac{\theta}{2}} = \tan^2\dfrac{\theta}{2} = $ R.H.S.

Answers

i L.H.S. = $\dfrac{1-\tan x}{1+\tan x} = \dfrac{\cos x - \sin x}{\cos x + \sin x}$

$= \dfrac{(\cos x - \sin x)(\cos x - \sin x)}{\cos^2 x - \sin^2 x}$

$= \dfrac{\cos^2 x + \sin^2 x - 2\sin x \cos x}{\cos^2 x - \sin^2 x} = \dfrac{1 - \sin 2x}{\cos 2x}$ = R.H.S.

2 a L.H.S. = $\sin(A + 60°) + \sin(A - 60°) = \sin A \cos 60°$
  $+ \cos A \sin 60° + \sin A \cos 60° - \cos A \sin 60°$
  $= 2 \sin A \cos 60° \equiv \sin A$ = R.H.S.

  b L.H.S. = $\dfrac{\cos A}{\sin B} - \dfrac{\sin A}{\cos B} = \dfrac{\cos A \cos B - \sin A \sin B}{\sin B \cos B}$

  $\equiv \dfrac{\cos(A+B)}{\sin B \cos B}$ = R.H.S.

  c L.H.S. = $\dfrac{\sin(x+y)}{\cos x \cos y} = \dfrac{\sin x \cos y + \cos x \sin y}{\cos x \cos y}$

  $= \dfrac{\sin x}{\cos x} + \dfrac{\sin y}{\cos y} \equiv \tan x + \tan y$ = R.H.S.

  d L.H.S. = $\dfrac{\cos(x+y)}{\sin x \sin y} + 1 = \dfrac{\cos x \cos y - \sin x \sin y}{\sin x \sin y} + 1$

  $= \dfrac{\cos x \cos y}{\sin x \sin y} - \dfrac{\sin x \sin y}{\sin x \sin y} + 1 = \dfrac{\cos x \cos y}{\sin x \sin y}$

  $\equiv \cot x \cot y$ = R.H.S.

  e L.H.S. = $\cos\left(\theta + \dfrac{\pi}{3}\right) + \sqrt{3}\sin\theta = \cos\theta\cos\dfrac{\pi}{3}$
  $- \sin\theta\sin\dfrac{\pi}{3} + \sqrt{3}\sin\theta = \dfrac{1}{2}\cos\theta - \dfrac{\sqrt{3}}{2}\sin\theta + \sqrt{3}\sin\theta$

  $= \dfrac{1}{2}\cos\theta + \dfrac{\sqrt{3}}{2}\sin\theta \equiv \sin\left(\theta + \dfrac{\pi}{6}\right)$ = R.H.S.

  f L.H.S. = $\cot(A+B) = \dfrac{\cos(A+B)}{\sin(A+B)}$

  $= \dfrac{\cos A \cos B - \sin A \sin B}{\sin A \cos B + \cos A \sin B}$

  $= \dfrac{\dfrac{\cos A \cos B}{\sin A \sin B} - \dfrac{\sin A \sin B}{\sin A \sin B}}{\dfrac{\sin A \cos B}{\sin A \sin B} + \dfrac{\cos A \sin B}{\sin A \sin B}} \equiv \dfrac{\cot A \cot B - 1}{\cot A + \cot B}$ = R.H.S.

  g L.H.S. = $\sin^2(45° + \theta) + \sin^2(45° - \theta) = (\sin(45° + \theta))^2$
  $+ (\sin(45° - \theta))^2 = (\sin 45° \cos\theta + \cos 45° \sin\theta)^2$
  $+ (\sin 45° \cos\theta - \cos 45° \sin\theta)^2$

  $= \left(\dfrac{\sqrt{2}}{2}\cos\theta + \dfrac{\sqrt{2}}{2}\sin\theta\right)^2 + \left(\dfrac{\sqrt{2}}{2}\cos\theta - \dfrac{\sqrt{2}}{2}\sin\theta\right)^2$

  $= \dfrac{1}{2}\cos^2\theta + \cos\theta\sin\theta + \dfrac{1}{2}\sin^2\theta + \dfrac{1}{2}\cos^2\theta$
  $- \cos\theta\sin\theta + \dfrac{1}{2}\sin^2\theta = \cos^2\theta + \sin^2\theta \equiv 1$ = R.H.S.

  h L.H.S. = $\cos(A+B)\cos(A-B)$
  $= (\cos A \cos B - \sin A \sin B) \times (\cos A \cos B + \sin A \sin B)$
  $= (\cos^2 A \cos^2 B) - (\sin^2 A \sin^2 B) = (\cos^2 A(1 - \sin^2 B))$
  $- ((1 - \cos^2 A)\sin^2 B) = \cos^2 A - \cos^2 A \sin^2 B$
  $- \sin^2 B + \cos^2 A \sin^2 B \equiv \cos^2 A - \sin^2 B$ = R.H.S.

3 a L.H.S. = $\dfrac{\sin\theta}{\cos\theta} + \dfrac{\cos\theta}{\sin\theta} = \dfrac{\sin^2\theta + \cos^2\theta}{\sin\theta\cos\theta}$

  $= \dfrac{1}{\left(\frac{1}{2}\right)\sin 2\theta} = 2\csc 2\theta$ = R.H.S.

  b 4

4 a Use $\sin 3\theta \equiv \sin(2\theta + \theta)$ and substitute $\cos 2\theta \equiv \cos^2\theta - \sin^2\theta$.

  b Use $\cos 3\theta \equiv \cos(2\theta + \theta)$ and substitute $\cos 2\theta \equiv \cos^2\theta - \sin^2\theta$.

  c $\tan 3\theta \equiv \dfrac{\sin 3\theta}{\cos 3\theta} = \dfrac{3\sin\theta\cos^2\theta - \sin^3\theta}{\cos^3\theta - 3\sin^2\theta\cos\theta}$

  $= \dfrac{3\tan\theta - \tan^3\theta}{1 - 3\tan^2\theta}$

  d $\tan\theta = 2\sqrt{2}$

  so $\tan 3\theta = \dfrac{6\sqrt{2} - 16\sqrt{2}}{1 - 24} = \dfrac{-10\sqrt{2}}{-23} = \dfrac{10\sqrt{2}}{23}$

5 a i $\cos x \equiv 2\cos^2\dfrac{x}{2} - 1$

  $\Rightarrow 2\cos^2\dfrac{x}{2} \equiv 1 + \cos x \Rightarrow \cos^2\dfrac{x}{2} \equiv \dfrac{1+\cos x}{2}$

  ii $\cos x \equiv 1 - 2\sin^2\dfrac{x}{2}$

  $\Rightarrow 2\sin^2\dfrac{x}{2} \equiv 1 - \cos x \Rightarrow \sin^2\dfrac{x}{2} \equiv \dfrac{1-\cos x}{2}$

  b i $\dfrac{2\sqrt{5}}{5}$ ii $\dfrac{\sqrt{5}}{5}$ iii $\dfrac{1}{2}$

  c $\cos^4\dfrac{A}{2} \equiv \left(\dfrac{1+\cos A}{2}\right)^2 \equiv \dfrac{1 + 2\cos A + \cos^2 A}{4}$

  $\equiv \dfrac{1 + 2\cos A + \left(\dfrac{1+\cos 2A}{2}\right)}{4}$

  $\equiv \dfrac{2 + 4\cos A + 1 + \cos 2A}{8} \equiv \dfrac{3 + 4\cos A + \cos 2A}{8}$

6 L.H.S. $\equiv \cos^4\theta \equiv (\cos^2\theta)^2 \equiv \left(\dfrac{1+\cos 2\theta}{2}\right)^2$

  $\equiv \dfrac{1}{4}(1 + 2\cos 2\theta + \cos^2 2\theta) \equiv \dfrac{1}{4} + \dfrac{1}{2}\cos 2\theta$

  $+ \dfrac{1}{4}\left(\dfrac{1+\cos 4\theta}{2}\right) \equiv \dfrac{1}{4} + \dfrac{1}{2}\cos 2\theta + \dfrac{1}{8} + \dfrac{1}{8}\cos 4\theta$

  $\equiv \dfrac{3}{8} + \dfrac{1}{2}\cos 2\theta + \dfrac{1}{8}\cos 4\theta$ = R.H.S.

7 $[\sin(x+y) + \sin(x-y)][\sin(x+y) - \sin(x-y)]$
  $\equiv [2\sin x \cos y][2\cos x \sin y]$
  $\equiv [2\sin x \cos x][2\cos y \sin y]$
  $\equiv \sin 2x \sin 2y$

8 $2\cos\left(2\theta + \dfrac{\pi}{3}\right) \equiv 2\left(\cos 2\theta\cos\dfrac{\pi}{3} - \sin 2\theta\sin\dfrac{\pi}{3}\right)$

  $\equiv 2\left(\cos 2\theta\dfrac{1}{2} - \sin 2\theta\dfrac{\sqrt{3}}{2}\right) \equiv \cos 2\theta - \sqrt{3}\sin 2\theta$

9 $4\cos\left(2\theta - \dfrac{\pi}{6}\right) \equiv 4\cos 2\theta\cos\dfrac{\pi}{6} + 4\sin 2\theta\sin\dfrac{\pi}{6}$
  $\equiv 2\sqrt{3}\cos 2\theta + 2\sin 2\theta \equiv 2\sqrt{3}(1 - 2\sin^2\theta) + 4\sin\theta\cos\theta$
  $\equiv 2\sqrt{3} - 4\sqrt{3}\sin^2\theta + 4\sin\theta\cos\theta$

10 a R.H.S. = $\sqrt{2}\left\{\sin\theta\cos\dfrac{\pi}{4} + \cos\theta\sin\dfrac{\pi}{4}\right\}$

  $= \sqrt{2}\left\{\sin\theta\dfrac{1}{\sqrt{2}} + \cos\theta\dfrac{1}{\sqrt{2}}\right\} = \sin\theta + \cos\theta$ = L.H.S.

  b R.H.S. = $2\left\{\sin 2\theta\cos\dfrac{\pi}{6} - \cos 2\theta\sin\dfrac{\pi}{6}\right\}$

  $= 2\left\{\sin 2\theta\dfrac{\sqrt{3}}{2} - \cos 2\theta\dfrac{1}{2}\right\} = \sqrt{3}\sin 2\theta - \cos 2\theta$ = L.H.S.

Answers

Challenge

1. **a** $\cos(A + B) - \cos(A - B)$
   $\equiv \cos A \cos B - \sin A \sin B - (\cos A \cos B + \sin A \sin B)$
   $\equiv -2 \sin A \sin B$

   b Let $A + B = P$ and $A - B = Q$. Solve to get $A = \dfrac{P + Q}{2}$ and $B = \dfrac{P - Q}{2}$. Then use result from part **a** to get
   $\cos P - \cos Q = -2 \sin\left(\dfrac{P + Q}{2}\right) \sin\left(\dfrac{P - Q}{2}\right)$

   c $-\frac{3}{2}(\cos 8x - \cos 6x)$

2. **a** $\sin(A + B) + \sin(A - B)$
   $= \sin A \cos B + \cos A \sin B + \sin A \cos B - \cos A \sin B$
   $= 2 \sin A \cos B$
   Let $A + B = P$ and $A - B = Q$
   $\therefore A = \dfrac{P + Q}{2}$ and $B = \dfrac{P - Q}{2}$
   $\therefore \sin P + \sin Q = 2 \sin\left(\dfrac{P + Q}{2}\right) \cos\left(\dfrac{P - Q}{2}\right)$

   b $\dfrac{11\pi}{24} = \dfrac{P + Q}{2}, \dfrac{5\pi}{24} = \dfrac{P - Q}{2}$
   $\dfrac{22\pi}{24} = P + Q, \dfrac{10\pi}{24} = P - Q$
   $\dfrac{32\pi}{24} = 2P \Rightarrow P = \dfrac{2\pi}{3}, Q = \dfrac{\pi}{4}$,
   $\sin\left(\dfrac{2\pi}{3}\right) + \sin\left(\dfrac{\pi}{4}\right) = \dfrac{\sqrt{3} + \sqrt{2}}{2}$

Exercise 7G

1. **a** 0.25 m **b** 0.013 minutes, 0.8 seconds
   c 0.2 minutes or 12 seconds
2. **a** 0.03 radians **b** 0.0085 radians
   c 0.251 seconds
   d 0.0492, 0.2021, 0.3006, 0.4534 seconds
3. **a** £17.12, £17.08
   b £19.40, 6.53 hours or 6 h 32 min
   c After 4.37 hours (4 h 22 min after market opens)
4. **a** 224.7°C
   b 2 m 17 s, 5 m 26 s, 8 m 34 s
   c 17.6 seconds.
5. **a** $R = 0.5, \alpha = 53.13°$
   b **i** 0.5   **ii** $\theta = 143.1°$
   c Minimum value is 22.5, occurs at 17.95 minutes
   d 3, 13, 23, 33, 43, 53 minutes
6. **a** $R = 68.0074, \alpha = 0.2985$ **b** 138.0 m
   c 31.4 minutes **d** 11.1 minutes
7. **a** $R = 250, \alpha = 0.6435$
   b **i** 1950 V/m   **ii** $x = 4.41$ cm, $x = 16.91$ cm
   c $2.10 \leq x \leq 6.71, 14.60 \leq x \leq 19.21$

Challenge

a 0 cm $\leq x < 0.39$ cm, 8.42 cm $< x < 12.89$ cm, 20.92 cm $< x < 25$ cm
b Identifying the exact point of maximum field strength; microwave oven would not work exactly the same every time it is used.

Mixed exercise 7

1. **a** $\frac{1}{2}$ **b** $\frac{1}{2}$ **c** $\dfrac{\sqrt{3}}{3}$

2. $\sin x = \dfrac{1}{\sqrt{5}}$, so $\cos x = \dfrac{2}{\sqrt{5}}$
   $\cos(x - y) = \sin y \Rightarrow \dfrac{2}{\sqrt{5}} \cos y + \dfrac{1}{\sqrt{5}} \sin y = \sin y$
   $\Rightarrow (\sqrt{5} - 1) \sin y = 2 \cos y \Rightarrow \tan y = \dfrac{2}{\sqrt{5} - 1} = \dfrac{\sqrt{5} + 1}{2}$

3. **a** $\tan A = 2, \tan B = \frac{1}{3}$ **b** 45°

4. Use the sine rule and addition formulae to get
   $\dfrac{1}{20} \sin\theta \times \dfrac{\sqrt{3}}{2} = \dfrac{9}{20} \cos\theta \times \dfrac{1}{2}$
   Then rearrage to get $\tan\theta = 3\sqrt{3}$.

5. 75°

6. **a** **i** $\frac{56}{65}$ **ii** $\frac{120}{119}$
   b Use $\cos\{180° - (A + B)\} \equiv -\cos(A + B)$ and expand. You can work out all the required trig. ratios (A and B are acute).

7. **a** Use $\cos 2x \equiv 1 - 2 \sin^2 x$ **b** $\frac{4}{5}$
   c **i** Use $\tan x = 2, \tan y = \frac{1}{3}$ in the expansion of $\tan(x + y)$.
   ii Find $\tan(x - y) = 1$ and note that $x - y$ has to be acute.

8. **a** Show that both sides are equal to $\frac{5}{6}$.
   b $\dfrac{3k}{2}$ **c** $\dfrac{12k}{4 - 9k^2}$

9. **a** $\sqrt{3} \sin 2\theta = 1 - 2\sin^2\theta = \cos 2\theta$
   $\Rightarrow \sqrt{3} \tan 2\theta = 1 \Rightarrow \tan 2\theta = \dfrac{1}{\sqrt{3}}$
   b $\dfrac{\pi}{12}, \dfrac{7\pi}{12}$

10. **a** $a = 2, b = 5, c = -1$ **b** 0.187, 2.95
11. **a** $\cos(x - 60°) = \cos x \cos 60° + \sin x \sin 60°$
    $= \dfrac{1}{2} \cos x + \dfrac{\sqrt{3}}{2} \sin x$
    So $\left(2 - \dfrac{\sqrt{3}}{2}\right) \sin x = \dfrac{1}{2} \cos x \Rightarrow \tan x = \dfrac{\frac{1}{2}}{2 - \frac{\sqrt{3}}{2}} = \dfrac{1}{4 - \sqrt{3}}$
    b 23.8°, 203.8°

12. **a** Using addition formulae:
    $\cos x \cos 20° - \sin x \sin 20°$
    $= \sin 70° \cos x - \cos 70° \sin x$
    Rearrange to get: $\sin x (5 \cos 70°) + \cos x (3 \sin 70°) = 0$
    $\Rightarrow \tan x = \dfrac{\sin x}{\cos x} = -\dfrac{3 \sin 70°}{5 \cos 70°} = -\dfrac{3}{5} \tan 70°$
    b 121.2°

13. **a** Find $\sin\alpha = \frac{3}{5}$ and $\cos\alpha = \frac{4}{5}$ and insert in expansions on L.H.S. Result follows.
    b 0.6, 0.8

14. **a** Example: $A = 60°, B = 0°$; $\sec(A + B) = 2$, $\sec A + \sec B = 2 + 1 = 3$
    b L.H.S. $= \dfrac{\sin\theta}{\cos\theta} + \dfrac{\cos\theta}{\sin\theta} \equiv \dfrac{\sin^2\theta + \cos^2\theta}{\sin\theta \cos\theta}$
    $\equiv \dfrac{1}{\frac{1}{2}\sin 2\theta} \equiv 2 \operatorname{cosec} 2\theta =$ R.H.S.

15. **a** Setting $\theta = \dfrac{\pi}{8}$ gives resulting quadratic equation in t,
    $t^2 + 2t - 1 = 0$, where $t = \tan\left(\dfrac{\pi}{8}\right)$.
    Solving this and taking +ve value for t gives result.
    b Expanding $\tan\left(\dfrac{\pi}{4} + \dfrac{\pi}{8}\right)$ gives answer: $\sqrt{2} + 1$

Answers

16 a $2\sin(x-60)°$

  b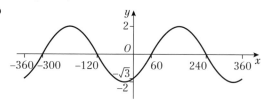

  Graph crosses y-axis at $(0, -\sqrt{3})$
  Graph crosses x-axis at $(-300°, -0)$, $(-120°, 0)$, $(60°, 0)$, $(240°, 0)$

17 a $R = 25$, $\alpha = 1.29$ **b** 32   **c** $\theta = 0.12, 1.17$

18 a $2.5\sin(2x + 0.927)$ **b** $\frac{3}{2}\sin 2x + 2\cos 2x + 2$ **c** 4.5

19 a $\alpha = 14.0°$ **b** $0°, 151.9°, 360°$

20 a $R = \sqrt{13}$, $\alpha = 56.3°$ **b** $\theta = 17.6°, 229.8°$

21 a L.H.S. $= \dfrac{1}{\cos\theta} \cdot \dfrac{1}{\sin\theta} \equiv \dfrac{1}{\frac{1}{2}\sin 2\theta} \equiv 2\csc 2\theta =$ R.H.S.

  b L.H.S. $= \dfrac{1 + \tan x}{1 - \tan x} - \dfrac{1 - \tan x}{1 + \tan x}$

  $\equiv \dfrac{(1 + \tan x)^2 - (1 - \tan x)^2}{(1 + \tan x)(1 - \tan x)}$

  $\equiv \dfrac{(1 + 2\tan x + \tan^2 x) - (1 - 2\tan x + \tan^2 x)}{1 - \tan^2 x}$

  $\equiv \dfrac{4\tan x}{1 - \tan^2 x} = \dfrac{2(2\tan x)}{1 - \tan^2 x} = 2\tan 2x =$ R.H.S.

  c L.H.S. $= (\sin x \cos y + \cos x \sin y)(\sin x \cos y - \cos x \sin y)$
  $= \sin^2 x \cos^2 y - \cos^2 x \sin^2 y$
  $= (1 - \cos^2 x)\cos^2 y - \cos^2 x(1 - \sin^2 y) =$ R.H.S.

  d L.H.S. $= 2\cos 2\theta + 1 + (2\cos^2 2\theta - 1)$
  $\equiv 2\cos 2\theta(1 + \cos 2\theta) \equiv 2\cos 2\theta(2\cos^2 \theta)$
  $\equiv 4\cos^2 \theta \cos 2\theta \equiv$ R.H.S.

22 a $\dfrac{1 - (1 - 2\sin^2 x)}{1 + (2\cos^2 x - 1)} \equiv \dfrac{2\sin^2 x}{2\cos^2 x}$
  $\equiv \tan^2 x$

  b $\pm\dfrac{\pi}{3}, \pm\dfrac{2\pi}{3}$

23 a L.H.S. $= \cos^4 2\theta - \sin^4 2\theta$
  $\equiv (\cos^2 2\theta - \sin^2 2\theta)(\cos^2 2\theta + \sin^2 2\theta)$
  $\equiv (\cos^2 2\theta - \sin^2 2\theta)(1)$
  $\equiv \cos 4\theta =$ R.H.S.

  b $15°, 75°, 105°, 165°$

24 a Use $\cos 2\theta = 1 - 2\sin^2 \theta$ and $\sin 2\theta = 2\sin\theta\cos\theta$.
  b $\sin 360° = 0$, $2 - 2\cos(360°) = 2 - 2 = 0$
  c $26.6°, 206.6°$

25 a $R = 3$, $\alpha = 0.841$ **b** $x = 1.07, 3.53$

26 a $R = 5.772$, $\alpha = 75.964°$ **b** 5.772 when $\theta = 166.0°$
  c 6.228 hours   **d** 350.8 days

27 a $13\sin(x + 22.6°)$ **b** 3.8 m/s
  c 168.5 minutes

Challenge

1 a $\dfrac{\cos 2\theta + \cos 4\theta}{\sin 2\theta - \sin 4\theta} \equiv \dfrac{2\cos 3\theta \cos\theta}{-2\cos 3\theta \sin\theta} \equiv -\cot\theta$

  b $\cos 5x + \cos x + 2\cos 3x$
  $\equiv 2\cos 3x \cos 2x + 2\cos 3x$
  $\equiv 2\cos 3x (\cos 2x + 1)$
  $\equiv 2\cos 3x (2\cos^2 x)$
  $\equiv 4\cos^2 x \cos 3x$

2 a $\theta = \angle OAB = \angle OBA \Rightarrow \angle AOB = \pi - 2\theta$, so $\angle BOD = 2\theta$
  $OB = 1$, $OD = \cos 2\theta$
  $BD = \sin 2\theta$, $AB = 2\cos\theta$
  $\sin\theta = \dfrac{BD}{AB} = \dfrac{BD}{2\cos\theta}$
  So $BD = 2\sin\theta\cos\theta$
  But $BD = \sin 2\theta$
  So $\sin 2\theta \equiv 2\sin\theta\cos\theta$

  b $AB = 2\cos\theta$
  $AD = (2\cos\theta)\cos\theta = 2\cos^2\theta$
  $OD = 2\cos^2\theta - 1$
  From part **a**, $OD = \cos 2\theta$, so $\cos 2\theta \equiv 2\cos^2\theta - 1$.

CHAPTER 8

Prior knowledge 8

1 a $t = \dfrac{x}{4 - k}$ **b** $t = \pm\sqrt{\dfrac{y}{3}}$

  c $t = e^{\frac{2-y}{4}}$ **d** $t = -\dfrac{1}{3}\ln\left(\dfrac{x-1}{2}\right)$

2 a $7 - 3\cos^2 x$ **b** $2\cos x\sqrt{1 - \cos^2 x}$

  c $\dfrac{\cos x}{\sqrt{1 - \cos^2 x}}$ **d** $2\cos x + 2\cos^2 x - 1$

3 a $y > 0$ **b** $0 < y < 2$
  c $-6 \le y < 3$ **d** $0 < y < 1$

4 $(4, 7)$ and $(-4.8, 2.6)$

Exercise 8A

1 a $y = (x + 2)^2 + 1$, $-6 \le x \le 2$, $1 \le y \le 17$
  b $y = (5 - x)^2 - 1$, $x \in \mathbb{R}$ $y \ge -1$
  c $y = 3 - \dfrac{1}{x}$, $x \ne 0$, $y \ne 3$
  d $y = \dfrac{2}{x - 1}$, $x > 1$, $y > 0$
  e $y = \left(\dfrac{1 + 2x}{x}\right)^2$, $x > 0$, $y > 4$
  f $y = \dfrac{x}{1 - 3x}$, $0 < x < \dfrac{1}{3}$, $y > 0$

2 a i $y = 20 - 10e^{\frac{1}{2}x} + e^x$, $x > 0$ **ii** $y \ge -5$
  b i $y = \dfrac{1}{e^x + 2}$, $x > 0$ **ii** $0 < y < \dfrac{1}{3}$
  c i $y = x^3$, $x > 0$ **ii** $y > 0$

3 a $y = 9x^2 - x^4$, $0 \le x \le \sqrt{5}$, $0 \le y \le \dfrac{81}{4}$
  b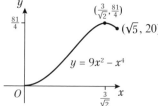

4 a i $y = \dfrac{15}{2} - \dfrac{1}{2}x$ **ii** $x > -3$, $y < 9$
  iii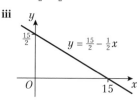

Answers

b i $y = \frac{1}{9}(x-2)(x+7)$
ii $-13 < x < 11$, $-\frac{9}{4} < y < 18$
iii

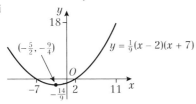

c i $y = \frac{1}{x-2}$
ii $x \in \mathbb{R}, x \neq 2, y \in \mathbb{R}, y \neq 0$
iii

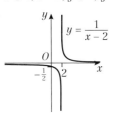

d i $y = 3x + 3$
ii $x > -1, y > 0$
iii

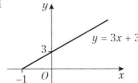

e i $y = 2 - e^x$
ii $x > 0, y < 1$
iii

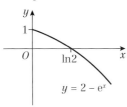

5 a $C_1: x = 1 + 2t, t = \frac{x-1}{2}$

Sub t into $y = 2 + 3t$:
$$y = 2 + 3\left(\frac{x-1}{2}\right) = 2 + \frac{3}{2}x - \frac{3}{2} = \frac{3}{2}x + \frac{1}{2}$$

$C_2: x = \frac{1}{2t-3}, t = \frac{1+3x}{2x}$

Sub into $y = \frac{t}{2t-3}$.

$$y = \frac{\frac{1+3x}{2x}}{2\left(\frac{1+3x}{2x}\right) - 3} = \frac{\frac{1+3x}{2x}}{\frac{1}{x}} = \frac{1+3x}{2} = \frac{3}{2}x + \frac{1}{2}$$

Therefore C_1 and C_2 represent a segment of the same straight line.

b Length of $C_1 = 3\sqrt{13}$, length of $C_2 = \frac{\sqrt{13}}{3}$

6 a $x \neq 2; y \leq -2, y \neq -3$
b $x = \frac{3}{t} + 2, t = \frac{3}{x-2}$

Sub into $y = 2t - 3 - t^2$
$$y = 2\left(\frac{3}{x-2}\right) - 3 - \left(\frac{3}{x-2}\right)^2 = \frac{6}{x-2} - 3 - \frac{9}{(x-2)^2}$$
$$= \frac{6(x-2) - 3(x-2)^2 - 9}{(x-2)^2}$$
$$= \frac{6x - 12 - 3x^2 + 12x - 12 - 9}{(x-2)^2} = \frac{-3x^2 + 18x - 33}{(x-2)^2}$$
$$= \frac{-3(x^2 - 6x + 11)}{(x-2)^2} \text{ so } A = -3, b = -6, c = 11$$

7 a $x = \ln(t+3)$ $t = e^x - 3$ Sub into $y = \frac{1}{t+5}$
$$y = \frac{1}{e^x - 3 + 5} = \frac{1}{e^x + 2}, \quad x > 0$$
b $0 < y < \frac{1}{3}$

8 a $y = \frac{x^6}{729} - \frac{2x^2}{9}, 0 \leq x \leq 3\sqrt{2}$
b $y = t^3 - 2t, \frac{dy}{dt} = 3t^2 - 2$
$0 = 3t^2 - 2$ $t^2 = \frac{2}{3}$ $t = \sqrt{\frac{2}{3}}$
c $-\frac{4\sqrt{6}}{9} \leq f(x) \leq 4$

9 a $y = 4 - t^2 \Rightarrow t = \sqrt{4 - y}$
Sub into $x = t^3 - t = t(t^2 - 1)$
$x = \sqrt{4-y}(4-y-1) = \sqrt{4-y}(3-y)$
$x^2 = (4-y)(3-y)^2$
$a = 4, b = 3$
b Max y is 4

Challenge

a $x^2 = \frac{(1-t^2)^2}{(1+t^2)^2}, y^2 = \frac{4t^2}{(1+t^2)^2}$

$x^2 + y^2 = \frac{(1-t^2)^2}{(1+t^2)^2} + \frac{4t^2}{(1+t^2)^2} = \frac{1 - 2t^2 + t^4}{(1+t^2)^2} + \frac{4t^2}{(1+t^2)^2}$

$= \frac{1 + 2t^2 + t^4}{(1+t^2)^2} = \frac{(1+t^2)^2}{(1+t^2)^2} = 1$

So $x^2 + y^2 = 1$
b Circle, centre (0,0), radius 1.

Exercise 8B

1 a $25(x+1)^2 + 4(y-4)^2 = 100$ **b** $y^2 = 4x^2(1-x^2)$
c $y = 4x^2 - 2$ **d** $y = \frac{2x\sqrt{1-x^2}}{1-2x^2}$
e $y = \frac{4}{x-2}$ **f** $y^2 = 1 + \left(\frac{x}{3}\right)^2$

2 a $(x+5)^2 + (y-2)^2 = 1$
b Centre $(-5, 2)$, radius 1
c $0 \leq t < 2\pi$

3 Centre $(3, -1)$, radius 4
4 $(x+2)^2 + (y-3)^2 = 1$

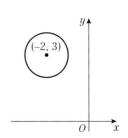

399

Answers

5 a $y = \frac{\sqrt{2}}{2}x + \frac{\sqrt{2(1-x^2)}}{2}, -1 < x < 1$

  b $y = \frac{\sqrt{3}}{3}x - \frac{\sqrt{9-x^2}}{3}, \frac{3}{2} < x < 3$

  c $y = -3x, -1 \leq x \leq 1$

6 a $y = \frac{16}{x^2}, 0 < x \leq 8$

  b

7 $y = \frac{9}{3+x}$ Domain: $x > 0$

8 a $y = 9x(1 - 12x^2) \Rightarrow a = 9, b = 12$
  b Domain: $0 < x < \frac{1}{3}$, Range: $-1 < y \leq 1$

9 $y = \sin t \cos\left(\frac{\pi}{6}\right) - \cos t \sin\left(\frac{\pi}{6}\right)$

$= \frac{\sqrt{3}}{2}\sin t - \frac{1}{2}\cos t = \frac{\sqrt{3\left(1-\frac{x^2}{4}\right)}}{2} - \frac{1}{4}x$

$= \frac{1}{4}\left(2\sqrt{3-\frac{3}{4}x^2} - x\right) = \frac{1}{4}(\sqrt{12-3x^2} - x)$

$t = 0 \Rightarrow x = 2, t = \pi \Rightarrow x = -2$, so $-2 < x < 2$.

10 a $y^2 = 25\left(1 - \frac{1}{x-4}\right)$ b $x > 5, 0 < y < 5$

11 $x = -\frac{y}{\sqrt{9-y^2}}, x > 0$

Challenge
$(4y^2 - 2 + 2x)^2 + 12x^2 - 3 = 0$

Exercise 8C

1

t	−5	−4	−3	−2	−1	−0.5
$x = 2t$	−10	−8	−6	−4	−2	−1
$y = \frac{5}{t}$	−1	−1.25	−1.67	−2.5	−5	−10

t	0.5	1	2	3	4	5
$x = 2t$	1	2	4	6	8	10
$y = \frac{5}{t}$	10	5	2.5	1.67	1.25	1

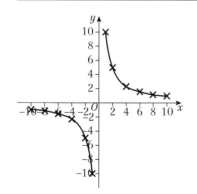

2

t	−4	−3	−2	−1	0
$x = t^2$	16	9	4	1	0
$y = \frac{t^3}{5}$	−12.8	−5.4	−1.6	−0.2	0

t	1	2	3	4
$x = t^2$	1	4	9	16
$y = \frac{t^3}{5}$	0.2	1.6	5.4	12.8

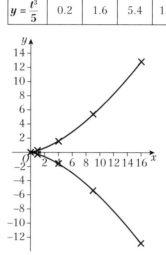

3

t	$-\frac{\pi}{4}$	$-\frac{\pi}{6}$	$-\frac{\pi}{12}$	0
$x = \tan t + 1$	0	0.423	0.732	1
$y = \sin t$	−0.707	−0.5	−0.259	0

t	$\frac{\pi}{12}$	$\frac{\pi}{6}$	$\frac{\pi}{4}$	$\frac{\pi}{3}$
$x = \tan t + 1$	1.268	1.577	2	2.732
$y = \sin t$	0.259	0.5	0.707	0.866

4 a   b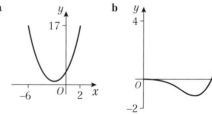

  c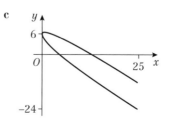

d 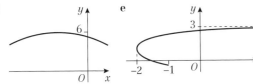   **e**

f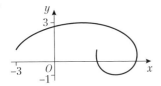

5 a $y = (3 - x)^2 - 2$
b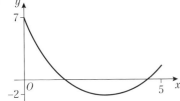

6 a $(x + 2)^2 + (y - 1)^2 = 81$
b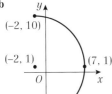
$(-2, 10)$
$(-2, 1)$, $(7, 1)$
c 6π

Challenge

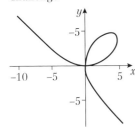

As t approaches -1 from the positive direction, the curve heads off to infinity in the 2nd quadrant.
As t approaches -1 from the negative direction, the curve heads off to infinity in the 4th quadrant.

Exercise 8D
1 a $(11, 0)$ **b** $(7, 0)$ **c** $(1, 0), (9, 0)$
  d $(1, 0), (2, 0)$ **e** $\left(\frac{9}{5}, 0\right)$
2 a $(0, -5)$ **b** $\left(0, \frac{9}{16}\right)$ **c** $(0, 0), (0, 12)$
  d $\left(0, \frac{1}{2}\right)$ **e** $(0, 1)$
3 4   **4** 4   **5** $\left(\frac{1}{2}, \frac{3}{2}\right)$
6 $t = \frac{5}{2}, t = -\frac{3}{2}$; $\left(\frac{25}{4}, 5\right), \left(\frac{9}{4}, -3\right)$
7 $(1, 2), (1, -2), (4, 4), (4, -4)$
8 a $\left(\frac{\pi^2}{4} - 1, 0\right), (0, \cos 1)$
  b $\left(\frac{\sqrt{3}}{2}, 0\right), (0, 1)$
  c $(1, 0)$
9 a $(e + 5, 0)$ **b** $(\ln 8, 0), (0, -63)$ **c** $\left(\frac{5}{4}, 0\right)$

10 $t = \frac{2}{3}, t = -1, \left(\frac{4}{9}, \frac{2}{3}\right), (1, -1)$
11 $t = \frac{14}{5}, \left(\ln \frac{9}{5}, \ln \frac{3}{5}\right)$
12 a $\left(6 \cos\left(\frac{\pi}{12}\right), 0\right), \left(6 \cos\left(\frac{5\pi}{12}\right), 0\right)$
  b $4 \sin 2t + 2 = 4 \Rightarrow 4 \sin 2t = 2 \Rightarrow \sin 2t = 0.5$
  $2t = \frac{\pi}{6}, \frac{5\pi}{6} \Rightarrow t = \frac{\pi}{12}, \frac{5\pi}{12}$
  c $\left(6 \cos\left(\frac{\pi}{12}\right), 4\right), \left(6 \cos\left(\frac{5\pi}{12}\right), 4\right)$
13 $y = 2x - 5 \Rightarrow 4t(t - 1) = 2(2t) - 5 \Rightarrow 4t^2 - 8t + 5 = 0$
  Discriminant $= 8^2 - 4 \times 4 \times 5 = 64 - 80 = -16 < 0$
  Since the discriminant is less than 0, the quadratic has no solutions, therefore the two equations do not intersect.
14 a $\cos 2t + 1 = k$
  max of $\cos 2t = 1$, so $k = 1 + 1 = 2$
  min of $\cos 2t = -1$, so $k = -1 + 1 = 0$
  Therefore, $0 \leq k \leq 2$
  b $y = 1 - 2\sin^2 t + 1 = 2 - 2\sin^2 t = 2 - 2x^2$
  $k = 2 - 2x^2 \Rightarrow 2x^2 + k - 2 = 0$
  Tangent when discriminant $= 0$
  Discriminant $= 0^2 - 4 \times 2 \times (k - 2) = 0$
  $-8(k - 2) = 0 \Rightarrow k - 2 = 0 \Rightarrow k = 2$
  Therefore, $y = k$ is a tangent to the curve when $k = 2$.
15 a $A(4, 1), B(9, 2)$ **b** Gradient of $l = \frac{2 - 1}{9 - 4} = \frac{1}{5}$
  c $x - 5y + 1 = 0$
16 $y + \sqrt{3}x - \sqrt{3} = 0$
17 a $A(0, -3), B\left(\frac{3}{4}, 0\right)$
  b Gradient of $l_1 = 4$
  Equation of l_2 and l_3: $y = 4x + c$
  $t - 4 = \frac{4(t - 1)}{t} + c \Rightarrow t^2 - 4t = 4t - 4 + ct$
  $\Rightarrow t^2 - (8 + c)t + 4 = 0$
  Tangent when discriminant $= 0$
  $(-(8+c))^2 - 4 \times 1 \times 4 = 0$
  $64 + 16c + c^2 - 16 = 0$
  $c^2 + 16c + 48 = 0$
  $(c + 4)(c + 12) = 0 \Rightarrow c = -4$ or $c = -12$
  So, two possible equations for l_2 and l_3 are
  $y = 4x - 4$ and $y = 4x - 12$
  c $\left(\frac{1}{2}, -2\right), \left(\frac{3}{2}, -6\right)$

Challenge
$(1, 1), (e, 2)$

Exercise 8E
1 a 83.3 seconds **b** 267 m
  c $t = \frac{x}{0.9} \Rightarrow y = -3.2 \frac{x}{0.9} \Rightarrow y = -\frac{32}{9}x$
  which is in the form, $y = mx + c$ and is therefore a straight line.
  d 3.32 ms^{-1}
2 a 3000 m
  b Initial point is when $t = 0$. For $t \geq 330$, y is negative ie. the plane is underground or below sea level.
  c 26 400 m (3 s.f.)
3 a 35.3 m
  b Between 1.75 and 1.88 seconds (3 s.f.)
  c 30.3 (3 s.f.).
4 a $\frac{100}{49}$ seconds **b** $\frac{200}{49}$ m

Answers

c $t = \frac{x}{2} \Rightarrow y = -4.9\left(\frac{x}{2}\right)^2 + 10\left(\frac{x}{2}\right) = -\frac{49}{40}x^2 + 5x$

Therefore, the dolphin's path is a quadratic curve

d $\frac{250}{49}$ m

5 a $\sin t = \frac{x}{12}$, $\cos t = \frac{y-12}{-12}$

$\left(\frac{x}{12}\right)^2 + \left(\frac{y-12}{-12}\right)^2 = 1 \Rightarrow x^2 + (y-12)^2 = 144$

Therefore, motion is a circle with centre (0,12) and radius $\sqrt{144} = 12$.

b 24 m **c** 2π minutes, 12 m/min

6 a 4.86 (3 s.f.) **b** Depth = 2

7 a $\frac{\sqrt{13}}{2}$ **b** (0, 2), (0, 4)

c $2t = 2\left(\frac{t^2 - 3t + 2}{t}\right) + 10$

$2t^2 = 2t^2 - 6t + 4 + 10t \Rightarrow 0 = 4t + 4 \Rightarrow t = -1$
Since, $t > 0$, the paths do not intersect.

8 a 10 m

b $k = 1.89$ (3 s.f.). Therefore, time taken is 1.89 seconds.

c 34.1 m (3 s.f.)

d $t = \frac{x}{18}$

$y = -4.9\left(\frac{x}{18}\right)^2 + 4\left(\frac{x}{18}\right) + 10 = -\frac{49}{3240}x^2 + \frac{2}{9}x + 10$

Therefore, the ski jumper's path is a quadratic equation. Maximum height = 10.8 m (3 s.f.)

9 a $t = \frac{\pi}{4}$ **b** (50, 20)

c (77.87, 18.19)

$\frac{\pi}{4} < 1 < \frac{\pi}{2}$, which is when $\sin 2t$ is decreasing, hence when y is decreasing, hence the cyclist is descending.

10 a (4.35, 4.33) (3 s.f.) **b** 25 m

c 3.47 m (3 s.f.) **d** −7.21

Mixed exercise 8

1 a $A(4, 0)$, $B(0, 3)$ **b** $C\left(2\sqrt{3}, \frac{3}{2}\right)$ **c** $\left(\frac{x}{4}\right)^2 + \left(\frac{y}{3}\right)^2 = 1$

2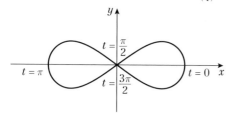

3 a $y = \ln(2\sqrt{x-1}) - \frac{1}{2}$, $x > e^3 + 1$

b $y > 1 + \ln 2$

4 $y = -\ln(4) - 2\ln(x)$, $0 < x < \frac{1}{2}$, $y > 0$

5 a $y = 1 - 2x^2$ **b** $\frac{\sqrt{2}}{2}, -\frac{\sqrt{2}}{2}$

6 $t = \frac{1}{x} - 1$

Sub into $y = \frac{1}{(1+t)(1-t)}$

$y = \frac{1}{\left(1 + \frac{1}{x} - 1\right)\left(1 - \frac{1}{x} + 1\right)} = \frac{1}{\left(\frac{1}{x}\right)\left(2 - \frac{1}{x}\right)}$

$y = \frac{x^2}{x^2\left(\frac{1}{x}\right)\left(2 - \frac{1}{x}\right)} = \frac{x^2}{2x - 1}$

7 a $(x + 3)^2 + (y - 5)^2 = 16$ **b**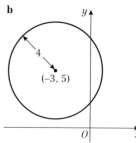

c $(0, 5 + \sqrt{7})$, $(0, 5 - \sqrt{7})$

8 a $x(1 + t) = 2 - 3t \Rightarrow xt + 3t = 2 - x \Rightarrow t(x + 3) = 2 - x$

$\Rightarrow t = \frac{2 - x}{x + 3}$

Sub into $y = \frac{3 + 2t}{1 + t}$

$y = \frac{3 + 2\left(\frac{2-x}{x+3}\right)}{1 + \left(\frac{2-x}{x+3}\right)} = \frac{3(x+3) + 2(2-x)}{x + 3 + 2 - x} = \frac{3x + 9 + 4 - 2x}{5}$

$= \frac{x + 13}{5} \Rightarrow y = \frac{1}{5}x + \frac{13}{5}$

This is in the form $y = mx + c$, therefore the curve C is a straight line.

b $\frac{4\sqrt{26}}{5}$

9 a $y = 2\sqrt{x + 2}$

b Domain: $-2 \leq x \leq 2$, Range: $0 \leq y \leq 4$

c

10 a $\cos t = \frac{x}{2}$, $\sin t = \frac{y + 5}{2}$

$\left(\frac{x}{2}\right)^2 + \left(\frac{y+5}{2}\right)^2 = 1 \Rightarrow x^2 + (y+5)^2 = 4$

Since $0 \leq t \leq \pi$, the curve C forms half of a circle.

b **c** 2π

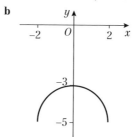

11 a $y = x^3 + 4x^2 + 4x$ **b**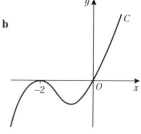

12 $4 - t^2 = 4(t - 3) + 20 \Rightarrow 0 = t^2 + 4t + 4$
Discriminant $= 4^2 - 4 \times 1 \times 4 = 16 - 16 = 0$
So, the line and the curve only intersect once.
Therefore, $y = 4x + 20$ is a tangent to the curve.

13 a $(5, e^5 - 1)$ **b** $k > -1$

14 a $A(0, -\frac{1}{2}), B(1, 0)$ **b** $x - 2y - 1 = 0$

15 $x + y \ln 2 - \ln 2 = 0$

16 a $t = \frac{x}{80}$, sub into $y = 3000 - 30t$
$y = 3000 - 30\left(\frac{x}{80}\right) \Rightarrow y = 3000 - \frac{3}{8}x$
This is in the form $y = mx + c$, therefore the plane's descent is a straight line.
  b $k = 99$ **c** $8458.56\,\text{m}$

17 a $1022\,\text{m}$
  b $1000 = 50\sqrt{2}\,t \Rightarrow t = 10\sqrt{2}$
Sub into $y = 1.5 - 4.9t^2 + 50\sqrt{2}\,t$
$y = 1.5 - 4.9(10\sqrt{2})^2 + 50\sqrt{2}(10\sqrt{2})$
$y = 21.5\,\text{m}$
$21.5 > 10$, therefore the arrow will be too high
  c $11.8\,\text{m}$ (3 s.f.)

18 a $976\,\text{m}$, 2 hours **b** $600\,\text{m}$

19 a $10\,\text{m}$ **b** $80\,\text{m}$

20 a $10\,\text{m}$ **b** 1 second **c** $0.9\,\text{m}$

Challenge
a $k = \frac{3}{2}$ **b** $\left(4, \frac{5}{2}\right)$

Review exercise 2

1 x-axis: $\left(-\frac{7\pi}{4}, 0\right), \left(-\frac{3\pi}{4}, 0\right), \left(\frac{\pi}{4}, 0\right), \left(\frac{5\pi}{4}, 0\right)$
y-axis: $\left(0, \frac{1}{\sqrt{2}}\right)$

2 a

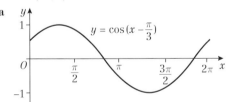

$y = \cos\left(x - \frac{\pi}{3}\right)$

  b y-axis at $(0, 0.5)$. x-axis at $\left(\frac{5\pi}{6}, 0\right)$ and $\left(\frac{11\pi}{6}, 0\right)$
  c $x = 2.89, x = 5.49$

3 a 1.287 radians **b** $6.44\,\text{cm}$

4 $12 + 2\pi\,\text{cm}$

5 a $\frac{1}{2}(r + 10)^2\theta - \frac{1}{2}r^2\theta = 40 \Rightarrow 20r\theta + 100\theta = 80$
$\Rightarrow r\theta + 5\theta = 4 \Rightarrow r = \frac{4}{\theta} - 5$
  b $28\,\text{cm}$

6 a $6\,\text{cm}$ **b** $6.7\,\text{cm}^2$

7 a $119.7\,\text{cm}^2$ **b** $40.3\,\text{cm}$

8 Split each half of the rectangle as shown.
Area $S = \frac{\pi}{12}r^2$
Area $T = \frac{\sqrt{3}}{8}r^2$
$\Rightarrow$ Area $R = \left(\frac{1}{2} - \frac{\sqrt{3}}{8} - \frac{\pi}{12}\right)r^2$

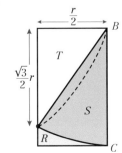

$U = \left(r^2 - \frac{\pi}{4}r^2\right) - 2R$
$= \left(1 - \frac{\pi}{4} - 1 + \frac{\sqrt{3}}{4} + \frac{\pi}{6}\right)r^2$
$= r^2\left(\frac{\sqrt{3}}{4} - \frac{\pi}{12}\right)$

∴ Shaded area
$= \frac{r^2}{6}(3\sqrt{3} - \pi)$

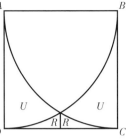

9 a $3\sin^2 x + 7\cos x + 3 = 3(1 - \cos^2 x) + 7\cos x + 3$
$= -3\cos^2 x + 7\cos x + 6 = 0$
Therefore $0 = 3\cos^2 x - 7\cos x - 6$
  b $x = 2.30, 3.98$

10 a For small values of θ:
$\sin 4\theta \approx 4\theta, \cos 4\theta \approx 1 - \frac{1}{2}(4\theta)^2, \tan 3\theta \approx 3\theta$
$\sin 4\theta - \cos 4\theta + \tan 3\theta \approx 4\theta - \left(1 - \frac{(4\theta)^2}{2}\right) + 3\theta$
$= 8\theta^2 + 7\theta - 1$
  b -1

11 a

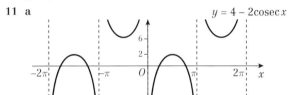

$y = 4 - 2\cosec x$

  b $2 < k < 6$

12 a $\frac{\pi}{3}$ **b** $k = 2$ **c** $-\frac{11\pi}{12}, -\frac{5\pi}{12}$

13 a $\frac{\cos x}{1 - \sin x} + \frac{1 - \sin x}{\cos x} = \frac{\cos^2 x + (1 - \sin x)^2}{\cos x(1 - \sin x)}$
$= \frac{\cos^2 x + 1 - 2\sin x + \sin^2 x}{\cos x(1 - \sin x)} = \frac{2 - 2\sin x}{\cos x(1 - \sin x)}$
$= \frac{2}{\cos x} = 2\sec x$

  b $x = \frac{3\pi}{4}, \frac{5\pi}{4}, \frac{11\pi}{4}, \frac{13\pi}{4}$

14 a $\frac{\sin\theta}{\cos\theta} + \frac{\cos\theta}{\sin\theta} = \frac{\sin^2\theta + \cos^2\theta}{\cos\theta\sin\theta}$
$= \frac{1}{\frac{1}{2}\sin 2\theta} = \frac{2}{\sin 2\theta} = 2\cosec 2\theta$

  b

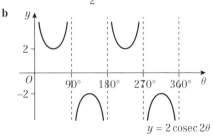

$y = 2\cosec 2\theta$

  c $20.9°, 69.1°, 200.9°, 249.1°$

15 a Note the angle $BDC = \theta$
$\cos\theta = \frac{BC}{10} \Rightarrow BC = 10\cos\theta$
$\sin\theta = \frac{BC}{BD} \Rightarrow BD = 10\cot\theta$

  b $10\cot\theta = \frac{10}{\sqrt{3}} \Rightarrow \cot\theta = \frac{1}{\sqrt{3}}, \theta = \frac{\pi}{3}$
$DC = 10\cos\theta\cot\theta = 10\left(\frac{1}{2}\right)\left(\frac{1}{\sqrt{3}}\right) = \frac{5}{\sqrt{3}}$

Answers

16 a $\sin^2\theta + \cos^2\theta = 1$
$\dfrac{\sin^2\theta}{\cos^2\theta} + \dfrac{\cos^2\theta}{\cos^2\theta} = \dfrac{1}{\cos^2\theta} \Rightarrow \tan^2\theta + 1 = \sec^2\theta$
b $0.0°, 131.8°, 228.2°$

17 a $ab = 2, a = \dfrac{2}{b}$

b $\dfrac{4-b^2}{a^2-1} = \dfrac{4-b^2}{\dfrac{4}{b^2}-1} = \dfrac{4-b^2}{\dfrac{4-b^2}{b^2}} = b^2$

18 a $\dfrac{\pi}{2} - y = \arccos x$ **b** $\dfrac{\pi}{2}$

19 a $\arccos\dfrac{1}{x} = p \Rightarrow \cos p = \dfrac{1}{x}$

Use Pythagorean Theorem to show that opposite side of right triangle is $\sqrt{x^2 - 1}$

$\sin p = \dfrac{\sqrt{x^2-1}}{x} \Rightarrow p = \arcsin\dfrac{\sqrt{x^2-1}}{x}$

b Possible answer: cannot take the square root of a negative number and for $0 \leqslant x < 1$, $x^2 - 1$ is negative.

20 a $\left(-1, \dfrac{3\pi}{2}\right)$

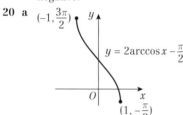

b $\left(\dfrac{1}{\sqrt{2}}, 0\right)$

21 $\tan\left(x + \dfrac{\pi}{6}\right) = \dfrac{1}{6} \Rightarrow \dfrac{\tan x + \dfrac{\sqrt{3}}{3}}{1 - \dfrac{\sqrt{3}}{3}\tan x} = \dfrac{1}{6}$

$6\tan x + 2\sqrt{3} = 1 - \dfrac{\sqrt{3}}{3}\tan x$

$\left(\dfrac{18+\sqrt{3}}{3}\right)\tan x = 1 - 2\sqrt{3}$

$\tan x = \dfrac{3 - 6\sqrt{3}}{18 + \sqrt{3}} \times \dfrac{18 - \sqrt{3}}{18 - \sqrt{3}} = \dfrac{72 - 111\sqrt{3}}{321}$

22 a $\sin(x+30°) = 2\sin(x-60°)$
$\sin x \cos 30° + \cos x \sin 30°$
$= 2(\sin x \cos 60° - \cos x \sin 60°)$
$\dfrac{\sqrt{3}}{2}\sin x + \dfrac{1}{2}\cos x = 2\left(\dfrac{1}{2}\sin x - \dfrac{\sqrt{3}}{2}\cos x\right)$
$\sqrt{3}\sin x + \cos x = 2\sin x - 2\sqrt{3}\cos x$
$(-2+\sqrt{3})\sin x = (-1-2\sqrt{3})\cos x$
$\dfrac{\sin x}{\cos x} = \dfrac{-1-2\sqrt{3}}{-2+\sqrt{3}} = \dfrac{-1-2\sqrt{3}}{-2+\sqrt{3}} \times \dfrac{-2-\sqrt{3}}{-2-\sqrt{3}}$
$= \dfrac{2 + 4\sqrt{3} + \sqrt{3} + 6}{4 + 2\sqrt{3} - 2\sqrt{3} - 3} = 8 + 5\sqrt{3}$

b $8 - 5\sqrt{3}$

23 a $\sin 165° = \sin(120° + 45°)$
$= \sin 120° \cos 45° + \cos 120° \sin 45°$
$= \dfrac{\sqrt{3}}{2} \times \dfrac{1}{\sqrt{2}} + \dfrac{-1}{2} \times \dfrac{1}{\sqrt{2}} = \dfrac{\sqrt{3}-1}{2\sqrt{2}} = \dfrac{\sqrt{6}-\sqrt{2}}{4}$

b $\text{cosec } 165° = \dfrac{1}{\sin 165°}$
$= \dfrac{4}{(\sqrt{6}-\sqrt{2})} \times \dfrac{(\sqrt{6}+\sqrt{2})}{(\sqrt{6}+\sqrt{2})} = \dfrac{4(\sqrt{6}+\sqrt{2})}{6-2} = \sqrt{6}+\sqrt{2}$

24 a $\cos A = \dfrac{3}{4} \Rightarrow \sin A = \dfrac{-\sqrt{7}}{4}$

$\sin 2A = 2\sin A \cos A = 2\left(\dfrac{-\sqrt{7}}{4}\right)\left(\dfrac{3}{4}\right) = \dfrac{-3\sqrt{7}}{8}$

b $\cos 2A = 2\cos^2 A - 1 = \dfrac{1}{8}$

$\tan 2A = \dfrac{\sin 2A}{\cos 2A} = \dfrac{\left(\dfrac{-3\sqrt{7}}{8}\right)}{\left(\dfrac{1}{8}\right)} = -3\sqrt{7}$

25 a $-180°, 0°, 30°, 150°, 180°$
b $-148.3°, -58.3°, 31.7°, 121.7°$ (1 d.p.)

26 a $3\sin x + 2\cos x = \sqrt{13}\sin(x + 0.588...)$
b 169
c $\Rightarrow x = 2.273, 5.976$ (3 d.p.)

27 a $\cot\theta - \tan\theta = \dfrac{\cos\theta}{\sin\theta} - \dfrac{\sin\theta}{\cos\theta} = \dfrac{\cos^2\theta - \sin^2\theta}{\sin\theta\cos\theta}$

$= \dfrac{\cos 2\theta}{\dfrac{1}{2}\sin 2\theta} = \dfrac{2\cos 2\theta}{\sin 2\theta} = 2\cot 2\theta$

b $\theta = -2.95, -1.38, 0.190, 1.76$ (3 s.f.)

28 a $\cos 3\theta = \cos(2\theta + \theta) = \cos 2\theta \cos\theta - \sin 2\theta \sin\theta$
$= (\cos^2\theta - \sin^2\theta)\cos\theta - (2\sin\theta\cos\theta)\sin\theta$
$= \cos^3\theta - 3\sin^2\theta\cos\theta$
$= \cos^3\theta - 3(1 - \cos^2\theta)\cos\theta$
$= 4\cos^3\theta - 3\cos\theta$

b $\sec 3\theta = \dfrac{-27}{19\sqrt{2}} = \dfrac{-27\sqrt{2}}{38}$

29 $\sin^4\theta = (\sin^2\theta)(\sin^2\theta)$
$\cos 2\theta = 1 - 2\sin^2\theta \Rightarrow \sin^2\theta = \dfrac{1 - \cos 2\theta}{2}$
$\sin^4\theta = \left(\dfrac{1-\cos 2\theta}{2}\right)\left(\dfrac{1-\cos 2\theta}{2}\right)$
$\sin^4\theta = \dfrac{1}{4}(1 - 2\cos 2\theta + \cos^2 2\theta)$
$\sin^4\theta = \dfrac{1}{4}\left(1 - 2\cos 2\theta + \dfrac{1+\cos 4\theta}{2}\right)$
$\sin^4\theta = \dfrac{3}{8} - \dfrac{1}{2}\cos 2\theta + \dfrac{1}{8}\cos 4\theta$

30 a $\sqrt{40}\sin(\theta + 0.32)$
b i $\sqrt{40}$ **ii** $\theta = 1.25$
c Minimum of $2.68°C$, occurs 16.77 hours after 9 am $\approx$ 1:46 am
d $t = 2.25, t = 7.29$. So 11:15 am and 4:17 pm

31 a $x \neq 1, y \geqslant -1.25$

b $t = \dfrac{-4}{x-1} = \dfrac{4}{1-x}$

$$y = \left(\frac{4}{1-x}\right)^2 - 3\left(\frac{4}{1-x}\right) + 1$$

$$y = \frac{16}{(1-x)^2} - \frac{12(1-x)}{(1-x)^2} + \frac{(1-x)^2}{(1-x)^2}$$

$$y = \frac{16 - 12 + 12x + 1 - 2x + x^2}{(1-x)^2}$$

$$y = \frac{x^2 + 10x + 5}{(1-x)^2} \Rightarrow a = 1, b = 10, c = 5$$

32 a $t = e^x - 2$

$$y = \frac{3t}{t+3} = \frac{3e^x - 6}{e^x + 1}$$

$t > 4 \Rightarrow e^x - 2 > 4 \Rightarrow e^x > 6 \Rightarrow x > \ln 6$

b $t = 4 \Rightarrow y = \frac{12}{7}, x \to \infty, y \to 3, \frac{12}{7} < y < 3$

33 $x = \frac{1}{1+t} \Rightarrow t = \frac{1-x}{x}$

$$y = \frac{1}{1 - \frac{1-x}{x}} = \frac{x}{x - (1-x)} = \frac{x}{2x-1}$$

34 a $y = \cos 3t = \cos(2t + t) = \cos 2t \cos t - \sin 2t \sin t$
$= (2\cos^2 t - 1)\cos t - 2\sin^2 t \cos t$
$= 2\cos^3 t - \cos t - 2(1 - \cos^2 t)\cos t$
$= 4\cos^3 t - 3\cos t$

$x = 2\cos t \Rightarrow \cos t = \frac{x}{2}$

$$y = 4\left(\frac{x}{2}\right)^3 - 3\left(\frac{x}{2}\right) = \frac{x}{2}(x^2 - 3)$$

b $0 \leq x \leq 2, -1 \leq y \leq 1$

35 a $y = \sin\left(t + \frac{\pi}{6}\right) = \sin t \cos\frac{\pi}{6} + \cos t \sin\frac{\pi}{6}$

$= \frac{\sqrt{3}}{2}\sin t + \frac{1}{2}\cos t$

$= \frac{\sqrt{3}}{2}\sin t + \frac{1}{2}\sqrt{1 - \sin^2 t}$

$= \frac{\sqrt{3}}{2}x + \frac{1}{2}\sqrt{1 - x^2}$

$-1 \leq \sin t \leq 1 \Rightarrow -1 \leq x \leq 1$

b $A = (-0.5, 0), B = (0, 0.5)$

36 a $y = 2\left(\frac{x}{3}\right)^2 - 1, -3 \leq x \leq 3$

b
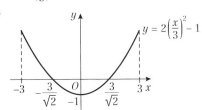

37 $y = 3x + c \Rightarrow 8t(2t - 1) = 3(4t) + c \Rightarrow 16t^2 - 20t - c = 0$
$(-20)^2 - 4(16)(-c) < 0$ so $64c < -400 \Rightarrow c < -\frac{25}{4}$

38 a $\left(-\frac{3\sqrt{3}}{2}, 0\right)$ and $\left(\frac{3\sqrt{3}}{2}, 0\right)$

b $3\sin 2t = 1.5 \Rightarrow \sin 2t = \frac{1}{2}$

$2t = \frac{\pi}{6}, \frac{5\pi}{6}, \frac{13\pi}{6}, \frac{17\pi}{6}, \ldots \Rightarrow t = \frac{\pi}{12}, \frac{5\pi}{12}, \frac{13\pi}{12}, \frac{17\pi}{12}, \ldots$

$t = \frac{13\pi}{12}, \frac{17\pi}{12}$

39 a $-4.9t^2 + 25t + 50 = 0$

$$t = \frac{-25 \pm \sqrt{25^2 - 4(-4.9)(50)}}{2(-4.9)}$$

$t \neq -1.54, t = 6.64s \Rightarrow k = 6.64$

b $t = \frac{x}{25\sqrt{3}}$

$$y = 25\left(\frac{x}{25\sqrt{3}}\right) - 4.9\left(\frac{x}{25\sqrt{3}}\right)^2 + 50$$

$$= \frac{x}{\sqrt{3}} - \frac{49}{18\,750}x^2 + 50$$

$t = 6.64$ $x = 25\sqrt{3}t = 25\sqrt{3}(6.64) = 287.5$
Domain of f(x) is $0 \leq x \leq 287.5$

Challenge

1 $\frac{\pi - 2}{2 + 3\pi} : 1$

2 a $\sin x$ **b** $\cos x$ **c** $\text{cosec}\, x$
  d $\cot x$ **e** $\tan x$ **f** $\sec x$

3 a $\sin^2 t + \cos^2 t = 1$

$$\left(\frac{x-3}{4}\right)^2 + \left(\frac{y+1}{4}\right)^2 = 1 \Rightarrow (x-3)^2 + (y+1)^2 = 16$$

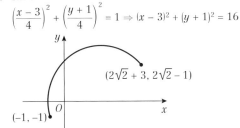

b $\frac{3}{8}(2\pi \times 4) = 3\pi$

CHAPTER 9

Prior knowledge 9

1 a $6x - 5$ **b** $-\frac{2}{x^2} - \frac{1}{2\sqrt{x}}$ **c** $8x - 16x^3$

2 $y = -6x + 17$ **3** $(0, 2), \left(0, \frac{179}{27}\right), (11.1, 0)$
4 $0.588, 3.73$

Exercise 9A

1 a $f'(x) = \lim_{h \to 0}\left(\frac{f(x+h) - f(x)}{h}\right)$

$= \lim_{h \to 0}\left(\frac{\cos(x+h) - \cos x}{h}\right)$

$= \lim_{h \to 0}\left(\frac{\cos x \cos h - \sin x \sin h - \cos x}{h}\right)$

$= \lim_{h \to 0}\left(\frac{\cos x(\cos h - 1) - \sin x \sin h}{h}\right)$

$= \lim_{h \to 0}\left(\left(\frac{\cos h - 1}{h}\right)\cos x - \left(\frac{\sin h}{h}\right)\sin x\right)$

b As $h \to 0, \cos h \to 1$, so $\left(\frac{\cos h - 1}{h}\right) \to 0$

and $\left(\frac{\sin h}{h}\right) \to 1$

So $f'(x) = \lim_{h \to 0}\left(\left(\frac{\cos h - 1}{h}\right)\cos x - \left(\frac{\sin h}{h}\right)\sin x\right)$

$= 0 \cos x - \sin x = -\sin x$

2 a $-2\sin x$ **b** $\cos\left(\frac{1}{2}x\right)$
  c $8\cos 8x$ **d** $4\cos\left(\frac{2}{3}x\right)$

3 a $-2\sin x$ **b** $-5\sin\left(\frac{5}{6}x\right)$
  c $-2\sin\left(\frac{1}{2}x\right)$ **d** $-6\sin 2x$

Answers

4   **a** $2\cos 2x - 3\sin 3x$ **b** $-8\sin 4x + 4\sin x - 14\sin 7x$
      c $2x - 12\sin 3x$ **d** $-\dfrac{1}{x^2} + 10\cos 5x$

5   (0.41, −0.532), (1.68, 2.63), (2.50, 1.56)
6   8   7   (0.554, 2.24), (2.12, −2.24)
8   $y = -5x + 5\pi - 1$
9   $\dfrac{dy}{dx} = 4x - \cos x$

At $x = \pi$, $y = 2\pi^2$, $\dfrac{dy}{dx} = 4\pi - \cos \pi = 4\pi + 1$

Gradient of normal $= -\dfrac{1}{4\pi + 1}$

Equation of normal:

$y - 2\pi^2 = -\dfrac{1}{4\pi + 1}(x - \pi)$

$(4\pi + 1)y - 2\pi^2(4\pi + 1) = -x + \pi$
$x + (4\pi + 1)y - 8\pi^3 - 2\pi^2 - \pi = 0$
$x + (4\pi + 1)y - \pi(8\pi^2 + 2\pi + 1) = 0$

10   Let $f(x) = \sin x$

$f'(x) = \lim_{h \to 0} \dfrac{f(x + h) - f(x)}{h} = \lim_{h \to 0} \dfrac{\sin(x + h) - \sin x}{h}$

$= \lim_{h \to 0} \dfrac{\sin x \cos h + \cos x \sin h - \sin x}{h}$

$= \lim_{h \to 0} \left[\left(\dfrac{\cos h - 1}{h}\right)\sin x + \left(\dfrac{\sin h}{h}\right)\cos x\right]$

Since $\dfrac{\cos h - 1}{h} \to 0$ and $\dfrac{\sin h}{h} \to 1$ the expression inside the limit $\to (0 \times \sin x + 1 \times \cos x)$

So $\lim_{h \to 0} \dfrac{\sin(x + h) - \sin x}{h} = \cos x$

Hence the derivative of $\sin x$ is $\cos x$.

Challenge
Let $f(x) = \sin(kx)$

$f'(x) = \lim_{h \to 0}\left(\dfrac{f(x+h) - f(x)}{h}\right) = \lim_{h \to 0}\left(\dfrac{\sin(kx + kh) - \sin kx}{h}\right)$

$= \lim_{h \to 0}\left(\dfrac{\sin kx \cos kh + \cos kx \sin kh - \sin kx}{h}\right)$

$= \lim_{h \to 0}\left(\left(\dfrac{\cos kh - 1}{h}\right)\sin kx + \left(\dfrac{\sin kh}{h}\right)\cos kx\right)$

As $h \to 0$, $\left(\dfrac{\sin kh}{h}\right) \to k$ and $\left(\dfrac{\cos kh - 1}{h}\right) \to 0$ as given,

So $f'(x) = 0 \sin kx + k \cos kx = k \cos kx$

Exercise 9B

1   **a** $28e^{7x}$ **b** $3^x \ln 3$ **c** $\left(\dfrac{1}{2}\right)^x \ln \dfrac{1}{2}$ **d** $\dfrac{1}{x}$

    e $4\left(\dfrac{1}{3}\right)^x \ln \dfrac{1}{3}$ **f** $\dfrac{3}{x}$ **g** $3e^{3x} + 3e^{-3x}$ **h** $-e^{-x} + e^x$

2   **a** $3^{4x} 4 \ln 3$ **b** $\left(\dfrac{3}{2}\right)^{2x} 2 \ln \dfrac{3}{2}$

    c $2^{4x} 8 \ln 2$ **d** $2^{3x} 3 \ln 2 - 2^{-x} \ln 2$
3   323.95 (2 d.p.)   4 $4y = 15 \ln 2(x-2) + 17$
5   $\dfrac{dy}{dx} = 2e^{2x} - \dfrac{1}{x}$ At $x = 1$, $y = e^2$, $\dfrac{dy}{dx} = 2e^2 - 1$

Equation of tangent: $y - e^2 = (2e^2 - 1)(x - 1)$
$\Rightarrow y = (2e^2 - 1)x - 2e^2 + 1 + e^2 \Rightarrow y = (2e^2 - 1)x - e^2 + 1$

6   −9.07 millicuries/day
7   **a** $P_0 = 37\,000$, $k = 1.01$ (2 d.p.)   **b**   1178
    c   The rate of change of the population in the year 2000

8   The student has treated "ln kx" as if it is "e^{kx}" – they have applied the incorrect standard differential.
Correct differential is: $\dfrac{1}{x}$

9   Let $y = a^{kx} \Rightarrow y = e^{\ln a^{kx}} = e^{kx \ln a}$
$\dfrac{dy}{dx} = k \ln a \, e^{kx \ln a} = k \ln a \, e^{\ln a^{kx}} = a^{kx} k \ln a$

10   **a** $2e^{2x} - \dfrac{2}{x}$

     b $2e^{2a} - \dfrac{2}{a} = 2 \Rightarrow 2ae^{2a} - 2 = 2a \Rightarrow a(e^{2a} - 1) = 1$

11   **a** $5\sin(3 \times 0) + 2\cos(3 \times 0) = 0 + 2 = 2 = y$
When $x = 0$, $y = 2$, therefore (0, 2) lies on C.
     b $y = -\dfrac{1}{15}x + 2$

12   $y = -\dfrac{1}{648 \ln 3}x + \dfrac{1}{648 \ln 3} + 162$

Challenge
$y = 3x - 2\ln 2 + 2$

Exercise 9C

1   **a** $8(1+2x)^3$ **b** $20x(3-2x^2)^{-6}$
    c $2(3 + 4x)^{-\frac{1}{2}}$ **d** $7(6 + 2x)(6x + x^2)^6$
    e $-\dfrac{2}{(3 + 2x)^2}$ **f** $-\dfrac{1}{2\sqrt{7 - x}}$
    g $128(2 + 8x)^3$ **h** $18(8 - x)^{-7}$

2   **a** $-\sin x \, e^{\cos x}$ **b** $-2\sin(2x - 1)$ **c** $\dfrac{1}{2x\sqrt{\ln x}}$

    d $5(\cos x - \sin x)(\sin x + \cos x)^4$
    e $(6x - 2)\cos(3x^2 - 2x + 1)$
    f $\cot x$ **g** $-8\sin 4x \, e^{\cos 4x}$ **h** $-2e^{2x}\sin(e^{2x} + 3)$

3   −1   4 $y = -54x + 81$ 5 $12e^{-3}$

6   **a** $\dfrac{1}{2y + 1}$ **b** $\dfrac{1}{e^y + 4}$ **c** $\dfrac{1}{2}\sec 2y$ **d** $\dfrac{4y}{1 + 3y^3}$

7   $\dfrac{1}{10}$ 8 $\dfrac{16}{3}$

9   **a** $e^y = \dfrac{dx}{dy}$
    b $y = \ln x$, $e^y = x$
Differentiate with respect to y using part **a**
$e^y = \dfrac{dx}{dy} \Rightarrow \dfrac{1}{e^y} = \dfrac{dy}{dx}$
Since $x = e^y$, $\dfrac{dy}{dx} = \dfrac{1}{x}$

10   **a** $4\cos 2\left(\dfrac{\pi}{6}\right) = 4\left(\dfrac{1}{2}\right) = 2$

When $y = \dfrac{\pi}{6}$, $x = 2$, therefore $\left(2, \dfrac{\pi}{6}\right)$ lies on C.

     b $\dfrac{dx}{dy} = -8\sin 2y$

At $Q\left(2, \dfrac{\pi}{6}\right)$: $\dfrac{dx}{dy} = -8\sin 2\left(\dfrac{\pi}{6}\right) = -8\left(\dfrac{\sqrt{3}}{2}\right) = -4\sqrt{3}$

So, $\dfrac{dy}{dx} = -\dfrac{1}{4\sqrt{3}}$

     c $4\sqrt{3}\,x - y - 8\sqrt{3} + \dfrac{\pi}{6} = 0$

11   **a** $6\sin 3x \cos 3x$ **b** $2(x + 1)e^{(x+1)^2}$ **c** $-2\tan x$

     d $\dfrac{2\sin 2x}{(3 + \cos 2x)^2}$ **e** $-\dfrac{1}{x^2}\cos\left(\dfrac{1}{x}\right)$

12   $3125x - 100y - 9371 = 0$ 13 $9\ln 3$

Challenge

a $\dfrac{\cos\sqrt{x}}{4\sqrt{x}\sin\sqrt{x}}$ **b** $9e^{\sin^3(3x+4)}\cos(3x + 4)\sin^2(3x + 4)$

Answers

Exercise 9D
1. **a** $(3x+1)^4(18x+1)$ **b** $2(3x^2+1)^2(21x^2+1)$
   c $16x^2(x+3)^3(7x+9)$ **d** $3x(5x-2)(5x-1)^{-2}$
2. **a** $-4(x-3)(2x-1)^4 e^{-2x}$
   b $2\cos 2x \cos 3x - 3\sin 2x \sin 3x$
   c $e^x(\sin x + \cos x)$ **d** $5\cos 5x \ln(\cos x) - \tan x \sin 5x$
3. **a** 52 **b** 13 **c** $\frac{3}{25}$
4. $(2, 0), \left(-\frac{1}{3}, \frac{343}{27}\right)$ 5 $\frac{5\pi^4}{256}$
6. $\sqrt{2\pi}\,(\pi - 4)x + 8y - \pi\sqrt{2}\left(\frac{\pi-2}{2}\right) = 0$
7. $6x(5x-3)^3 + 3x^2[3(5x-3)^2(5)] = 6x(5x-3)^3 + 45x^2(5x-3)^2$
   $= 3x(5x-3)^2(2(5x-3) + 15x) = 3x(5x-3)^2(10x - 6 + 15x)$
   $= 3x(5x-3)^2(25x - 6) \Rightarrow n = 2, A = 3, B = 25, C = -6$
8. **a** $(x+3)(3x+11)e^{3x}$ **b** $85e^6$
9. **a** $(3\sin x + 2\cos x)\ln(3x) + \frac{2\sin x - 3\cos x}{x}$
   b $x^3(7x+4)e^{7x-3}$
10. 21.25

Challenge
a $-e^x \sin x (\sin^2 x - \cos x \sin x - 2\cos^2 x)$
b $-(4x-3)^5(4x-1)^8(256x^2 - 148x + 3)$

Exercise 9E
1. **a** $\frac{5}{(x+1)^2}$ **b** $-\frac{4}{(3x-2)^2}$ **c** $-\frac{5}{(2x+1)^2}$
   d $-\frac{6x}{(2x-1)^3}$ **e** $\frac{15x+18}{(5x+3)^{\frac{3}{2}}}$
2. **a** $\frac{e^{4x}(\sin x + 4\cos x)}{\cos^2 x}$ **b** $\frac{1}{x(x+1)} - \frac{\ln x}{(x+1)^2}$
   c $\frac{e^{-2x}(2x(e^{4x}-1)\ln x - e^{4x} - 1)}{x(\ln x)^2}$
   d $\frac{(e^x+3)^2((e^x+3)\sin x + 3e^x \cos x)}{\cos^2 x}$
   e $\frac{2\sin x \cos x}{\ln x} - \frac{\sin^2 x}{x(\ln x)^2}$
3. $\frac{1}{16}$ 4 $\frac{2}{25}$ 5 $(0.5, 2e^4)$
6. $y = \frac{1}{3}e$ 7 $\frac{6\sqrt{3} - 2\pi \ln\left(\frac{\pi}{9}\right)}{\pi}$
8. **a** $\left(\frac{1}{3}, 0\right)$ **b** $y = -\frac{1}{9}x + \frac{1}{27}$
9. $\frac{x^3(3x \sin 3x + 4\cos 3x)}{\cos^2 3x}$
10. **a** $\frac{(x-2)^2(2e^{2x}) - e^{2x}[2(x-2)]}{(x-2)^4} = \frac{2(x-2)^2 e^{2x} - 2e^{2x}(x-2)}{(x-2)^4}$
    $= \frac{2(x-2)e^{2x} - 2e^{2x}}{(x-2)^3} = \frac{2e^{2x}(x-2-1)}{(x-2)^3} = \frac{2e^{2x}(x-3)}{(x-2)^3}$
    $A = 2, B = 1, C = 3$
    b $y = 4e^2 x - 3e^2$
11. **a** $\frac{2x}{x+5} + \frac{6x}{(x+5)(x+2)} = \frac{2x(x+2)}{(x+5)(x+2)} + \frac{6x}{(x+5)(x+2)}$
    $= \frac{2x(x+2+3)}{(x+5)(x+2)} = \frac{2x(x+5)}{(x+5)(x+2)} = \frac{2x}{(x+2)}$
    b $\frac{4}{25}$
12. **a** $f'(x) = -2e^{x-2}(2\sin 2x - \cos 2x) = 0$
    $2\sin 2x - \cos 2x = 0 \Rightarrow \tan 2x = \frac{1}{2}$
    b $-1.47 < f(x) < 6.26$

Exercise 9F
1. **a** $3\sec^2 3x$ **b** $12\tan^2 x \sec^2 x$ **c** $\sec^2(x-1)$
   d $\frac{1}{2}x^2 \sec^2 \frac{1}{2}x + 2x \tan \frac{1}{2}x + \sec^2\left(x - \frac{1}{2}\right)$
2. **a** $-4\csc^2 4x$ **b** $5\sec 5x \tan 5x$
   c $-4\csc 4x \cot 4x$ **d** $6\sec^2 3x \tan 3x$
   e $\cot 3x - 3x \csc^2 3x$ **f** $\frac{\sec^2 x(2x\tan x - 1)}{x^2}$
   g $-6\csc^3 2x \cot 2x$
   h $-4\cot(2x-1)\csc^2(2x-1)$
3. **a** $\frac{1}{2}(\sec x)^{\frac{1}{2}} \tan x$ **b** $-\frac{1}{2}(\cot x)^{-\frac{1}{2}} \csc^2 x$
   c $-2\csc^2 x \cot x$ **d** $2\tan x \sec^2 x$
   e $3\sec^3 x \tan x$ **f** $-3\cot^2 x \csc^2 x$
4. **a** $2x\sec 3x + 3x^2 \sec 3x \tan 3x$
   b $\frac{2x\sec^2 2x - \tan 2x}{x^2}$ **c** $\frac{2x\tan x - x^2 \sec^2 x}{\tan^2 x}$
   d $e^x \sec 3x (1 + 3\tan 3x)$ **e** $\frac{\tan x - x\sec^2 x \ln x}{x\tan^2 x}$
   f $e^{\tan x} \sec x (\tan x + \sec^2 x)$
5. **a** $\frac{1}{\cos^2 x} - \frac{1}{\sin^2 x}$ **b** 2
   c $24x - 9y + 12\sqrt{3} - 8\pi = 0$
6. $y = \frac{1}{\cos x}, \frac{dy}{dx} = \frac{\cos x \times 0 - 1 \times -\sin x}{\cos^2 x} = \frac{\sin x}{\cos^2 x}$
   $= \sec x \tan x$
7. $y = \frac{1}{\tan x}$
   $\frac{dy}{dx} = \frac{\tan x \times 0 - 1 \times \sec^2 x}{\tan^2 x} = -\frac{\sec^2 x}{\tan^2 x} = -\frac{\frac{1}{\cos^2 x}}{\frac{\sin^2 x}{\cos^2 x}} = -\csc^2 x$
8. $\frac{dx}{dy} = -2\sin 2y, \frac{dx}{dy} = \frac{-1}{2\sin 2y}$
   $\sin^2 2y + \cos^2 2y = 1$
   $\sin 2y = \sqrt{1 - \cos^2 2y} = \sqrt{1 - x^2}$
   $\frac{dx}{dy} = \frac{-1}{2\sqrt{1-x^2}}$
9. **a** $\frac{-1}{5\cot 5y \csc 5y}$ **b** $-\frac{1}{5x\sqrt{x^2-1}}$

Challenge
a Let $y = \arccos x \Rightarrow \cos y = x \Rightarrow \frac{dx}{dy} = -\sin y$
$\frac{dy}{dx} = -\frac{1}{\sin y} = -\frac{1}{\sqrt{1-\cos^2 y}} = -\frac{1}{\sqrt{1-x^2}}$
b Let $y = \arctan x$
Then, $\tan y = x$
$\frac{dx}{dy} = \sec^2 y$
$\frac{dy}{dx} = \frac{1}{\sec^2 y} = \frac{1}{1+\tan^2 y} = \frac{1}{1+x^2}$

Answers

Exercise 9G

1. **a** $\dfrac{2t-3}{2}$ **b** $\dfrac{6t^2}{6t}=t$ **c** $\dfrac{4}{1+6t}$ **d** $\dfrac{15t^3}{2}$
   e $-3t^3$ **f** $t(1-t)$ **g** $\dfrac{2t}{t^2-1}$ **h** $\dfrac{2}{(t^2+2t)e^t}$
   i $-\dfrac{3}{4}\tan 3t$ **j** $4\tan t$ **k** $\text{cosec}\, t$ **l** $\dfrac{2\sin 2t}{2-2\cos 2t}$
   m $\dfrac{1}{te^t}$ **n** $2t^2$ **o** $\dfrac{1}{e^t}$

2. **a** $y=-\dfrac{1}{2}x+\dfrac{3}{2}-\pi$ **b** $2y+5x=57$

3. **a** $x=1$ **b** $y+\sqrt{3}x=\sqrt{3}$

4. $(0, 0)$ and $(-2, -4)$

5. **a** $y=\dfrac{1}{4}x$
   b $\dfrac{dy}{dx}=\dfrac{1}{2}e^{-t}=0 \Rightarrow e^{-t}=0$
   No solution, therefore no stationary points.

6. $y=x+7$

7. **a** $-\dfrac{1}{2}\sec t \, \text{cosec}^3 t$ **b** $8x+\sqrt{3}y-10=0$

8. **a** $\dfrac{\pi}{3}$
   b $\dfrac{dy}{dt}=-4\cot 2t\,\text{cosec}\,2t,\;\dfrac{dx}{dt}=4\cos t$
   $\dfrac{dy}{dx}=\dfrac{-4\cot 2t\,\text{cosec}\,2t}{4\cos t}=\dfrac{-\cot 2t\,\text{cosec}\,2t}{\cos t}$
   At $t=\dfrac{\pi}{3}$, $\dfrac{dy}{dx}=\dfrac{4}{3}$
   Gradient of normal: $-\dfrac{3}{4}$
   Equation of normal:
   $y-\dfrac{4\sqrt{3}}{3}=-\dfrac{3}{4}(x-2\sqrt{3})\Rightarrow 9x+12y-34\sqrt{3}=0$

9. **a** $(30, 101)$ **b** $y=2x+41$
   c $t^2-10t+5=2(t^2+t)+41$
   $t^2-10t+5=2t^2+2t+41$
   $0=t^2+12t+36$
   Discriminant $=12^2-4\times 1\times 36=144-144=0$
   Therefore the curve and the line only intersects once.
   Therefore it does not intersect the curve again.

10. **a** $-2\sqrt{2}\sin t$ **b** $x-\sqrt{6}y-2\sqrt{3}=0$
    c $2\sin t-\sqrt{12}\cos 2t-2\sqrt{3}=0$
    $\sin t-\sqrt{3}\cos 2t-\sqrt{3}=0$
    $2\sqrt{3}\sin^2 t+\sin t-2\sqrt{3}=0$
    $(2\sin t-\sqrt{3})(\sqrt{3}\sin t+2)=0$
    $\sin t=\dfrac{\sqrt{3}}{2}\left(\sin t\neq -\dfrac{2}{\sqrt{3}}\right)\Rightarrow t=\dfrac{\pi}{3}$ or $\dfrac{2\pi}{3}$
    B is when $t=\dfrac{2\pi}{3}$: $\left(2\sin\dfrac{2\pi}{3},\sqrt{2}\cos\dfrac{4\pi}{3}\right)=\left(\sqrt{3},-\dfrac{1}{\sqrt{2}}\right)$
    Same point as A, so l only intersects C once.

11. **a** $-\dfrac{\cos 2t}{\sin t}$ **b** $y=-x+\dfrac{3\sqrt{3}}{4}$
    c $y=-x$ and $y=-x-\dfrac{3\sqrt{3}}{4}$

Exercise 9H

1. Letting $u=y^n$, $\dfrac{du}{dy}=ny^{n-1}$
   $\dfrac{d}{dx}(y^n)=\dfrac{du}{dx}=\dfrac{du}{dy}\times\dfrac{dy}{dx}=ny^{n-1}\dfrac{dy}{dx}$

2. $\dfrac{d}{dx}(xy)=x\dfrac{d}{dx}(y)+\dfrac{d}{dx}(x)y=x\dfrac{dy}{dx}+1\times y=x\dfrac{dy}{dx}+y$

3. **a** $-\dfrac{2x}{3y^2}$ **b** $-\dfrac{x}{5y}$ **c** $\dfrac{-3-x}{5y-4}$
   d $\dfrac{4-6xy}{3x^2+3y^2}$ **e** $\dfrac{3x^2-2y}{6y-2+2x}$ **f** $\dfrac{3x^2-y}{2+x}$
   g $\dfrac{4(x-y)^3-1}{1+4(x-y)^3}$ **h** $\dfrac{e^xy-e^y}{xe^y-e^x}$ **i** $\dfrac{-2\sqrt{xy}-y}{4y\sqrt{xy}+x}$

4. $y=-\dfrac{7}{9}x+\dfrac{23}{9}$

5. $y=2x-2$

6. $(3, 1)$ and $(3, 3)$

7. $3x+2y+1=0$

8. $2-3\ln 3$

9. $\dfrac{1}{4}(4+3\ln 3)$

10. **a** $\dfrac{\cos x}{\sin y}$ **b** $\left(\dfrac{\pi}{2},\dfrac{2\pi}{3}\right)$ and $\left(\dfrac{\pi}{2},-\dfrac{2\pi}{3}\right)$

11. **a** $\dfrac{3+3ye^{-3x}}{e^{-3x}-2y}$
    b At O, $\dfrac{dy}{dx}=\dfrac{3+0}{e^0-0}=3$
    So the tangent is $y-0=3(x-0)$, or $y=3x$.

Challenge

a $6+2y\dfrac{dy}{dx}+2y+2x\dfrac{dy}{dx}=2x\Rightarrow\dfrac{dy}{dx}=\dfrac{x-y-3}{y+x}$
So $\dfrac{dy}{dx}=0\Leftrightarrow x-y=3$
Substitute: $6x+(x-3)^2+2x(x-3)=x^2$
So $2x^2-6x+9=0$
Discriminant $=-36$, so no real solutions to quadratic.
Therefore no points on C s.t. $\dfrac{dy}{dx}=0$.

b $(0, 0)$ and $(3, -3)$

Exercise 9I

1. **a i** $[1, \infty)$ **ii** $(-\infty, 1]$
   b i $(-\infty, 0]\cup\left[\dfrac{3}{2},\infty\right)$ **ii** $\left[0,\dfrac{3}{2}\right]$
   c i $[\pi, 2\pi)$ **ii** $(0, \pi]$
   d i nowhere **ii** $(-\infty, \infty)$
   e i $[\ln 2, \infty)$ **ii** $(-\infty, \ln 2]$
   f i nowhere **ii** $(0, \infty)$

Answers

2 a Let $y = f(x)$. Then $x = \sin y$.
$$\frac{dx}{dy} = \cos y \Rightarrow \frac{dy}{dx} = \frac{1}{\cos y} = \frac{1}{\sqrt{1 - \sin^2 y}}$$
so $f'(x) = \frac{1}{\sqrt{1-x^2}}$

b $f'(x) = \frac{1}{\sqrt{1-x^2}}$, $f''(x) = \frac{x}{(1-x^2)^{\frac{3}{2}}}$
$f''(x) \leq 0 \Rightarrow x \leq 0$, so $f(x)$ concave for $x \in (-1, 0)$

c $f''(x) \geq 0 \Rightarrow x \geq 0$, so $f(x)$ convex for $x \in (0, 1)$

d $(0, 0)$

3 a $\left(\frac{\pi}{6}, -\frac{1}{4}\right), \left(\frac{5\pi}{6}, -\frac{1}{4}\right)$
b $(1, -1)$ **c** $(0, 0)$

4 $f'(x) = 2x + 4x \ln x = 2x(1 + 2 \ln x)$, $f''(x) = 6 + 4 \ln x$
$f''(x) = 0 \Rightarrow 4 \ln x = -6 \Rightarrow \ln x = -\frac{3}{2} \Rightarrow x = e^{-\frac{3}{2}}$
There is one point of inflection where $x = e^{-\frac{3}{2}}$

5 a $(0, 2)$, point of inflection   **b** $\left(-2, \frac{10}{e^2}\right)$

6 a $\left(-1, -\frac{1}{e}\right)$, minimum   **b** $\left(-2, -\frac{2}{e^2}\right)$

c

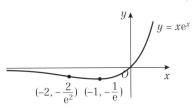

7 A **i** negative **ii** positive
B **i** zero **ii** positive
C **i** positive **ii** negative
D **i** zero **ii** zero

8 $f'(x) = \sec^2 x$, $f''(x) = 2 \sin x \sec^3 x$
$f''(x) = 0 \Leftrightarrow \sin x = 0$ (as $\sec x \neq 0$) $\Leftrightarrow x = 0$
So there is one point of inflection at $(0, \tan 0) = (0, 0)$.

9 a $\frac{dy}{dx} = 15x(3x-1)^4 + (3x-1)^5$
$\frac{d^2y}{dx^2} = 30(3x-1)^4 + 180x(3x-1)^3$

b $\left(\frac{1}{9}, -\frac{32}{2187}\right), \left(\frac{1}{3}, 0\right)$

10 a Although $\frac{d^2y}{dx^2} = 0$, the sign does not change, so there is not a point of inflection when $x = 5$.

b $(5, 0)$; minimum

11 $\frac{dy}{dx} = \frac{2}{3} x \ln x + \frac{1}{3} x - 2$, $\frac{d^2y}{dx^2} = \frac{2}{3} \ln x + 1$
$\frac{d^2y}{dx^2} \geq 0 \Leftrightarrow \frac{2}{3} \ln x \geq -1 \Leftrightarrow x \geq e^{-\frac{3}{2}}$

Challenge

1 A general cubic can be written as $f(x) = ax^3 + bx^2 + cx + d$.
$f''(x) = 6ax + 2b$. $f''(x) = 0 \Leftrightarrow x = -\frac{b}{3a}$
Let $\varepsilon \in \mathbb{R}$, $\varepsilon > 0$:
$f''\left(-\frac{b}{3a} + \varepsilon\right) = 6a\varepsilon > 0$, $f''\left(-\frac{b}{3a} - \varepsilon\right) = -6a\varepsilon < 0$
So the sign of $f''(x)$ changes either side of $x = -\frac{b}{3a}$, and this is a point of inflection.

2 a $f''(x) = 12ax^2 + 6bx + 2c$ is quadratic, so there are at most two values of x at which $f''(x) = 0$.

b $\frac{d^2y}{dx^2} = 12ax^2 + 6bx + 2c$

Discriminant $= 36b^2 - 96ac < 0 \Leftrightarrow 3b^2 < 8ac$
So when $3b^2 < 8ac$, $\frac{d^2y}{dx^2} = 0$ has no solutions.
Therefore C has no points of inflection.

Exercise 9J

1 6π **2** $15e^2$ **3** $-\frac{9}{2}$ **4** $\frac{8}{9\pi}$ **5** $\frac{dP}{dt} = kP$

6 $\frac{dy}{dx} = kxy$; at $(4, 2)$ $\frac{dy}{dx} = \frac{1}{2}$, so $8k = \frac{1}{2}$, $k = \frac{1}{16}$
Therefore $\frac{dy}{dx} = \frac{xy}{16}$

7 $\frac{dV}{dt} =$ rate in $-$ rate out $= 30 - \frac{2}{15}V \Rightarrow 15\frac{dV}{dt} = 450 - 2V$
So $-15\frac{dV}{dt} = 2V - 450$

8 $\frac{dQ}{dt} = -kQ$ **9** $\frac{dx}{dt} = \frac{k}{x^2}$

10 a Circumference, $C, = 2\pi r$, so $\frac{dC}{dt} = 2\pi \times 0.4$
$= 0.8\pi$ cm s^{-1}
Rate of increase of circumference with respect to time.

b 8π cm^2 s^{-1} **c** $\frac{25}{\pi}$ cm

11 a 0.070 cm per second   **b** 20.5 cm^3

12 $\frac{dV}{dt} \propto \sqrt{V} \Rightarrow \frac{dV}{dt} = -k_1\sqrt{V}$, $V \propto h \Rightarrow h = k_2 V$
$\frac{dh}{dt} = \frac{dh}{dV} \times \frac{dV}{dt} = k_2 \times (-k_1\sqrt{V}) = -k_1 k_2 \sqrt{\frac{h}{k_2}}$
$= \frac{-k_1 k_2}{\sqrt{k_2}}\sqrt{h} = -k\sqrt{h}$

13 a $V = \left(\frac{A}{6}\right)^{\frac{3}{2}}$ **b** $\frac{1}{4}\left(\frac{A}{6}\right)^{\frac{1}{2}}$

c $\frac{dV}{dt} = \frac{dV}{dA} \times \frac{dA}{dt} = \frac{1}{4}\left(\frac{A}{6}\right)^{\frac{1}{2}} \times 2 = \frac{1}{2}(V^{\frac{2}{3}})^{\frac{1}{2}} = \frac{1}{2}V^{\frac{1}{3}}$

14 $V = \frac{\pi}{3} r^2 h = \frac{\pi}{3}(h \tan 30°)^2 h = \frac{\pi}{9} h^3$
$\frac{dh}{dt} = \frac{dh}{dV} \times \frac{dV}{dt} = \frac{1}{\frac{dV}{dh}} \times \frac{dV}{dt} = \frac{1}{\frac{\pi}{3}h^2} \times (-6) = -\frac{18}{\pi h^2}$
So $\frac{dh}{dt} \propto \frac{1}{h^2}$

Mixed exercise 9

1 a $\frac{2}{x}$ **b** $2x \sin 3x + 3x^2 \cos 3x$

2 a $2\frac{dy}{dx} = 1 - \sin x \frac{d}{dx}(\cos x) - \frac{d}{dx}(\sin x) \cos x$
$= 1 + \sin^2 x - \cos^2 x = 2 \sin^2 x$
So $\frac{dy}{dx} = \sin^2 x$

b $\left(\frac{\pi}{2}, \frac{\pi}{4}\right), \left(\pi, \frac{\pi}{2}\right), \left(\frac{3\pi}{2}, \frac{3\pi}{4}\right)$

3 a $\frac{x \cos x - \sin x}{x^2}$ **b** $-\frac{2x}{x^2 + 9}$

4 a $k = \sqrt{2}$ **b** $(0, 0), \left(\pm\sqrt{6}, \pm\frac{\sqrt{3}}{4\sqrt{2}}\right)$

5 a $x > 0$ **b** $(\sqrt[3]{256}, 32 \ln 2 + 16)$

6 $\left(\frac{\pi}{6}, \frac{5}{4}\right), \left(\frac{\pi}{2}, 1\right), \left(\frac{5\pi}{6}, \frac{5}{4}\right), \left(\frac{3\pi}{2}, -1\right)$

7 Maximum is when $\frac{dy}{dx} = 0$
$\frac{dy}{dx} = \sqrt{\sin x} + x\left(\cos x \times \frac{1}{2\sqrt{\sin x}}\right) = \frac{2 \sin x + x \cos x}{2\sqrt{\sin x}} = 0$
So $2 \sin x + x \cos x = 0 \Rightarrow 2 \sin x = -x \cos x \Rightarrow 2 \tan x = -x$
$\therefore 2 \tan x + x = 0$

8 a $f'(x) = 0.5 e^{0.5x} - 2x$
b $f'(6) = -1.957... < 0$, $f'(7) = 2.557... > 0$
So there exists $p \in (6, 7)$ such that $f'(p) = 0$.
$\therefore$ there is a stationary point for some $x = p \in (6, 7)$.

Answers

9 a $\left(\dfrac{3\pi}{8}, \dfrac{e^{\frac{3\pi}{4}}}{\sqrt{2}}\right), \left(\dfrac{7\pi}{8}, -\dfrac{e^{\frac{7\pi}{4}}}{\sqrt{2}}\right)$

b $f''(x) = 2e^{2x}(-2\sin 2x + 2\cos 2x) + 4e^{2x}(\cos 2x + \sin 2x)$
$= 4e^{2x}(-\sin 2x + \cos 2x + \cos 2x + \sin 2x)$
$= 8e^{2x}\cos 2x$

c $\left(\dfrac{3\pi}{8}, \dfrac{e^{\frac{3\pi}{4}}}{\sqrt{2}}\right)$ is a maximum; $\left(\dfrac{7\pi}{8}, -\dfrac{e^{\frac{7\pi}{4}}}{\sqrt{2}}\right)$ is a minimum.

d $\left(\dfrac{\pi}{4}, e^{\frac{\pi}{2}}\right), \left(\dfrac{3\pi}{4}, -e^{\frac{3\pi}{2}}\right)$

10 $x + 2y - 8 = 0$

11 a $x = \dfrac{1}{3}$ **b** $y = -\dfrac{1}{2}x + 1\dfrac{1}{2}$

12 a $f'(x) = e^{2x}(2\cos x - \sin x)$
$2\cos x - \sin x = 0 \Rightarrow \tan x = 2$ **b** $y = 2x + 1$

13 a $y + 2y\ln y$ **b** $\dfrac{1}{3e}$

14 a $e^{-x}(-x^3 + 3x^2 + 2x - 2)$
b $f'(0) = -2 \Rightarrow$ gradient of normal $= \dfrac{1}{2}$
Equation of normal is $y = \dfrac{1}{2}x$
$(x^3 - 2x)e^{-x} = \dfrac{1}{2}x \Rightarrow 2x^3 - 4x = xe^x \Rightarrow 2x^2 = e^x + 4$

15 a $1 + x + (1 + 2x)\ln x$
b $1 + x + (1 + 2x)\ln x = 0 \Rightarrow x = e^{-\frac{1+x}{1+2x}}$

16 a $\dfrac{dy}{dx} = -\dfrac{4}{t^3}$ **b** $y = 2x - 8$

17 $3y + x = 33$ **18** $y = \dfrac{2}{3}x + \dfrac{1}{3}$

19 a $\dfrac{dx}{dt} = -2\sin t + 2\cos 2t$; $\dfrac{dy}{dt} = -\sin t - 4\cos 2t$
b $\dfrac{1}{2}$ **c** $y + 2x = \dfrac{5\sqrt{2}}{2}$

20 a $\dfrac{dy}{dt} = 3t^2 - 4, \dfrac{dx}{dt} = 2, \dfrac{dy}{dx} = \dfrac{3t^2 - 4}{2}$
At $t = -1, \dfrac{dy}{dx} = -\dfrac{1}{2}, x = 1, y = 3$.
Equation of l is $2y + x = 7$.
b 2

21 $\dfrac{dV}{dt} = -kV$ **22** $\dfrac{dM}{dt} = -kM$ **23** $\dfrac{dP}{dt} = kP - Q$

24 $\dfrac{dr}{dt} = \dfrac{k}{r}$ **25** $\dfrac{d\theta}{dt} = -k(\theta - \theta_0)$

26 a $\dfrac{\pi}{6}$ **b** $-\dfrac{3}{16}\csc t$

c $\dfrac{dy}{dx} = -\dfrac{3}{8} \Rightarrow$ gradient of normal $= \dfrac{8}{3}$
$y - \dfrac{3}{2} = \dfrac{8}{3}(x - 2) \Rightarrow 6y - 16x + 23 = 0$

d $-\dfrac{123}{64}$

27 a $-\dfrac{1}{2}\sec t$ **b** $4y + 4x = 5a$
c Tangent crosses the x-axis at $x = \dfrac{5}{4}a$, and crosses the y-axis at $y = \dfrac{5}{4}a$.
So area $AOB = \dfrac{1}{2}\left(\dfrac{5}{4}a\right)^2 = \dfrac{25}{32}a^2$, $k = \dfrac{25}{32}$

28 $y + x = 16$ **29** $\dfrac{1}{7}$

30 $\dfrac{y - 2e^{2x}}{2e^{2y} - x}$ **31** $(1, 1)$ and $(-\sqrt[3]{3}, \sqrt[3]{-3})$.

32 a $\dfrac{2x - 2 - y}{1 + x - 2y}$ **b** $\dfrac{4}{3}, -\dfrac{1}{3}$

c $\left(\dfrac{5 + 2\sqrt{13}}{3}, \dfrac{4 + \sqrt{13}}{3}\right)$ and $\left(\dfrac{5 - 2\sqrt{13}}{3}, \dfrac{4 - \sqrt{13}}{3}\right)$

33 $14x + 48y + 48x\dfrac{dy}{dx} - 14y\dfrac{dy}{dx} = 0 \Rightarrow \dfrac{dy}{dx} = \dfrac{-7x - 24y}{24x - 7y}$

So $\dfrac{-7x - 24y}{24x - 7y} = \dfrac{2}{11}$

$\Rightarrow -77x - 264y = 48x - 14y \Rightarrow x + 2y = 0$

34 $\ln y = x\ln x \Rightarrow \dfrac{1}{y}\times\dfrac{dy}{dx} = x\dfrac{d}{dx}(\ln x) + \dfrac{d}{dx}(x)\ln x = 1 + \ln x$

So $\dfrac{dy}{dx} = y(1 + \ln x) = x^x(1 + \ln x)$

35 a $\ln a^x = \ln e^{kx} \Rightarrow x\ln a = kx\ln e = kx \Rightarrow k = \ln a$

b $y = e^{(\ln 2)x} \Rightarrow \dfrac{dy}{dx} = \ln 2\, e^{(\ln 2)x} = 2^x\ln 2$

c $\dfrac{dy}{dx} = 2^2\ln 2 = 4\ln 2 = \ln 2^4 = \ln 16$

36 a $\dfrac{\ln P - \ln P_0}{\ln 1.09}$ **b** 8.04 years   **c** $0.172P_0$

37 a $\left(\dfrac{\pi}{2}, 0\right)$

b $\dfrac{d^2y}{dx^2} = -\csc^2 x \cdot \csc^2 x > 0$ for all x,
hence $-\csc^2 x < 0$, so $\dfrac{d^2y}{dx^2} < 0$ for all x.
Thus C is concave for all values of x.

38 a $40e^{-0.183} = 33.31...$ **b** $-9.76e^{-0.244t}$
c The mass is decreasing

39 a $f'(x) = -\dfrac{2\sin 2x + \cos 2x}{e^x}$
$f'(x) = 0 \Leftrightarrow 2\sin 2x + \cos 2x = 0 \Leftrightarrow \tan 2x = -0.5$
$A(1.34, -0.234), B(2.91, 0.0487)$
b Maximum $(6.91, 2.19)$; minimum $(5.34, 1.06)$ to 3 s.f.
c $0 < x \leqslant 0.322$, $1.89 \leqslant x < \pi$ (decimals to 3 s.f.)

Challenge

a $-\dfrac{4\cos 2t}{5\sin\left(t + \dfrac{\pi}{12}\right)}$

b $\left(\dfrac{5}{2}, 2\right), \left(-\dfrac{5\sqrt{3}}{2}, -2\right), \left(-\dfrac{5}{2}, 2\right), \left(\dfrac{5\sqrt{3}}{2}, -2\right)$

c Cuts the x-axis at:
$(4.83, 0)$ gradient -3.09; $(-1.29, 0)$ gradient 0.828
$(-4.83, 0)$ gradient 3.09; $(1.29, 0)$ gradient -0.828
Cuts the y-axis twice at $(0, 1)$ gradients 0.693 and -0.693

d $(-5, -1)$ and $(5, -1)$

e

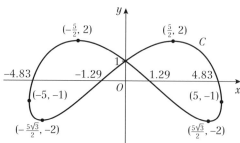

CHAPTER 10

Prior knowledge 10
1 a 3.25   b 11.24
2 a $f'(x) = \dfrac{3}{2\sqrt{x}} + 8x + \dfrac{15}{x^4}$ b $f'(x) = \dfrac{5}{x+2} - 7e^{-x}$
   c $f'(x) = x^2 \cos x + 2x \sin x + 4 \sin x$
3 $u_1 = 2, u_2 = 2.5, u_3 = 2.9$

Exercise 10A
1 a $f(-2) = -1 < 0, f(-1) = 5 > 0$. Sign change implies root.
   b $f(3) = -2.732 < 0, f(4) = 4 > 0$. Sign change implies root.
   c $f(-0.5) = -0.125 < 0, f(-0.2) = 2.992 > 0$. Sign change implies root.
   d $f(1.65) = -0.294 < 0, f(1.75) = 0.195 > 0$. Sign change implies root.
2 a $f(1.8) = 0.408 > 0, f(1.9) = -0.249$. Sign change implies root.
   b $f(1.8635) = 0.00138... > 0, f(1.8645) = -0.00531... < 0$. Sign change implies root.
3 a $h(1.4) = -0.0512... < 0, h(1.5) = 0.0739.... > 0$. Sign change implies root.
   b $h(1.4405) = -0.00055... < 0, h(1.4415) = 0.00069... > 0$. Sign change implies root.
4 a $f(2.2) = 0.020 > 0, f(2.3) = -0.087$. Sign change implies root.
   b $f(2.2185) = 0.00064... > 0, f(2.2195) = -0.00041... < 0$. There is a sign change in the interval $2.2185 < x < 2.2195$, so $\alpha = 2.219$ correct to 3 decimal places.
5 a $f(1.5) = 16.10... > 0, f(1.6) = -32.2... < 0$. Sign change implies root.
   b There is an asymptote in the graph of $y = f(x)$ at $x = \dfrac{\pi}{2} \approx 1.57$. So there is not a root in this interval.
6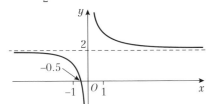
   Alternatively: $\dfrac{1}{x} + 2 = 0 \Rightarrow \dfrac{1}{x} = -2 \Rightarrow x = -\dfrac{1}{2}$
7 a $f(0.2) = -0.4421..., f(0.8) = -0.1471...$
   b There are either no roots or an even numbers of roots in the interval $0.2 < x < 0.8$.
   c $f(0.3) = 0.01238... > 0, f(0.4) = -0.1114... < 0, f(0.5) = -0.2026... < 0, f(0.6) = 0, f(0.7) = 0.2710... > 0$
   d There exists at least one root in the interval $0.2 < x < 0.3, 0.3 < x < 0.4$ and $0.7 < x < 0.8$. Additionally $x = 0.6$ is a root. Therefore there are at least four roots in the interval $0.2 < x < 0.8$.
8 a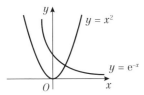
   b One point of intersection, so one root.
   c $f(0.7) = 0.0065... > 0, f(0.71) = -0.0124... < 0$. Sign change implies root.
9 a 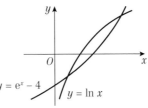   b 2
   c $f(x) = \ln x - e^x + 4. f(1.4) = 0.2812... < 0, f(1.5) = -0.0762... < 0$. Sign change implies root.
10 a $h'(x) = 2\cos 2x + 4e^{4x}. h'(-0.9) = -0.3451... < 0. h'(-0.8) = 0.1046... > 0$. Sign change implies slope changes from decreasing to increasing over interval, which implies turning point.
   b $h'(-0.8235) = -0.003839.... < 0, h'(-0.8225) = 0.00074... > 0$. Sign change implies α lies in the range $-0.8235 < \alpha < -0.8225$, so $\alpha = -0.823$ correct to 3 decimal places.
11 a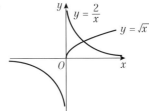
   b 1 point of intersection $\Rightarrow$ 1 root
   c $f(1) = -1, f(2) = 0.414...$ d $p = 3, q = 4$ e $4^{\frac{1}{3}}$
12 a $f(-0.9) = 1.5561 > 0, f(-0.8) = -0.7904 < 0$. There is a change of sign in the interval $[-0.9, -0.8]$, so there is at least one root in this interval.
   b $(1.74, -45.37)$ to 2 d.p.   c $a = 3, b = 9$ and $c = 6$
   d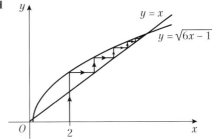

Exercise 10B
1 a i $x^2 - 6x + 2 = 0 \Rightarrow 6x = x^2 + 2 \Rightarrow x = \dfrac{x^2+2}{6}$
        ii $x^2 - 6x + 2 = 0 \Rightarrow x^2 = 6x - 2 \Rightarrow x = \sqrt{6x-2}$
        iii $x^2 - 6x + 2 = 0 \Rightarrow x - 6 + \dfrac{2}{x} = 0 \Rightarrow x = 6 - \dfrac{2}{x}$
   b i $x = 0.354$ ii $x = 5.646$ iii $x = 5.646$
   c $a = 3, b = 7$
2 a i $x^2 - 5x - 3 = 0 \Rightarrow x^2 = 5x + 3 \Rightarrow x = \sqrt{5x+3}$
        ii $x^2 - 5x - 3 = 0 \Rightarrow x^2 - 3 = 5x \Rightarrow x = \dfrac{x^2-3}{5}$
   b i 5.5 (1 d.p.)   ii -0.5 (1 d.p.)
3 a $x^2 - 6x + 1 = 0 \Rightarrow x^2 = 6x - 1 \Rightarrow x = \sqrt{6x-1}$
   c The graph shows there are two roots of $f(x) = 0$
   b, d

Answers

e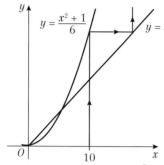

4 a $xe^{-x} - x + 2 = 0 \Rightarrow e^{-x} = \frac{x-2}{x} \Rightarrow e^x = \frac{x}{x-2}$
$\Rightarrow x = \ln\left|\frac{x}{x-2}\right|$

b $x_1 = -1.10$, $x_2 = -1.04$, $x_3 = -1.07$

5 a i $x^3 + 5x^2 - 2 = 0 \Rightarrow x^3 = 2 - 5x^2 \Rightarrow x = \sqrt[3]{2 - 5x^2}$

ii $x^3 + 5x^2 - 2 = 0 \Rightarrow x + 5 - \frac{2}{x^2} = 0 \Rightarrow x = \frac{2}{x^2} - 5$

iii $x^3 + 5x^2 - 2 = 0 \Rightarrow 5x^2 = 2 - x^3 \Rightarrow x^2 = \frac{2 - x^3}{5}$
$\Rightarrow x = \sqrt{\frac{2 - x^3}{5}}$

b $x = -4.917$ **c** $x = 0.598$
d It is not possible to take the square root of a negative number over $\mathbb{R}$.

6 a $x^4 - 3x^3 - 6 = 0 \Rightarrow \frac{1}{3}x^4 - x^3 - 2 = 0$
$\Rightarrow \frac{1}{3}x^4 - 2 = x^3 \Rightarrow x = \sqrt[3]{\frac{1}{3}x^4 - 2} \Rightarrow p = \frac{1}{3}, q = -2$

b $x_1 = -1.260$, $x_2 = -1.051$, $x_3 = -1.168$
c $f(-1.1315) = -0.014... < 0$, $f(-1.1325) = 0.0024... > 0$
There is a sign change in this interval, which implies $\alpha = -1.132$ correct to 3 decimal places.

7 a $3\cos(x^2) + x - 2 = 0 \Rightarrow \cos(x^2) = \frac{2-x}{3}$
$\Rightarrow x^2 = \arccos\left(\frac{2-x}{3}\right) \Rightarrow x = \left[\arccos\left(\frac{2-x}{3}\right)\right]^{1/2}$

b $x_1 = 1.109$, $x_2 = 1.127$, $x_3 = 1.129$
c $f(1.12975) = 0.000423... > 0$,
$f(1.12985) = -0.0001256... < 0$. There is a sign change in this interval, which implies $\alpha = 1.1298$ correct to 4 decimal places.

8 a $f(0.8) = 0.484...$, $f(0.9) = -1.025...$. There is a change of sign in the interval, so there must exist a root in the interval, since f is continuous over the interval.

b $\frac{4\cos x}{\sin x} - 8x + 3 = 0 \Rightarrow 8x = \frac{4\cos x}{\sin x} + 3$
$\Rightarrow x = \frac{\cos x}{2\sin x} + \frac{3}{8}$

c $x_1 = 0.8142$, $x_2 = 0.8470$, $x_3 = 0.8169$
d $f(0.8305) = 0.0105... > 0$, $f(0.8315) = -0.0047... < 0$. There is a change of sign in the interval, so there must exist a root in the interval.

9 a $e^{x-1} + 2x - 15 = 0 \Rightarrow e^{x-1} = 15 - 2x$
$\Rightarrow x - 1 = \ln(15 - 2x)$
$\Rightarrow x = \ln(15 - 2x) + 1$ where $x < \frac{15}{2}$

b $x_1 = 3.1972$, $x_2 = 3.1524$, $x_3 = 3.1628$
c $f(3.155) = -0.062... < 0$, $f(3.165) = 0.044... > 0$.
There is a sign change in this interval, which implies $\alpha = 3.16$ correct to 2 decimal places.

10 a $A(0, 0)$ and $B(\ln 4, 0)$
b $f'(x) = xe^x + e^x - 4 = e^x(x + 1) - 4$

c $f'(0.7) = -0.5766... < 0$, $f'(0.8) = 0.0059... > 0$.
There is a sign change in this interval, which implies $f'(x) = 0$ in this range. $f'(x) = 0$ at a turning point.
d $e^x(x + 1) - 4 = 0 \Rightarrow e^x = \frac{4}{x+1} \Rightarrow x = \ln\left(\frac{4}{x+1}\right)$
e $x_1 = 1.386$, $x_2 = 0.517$, $x_3 = 0.970$, $x_4 = 0.708$

Exercise 10C

1 a $f(1) = -2$, $f(2) = 3$. There is a sign change in the interval $1 < \alpha < 2$, so there is a root in this interval.
b $x_1 = 1.632$

2 a $f'(x) = 2x + \frac{4}{x^2} + 6$ **b** -0.326

3 a It's a turning point, so $f'(x) = 0$, and you cannot divide by zero in the Newton–Raphson formula.
b 1.247

4 a $f(1.4) = -0.020...$, $f(1.5) = 0.12817...$ As there is a change of sign in the interval, there must be a root α in this interval.
b $x_1 = 1.413$
c $f(1.4125) = -0.00076...$, $f(1.4135) = 0.0008112...$.

5 a $f(1.3) = -0.085...$, $f(1.4) = 0.429...$ As there is a change of sign in the interval, there must be a root α in this interval.
b $f'(x) = 2x + \frac{6}{x^3}$ **c** 1.316

6 a $f(0.6) = 0.0032... > 0$, $f(0.7) = -0.0843... < 0$.
Sign change implies root in the interval.
$f(1.2) = -0.0578... < 0$, $f(1.3) = 0.0284... > 0$.
Sign change implies root in the interval.
$f(2.4) = 0.0906... > 0$, $f(2.5) = -0.2595... < 0$.
Sign change implies root in the interval.
b It's a turning point, so $f'(x) = 0$, and you cannot divide by zero in the Newton–Raphson formula.
c 2.430

7 a $f(3.4) = 0.2645... > 0$, $f(3.5) = -0.3781.... < 0$.
Sign change implies root in the interval.
b $f'(x) = \frac{3}{3x-4} - 2x$ **c** 3.442

Challenge

a From the graph, $f(x) > 0$ for all values of $x > 0$. Note also that $xe^{-x^2} > 0$ when $x > 0$. So the same must be true for $x > \frac{1}{\sqrt{2}}$.

$f'(x) = e^{-x^2}(1 - 2x^2) = 0 \Rightarrow x = \frac{1}{\sqrt{2}}$

So $f'(x) < 0$ for $x > \frac{1}{\sqrt{2}}$.

$x_{n+1} = x_n - \frac{f(x)}{f'(x)}$ is an increasing sequence as

$f(x) > 0$ and $f'(x) < 0$, for $x > \frac{1}{\sqrt{2}}$. Therefore the

Newton–Raphson method will fail to converge.
b -0.209

Exercise 10D

1 a $\frac{\pi}{6} = E - 0.1 \sin E$, if E is a root then $f(E) = 0$
$E - 0.1 \sin E - k = 0 \Rightarrow E - 0.1 \sin E = k \Rightarrow \frac{\pi}{6} = k$
b $0.5782...$
c $f(0.5775) = -0.00069...< 0$, $f(0.5785) = 0.00022 > 0$.
Change of sign implies root in interval $[0.5775, 0.5785]$, so root is 0.578 to 3 d.p.

2 a $A(0, 0)$ and $B(19, 0)$
b $f'(t) = \frac{10}{t+1} - \left(\frac{\ln(t+1)}{2} + \frac{1}{2}\right)$

c $f'(5.8) = \dfrac{10}{5.8 + 1} - \left(\dfrac{\ln(5.8 + 1)}{2} + \dfrac{1}{2}\right) = 0.0121... > 0$

$f'(5.9) = \dfrac{10}{5.9 + 1} - \left(\dfrac{\ln(5.9 + 1)}{2} + \dfrac{1}{2}\right) = -0.0164... < 0$

The sign change implies that the speed changes from increasing to decreasing, so the greatest speed of the skier lies between 5.8 and 5.9.

d $f'(t) = \dfrac{10}{t+1} - \left(\dfrac{\ln(t+1)}{2} + \dfrac{1}{2}\right) = 0$

$\dfrac{\ln(t+1) + 1}{2} = \dfrac{10}{t+1}$

$(t+1)(\ln(t+1) + 1) = 20$

$t + 1 = \dfrac{20}{\ln(t+1) + 1}$

$t = \dfrac{20}{\ln(t+1) + 1} - 1$

e $t_1 = 6.164$, $t_2 = 5.736$, $t_3 = 5.879$

3 a $d(x) = 0 \Rightarrow x^2 - 3x = 0$
$x(x - 3) = 0 \Rightarrow x = 0, x = 3$
The river bed is 3 m wide so the function is only valid for $0 \leq x \leq 3$.

b $d'(x) = 2xe^{-0.6x} - \dfrac{3}{5}x^2 e^{-0.6x} - 3e^{-0.6x} + \dfrac{9}{5}xe^{-0.6x}$

$d'(x) = e^{-0.6x}\left(-\dfrac{3}{5}x^2 + \dfrac{19}{5}x - \dfrac{15}{5}\right)$

$d'(x) = -\dfrac{1}{5}e^{-0.6x}(3x^2 - 19x + 15)$

So $a = 3$, $b = -19$ and $c = 15$

c $-\dfrac{1}{5}e^{-0.6x} \neq 0$ so $d'(x) = 0 \Rightarrow 3x^2 - 19x + 15 = 0$

i $3x^2 - 19x + 15 = 0 \Rightarrow 3x^2 = 19x - 15$

$\Rightarrow x^2 = \dfrac{19x - 15}{3} \Rightarrow x = \sqrt{\dfrac{19x - 15}{3}}$

ii $3x^2 - 19x + 15 = 0 \Rightarrow 3x^2 + 15 = 19x$

$\Rightarrow x = \dfrac{3x^2 + 15}{19}$

iii $3x^2 - 19x + 15 = 0 \Rightarrow 3x^2 = 19x - 15$

$\Rightarrow x = \dfrac{19x - 15}{3x}$

d Part **i** and **iii** tend to 5.408... which is outside the required range. Part **ii** tends to $x = 0.924$.

e 1.10 m.

4 a $h(t) = 0$

$40\sin\left(\dfrac{t}{10}\right) - 9\cos\left(\dfrac{t}{10}\right) - 0.5t^2 + 9 = 0$

$40\sin\left(\dfrac{t}{10}\right) - 9\cos\left(\dfrac{t}{10}\right) + 9 = 0.5t^2$

$80\sin\left(\dfrac{t}{10}\right) - 18\cos\left(\dfrac{t}{10}\right) + 18 = t^2$

$\Rightarrow t = \sqrt{18 + 80\sin\left(\dfrac{t}{10}\right) - 18\cos\left(\dfrac{t}{10}\right)}$

b $t_1 = 7.928$, $t_2 = 7.896$, $t_3 = 7.882$, $t_4 = 7.876$

c $h'(t) = 4\cos\left(\dfrac{t}{10}\right) + 0.9\sin\left(\dfrac{t}{10}\right) - t$

d 7.874 (3 d.p.)

e Restrict the range of validity to $0 \leq t \leq A$

5 a $c'(x) = -5e^{-x} + 2\cos\left(\dfrac{x}{2}\right) + \dfrac{1}{2}$

b i $-5e^{-x} + 2\cos\left(\dfrac{x}{2}\right) + \dfrac{1}{2} = 0$

$\Rightarrow \cos\left(\dfrac{x}{2}\right) = \dfrac{5}{2}e^{-x} - \dfrac{1}{4} \Rightarrow x = 2\arccos\left[\dfrac{5}{2}e^{-x} - \dfrac{1}{4}\right]$

ii $-5e^{-x} + 2\cos\left(\dfrac{x}{2}\right) + \dfrac{1}{2} = 0 \Rightarrow e^{-x} = \dfrac{4\cos\left(\dfrac{x}{2}\right) + 1}{10}$

$\Rightarrow e^x = \dfrac{10}{4\cos\left(\dfrac{x}{2}\right) + 1} \Rightarrow x = \ln\left(\dfrac{10}{4\cos\left(\dfrac{x}{2}\right) + 1}\right)$

c $x_1 = 3.393$, $x_2 = 3.475$, $x_3 = 3.489$, $x_4 = 3.491$

d $x_1 = 0.796$, $x_2 = 0.758$, $x_3 = 0.752$, $x_4 = 0.751$

e The model does support the assumption that the crime rate was increasing. The model shows that there is a minimum point $\dfrac{3}{4}$ of the way through 2000 and a maximum point mid-way through 2003. So, the crime rate is increasing in the interval between October 2000 and June 2003.

Mixed exercise 10

1 a $x^3 - 6x - 2 = 0 \Rightarrow x^3 = 6x + 2$

$\Rightarrow x^2 = 6 + \dfrac{2}{x} \Rightarrow x = \pm\sqrt{6 + \dfrac{2}{x}}$; $a = 6$, $b = 2$

b $x_1 = 2.6458$, $x_2 = 2.5992$, $x_3 = 2.6018$, $x_4 = 2.6017$

c $f(2.6015) = (2.6015)^3 - 6(2.6015) - 2 = -0.0025... < 0$
$f(2.6025) = (2.6025)^3 - 6(2.6025) - 2 = 0.0117 > 0$
There is a sign change in the interval $2.6015 < x < 2.6025$, so this implies there is a root in the interval.

2 a $f(3.9) = 13$, $f(4.1) = -7$

b There is an asymptote at $x = 4$ which causes the change of sign, not a root.

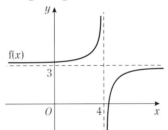

c $\alpha = \dfrac{13}{3}$

3 a

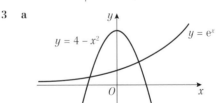

b 2 roots – 1 positive and 1 negative

c $x^2 + e^x - 4 = 0 \Rightarrow x^2 = 4 - e^x \Rightarrow x = \pm(4 - e^x)^{1/2}$

d $x_1 = -1.9659$, $x_2 = -1.9647$, $x_3 = -1.9646$, $x_4 = -1.9646$

e You would need to take the square root of a negative number.

4 a $g(1) = -10 < 0$, $g(2) = 16 > 0$. The sign change implies there is a root in this interval.

b $g(x) = 0 \Rightarrow x^5 - 5x - 6 = 0$
$\Rightarrow x^5 = 5x + 6 \Rightarrow x = (5x + 6)^{\frac{1}{5}}$
$p = 5$, $q = 6$, $r = 5$

c $x_1 = 1.6154$, $x_2 = 1.6971$, $x_3 = 1.7068$

d $g(1.7075) = -0.0229... < 0$, $g(1.7085) = 0.0146... > 0$.
The sign change implies there is a root in this interval.

5 a $g(x) = 0 \Rightarrow x^2 - 3x - 5 = 0$
$\Rightarrow x^2 = 3x + 5 \Rightarrow x = \sqrt{3x + 5}$

b, c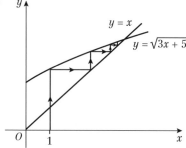

d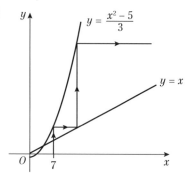

6 **a** f(1.1) = −0.0648... < 0, f(1.15) = 0.0989... > 0.
The sign change implies there is a root in this interval.
b $5x - 4\sin x - 2 = 0 \Rightarrow 5x = 4\sin x + 2$
$\Rightarrow x = \frac{4}{5}\sin x + \frac{2}{5} \Rightarrow p = \frac{4}{5}, q = \frac{2}{5}$
c $x_1 = 1.113, x_2 = 1.118, x_3 = 1.119, x_4 = 1.120$

7 **a** 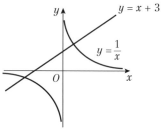 $y = x + 3$ **b** 2

c $\frac{1}{x} = x + 3 \Rightarrow 0 = x + 3 - \frac{1}{x}$, let $f(x) = x + 3 - \frac{1}{x}$
f(0.30) = −0.0333...< 0, f(0.31) = 0.0841... > 0.
Sign change implies root.
d $\frac{1}{x} = x + 3 \Rightarrow 1 = x^2 + 3x \Rightarrow 0 = x^2 + 3x - 1$
e 0.303

8 **a** $g'(x) = 3x^2 - 14x + 2$ **b** 6.606
c $(x - 1)(x^2 - 6x - 4) \Rightarrow x^2 - 6x - 4 = 0 \Rightarrow x = 3 \pm \sqrt{13}$
d 0.007%

9 **a** f(0.4) = −0.0285... < 0, f(0.5) = 0.2789... > 0.
Sign change implies root.
b 0.410
c f(−1.1905) = 0.0069... > 0,
f(−1.1895) = −0.0044... < 0.
Sign change implies root.

10 **a** $e^{0.8x} - \frac{1}{3 - 2x} = 0 \Rightarrow (3 - 2x)e^{0.8x} - 1 = 0$
$\Rightarrow (3 - 2x)e^{0.8x} = 1 \Rightarrow 3 - 2x = e^{-0.8x}$
$\Rightarrow 3 - e^{-0.8x} = 2x \Rightarrow x = 1.5 - 0.5e^{-0.8x}$
b $x_1 = 1.32327..., x_2 = 1.32653..., x_3 = 1.32698...$,
root = 1.327 (3 d.p.)
c $e^{0.8x} - \frac{1}{3 - 2x} = 0 \Rightarrow e^{0.8x} = \frac{1}{3 - 2x} \Rightarrow 3 - 2x = e^{-0.8x}$

$\Rightarrow -0.8x = \ln(3 - 2x) \Rightarrow x = -1.25\ln(3 - 2x)$
$p = -1.25$
d $x_1 = -2.6302, x_2 = -2.6393, x_3 = -2.6421$,
root = −2.64 (2 d.p.)

11 **a** $\ln y = x\ln x \Rightarrow \frac{1}{y}\frac{dy}{dx} = (1)(\ln x) + (x)\left(\frac{1}{x}\right)$

$\Rightarrow \frac{1}{y}\frac{dy}{dx} = \ln x + 1 \Rightarrow \frac{dy}{dx} = y(\ln x + 1) = x^x(1 + \ln x)$

b f(1.4) = −0.3983... < 0, f(1.6) = 0.1212... > 0.
Sign change implies root in the interval.
c $x_1 = 1.5631$ (4 d.p.)
d f(1.55955) = −0.00017... < 0,
f(1.55965) = 0.00011... > 0. Sign change implies root in the interval.

12 **a** f(1.3) =−0.18148..., f(1.4) =0.07556..... There is a sign change in the interval [1.3, 1.4], so there is a root in this interval.
b (0.817, −1.401)
c $x_1 = 0.3424, x_2 = 0.3497, x_3 = 0.3488, x_4 = 0.3489$
d $x_1 = 1.708$
e f(1.7075) = 0.000435...., f(1.7085) = −0.002151....
There is a sign change in the interval [1.7075, 1.7085], so there is a root in this interval.

Challenge
a $f(x) = x^6 + x^3 - 7x^2 - x + 3$
$f'(x) = 6x^5 + 3x^2 - 14x - 1$
$f''(x) = 30x^4 + 6x - 14$
$f''(x) = 0 \Rightarrow 15x^4 + 3x - 7 = 0$
i $15x^4 + 3x - 7 = 0 \Rightarrow 3x = 7 - 15x^4 \Rightarrow x = \frac{7 - 15x^4}{3}$

ii $15x^4 + 3x - 7 = 0 \Rightarrow 15x^4 + 3x = 7$
$\Rightarrow x(15x^3 + 3) = 7 \Rightarrow x = \frac{7}{15x^3 + 3}$

iii $15x^4 + 3x - 7 = 0 \Rightarrow 15x^4 = 7 - 3x$
$\Rightarrow x^4 = \frac{7 - 3x}{15} \Rightarrow x = \sqrt[4]{\frac{7 - 3x}{15}}$

b Using formula **iii**, root = 0.750 (3 d.p.)
c Formula **iii** gives the positive fourth root, so cannot be used to find a negative root.
d −0.897 (3 d.p.)

CHAPTER 11

Prior knowledge 11

1 **a** $12(2x - 7)^5$ **b** $5\cos 5x$ **c** $\frac{1}{3}e^{\frac{x}{3}}$

2 **a** $y = \frac{16}{3}x^{\frac{3}{2}} - 12x^{\frac{1}{2}}$ **b** $\frac{268}{3}$

3 $\frac{7}{4x - 1} - \frac{1}{x + 3}$ 4 6 units^2

Exercise 11A

1 **a** $3\tan x + 5\ln|x| - \frac{2}{x} + c$ **b** $5e^x + 4\cos x + \frac{x^4}{2} + c$
c $-2\cos x - 2\sin x + x^2 + c$ **d** $3\sec x - 2\ln|x| + c$
e $5e^x + 4\sin x + \frac{2}{x} + c$ **f** $\frac{1}{2}\ln|x| - 2\cot x + c$
g $\ln|x| - \frac{1}{x} - \frac{1}{2x^2} + c$ **h** $e^x - \cos x + \sin x + c$
i $-2\csc x - \tan x + c$ **j** $e^x + \ln|x| + \cot x + c$

2 **a** $\tan x - \frac{1}{x} + c$ **b** $\sec x + 2e^x + c$
c $-\cot x - \csc x - \frac{1}{x} + \ln|x| + c$
d $-\cot x + \ln|x| + c$ **e** $-\cos x + \sec x + c$
f $\sin x - \csc x + c$ **g** $-\cot x + \tan x + c$

Answers

h $\tan x + \cot x + c$
i $\tan x + e^x + c$
j $\tan x + \sec x + \sin x + c$

3 a $2e^7 - 2e^3$ **b** $\frac{95}{72}$ **c** -5 **d** $2 - \sqrt{2}$
4 $a = 2$ **5** $a = 7$ **6** $b = 2$
7 a $x = 4$ **b** $\frac{1}{20}x^{\frac{5}{2}} - 4\ln|x| + c$
c $\frac{31}{20} - 4\ln 4$

Exercise 11B

1 a $-\frac{1}{2}\cos(2x+1) + c$ **b** $\frac{3}{2}e^{2x} + c$
c $4e^{x+5} + c$ **d** $-\frac{1}{2}\sin(1-2x) + c$
e $-\frac{1}{3}\cot 3x + c$ **f** $\frac{1}{4}\sec 4x + c$
g $-6\cos\left(\frac{1}{2}x+1\right) + c$ **h** $-\tan(2-x) + c$
i $-\frac{1}{2}\operatorname{cosec} 2x + c$ **j** $\frac{1}{3}(\sin 3x + \cos 3x) + c$

2 a $\frac{1}{2}e^{2x} + \frac{1}{4}\cos(2x-1) + c$ **b** $\frac{1}{2}e^{2x} + 2e^x + x + c$
c $\frac{1}{2}\tan 2x + \frac{1}{2}\sec 2x + c$
d $-6\cot\left(\frac{1}{2}x\right) + 4\operatorname{cosec}\left(\frac{1}{2}x\right) + c$
e $-e^{3-x} + \cos(3-x) - \sin(3-x) + c$

3 a $\frac{1}{2}\ln|2x+1| + c$ **b** $-\frac{1}{2(2x+1)} + c$
c $\frac{(2x+1)^3}{6} + c$ **d** $\frac{3}{4}\ln|4x-1| + c$
e $-\frac{3}{4}\ln|1-4x| + c$ **f** $\frac{3}{4(1-4x)} + c$
g $\frac{(3x+2)^6}{18} + c$ **h** $\frac{3}{4(1-2x)^2} + c$

4 a $-\frac{3}{2}\cos(2x+1) + 2\ln|2x+1| + c$
b $\frac{1}{5}e^{5x} - \frac{(1-x)^6}{6} + c$
c $-\frac{1}{2}\cot 2x + \frac{1}{2}\ln|1+2x| - \frac{1}{2(1+2x)} + c$
d $\frac{(3x+2)^3}{9} - \frac{1}{3(3x+2)} + c$

5 a 1 **b** $\frac{7}{4}$ **c** $\frac{2\sqrt{3}}{9}$ **d** $\frac{5}{2}\ln 3$
6 $b = 6$ **7** $k = 24$ **8** $k = \frac{1}{12}$

Challenge
$a = 4, b = -3$ or $a = 8, b = -6$

Exercise 11C

1 a $-\cot x - x + c$ **b** $\frac{1}{2}x + \frac{1}{4}\sin 2x + c$
c $-\frac{1}{8}\cos 4x + c$ **d** $\frac{3}{2}x - 2\cos x - \frac{1}{4}\sin 2x + c$
e $\frac{1}{3}\tan 3x - x + c$ **f** $-2\cot x - x + 2\operatorname{cosec} x + c$
g $x - \frac{1}{2}\cos 2x + c$ **h** $\frac{1}{8}x - \frac{1}{32}\sin 4x + c$
i $-2\cot 2x + c$ **j** $\frac{3}{2}x + \frac{1}{8}\sin 4x - \sin 2x + c$

2 a $\tan x - \sec x + c$ **b** $-\cot x - \operatorname{cosec} x + c$
c $2x - \tan x + c$ **d** $-\cot x - x + c$
e $-2\cot x - x - 2\operatorname{cosec} x + c$
f $-\cot x - 4x + \tan x + c$ **g** $x + \frac{1}{2}\cos 2x + c$
h $-\frac{3}{2}x + \frac{1}{4}\sin 2x + \tan x + c$ **i** $-\frac{1}{2}\operatorname{cosec} 2x + c$

3 $\int_{\frac{\pi}{4}}^{\frac{\pi}{2}} \sin^2 x \, dx = \int_{\frac{\pi}{4}}^{\frac{\pi}{2}} \left(\frac{1}{2} - \frac{1}{2}\cos 2x\right) dx$

$= \left[\frac{1}{2}x - \frac{1}{4}\sin 2x\right]_{\frac{\pi}{4}}^{\frac{\pi}{2}} = \frac{\pi}{8} + \frac{1}{4} = \frac{2+\pi}{8}$

4 a $\frac{4\sqrt{3}}{3}$ **b** $\frac{9\sqrt{3} - 10 - \pi}{8}$ **c** $2\sqrt{2} - \frac{\pi}{4}$ **d** $\frac{\sqrt{2}-1}{2}$

5 a $\sin(3x + 2x) = \sin 3x \cos 2x + \cos 3x \sin 2x$
$\sin(3x - 2x) = \sin 3x \cos 2x - \cos 3x \sin 2x$
Adding gives $\sin 5x + \sin x = 2\sin 3x \cos 2x$
b So $\int \sin 3x \cos 2x \, dx = \int \frac{1}{2}(\sin 5x + \sin x) dx$
$= \frac{1}{2}\left(-\frac{1}{5}\cos 5x - \cos x\right) + c = -\frac{1}{10}\cos 5x - \frac{1}{2}\cos x + c$

6 a $5\sin^2 x + 7\cos^2 x = 5 + 2\cos^2 x = 6 + (2\cos^2 x - 1)$
$= \cos 2x + 6$
b $\frac{1}{2}(1 + 3\pi)$

7 a $\cos^4 x = (\cos^2 x)^2 = \left(\frac{1 + \cos 2x}{2}\right)^2 = \frac{1}{4} + \frac{1}{2}\cos 2x$
$+ \frac{1}{4}\cos^2 2x = \frac{1}{4} + \frac{1}{2}\cos 2x + \frac{1}{4}\left(\frac{1 + \cos 4x}{2}\right)$
$= \frac{3}{8} + \frac{1}{2}\cos 2x + \frac{1}{8}\cos 4x$
b $\frac{1}{32}\sin 4x + \frac{1}{4}\sin 2x + \frac{3}{8}x + c$

Exercise 11D

1 a $\frac{1}{2}\ln|x^2 + 4| + c$ **b** $\frac{1}{2}\ln|e^{2x} + 1| + c$
c $-\frac{1}{4}(x^2 + 4)^{-2} + c$ **d** $-\frac{1}{4}(e^{2x} + 1)^{-2} + c$
e $\frac{1}{2}\ln|3 + \sin 2x| + c$ **f** $\frac{1}{4}(3 + \cos 2x)^{-2} + c$
g $\frac{1}{2}e^{x^2} + c$ **h** $\frac{1}{10}(1 + \sin 2x)^5 + c$
i $\frac{1}{3}\tan^3 x + c$ **j** $\tan x + \frac{1}{3}\tan^3 x + c$

2 a $\frac{1}{10}(x^2 + 2x + 3)^5 + c$ **b** $-\frac{1}{4}\cot^2 2x + c$
c $\frac{1}{18}\sin^6 3x + c$ **d** $e^{\sin x} + c$
e $\frac{1}{2}\ln|e^{2x} + 3| + c$ **f** $\frac{1}{5}(x^2 + 1)^{\frac{5}{2}} + c$
g $\frac{2}{3}(x^2 + x + 5)^{\frac{3}{2}} + c$ **h** $2(x^2 + x + 5)^{\frac{1}{2}} + c$
i $-\frac{1}{2}(\cos 2x + 3)^{\frac{1}{2}} + c$ **j** $-\frac{1}{4}\ln|\cos 2x + 3| + c$

3 a 468 **b** $2\ln 3$ **c** $\frac{1}{2}\ln\left(\frac{16}{5}\right)$ **d** $\frac{1}{4}(e^4 - 1)$
4 $k = 2$ **5** $\theta = \frac{\pi}{2}$
6 a $\ln|\sin x| + c$
b $\int \tan x \, dx = -\ln|\cos x| + c = \ln\left|\frac{1}{\cos x}\right| + c$
$= \ln|\sec x| + c$

Exercise 11E

1 a $\frac{2}{5}(1+x)^{\frac{5}{2}} - \frac{2}{3}(1+x)^{\frac{3}{2}} + c$ **b** $-\ln|1 - \sin x| + c$
c $\frac{\cos^3 x}{3} - \cos x + c$ **d** $\ln\left|\frac{\sqrt{x}-2}{\sqrt{x}+2}\right| + c$
e $\frac{2}{5}(1+\tan x)^{\frac{5}{2}} - \frac{2}{3}(1+\tan x)^{\frac{3}{2}} + c$
f $\tan x + \frac{1}{3}\tan^3 x + c$

2 a $\frac{506}{15}$ **b** $\frac{392}{5}$ **c** $\frac{14}{9}$ **d** $\frac{16}{3} - 2\sqrt{3}$ **e** $\frac{1}{2}\ln\frac{9}{5}$

3 a $\frac{(3+2x)^7}{28} - \frac{(3+2x)^6}{8} + c$ **b** $\frac{2}{3}(1+x)^{\frac{3}{2}} - 2\sqrt{1+x} + c$
c $\sqrt{x^2+4} + \ln\left|\frac{\sqrt{x^2+4}-2}{\sqrt{x^2+4}+2}\right| + c$

4 a $\frac{886}{15}$ **b** $2 + 2\ln\frac{2}{3}$ **c** $2 - 2\ln 2$

5 $\frac{592}{3}$

6 $\int_{\ln 3}^{\ln 4} \frac{e^{4x}}{e^x - 2} dx = \int_1^{\sqrt{2}} \frac{2(u^2+2)^3}{u} du$
$= \int_1^{\sqrt{2}} \left(2u^5 + 12u^3 + 24u + \frac{16}{u}\right) du$
$= \left[\frac{1}{3}u^6 + 3u^4 + 12u^2 + 16\ln u\right]_1^{\sqrt{2}}$

Answers

$= \left(\dfrac{116}{3} + 16\ln\sqrt{2}\right) - \left(\dfrac{46}{3} + 16\ln 1\right)$

$= \dfrac{70}{3} + 16\ln\sqrt{2} = \dfrac{70}{3} + 8\ln 2 \Rightarrow$

$a = 70, b = 3, c = 8, d = 2$

7 $x = \cos\theta, \dfrac{dx}{d\theta} = -\sin\theta$

$\displaystyle\int -\dfrac{1}{\sqrt{1-x^2}}\,dx = \int -\dfrac{1}{\sin\theta}(-\sin\theta)\,d\theta$

$= \int 1\,d\theta = \theta + c = \arccos x + c$

8 $\displaystyle\int_0^{\frac{\pi}{3}} \sin^3 x \cos^2 x\,dx = \int_1^{\frac{1}{2}} (u^2 - 1)u^2\,du = \int_{\frac{1}{2}}^{1}(u^4 - u^2)\,du$

$= \left[\dfrac{1}{5}u^5 - \dfrac{1}{3}u^3\right]_{\frac{1}{2}}^{1} = \dfrac{47}{480}$

9 $\dfrac{2\pi + 3\sqrt{3}}{96}$

Challenge

$x = 3\sin u, \dfrac{dx}{du} = 3\cos u \Rightarrow dx = 3\cos u\,du$

$(3\sin u)^2 + (3\cos u)^2 = 9$

$\Rightarrow x^2 + (3\cos u)^2 = 9 \Rightarrow \cos u = \dfrac{\sqrt{9-x^2}}{3}$

$\displaystyle\int \dfrac{1}{x^2\sqrt{9-x^2}}\,dx = \int \dfrac{1}{9\sin^2 u \cos u}(3\cos u)\,du$

$= \dfrac{1}{9}\int \mathrm{cosec}^2 u\,du = -\dfrac{1}{9}\cot u + c = -\dfrac{\cos u}{9\sin u} + c$

$= -\dfrac{\frac{\sqrt{9-x^2}}{3}}{\frac{x}{3}} + c = -\dfrac{\sqrt{9-x^2}}{9x} + c$

Exercise 11F

1 a $-x\cos x + \sin x + c$ **b** $xe^x - e^x + c$
 c $x\tan x - \ln|\sec x| + c$ **d** $x\sec x - \ln|\sec x + \tan x| + c$
 e $-x\cot x + \ln|\sin x| + c$

2 a $3x\ln x - 3x + c$ **b** $\dfrac{x^2}{2}\ln x - \dfrac{x^2}{4} + c$
 c $-\dfrac{\ln x}{2x^2} - \dfrac{1}{4x^2} + c$ **d** $x(\ln x)^2 - 2x\ln x + 2x + c$
 e $\dfrac{x^3}{3}\ln x - \dfrac{x^3}{9} + x\ln x - x + c$

3 a $-e^{-x}x^2 - 2xe^{-x} - 2e^{-x} + c$
 b $x^2\sin x + 2x\cos x - 2\sin x + c$
 c $x^2(3 + 2x)^6 - \dfrac{x(3+2x)^7}{7} + \dfrac{(3+2x)^8}{112} + c$
 d $-x^2\cos 2x + x\sin 2x + \dfrac{1}{2}\cos 2x + c$
 e $x^2\sec^2 x - 2x\tan x + 2\ln|\sec x| + c$

4 a $2\ln 2 - \dfrac{3}{4}$ **b** 1 **c** $\dfrac{\pi}{2} - 1$
 d $\dfrac{1}{2}(1 - \ln 2)$ **e** 9.8 **f** $2\sqrt{2}\pi + 8\sqrt{2} - 16$
 g $\dfrac{1}{2}(1 - \ln 2)$

5 a $\dfrac{1}{16}(4x\sin 4x + \cos 4x) + c$
 b $\dfrac{1}{32}((1 - 8x^2)\cos 4x + 4x\sin 4x)) + c$

6 a $-\dfrac{2}{3}(8-x)^{\frac{3}{2}} + c$
 b $u = x - 2 \Rightarrow \dfrac{du}{dx} = 1; \dfrac{dv}{dx} = \sqrt{8-x} \Rightarrow v = -\dfrac{2}{3}(8-x)^{\frac{3}{2}}$

$I = (x-2)\left(-\dfrac{2}{3}(8-x)^{\frac{3}{2}}\right) - \int -\dfrac{2}{3}(8-x)^{\frac{3}{2}}\,dx$

$= -\dfrac{2}{3}(x-2)(8-x)^{\frac{3}{2}} + \dfrac{2}{3}\int(8-x)^{\frac{3}{2}}\,dx$

$= -\dfrac{2}{3}(x-2)(8-x)^{\frac{3}{2}} - \dfrac{4}{15}(8-x)^{\frac{5}{2}} + c$

$= -\dfrac{2}{3}(x-2)(8-x)^{\frac{3}{2}} - \dfrac{4}{15}(8-x)(8-x)^{\frac{3}{2}} + c$

$= (8-x)^{\frac{3}{2}}\left(-\dfrac{2}{3}(x-2) - \dfrac{4}{15}(8-x)\right) + c$

$= (8-x)^{\frac{3}{2}}\left(-\dfrac{2x}{5} - \dfrac{4}{5}\right) + c = -\dfrac{2}{5}(8-x)^{\frac{3}{2}}(x+2) + c$

 c 15.6

7 a $\dfrac{1}{3}\tan 3x + c$ **b** $\dfrac{1}{3}x\tan 3x - \dfrac{1}{9}\ln|\sec 3x| + c$

 c $\displaystyle\int_{\frac{\pi}{18}}^{\frac{\pi}{9}} x\sec^2 3x = \left[\dfrac{1}{3}x\tan 3x - \dfrac{1}{9}\ln|\sec 3x|\right]_{\frac{\pi}{18}}^{\frac{\pi}{9}}$

$= \left(\dfrac{\sqrt{3}\pi}{27} - \dfrac{1}{9}\ln 2\right) - \left(\dfrac{\sqrt{3}\pi}{162} - \dfrac{1}{9}\ln\dfrac{2}{\sqrt{3}}\right)$

$= \dfrac{5\sqrt{3}\pi}{162} - \dfrac{1}{9}\ln 2 + \dfrac{1}{9}\ln 2 - \dfrac{1}{9}\ln\sqrt{3}$

$= \dfrac{5\sqrt{3}\pi}{162} - \dfrac{1}{18}\ln 3 \Rightarrow p = \dfrac{5\sqrt{3}}{162}$ and $q = \dfrac{1}{18}$

Exercise 11G

1 a $\ln|(x+1)^2(x+2)| + c$ **b** $\ln|(x-2)\sqrt{2x+1}| + c$
 c $\ln\left|\dfrac{(x+3)^3}{x-1}\right| + c$ **d** $\ln\left|\dfrac{2+x}{1-x}\right| + c$

2 a $x + \ln|(x+1)^2\sqrt{2x-1}| + c$ **b** $\dfrac{x^2}{2} + x + \ln\left|\dfrac{x^2}{(x+1)^3}\right| + c$
 c $x + \ln\left|\dfrac{x-2}{x+2}\right| + c$ **d** $-x + \ln\left|\dfrac{(3+x)^2}{1-x}\right| + c$

3 a $A = 2, B = 2$ **b** $\ln\left|\dfrac{2x+1}{1-2x}\right| + c$ **c** $\ln\dfrac{5}{9}$, so $k = \dfrac{5}{9}$

4 a $f(x) = \dfrac{2}{3+2x} + \dfrac{1}{2-x} + \dfrac{1}{(2-x)^2}$ **b** $\dfrac{1}{2} + \ln\dfrac{10}{3}$

5 a $A = 1, B = 2, C = -2$ **b** $a = \dfrac{2}{3}, b = -\dfrac{4}{3}, c = 3$

6 a $f(x) = \dfrac{3}{x^2} - \dfrac{1}{x+2}$ **b** $a = \dfrac{3}{4}, b = \dfrac{2}{3}$

7 a $f(x) = 2 - \dfrac{3}{4x+1} + \dfrac{3}{4x-1}$, so $A = 2, B = -3$ and $C = 3$
 b $k = \dfrac{3}{4}, m = \dfrac{35}{27}$

Exercise 11H

1 a $2\ln 2$ **b** $\ln(2 + \sqrt{3})$ **c** $2\ln 2 - 1$
 d $\sqrt{2} - 1$ **e** $\dfrac{8}{3}$

2 a $\ln\dfrac{8}{5}$ **b** $\ln 3 - \dfrac{2}{3}$ **c** 1
 d $\dfrac{2\sqrt{2}-1}{3}$ **e** $\dfrac{1}{2}(1 - \ln 2)$

3 $\ln 4$ **4** $2e^2 - 2e + \ln 2$

5 a $A(0, 0), B(\pi, 0)$ and $C(2\pi, 0)$

 b Area $= \displaystyle\int_0^{\pi} x\sin x\,dx - \int_{\pi}^{2\pi} x\sin x\,dx$
 $= [-x\cos x + \sin x]_0^{\pi} - [-x\cos x + \sin x]_{\pi}^{2\pi}$
 $= \pi + 3\pi = 4\pi$

6 a $\dfrac{1}{3}x^3\ln x - \dfrac{1}{9}x^3 + c$ **b** $\dfrac{2}{3}(4\ln 2 - 1)$

7 a $A\left(-\dfrac{\pi}{2}, 0\right), B\left(\dfrac{\pi}{2}, 0\right), C\left(\dfrac{3\pi}{2}, 0\right)$ and $D(0, 3)$
 b $2(\sin x + 1)^{\frac{3}{2}} + c$ **c** $a = 32$

8 a 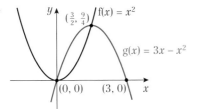   **b** $\frac{9}{8}$

c $t = \sqrt{2x+1} \Rightarrow \frac{dt}{dx} = (2x+1)^{-\frac{1}{2}}$

$\Rightarrow (2x+1)^{\frac{1}{2}} dt = dx \Rightarrow t\,dt = dx$

$\int_0^3 e^{(2x+1)^{1/2}} dx = \int_1^{\sqrt{7}} te^t\,dt \Rightarrow a = 1, b = \sqrt{7}, k = 1$

d 23.20

9 a $A\left(-\frac{\pi}{3}, 3\right), B\left(\frac{\pi}{3}, 3\right)$ and $C\left(\frac{5\pi}{3}, 3\right)$

b $a = 4, b = -4, c = 3$ (or $a = 4, b = 4, c = -3$)

c $R_2 = \int_{\frac{\pi}{3}}^{\frac{5\pi}{3}}(-2\cos x + 4)\,dx - \int_{\frac{\pi}{3}}^{\frac{5\pi}{3}}(2\cos x + 2)\,dx$

$= \int_{\frac{\pi}{3}}^{\frac{5\pi}{3}}(-4\cos x + 2)\,dx = [-4\sin x + 2x]_{\frac{\pi}{3}}^{\frac{5\pi}{3}}$

$= \left(2\sqrt{3} + \frac{10\pi}{3}\right) - \left(-2\sqrt{3} + \frac{2\pi}{3}\right) = 4\sqrt{3} + \frac{8\pi}{3}$

$R_2 : R_1 \Rightarrow 4\sqrt{3} + \frac{8\pi}{3} : 4\sqrt{3} - \frac{4\pi}{3} \Rightarrow 3\sqrt{3} + 2\pi : 3\sqrt{3} - \pi$

10 $y = \sin\theta$: Area $= \int_0^\pi 2\sin\theta\,d\theta = [-2\cos\theta]_0^\pi = (2) - (-2) = 4$

$y = \sin 2\theta$: Area $= \int_0^{\frac{\pi}{2}} 4\sin 2\theta\,d\theta = [-2\cos 2\theta]_0^{\frac{\pi}{2}}$
$= (2) - (-2) = 4$

11 a $\left(\frac{\pi}{4}, \frac{1}{\sqrt{2}}\right)$

b i $\sqrt{2} - 1$ **ii** $2 - \sqrt{2}$ **iii** $\sqrt{2}$

c $R_1 : R_2 \Rightarrow \sqrt{2} - 1 : 2 - \sqrt{2}$
$\Rightarrow (\sqrt{2} - 1)(2 + \sqrt{2}) : (2 - \sqrt{2})(2 + \sqrt{2}) \Rightarrow \sqrt{2} : 2$

12 Area $= \int y\frac{dx}{dt} dt = \int_0^{\sqrt[3]{4}} t^2(3t^2)\,dt = \frac{3}{5}(\sqrt[3]{4})^5 = \frac{3}{5}\,2^{\frac{10}{3}}$

$= \frac{3}{5}(2^3)(2^{\frac{1}{3}}) = \frac{24}{5}\sqrt[3]{2}$

13 $\frac{2}{3}$

14 a $x + y = 16$ **b** 58.9

Challenge

Area of region $R = \frac{2 - \sqrt{2}}{2}$.

Exercise 11I

1 a 1.1260, 1.4142   **b i** 1.352   **ii** 1.341
2 a 0.7071, 0.7071   **b** 0.758
  c The shape of the graph is concave, so the trapezium lines will underestimate the area.
  d 0.8   **e** 5.25%
3 a 0.427   **b** 1.04
4 a 1, 1.4581   **b i** 2.549   **ii** 2.402
  c Increasing the number of values decreases the interval. This leads to the approximation more closely following the curve.
  d $\int x\ln x\,dx - \int 2\ln x\,dx + \int 1\,dx$
  $= \left[\left(\frac{1}{2}x^2 - 2x\right)\ln x - \frac{1}{4}x^2 + 3x\right]_1^3$
  $= \left(-\frac{3}{2}\ln 3 + \frac{27}{4}\right) - \left(\frac{11}{4}\right) = -\frac{3}{2}\ln 3 + 4$
5 a 1.0607   **b** 1.337   **c** $p = \frac{16}{15}, q = \frac{1}{2}$ **d** 11.4%
6 a $4x - 5 = 0, x = \frac{5}{4}$ **b** 0.3556
  c 0.7313   **d** $\ln\left(\frac{49}{24}\right)$ **e** 2.5%
7 a 4.1133, 5.6522, 7.3891   **b** 23.25

Exercise 11J

1 a $y = Ae^{x-x^2} - 1$ **b** $y = k\sec x$
  c $y = \frac{-1}{\tan x - x + c}$ **d** $y = \ln|2e^x + c|$
2 a $\frac{1}{24} - \frac{\cos^3 x}{3}$ **b** $\sin 2y + 2y = 4\tan x - 4$
  c $\tan y = \frac{1}{2}\sin 2x + x + 1$ **d** $y = \arccos(e^{-\tan x})$
3 a $y = Axe^{-\frac{1}{x}}$ **b** $y = -e^3xe^{-\frac{1}{x}} = -xe^{(\frac{3x-1}{x})}$
4 $y = \sqrt{\frac{x}{x+1}}$ **5** $\ln|2 + e^y| = -xe^{-x} - e^{-x} + c$
6 $y = \frac{3}{1-x}$ **7** $y = \frac{3(1+x^2)+1}{3(1+x^2)-1}$
8 $y = \ln\left|\frac{x^2-12}{2}\right|$ **9** $\tan y = x + \frac{1}{2}\sin 2x + \frac{2-\pi}{4}$
10 $\ln|y| = -x\cos x + \sin x - 1$
11 a $3x + 4\ln|x| + c$ **b** $y = \left(\frac{3}{2}x + 2\ln|x| + \frac{5}{2}\right)^2$
12 a $\frac{5}{3x-8} + \frac{1}{x-2}$
  b $\ln|y| = \frac{5}{3}\ln|3x-8| + \ln|x-2| + c$
  c $y = 8(x-2)(3x-8)^{\frac{5}{3}}$
13 a $y = x^2 - 4x + c$
  b Graphs of the form $y = x^2 - 4x + c$, where c is any real number
14 a $y = \frac{1}{x+2} + c$
  b Graphs of the form $y = \frac{1}{x+2} + c$, where c is any real number
  c $y = \frac{1}{x+2} + 3$
15 a $\frac{dy}{dx} = -\frac{x}{y} \Rightarrow \int y\,dy = \int -x\,dx$
  $\Rightarrow \frac{1}{2}y^2 = -\frac{1}{2}x^2 + b \Rightarrow y^2 + x^2 = c$
  b Circles with centre (0, 0) and radius $\sqrt{c}$, where c is any positive real number.
  c $y^2 + x^2 = 49$

Exercise 11K

1 a $y = 200e^{kt}$ **b** 1 year
  c The population could not increase in size in this way forever due to limitations such as available food or space.
2 a $M = \frac{e^t}{1+e^t}$ **b** $\frac{2}{3}$ **c** M approaches 1
3 a $\frac{dx}{dt} = \frac{k}{x^2} \Rightarrow \frac{1}{3}x^3 = kt + c$
  $t = 0, x = 1 \Rightarrow c = \frac{1}{3} \Rightarrow t = 20, x = 2 \Rightarrow k = \frac{7}{60}$
  $\frac{1}{3}x^3 = \frac{7}{60}t + \frac{1}{3} \Rightarrow x = \sqrt[3]{\left(\frac{7}{20}t + 1\right)}$
  b $x = 3, t = 74.3$ days. So it takes 54.3 days to increase from 2 cm to 3 cm.
4 a The difference in temperature is $T - 25$. The tea is cooling, so there should be a negative sign. k has to be positive or the tea would be warming.
  b 46.2°C

Answers

5 a $A = \left(\dfrac{20t}{1 + 19t}\right)^2$

  b As $t \to \infty$, $A \to \left(\dfrac{20}{19}\right)^2 = \dfrac{400}{361}$ from below

6 a $V = 6000h \Rightarrow \dfrac{dV}{dh} = 6000$, $\dfrac{dV}{dt} = 12000 - 500h$,

  $\dfrac{dh}{dt} = \dfrac{dV}{dt} \div \dfrac{dV}{dh} = \dfrac{1}{6000}(12000 - 500h)$

  $60\dfrac{dh}{dt} = 120 - 5h$

  b $t = 12\ln\left(\dfrac{9}{7}\right)$

7 a $\dfrac{\left(\dfrac{1}{10000}\right)}{P} + \dfrac{\left(\dfrac{1}{10000}\right)}{10000 - P}$

  b $P = \dfrac{10000}{1 + 3e^{-0.5t}}$ so $a = 10000$, $b = 1$ and $c = 3$.

  c 10 000 deer

8 a $\dfrac{dV}{dt} = 40 - \dfrac{1}{4}V \Rightarrow -4\dfrac{dV}{dt} = V - 160$

  b $V = 160 + 4840e^{-\frac{1}{4}t}$, $a = 160$ and $b = 4840$

  c $V \to 160$

9 a $\dfrac{dR}{dt} = -kR \Rightarrow \int \dfrac{1}{R}dR = -k\int dt$

  $\Rightarrow \ln R = -kt + c \Rightarrow R = e^{-kt+c}$

  $\Rightarrow R = Ae^{-kt} \Rightarrow R_0 = Ae^0 \Rightarrow A = R_0 \Rightarrow R = R_0e^{-kt}$

  b $k = \dfrac{1}{5730}\ln 2$

  c $0.1R_0 = R_0 e^{\frac{1}{5730}\ln\left(\frac{1}{2}\right) \times t}$

  $\ln(0.1) = \dfrac{1}{5730}\ln\left(\dfrac{1}{2}\right) \times t \Rightarrow t \approx 19035$

Exercise 11L

1 a 567   **b** $\dfrac{196}{3}$ **c** $\dfrac{1456}{15}$

2 $-1 + \ln\left(\dfrac{27}{4}\right)$

3 4.523

Mixed exercise 11

1 a $\dfrac{1}{16}(2x - 3)^8 + c$ **b** $\dfrac{1}{40}(4x - 1)^{\frac{5}{2}} + \dfrac{1}{24}(4x - 1)^{\frac{3}{2}} + c$

  c $\dfrac{1}{3}\sin^3 x + c$ **d** $\dfrac{x^2}{2}\ln x - \dfrac{1}{4}x^2 + c$

  e $-\dfrac{1}{4}\ln|\cos 2x| + c$ **f** $-\dfrac{1}{4}\ln|3 - 4x| + c$

2 a $\dfrac{995085}{4}$ **b** $\dfrac{1}{4}\pi - \dfrac{1}{2}\ln 2$ **c** $\dfrac{992}{5} - 2\ln 4$

  d $\dfrac{\sqrt{3} - 1}{4}$ **e** $\dfrac{1}{4}\ln\left(\dfrac{35}{19}\right)$ **f** $\ln\left(\dfrac{4}{3}\right)$

3 a $\int \dfrac{1}{x^2}\ln x\, dx = (\ln x)\left(-\dfrac{1}{x}\right) - \int\left(-\dfrac{1}{x}\right)\left(\dfrac{1}{x}\right)dx$

  $= -\dfrac{\ln x}{x} + \int \dfrac{1}{x^2}dx = -\dfrac{\ln x}{x} - \dfrac{1}{x} + c$

  $\int_1^e \dfrac{1}{x^2}\ln x\, dx = \left[-\dfrac{\ln x}{x} - \dfrac{1}{x}\right]_1^e = \left(-\dfrac{1}{e} - \dfrac{1}{e}\right) - (0 - 1) = 1 - \dfrac{2}{e}$

  b $\dfrac{1}{(x + 1)(2x - 1)} = \dfrac{A}{x + 1} + \dfrac{B}{2x - 1} \Rightarrow A = -\dfrac{1}{3}, B = \dfrac{2}{3}$

  $\int_1^p \dfrac{1}{(x + 1)(2x - 1)}dx = \int_1^p\left(-\dfrac{1}{3(x + 1)} + \dfrac{2}{3(2x - 1)}\right)dx$

  $= \left[-\dfrac{1}{3}\ln(x + 1) + \dfrac{1}{3}\ln(2x - 1)\right]_1^p = \left[\dfrac{1}{3}\ln\left(\dfrac{2x - 1}{x + 1}\right)\right]_1^p$

  $= \left(\dfrac{1}{3}\ln\left(\dfrac{2p - 1}{p + 1}\right)\right) - \left(\dfrac{1}{3}\ln\left(\dfrac{1}{2}\right)\right) = \dfrac{1}{3}\ln\left(\dfrac{4p - 2}{p + 1}\right)$

4 $b = 2$ **5** $\theta = \dfrac{\pi}{3}$

6 a $\dfrac{2}{3}(x - 2)\sqrt{x + 1} + c$ **b** $\dfrac{8}{3}$

7 a $-\dfrac{1}{8}x\cos 8x + \dfrac{1}{64}\sin 8x + c$

  b $\dfrac{1}{8}x^2\sin 8x + \dfrac{1}{32}x\cos 8x - \dfrac{1}{256}\sin 8x + c$

8 a $A = \dfrac{1}{2}, B = 2, C = -1$

  b $\dfrac{1}{2}\ln|x| + 2\ln|x - 1| + \dfrac{1}{x - 1} + c$

  c $\int_4^9 f(x)\, dx = \left[\dfrac{1}{2}\ln|x| + 2\ln|x - 1| + \dfrac{1}{x - 1}\right]_4^9$

  $= \left(\dfrac{1}{2}\ln 9 + 2\ln 8 + \dfrac{1}{8}\right) - \left(\dfrac{1}{2}\ln 4 + 2\ln 3 + \dfrac{1}{3}\right)$

  $= \left(\ln 3 + \ln 64 + \dfrac{1}{8}\right) - \left(\ln 2 + \ln 9 + \dfrac{1}{3}\right)$

  $= \ln\left(\dfrac{3 \times 64}{2 \times 9}\right) - \dfrac{5}{24} = \ln\left(\dfrac{32}{3}\right) - \dfrac{5}{24}$

9 a $x = 4, y = 20$

  b $\dfrac{d^2y}{dx^2} = \dfrac{3}{4}x^{-\frac{1}{2}} + \dfrac{96}{x^3}$

  when $x = 4$, $\dfrac{d^2y}{dx^2} = \dfrac{15}{8} > 0 \Rightarrow$ minimum

  c $\dfrac{62}{5} + 48\ln 4$; $p = \dfrac{62}{5}, q = 48, r = 4$

10 a $\dfrac{1}{3}x^3\ln 2x - \dfrac{1}{9}x^3 + c$

  b $\left[\dfrac{1}{3}x^3\ln 2x - \dfrac{1}{9}x^3\right]_{\frac{1}{2}}^3 = (9\ln 6 - 3) - \left(0 - \dfrac{1}{72}\right)$

  $= 9\ln 6 - \dfrac{215}{72}$

11 a $(1 + \sin 2x)^2 \equiv 1 + 2\sin 2x + \sin^2 2x$

  $\equiv 1 + 2\sin 2x + \dfrac{1 - \cos 4x}{2} \equiv \dfrac{3}{2} + 2\sin 2x - \dfrac{\cos 4x}{2}$

  $\equiv \dfrac{1}{2}(3 + 4\sin 2x - \cos 4x)$

  b $\dfrac{9\pi}{8} + 1$ **c** $\left(\dfrac{\pi}{4}, 4\right)$

12 a $-xe^{-x} - e^{-x} + c$ **b** $\cos 2y = 2e^{-x}(x - e^x + 1)$

13 a $-\dfrac{1}{2}x\cos 2x + \dfrac{1}{4}\sin 2x + c$

  b $\tan y = -\dfrac{1}{2}x\cos 2x + \dfrac{1}{4}\sin 2x - \dfrac{1}{4}$

14 a $-\dfrac{1}{y} = \dfrac{1}{2}x^2 + c$

  b $x = 1: -\dfrac{1}{1} = \dfrac{1}{2} + c \Rightarrow c = -\dfrac{3}{2}$

  $-\dfrac{1}{y} = \dfrac{1}{2}x^2 - \dfrac{3}{2} \Rightarrow \dfrac{1}{y} = \dfrac{1}{2}(3 - x^2) \Rightarrow y = \dfrac{2}{3 - x^2}$

  c 1   **d** $y = x; (-2, -2)$

15 a $\ln|1 + 2x| + \dfrac{1}{1 + 2x} + c$

  b $2y - \sin 2y = \ln|1 + 2x| + \dfrac{1}{1 + 2x} + \dfrac{\pi}{2} - 2$

16 a $A_1 = \dfrac{1}{4} - \dfrac{1}{2e}, A_2 = \dfrac{1}{4}$

  b $A_1 : A_2 \Rightarrow \dfrac{1}{4} - \dfrac{1}{2e} : \dfrac{1}{4} \Rightarrow 1 - \dfrac{2}{e} : 1 \Rightarrow (e - 2) : e$

17 a $-e^{-x}(x^2 + 2x + 2) + c$

  b $y = -\dfrac{1}{3}\ln|3e^{-x}(x^2 + 2x + 2) - 5|$

18 a $\dfrac{1}{3}\ln 7$

  b $\int_0^{\frac{1}{3}\ln 7}(e^{3x} + 1)dx = \left[\dfrac{e^{3x}}{3} + x\right]_0^{\frac{1}{3}\ln 7}$

  $= \left(\dfrac{7}{3} + \dfrac{1}{3}\ln 7\right) - \left(\dfrac{1}{3} - 0\right) = 2 + \dfrac{1}{3}\ln 7$

19 a $A = 1, B = \frac{1}{2}, C = -\frac{1}{2}$
  b $x + \frac{1}{2}\ln|x-1| - \frac{1}{2}\ln|x+1| = 2t - \frac{1}{2}\ln 3$
20 a 4.00932 **b** 2.6254
  c The curve is convex, so it is an overestimate.
  d $\dfrac{e^3 - 3e + 2}{2e}$, $P = -3$, $Q = 2$ **e** 2.5%
21 a $\dfrac{dV}{dt} = -kV \Rightarrow \int \dfrac{1}{V}\,dV = \int -k\,dt \Rightarrow \ln V = -kt + c$
  $\Rightarrow V = Ae^{-kt}$
  b

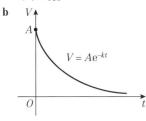

  c $\frac{1}{2}A = Ae^{-kT} \Rightarrow \frac{1}{2} = e^{-kT} \Rightarrow 2 = e^{kT} \Rightarrow \ln 2 = kT$
22 a $\int (k-y)\,dy = \int x\,dx \Rightarrow ky - \frac{1}{2}y^2 = \frac{1}{2}x^2 + c$
  $x^2 + (y-k)^2 = C$
  b Concentric circles with centre $(0, 2)$.
23 a 0.9775 **b** 3.074
  c Use more values, use smaller intervals. The lines would then more closely follow the curve.
  d $\int_1^4 \left(\frac{1}{5}x^2\right)\ln x - x + 2\,dx$
  $= \left[\frac{1}{15}x^3 \ln x - \frac{1}{45}x^3 - \frac{1}{2}x^2 + 2x\right]_1^4$
  $= \left(\frac{64}{15}\ln 4 - \frac{64}{45}\right) - \left(-\frac{1}{45} - \frac{1}{2} + 2\right) = \frac{-29}{10} + \frac{64}{15}\ln 4$
  e 2.0%
24 a $\dfrac{(1+2x^2)^6}{24} + c$ **b** $\tan 2y = \frac{1}{12}(1 + 2x^2)^6 + \frac{11}{12}$
25 $\arctan x + c$ **26** $y^2 = \dfrac{8x}{x+2}$
27 a $A = \pi r^2 \Rightarrow \dfrac{dA}{dr} = 2\pi r$
  $\dfrac{dr}{dt} = \dfrac{dr}{dA} \times \dfrac{dA}{dt} = \dfrac{1}{2\pi r} \times k \sin\left(\dfrac{t}{3\pi}\right) = \dfrac{k}{2\pi r} \sin\left(\dfrac{t}{3\pi}\right)$
  b $r^2 = -6\cos\left(\dfrac{t}{3\pi}\right) + 7$ **c** 6 days, 5 hours
28 a $\dfrac{\pi}{2}$ **b** $\dfrac{3\pi}{2}$
29 a $-\dfrac{4}{5}$ **b** $y - 2\sqrt{2} = -\dfrac{4}{5}\left(x - \dfrac{5}{\sqrt{2}}\right)$ **c** $10 - \dfrac{5\pi}{2}$
30 $\dfrac{41}{60}$

Challenge
a 15 **b** −3

CHAPTER 12
Prior knowledge 12
1 a 5**i** **b** −13**i** + 11**j**
2 a $\sqrt{34}$ **b** $\dfrac{5}{\sqrt{34}}\mathbf{i} - \dfrac{3}{\sqrt{34}}\mathbf{j}$
3 a $-2\mathbf{i} - \frac{1}{2}\mathbf{j}$ **b** $3\mathbf{i} + \frac{3}{4}\mathbf{j}$

Exercise 12A
1 $2\sqrt{21}$ **2** $7\sqrt{3}$
3 a $\sqrt{14}$ **b** 15 **c** $5\sqrt{2}$ **d** $\sqrt{30}$
4 $k = 5$ or $k = 9$ **5** $k = 10$ or $k = -4$

Challenge
a $(1, -3, 4), (1, -3, -2), (7, 3, 4), (7, 3, -2), (7, -3, -2)$
b $6\sqrt{5}$

Exercise 12B
1 a i $\begin{pmatrix} -3 \\ 5 \\ -9 \end{pmatrix}$ **ii** $\begin{pmatrix} 11 \\ -11 \\ 19 \end{pmatrix}$
  b **a** − **b** is parallel as −2(**a** − **b**) = 6**i** − 10**j** + 18**k**
    −**a** + 3**b** is not parallel as it is not a multiple of 6**i** − 10**j** + 18**k**.
2 $3\mathbf{a} + 2\mathbf{b} = 3\begin{pmatrix} 3 \\ 2 \\ -1 \end{pmatrix} + 2\begin{pmatrix} -3 \\ -2 \\ 4 \end{pmatrix} = 3\mathbf{i} + 2\mathbf{j} + 5\mathbf{k} = \frac{1}{2}(6\mathbf{i} + 4\mathbf{j} + 10\mathbf{k})$
3 $p = 2, q = 1, r = 2$
4 a $\sqrt{35}$ **b** $2\sqrt{5}$ **c** $\sqrt{3}$ **d** $\sqrt{170}$ **e** $5\sqrt{3}$
5 a $\begin{pmatrix} 7 \\ 1 \\ -1 \end{pmatrix}$ **b** $\begin{pmatrix} -5 \\ 5 \\ -5 \end{pmatrix}$ **c** $\begin{pmatrix} 14 \\ -3 \\ 1 \end{pmatrix}$ **d** $\begin{pmatrix} 8 \\ 4 \\ 4 \end{pmatrix}$ **e** $\begin{pmatrix} 8 \\ -6 \\ 10 \end{pmatrix}$
6 7**i** − 3**j** + 2**k** **7** 6 or −6 **8** $\sqrt{3}$ or $-\sqrt{3}$
9 a i $A: 2\mathbf{i} + \mathbf{j} + 4\mathbf{k}, B: 3\mathbf{i} - 2\mathbf{j} + 4\mathbf{k}, C: -\mathbf{i} + 2\mathbf{j} + 2\mathbf{k}$
    ii $-3\mathbf{i} + \mathbf{j} - 2\mathbf{k}$
  b i $\sqrt{14}$ **ii** 3
10 a $-4\mathbf{i} + 3\mathbf{j} - 12\mathbf{k}$ **b** 13 **c** $-\dfrac{4}{13}\mathbf{i} + \dfrac{3}{13}\mathbf{j} - \dfrac{12}{13}\mathbf{k}$
11 a $-6\mathbf{i} + 4\mathbf{j} + 3\mathbf{k}$ **b** $\sqrt{61}$ **c** $-\dfrac{6}{\sqrt{61}}\mathbf{i} + \dfrac{4}{\sqrt{61}}\mathbf{j} + \dfrac{3}{\sqrt{61}}\mathbf{k}$
12 a $\dfrac{3}{\sqrt{29}}\mathbf{i} - \dfrac{4}{\sqrt{29}}\mathbf{j} - \dfrac{2}{\sqrt{29}}\mathbf{k}$ **b** $\dfrac{\sqrt{2}}{5}\mathbf{i} - \dfrac{4}{5}\mathbf{j} - \dfrac{\sqrt{7}}{5}\mathbf{k}$
  c $\dfrac{\sqrt{5}}{4}\mathbf{i} - \dfrac{2\sqrt{2}}{4}\mathbf{j} - \dfrac{\sqrt{3}}{4}\mathbf{k}$
13 a $\overrightarrow{AB} = 4\mathbf{j} - \mathbf{k}, \overrightarrow{AC} = 4\mathbf{i} + \mathbf{j} - \mathbf{k}, \overrightarrow{BC} = 4\mathbf{i} - 3\mathbf{j}$
  b $|\overrightarrow{AB}| = \sqrt{17}, |\overrightarrow{AC}| = 3\sqrt{2}, |\overrightarrow{BC}| = 5$
  c scalene
14 a $\overrightarrow{AB} = -2\mathbf{i} - 6\mathbf{j} - 3\mathbf{k}, \overrightarrow{AC} = 4\mathbf{i} - 9\mathbf{j} - \mathbf{k},$
    $\overrightarrow{BC} = 6\mathbf{i} - 3\mathbf{j} + 2\mathbf{k}$
  b $|\overrightarrow{AB}| = 7, |\overrightarrow{AC}| = 7\sqrt{2}, |\overrightarrow{BC}| = 7$ **c** 45°
15 a i 98.0° **ii** 11.4° **iii** 82.0°
  b i 69.6° **ii** 62.3° **iii** 35.5°
  c i 56.3° **ii** 90° **iii** 146.3°
16 5.41
17 $|\overrightarrow{PQ}| = \sqrt{14}, |\overrightarrow{QR}| = \sqrt{29}, |\overrightarrow{PR}| = \sqrt{35}$
  Let $\theta = \angle PQR$. $14 + 29 - 2\sqrt{406}\cos\theta = 35$
  $\Rightarrow \cos\theta = 0.198... \Rightarrow \theta = 78.5°$ (1 d.p.)

Challenge
25.4°

Exercise 12C
1 a i $|\overrightarrow{OA}| = 9; |\overrightarrow{OB}| = 9 \Rightarrow |\overrightarrow{OA}| = |\overrightarrow{OB}|$
    ii $\overrightarrow{AC} = \begin{pmatrix} 9 \\ 4 \\ 22 \end{pmatrix}, |\overrightarrow{AC}| = \sqrt{581}; \overrightarrow{BC} = \begin{pmatrix} 6 \\ -4 \\ 23 \end{pmatrix}, |\overrightarrow{BC}| = \sqrt{581}$
    Therefore $|\overrightarrow{AC}| = |\overrightarrow{BC}|$
  b $OACB$ is a kite
2 a $\overrightarrow{AB} = 2\mathbf{i} + 3\mathbf{j} - 2\mathbf{k} \Rightarrow |\overrightarrow{AB}| = \sqrt{17}$
    $\overrightarrow{AC} = 6\mathbf{j} \Rightarrow |\overrightarrow{AC}| = 6$
    $\overrightarrow{BC} = -2\mathbf{i} + 3\mathbf{j} + 2\mathbf{k} \Rightarrow |\overrightarrow{BC}| = \sqrt{17}$
    $|\overrightarrow{AB}| = |\overrightarrow{BC}|$, so ABC is isosceles.
  b $6\sqrt{2}$ **c** $(0, 4, 7)$

Answers

3   **a** $\vec{AB} = 4\mathbf{i} - 10\mathbf{j} - 8\mathbf{k} = 2(2\mathbf{i} - 5\mathbf{j} - 4\mathbf{k})$
    $\vec{CD} = -6\mathbf{i} + 15\mathbf{j} + 12\mathbf{k} = -3(2\mathbf{i} - 5\mathbf{j} - 4\mathbf{k})$
    $\vec{CD} = -\frac{3}{2}\vec{AB}$, so AB is parallel to CD
    $AB : CD = 2 : 3$
   b $ABCD$ is a trapezium

4   $a = \frac{8}{3}, b = -1, c = \frac{3}{2}$

5   $(7, 14, -22), (-7, 14, -22)$ and $\left(\frac{1813}{20}, 14, -22\right)$

6   **a** 18.67 (2 d.p.)   **b** 168.07 (2 d.p.)

7   Let H = point of intersection of OF and AG.
    $\vec{OH} = r\vec{OF} = \vec{OA} + s\vec{AG}$
    $\vec{OF} = \mathbf{a} + \mathbf{b} + \mathbf{c}, \vec{AG} = -\mathbf{a} + \mathbf{b} + \mathbf{c}$
    So $r(\mathbf{a} + \mathbf{b} + \mathbf{c}) = \mathbf{a} + s(-\mathbf{a} + \mathbf{b} + \mathbf{c})$
    $r = 1 - s = s \Rightarrow r = s = \frac{1}{2}$, so $\vec{OH} = \frac{1}{2}\vec{OF}$ and $\vec{AH} = \frac{1}{2}\vec{AG}$.

8   Show that $\vec{FP} = \frac{2}{3}\mathbf{a}$ (multiple methods possible)
    Show that $\vec{PE} = \frac{1}{3}\mathbf{a}$ (multiple methods possible)
    Therefore FP and PE are parallel, so P lies on FE
    $FP : PE = 2 : 1$

Challenge

1   $p = \frac{24}{11}, q = \frac{32}{11}, r = -4$

2   $\vec{OM} = \frac{1}{2}\mathbf{a} + \mathbf{b} + \mathbf{c}, \vec{BN} = \mathbf{a} - \mathbf{b} + \frac{1}{2}\mathbf{c}, \vec{AF} = -\mathbf{a} + \mathbf{b} + \mathbf{c}$
    Let $\vec{OM}$ and $\vec{AF}$ intersect at X: $\vec{AX} = r\vec{AF} = r(-\mathbf{a} + \mathbf{b} + \mathbf{c})$
    $\vec{OX} = s\vec{OM} = s(\frac{1}{2}\mathbf{a} + \mathbf{b} + \mathbf{c})$ for scalars r and s
    $\vec{OX} = \vec{OA} + \vec{AX} = \mathbf{a} + r(-\mathbf{a} + \mathbf{b} + \mathbf{c})$
    $\Rightarrow s(\frac{1}{2}\mathbf{a} + \mathbf{b} + \mathbf{c}) = \mathbf{a} + r(-\mathbf{a} + \mathbf{b} + \mathbf{c})$
    Comparing coefficients in **a**, **b** and **c** gives $r = s = \frac{2}{3}$
    Let $\vec{BN}$ and $\vec{AF}$ intersect at Y: $\vec{AY} = p\vec{AF} = p(-\mathbf{a} + \mathbf{b} + \mathbf{c})$
    $\vec{BY} = q\vec{BN} = q(\mathbf{a} - \mathbf{b} + \frac{1}{2}\mathbf{c})$ for scalars p and q
    $\vec{BY} = \vec{BA} + \vec{AY} = \mathbf{a} - \mathbf{b} + p(-\mathbf{a} + \mathbf{b} + \mathbf{c})$
    $\Rightarrow q(\mathbf{a} - \mathbf{b} + \frac{1}{2}\mathbf{c}) = \mathbf{a} - \mathbf{b} + p(-\mathbf{a} + \mathbf{b} + \mathbf{c})$
    Comparing coefficients in **a**, **b** and **c** gives $p = \frac{1}{3}, q = \frac{2}{3}$
    $\vec{AX} = \frac{2}{3}\vec{AF}, \vec{AY} = \frac{1}{3}\vec{AF}$
    So the line segments OM and BN trisect the diagonal AF.

Exercise 12D

1   **a** $(5\mathbf{i} - \mathbf{j} + 4\mathbf{k})$ N   **b** $\sqrt{42}$ N
2   $2\sqrt{29}$ m
3   **a** $(\frac{1}{2}\mathbf{i} - \frac{5}{4}\mathbf{j} + \frac{3}{4}\mathbf{k})$ m s^{-2} **b** 1.54 m s^{-2}
4   $(5\mathbf{i} - 3\mathbf{j} - 7\mathbf{k})$ N
5   **a** $a = -2, b = 4$ **b** $(\mathbf{i} - 3\mathbf{j} - 4\mathbf{k})$ N
    c $(\frac{1}{2}\mathbf{i} - \frac{3}{2}\mathbf{j} - 2\mathbf{k})$ m s^{-2} **d** $\frac{1}{2}\sqrt{26}$ m s^{-2}
    e 126°

Mixed exercise 12

1   $\sqrt{22}$ 2 $a = 5$ or $a = 6$
3   $|\vec{AB}| = 5\sqrt{2} \Rightarrow 9 + t^2 + 25 = 50 \Rightarrow t^2 = 16 \Rightarrow t = 4$
    $6\mathbf{i} - 8\mathbf{j} - \frac{5}{2}t\mathbf{k} = 6\mathbf{i} - 8\mathbf{j} - 10\mathbf{k} = -2(-3\mathbf{i} + 4\mathbf{j} + 5\mathbf{k}) = -2\vec{AB}$
    So $\vec{AB}$ is parallel to $6\mathbf{i} - 8\mathbf{j} - \frac{5}{2}t\mathbf{k}$
4   **a** $\vec{PQ} = -3\mathbf{i} - 8\mathbf{j} + 3\mathbf{k}, \vec{PR} = -3\mathbf{i} - 9\mathbf{j} + 8\mathbf{k}, \vec{QR} = -\mathbf{j} + 5\mathbf{k}$
    b 20.0

5   **a** $\vec{DE} = 4\mathbf{i} + 3\mathbf{j} + 4\mathbf{k}, \vec{EF} = -3\mathbf{i} - 4\mathbf{j} + 4\mathbf{k}, \vec{FD} = -\mathbf{i} + \mathbf{j} - 8\mathbf{k}$
    b $|\vec{DE}| = \sqrt{41}, |\vec{EF}| = \sqrt{41}, |\vec{FD}| = \sqrt{66}$ **c** isosceles
6   **a** $\vec{PQ} = 9\mathbf{i} - 4\mathbf{j}, \vec{PR} = 7\mathbf{i} + \mathbf{j} - 3\mathbf{k}, \vec{QR} = -2\mathbf{i} + 5\mathbf{j} - 3\mathbf{k}$
    b $|\vec{PQ}| = \sqrt{97}, |\vec{PR}| = \sqrt{59}, |\vec{QR}| = \sqrt{38}$ **c** 51.3°
7   31.5°
8   184 (3 s.f.)
9   **a** $(2, -7, -2)$ **b** rhombus   **c** 36.1
10 $\vec{PQ} = \frac{1}{2}(\mathbf{a} + \mathbf{b} - \mathbf{c}), \vec{RS} = \frac{1}{2}(-\mathbf{a} + \mathbf{b} + \mathbf{c}), \vec{TU} = \frac{1}{2}(\mathbf{a} - \mathbf{b} + \mathbf{c})$
    Let $\vec{PQ}, \vec{RS}$ and $\vec{TU}$ intersect at X: $\vec{PX} = r\vec{PQ} = \frac{r}{2}(\mathbf{a} + \mathbf{b} - \mathbf{c})$
    $\vec{RX} = s\vec{RS} = \frac{s}{2}(-\mathbf{a} + \mathbf{b} + \mathbf{c})$
    $\vec{TX} = t\vec{TU} = \frac{t}{2}(\mathbf{a} - \mathbf{b} + \mathbf{c})$ for scalars r, s and t
    $\vec{RX} = \vec{RO} + \vec{OP} + \vec{PX} = \frac{1}{2}(-\mathbf{a} + \mathbf{c}) + \frac{r}{2}(\mathbf{a} + \mathbf{b} - \mathbf{c})$
    $\Rightarrow \frac{s}{2}(-\mathbf{a} + \mathbf{b} + \mathbf{c}) = \frac{1}{2}(-\mathbf{a} + \mathbf{c}) + \frac{r}{2}(\mathbf{a} + \mathbf{b} - \mathbf{c})$
    Comparing coefficients in **a**, **b** and **c** gives $r = s = \frac{1}{2}$
    $\vec{TX} = \vec{TO} + \vec{OP} + \vec{PX} = \frac{1}{2}(-\mathbf{b} + \mathbf{c}) + \frac{1}{4}(\mathbf{a} + \mathbf{b} - \mathbf{c})$
    $\Rightarrow \frac{t}{2}(\mathbf{a} - \mathbf{b} + \mathbf{c}) = \frac{1}{4}(\mathbf{a} - \mathbf{b} + \mathbf{c})$
    Comparing coefficients in **a**, **b** and **c** gives $t = \frac{1}{2}$
    So the line segments PQ, RS and TU meet at a point and bisect each other.
11 $b = 1$ or $\frac{17}{3}$
12 **a** Air resistance acts in opposition to the motion of the BASE jumper. The motion downwards will be greater than the motion in the other directions.
    b $(16\mathbf{i} + 13\mathbf{j} - 40\mathbf{k})$ N   **c** 20 seconds

Challenge

If $\mathbf{a} = \begin{pmatrix} 1 \\ 0 \\ 0 \end{pmatrix}, \mathbf{b} = \begin{pmatrix} 0 \\ 1 \\ 0 \end{pmatrix}, \mathbf{c} = \begin{pmatrix} 1 \\ 1 \\ 0 \end{pmatrix}$, then $\mathbf{a} + \mathbf{b} + \mathbf{c} = 2\mathbf{a} + 2\mathbf{b} + 0\mathbf{c}$.

Review exercise 3

1   $\frac{dy}{dx} = x - 4\sin x$
    $x = \frac{\pi}{2}, \frac{dy}{dx} = \frac{\pi}{2} - 4, y = \frac{\pi^2}{8}, m_n = -\frac{1}{\frac{\pi}{2} - 4}$
    $y - \frac{\pi^2}{8} = -\frac{1}{\frac{\pi}{2} - 4}\left(x - \frac{\pi}{2}\right)$
    $\Rightarrow 8y(8 - \pi) - 16x + \pi(\pi^2 - 8\pi + 8) = 0$

2   $\frac{dy}{dx} = 3e^{3x} - \frac{2}{x}, x = 2, y = e^6 - \ln 4, \frac{dy}{dx} = 3e^6 - 1$
    $y - e^6 + \ln 4 = (3e^6 - 1)(x - 2)$
    $\Rightarrow y - (3e^6 - 1)x - 2 + \ln 4 + 5e^6 = 0$

3   $8x + 36y + 19 = 0$

4   **a** $\frac{dy}{dx} = 4(2x - 3)(e^{2x}) + 2(2x - 3)^2(e^{2x})$
    $= 2(e^{2x})(2x - 3)(2x - 1)$
    b $\left(\frac{3}{2}, 0\right)$ and $\left(\frac{1}{2}, 4e\right)$

5   **a** $\frac{dy}{dx} = \frac{(x - 1)(2\sin x + \cos x - x \cos x)}{\sin^2 x}$
    b $x = \frac{\pi}{2}, y = \left(\frac{\pi}{2} - 1\right)^2, \frac{dy}{dx} = 2\left(\frac{\pi}{2} - 1\right)$
    $y - \left(\frac{\pi}{2} - 1\right)^2 = (\pi - 2)\left(x - \frac{\pi}{2}\right)$
    $\Rightarrow y = (\pi - 2)x + \left(1 - \frac{\pi^2}{4}\right)$

Online   Full worked solutions are available in SolutionBank.

420

Answers

6 a $y = \operatorname{cosec} x = \dfrac{1}{\sin x}$

$\dfrac{dy}{dx} = -\dfrac{\cos x}{\sin^2 x} = -\dfrac{1}{\sin x} \times \dfrac{\cos x}{\sin x} = -\operatorname{cosec} x \cot x$

b $\dfrac{dy}{dx} = -\dfrac{1}{6x\sqrt{x^2-1}}$

7 $y = \arcsin x \Rightarrow x = \sin y$

$\dfrac{dx}{dy} = \cos y \Rightarrow \dfrac{dy}{dx} = \dfrac{1}{\cos y}$

$\cos y = \sqrt{1-\sin^2 y} = \sqrt{1-x^2} \Rightarrow \dfrac{dy}{dx} = \dfrac{1}{\sqrt{1-x^2}}$

8 a $-2\sin^3 t \cos t$ **b** $y = -\tfrac{1}{2}x + 2$

c $y = \dfrac{8}{4+x^2}$

$x \geqslant 0$ is the domain of the function.

9 a $y = -9x + 8$ **b** $y = \dfrac{x}{2x-1}$

10 $7x + 2y - 2 = 0$

11 a $\dfrac{dy}{dx} = \dfrac{\cos x}{\sin y}$

b Stationary points at $\left(\dfrac{\pi}{2}, \dfrac{2\pi}{3}\right)$ and $\left(\dfrac{\pi}{2}, \dfrac{-2\pi}{3}\right)$ only in the given range.

12 $\dfrac{d^2 y}{dx^2} = e^{-x}(x^2 - 4x + 2)$ can show that roots of $x^2 - 4x + 2$ are $x = 2 \pm \sqrt{2}$ which means that $f''(x) \geqslant 0$ for all $x < 0$.

13 a $\dfrac{dV}{dr} = 4\pi r^2$ **b** $\dfrac{dr}{dt} = \dfrac{250}{\pi(2t+1)^2 r^2}$

14 a $g(1.4) = -0.216 < 0$, $g(1.5) = 0.125 > 0$. Sign change implies root.

b $g(1.4655) = -0.00025... < 0$, $g(1.4665) = 0.00326... > 0$. Sign change implies root.

15 a $p(1.7) = 0.0538... > 0$, $p(1.8) = -0.0619... < 0$. Sign change implies root.

b $p(1.7455) = 0.00074... > 0$, $p(1.7465) = -0.00041... < 0$. Sign change implies root.

16 a $e^{x-2} - 3x + 5 = 0 \Rightarrow e^{x-2} = 3x - 5$
$\Rightarrow x - 2 = \ln(3x-5) \Rightarrow x = \ln(3x-5) + 2$, $x > \tfrac{5}{3}$

b $x_0 = 4$, $x_1 = 3.9459$, $x_2 = 3.9225$, $x_3 = 3.9121$

17 a $f(0.2) = -0.01146... < 0$, $f(0.3) = 0.1564... > 0$. Sign change implies root.

b $\dfrac{1}{(x-2)^3} + 4x^2 = 0 \Rightarrow \dfrac{1}{(x-2)^3} = -4x^2$

$\Rightarrow \dfrac{-1}{4x^2} = (x-2)^3 \Rightarrow \sqrt[3]{\dfrac{-1}{4x^2}} + 2 = x$

c $x_0 = 1$, $x_1 = 1.3700, 75$, $x_2 = 1.4893$, $x_3 = 1.5170$, $x_4 = 1.5228$

d $f(1.5235) = 0.0412... > 0$, $f(1.5245) = -0.0050... < 0$

18 a It's a turning point, so the gradient is zero, which means dividing by zero in the Newton Raphson formula.

b $x_{n+1} = 2.9 - \dfrac{-0.5155...}{23.825...} = 2.922$

19 a i There is a sign change between $f(0.2)$ and $f(0.3)$. Sign change implies root.

ii There is a sign change between $f(2.6)$ and $f(2.7)$. Sign change implies root.

b $\dfrac{3}{10}x^3 - x^{\frac{2}{3}} + \dfrac{1}{x} - 4 = 0 \Rightarrow \dfrac{3}{10}x^3 = x^{\frac{2}{3}} - \dfrac{1}{x} + 4$

$\Rightarrow x^3 = \dfrac{10}{3}\left(4 + x^{\frac{2}{3}} - \dfrac{1}{x}\right) \Rightarrow x = \sqrt[3]{\dfrac{10}{3}\left(4 + x^{\frac{2}{3}} - \dfrac{1}{x}\right)}$

c $x_0 = 2.5$, $x_1 = 2.6275$, $x_2 = 2.6406$, $x_3 = 2.6419$, $x_4 = 2.6420$

d $0.3 - \dfrac{-1.10670714}{-12.02597883} = 0.208$

20 a $R = 0.37$, $\alpha = 1.2405$

b $v'(x) = -0.148 \sin\left(\dfrac{2x}{5} + 1.2405\right)$

c $v'(4.7) = -0.00312... < 0$, $v'(4.8) = 0.002798... > 0$. Sign change implies maximum or minimum.

d 12.607

e $v'(12.60665) = 0.0000037... > 0$, $v'(12.60675) = -0.0000021... < 0$. Sign change implies maximum or minimum.

21 $a = 1$

22 a $\cos 7x + \cos 3x = \cos(5x + 2x) + \cos(5x - 2x)$
$= \cos 5x \cos 2x - \sin 5x \sin 2x + \cos 5x \cos 2x + \sin 5x \sin 2x = 2\cos 5x \cos 2x$

b $\tfrac{3}{7}\sin 7x + \sin 3x + c$

23 $m = 3$ **24** 16

25 $\dfrac{2}{3} - \dfrac{3\sqrt{3}}{8}$ **26** $\tfrac{1}{9}(2e^3 + 10)$

27 a $\dfrac{5x+3}{(2x-3)(x-2)} \equiv \dfrac{3}{2x-3} + \dfrac{1}{x+2}$ **b** $\ln 54$

28 a $2\cos t = 1 \Rightarrow \cos t = \dfrac{1}{2} \Rightarrow t = \dfrac{\pi}{3}$ or $t = \dfrac{5\pi}{3}$

b $\int_{\frac{\pi}{3}}^{\frac{5\pi}{3}} y \dfrac{dx}{dt} dt = \int_{\frac{\pi}{3}}^{\frac{5\pi}{3}} (1 - 2\cos t)(1 - 2\cos t) dt$

$= \int_{\frac{\pi}{3}}^{\frac{5\pi}{3}} (1 - 2\cos t)^2 dt$

c $4\pi + 3\sqrt{3}$

29 a $\dfrac{\sqrt{3}}{16}a^2$ **b** 6.796 (4 s.f.)

30 a $\tfrac{1}{4}e^2 + \tfrac{1}{4}$

b

x	0	0.2	0.4	0.6	0.8	1
y	0	0.29836	0.89022	1.99207	3.96243	7.38906

c 2.168 (4 s.f.)   **d** 3.37%

31 a $\dfrac{2x-1}{(x-1)(2x-3)} \equiv \dfrac{-1}{x-1} + \dfrac{4}{2x-3}$

b $y = \dfrac{A(2x-3)^2}{(x-1)}$ **c** $y = \dfrac{10(2x-3)^2}{(x-1)}$

32 a $\dfrac{3k}{16\pi^2 r^5}$ **b** $r = \left[\dfrac{9k}{8\pi^2}t + A'\right]^{\frac{1}{6}}$

33 a Rate in = 20, rate out = $-kV$. So $\dfrac{dV}{dt} = 20 - kV$

b $A = \dfrac{20}{k}$ and $B = -\dfrac{20}{k}$ **c** 108 cm³ (3 s.f.)

34 a $\dfrac{dC}{dt} = -kC$, because k is the constant of proportionality. The negative sign and $k > 0$ indicates rate of decrease.

b $C = Ae^{-kt}$ **c** $k = \tfrac{1}{4}\ln 10$

35 $k = -4$, $k = 16$ **36** 130.3°

37 a $10\mathbf{i} - 5\mathbf{j} - 2\mathbf{k}$ **b** $\dfrac{10}{\sqrt{129}}\mathbf{i} - \dfrac{5}{\sqrt{129}}\mathbf{j} - \dfrac{2}{\sqrt{129}}\mathbf{k}$

c 100.1°   **d** Not parallel: $\overrightarrow{PQ} \neq m\overrightarrow{AB}$.

38 $k = 2$ **39** $p = -2$, $q = -8$, $r = -4$

Challenge

1 a $(0, 0)$ and $\left(\dfrac{-2a}{5}, \dfrac{a}{5}\right)$

b $\dfrac{dx}{dy} = 0 \Rightarrow y = 2x + \dfrac{a}{2} \Rightarrow 5x^2 + 2ax + \dfrac{a^2}{4} = 0$

$b^2 - 4ac = 4a^2 - 5a^2 = -a^2 < 0$

2 $3\sqrt{3} - 1$

Answers

Exam-style practice: Paper 1

1 $\dfrac{dy}{dx} = \dfrac{2\sec^2 t}{2\sin t \cos t} = \dfrac{1}{\sin t \cos^3 t} = \operatorname{cosec} t \sec^3 t$

2 **a** $x > -\dfrac{3}{2}$ **b** $x < -4, x > -1$ **c** $x > -1$

3 **a** $2x + y - 3 = 0 \rightarrow y = 3 - 2x$
   $x^2 + kx + y^2 + 4y = 4$
   $x^2 + kx + (3 - 2x)^2 + 4(3 - 2x) = 4$
   $5x^2 + kx - 20x + 17 = 0$
   $5x^2 + (k - 20)x + 17 > 0$ for no intersections.
   b $20 - 2\sqrt{85} < k < 20 + 2\sqrt{85}$

4 Let $f(\theta) = \cos\theta$
   $f'(\theta) = \lim\limits_{h \to 0} \dfrac{f(\theta + h) - f(\theta)}{h} = \lim\limits_{h \to 0} \dfrac{\cos(\theta + h) - \cos\theta}{h}$
   $= \lim\limits_{h \to 0} \dfrac{\cos\theta\cos h - \sin\theta\sin h - \cos\theta}{h}$
   $= \lim\limits_{h \to 0} \left[\left(\dfrac{\cos h - 1}{h}\right)\cos\theta - \left(\dfrac{\sin h}{h}\right)\sin\theta\right] = -\sin\theta$

5 **a** $p = 4, p = -4$ **b** Use $p = -4, -18432$

6 $\left(-\dfrac{49}{8}, \dfrac{705}{64}\right)$

7 **a** $u_1 = a, u_2 = 96 = ar, S_\infty = 600 = \dfrac{a}{1-r}$
   So $\dfrac{96}{1-r} \cdot \dfrac{1}{r} = 600 \Rightarrow 96 = 600r(1-r) \Rightarrow 96 = 600r - 600r^2$ and therefore $25r^2 - 25r + 4 = 0$
   b $r = 0.2, 0.8$ **c** $a = 120$ **d** $n = 39$

8 **a**
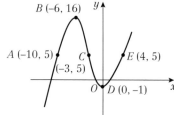

   b
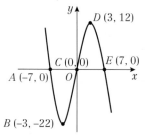

   c

9 $x = 0.77, x = 5.51$

10 **a** $\dfrac{dV}{dt} = -kV \Rightarrow \ln V = -kt + c \Rightarrow V = V_0 e^{-kt}$
   b $k = \dfrac{1}{3}\ln\left(\dfrac{5}{3}\right), V_0 = £35\,100$ **c** $t = 11.45$ years
   d k should be changed to a smaller value e.g. 0.1 (any value smaller than 0.17 acceptable)

11 **a** 14.9 miles
   b It is unlikely that a road could be built in a straight line, so the actual length of a road will be greater than 14.9 miles.

12 **a** $y = 2.79 - 0.01(x - 11)^2$, A = 2.79, B = 0.01, C = −11
   b 11 m from goal, height of 2.79 m.
   c 27.7 m (or 27.70 m)
   d $x = 0, y = 1.58$. The ball will enter the goal.

13 **a** Surface area of box $= 2x^2 + 2(2xh + xh) = 2x^2 + 6xh$
   Surface area of lid $= 2x^2 + 2(6x + 3x) = 2x^2 + 18x$
   Total surface area $= 4x^2 + 6xh + 18x = 5356$
   So $h = \dfrac{5356 - 18x - 4x^2}{6x} = \dfrac{2678 - 9x - 2x^2}{3x}$
   $V = 2x^2 h = \dfrac{2}{3}(2678x - 9x^2 - 2x^3)$
   b $6x^2 + 18x - 2678 = 0, x = 19.68$
   c $\dfrac{d^2V}{dx^2} < 0 \Rightarrow$ maximum   **d** $22\,648.7$ cm^3 **e** 21.1%

Exam-style practice: Paper 2

1 $a = -1, b = -4, c = 4$

2 **a** $y = -4x + 28$ **b** $y = \dfrac{1}{4}x + \dfrac{5}{2}$
   c $R(-10, 0)$ **d** 204 units^2

3 $y = \dfrac{1}{3}\ln(x + 1), x > -1$

4 **a** Student did not apply the laws of logarithms correctly in moving from the first line to the second line: $4^{A+B} \neq 4^A + 4^B$
   b $x = -2$. Note $x \neq -5$

5 **a**
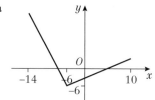

   b $-6 \leq y \leq 18$ **c** $a = -7, a = 0$

6 **a** $k = 4$ **b** $x = -2, x = 3 + \sqrt{7}, x = 3 - \sqrt{7}$

7 **a** Area $= \dfrac{1}{2}(x - 10)(x - 3)\sin 30° = \dfrac{1}{4}(x^2 - 13x + 30)$
   $= 11$
   So $x^2 - 13x + 30 = 44$, and $x^2 - 13x - 14 = 0$
   b $x = 14$ ($x \neq -1$, as $x - 10$ and $x - 3$ must be positive.)

8 **a** $h = -5, k = 2, c = 36$ **b** $\dfrac{13\pi}{2}$

9 **a** A = 4, B = 5, C = −6   **b** $-\dfrac{7}{8}x - \dfrac{23}{32}x^2$

10 $\overrightarrow{OA} = \mathbf{a}$ and $\overrightarrow{OB} = \mathbf{b}$
   $\overrightarrow{MN} = \overrightarrow{MB} + \overrightarrow{BN} = \dfrac{1}{2}\overrightarrow{OB} + \dfrac{1}{2}\overrightarrow{BA} = \dfrac{1}{2}\mathbf{b} + \dfrac{1}{2}(-\mathbf{b} + \mathbf{a}) = \dfrac{1}{2}\mathbf{a}$
   Therefore $\overrightarrow{OA}$ and $\overrightarrow{MN}$ are parallel and $\overrightarrow{MN} = \dfrac{1}{2}\overrightarrow{OA}$ as required.

11 **a** 2.46740   **b** 2.922   **c** 3.023   **d** 3.3%

12 **a** £3550   **b** £40 950   **c** £45 599.17

13 **a** $R = 0.41, \alpha = 1.3495$
   b 40 cm at time $t = 2.70$ seconds
   c 0.38 seconds and 5.02 seconds.

14 **a** $h'(t) = 3e^{-0.3(t-6.4)} - 8e^{0.8(t-6.4)}$
   b $\dfrac{3}{8}e^{-0.3(t-6.4)} = e^{0.8(t-6.4)} \Rightarrow \ln\left(\dfrac{3}{8}e^{-0.3(t-6.4)}\right) = 0.8(t - 6.4)$
   $\Rightarrow \dfrac{5}{4}\ln\left(\dfrac{3}{8}e^{-0.3(t-6.4)}\right) = t - 6.4 \Rightarrow t = \dfrac{5}{4}\ln\left(\dfrac{3}{8}e^{-0.3(t-6.4)}\right) + 6.4$
   c $t_0 = 5, t_1 = 5.6990, t_2 = 5.4369, t_3 = 5.5351, t_4 = 5.4983$
   d $h'(5.5075) = 0.00360... > 0$, $h'(5.5085) = -0.000702... < 0$. Sign change implies slope change, which implies a turning point.

Index

absolute value 23
addition, algebraic fractions 7–8
addition formulae 167–172
algebraic fractions 5–8
  addition 7–8
  division 6, 14–17
  improper 14
  integration 310–312
  multiplication 5
  subtraction 7–8
angles between vectors 341–342
arc length 118–120
arccos x 158–160
  differentiation 248
  domain 159
  range 159
arcsin x 158–160
  differentiation 248
  domain 158
  range 158
arctan x 158–160
  differentiation 248–249
  domain 159
  range 159
areas of regions, integration to find 313–314
argument, of modulus 24
arithmetic sequences 60–61, 63
arithmetic series 63–64

binomial expansion 92–102
  $(1 + bx)^n$ 92–95
  $(1 + x)^n$ 92–95
  $(a + bx)^n$ 97–99
  complex expressions 101–102
  using partial fractions 101–102
boundary conditions 323

Cartesian coordinates, in 3D 338–339
Cartesian equations, converting to/from parametric equations 198–199, 202–204
CAST diagram 117
chain rule 237–239, 261–263
chain rule reversed 296–297, 300–306
cobweb diagram 278
column vectors 340
common difference 63
common ratio 66
composite functions 32–34
  differentiation 237–239
compound angle formulae 167–172
concave functions 257–259
constant of integration 322
continuous functions 274
contradiction, proof by 2–3
convergent sequences 278–279
convex functions 257–259
cos θ
  any angle 117
  differentiation 232–233

small angle approximation 133–134, 232
cosec x
  calculation 143–144
  definition 143–144
  differentiation 247
  domain 146
  graph 146–147
  identities 153–156
  range 146
  using 149–151
cosines and sines, sums and differences 181–184
cot x
  calculation 143–144
  definition 143
  differentiation 247
  domain 147
  graph 146–147
  identities 153–156
  range 147
  using 149–151
curves
  defined using parametric equations 198–199
  sketching 206–207

degree of polynomial 14
differential equations 262–263
  families of solutions 322–323
  first order 322–324
  general solutions 322–323
  modelling with 326–328
  particular solutions 323
  second order 322
  solving by integration 322–324
differentiation 232–263
  chain rule 237–239, 261–263
  exponentials 235
  functions of a function 237–239
  implicit 254–255
  logarithms 235
  parametric 251–252
  product rule 241
  quotient rule 243–244
  rates of change 261–263
  second derivatives 257–259
  trigonometric functions 232–233, 246–249
distance between points 338–339
divergent sequences 278–280
division, algebraic fractions 6, 14–17
domain
  Cartesian function 198–199
  function 28–30, 36
  mapping 27–28
  parametric function 198–199
double-angle formulae 174–175

equating coefficients 9
exponentials, differentiation 235

functions
  composite 32–34, 237–239
  concave 257–259
  continuous 274
  convex 257–259
  domain 28–30, 36
  inverse 36–38
  many-to-one 27–30
  one-to-one 27–30
  piecewise-defined 29
  range 28–30, 36
  root location 274–276
  self-inverse 38
  see also modulus functions

geometric sequences 66–69, 70, 83
geometric series 70–72, 73–75, 83

implicit equations, differentiation 254–255
improper, algebraic fractions 14
inflection, points of 258–259
integration 294–328
  algebraic fractions 310–312
  areas of regions 313–314
  boundary conditions 323
  chain rule reversed 296–297, 300–306
  changing the variable 303–306
  constant of 322
  differential equations 322–324
  $f(ax + b)$ 296–297
  as the limit of a sum 329
  modelling with differential equations 326–328
  modulus sign in 294
  of a parametric curve 314
  partial fractions 310–312
  by parts 307–309
  standard functions 294–295
  by substitution 303–306
  trapezium rule 317–319
  trigonometric identities 298–299
intersection, points of 209–211
inverse functions 36–38
irrational numbers 2
iteration 278–280

key points summaries
  algebraic methods 21
  binomial expansion 106
  differentiation 271–272
  functions and graphs 58
  integration 336
  numerical methods 292
  parametric equations 224
  radians 141
  sequences and series 90
  trigonometric functions 164–165
  trigonometry and modelling 196
  vectors 351

limits
  of expression 73
  of sequence 66
  in sigma notation 76
line segments 345
logarithms, differentiation 235

many-to-one functions 27–30
mappings 27–30
  domain 27–28
  range 27
mechanics problems, modelling with vectors 348–349
minor arc 120
modelling
  with differential equations 326–328
  numerical methods, applications to 286
  with parametric equations 213–217
  with series 83–84
  with trigonometric functions 189–190
modulus functions 23–26
  graph of $y = |f(x)|$ 40–42
  graph of $y = f|x|$ 40–42
  problem solving 48–51
multiplication, algebraic fractions 5

natural numbers 63
negation 2
Newton–Raphson method 282–284
notation
  differential equations 322
  integration 294
  inverse functions 36
  limit 73
  major sector 122
  minor arc 120
  minor sector 122
  sequences and series 60
  sum to infinity 73
  vectors 340, 344, 345
  'x is small' 97
numerical methods 274–286
  applications to modelling 286
  iteration 278–280
  locating roots 274–276
  Newton–Raphson method 282–284

one-to-one functions 27–30
order, of sequence 81

parametric differentiation 251–252
parametric equations 198–217
  converting to/from Cartesian equations 198–199, 202–204
  curve sketching 206–207
  modelling with 213–217
  points of intersection 209–211

index

parametric integration 314
partial fractions 9–10
  binomial expansion using 101–102
  integration by 310–312
period 81
position vectors 340
product rule (differentiation) 241
Pythagoras' theorem, in 3D 338–339

quotient rule (differentiation) 243–244

radians 114–134
  angles in 115
  definition 114
  measuring angles using 114–118
  small angle approximations 133–134, 232
  solving trigonometric equations in 128–131
range
  Cartesian function 198–199
  function 28–30, 36
  mapping 27
  parametric function 198–199
rates of change 261–263
rational numbers 2
recurrence relations 79–82
reflection 44–47
repeated factors 12
reverse chain rule 296–297, 300–306
roots, locating 274–276

$\sec x$
  calculation 143–144

definition 143
differentiation 247
domain 146
graph 145–148
identities 153–155
range 146
using 149–151
sectors
  areas 122–125
  major 122
  minor 122
segments, areas 123–125
self-inverse functions 38
separation of variables 322–324
sequences 60–84
  alternating 66, 80
  arithmetic 60–61, 63
  decreasing 84
  geometric 66–69, 70, 83
  increasing 81
  order 81
  periodic 81
  recurrence relations 79–82
series 60–84
  arithmetic 63–64
  convergent 73
  divergent 73
  geometric 70–72, 73–75, 83
  modelling with 83–84
sigma notation 76–77
$\sin \theta$
  any angle 117
  differentiation 232–233
  small angle approximation 133–134, 232
sines and cosines, sums and differences 181–184
small angle approximations 133–134, 232

staircase diagram 278
stretch 44–46
substitution 9
subtraction, algebraic fractions 7–8
sum to infinity 73–75

$\tan \theta$
  any angle 117
  differentiation 246
  small angle approximation 133–134
transformations
  combining 44–47
  reflection 44–47
  stretch 44–46
  translation 44–46
translation 44–46
trapezium rule 317–319
trigonometric equations, solving 128–131, 177–179
trigonometric functions 143–160
  differentiation 232–233, 246–249
  graphs 145–148
  inverse 158–160
  modelling with 189–190
  reciprocal 143
  using reciprocal functions 149–151
trigonometric identities 130, 153–156
  integration using 298–299
  proving 186–187
  using to convert parametric equations into Cartesian equations 202–204

trigonometry
  $a \cos \theta + b \sin \theta$ expressions 181–184
  addition formulae 167–172
  double-angle formulae 174–175

unit vectors 340

vectors 338–348
  in 3D 340–342
  addition 340
  angles between 341–342
  Cartesian coordinates in 3D 338–339
  column 340
  comparing coefficients 345
  coplanar 345
  distance between points 338–339
  geometric problems involving 344–346
  modelling mechanics problems 348–349
  non-coplanar 345
  position 340
  scalar multiplication 340
  in three dimensions 340–342
  unit 340

$y = |f(x)|$, graph of 40–42
$y = f|x|$, graph of 40–42